Explorations in College Algebra

Explorations
in College Algebra

Discovering Algebra from Data Based Applications

Linda Almgren Kime
Judy Clark
University of Massachusetts, Boston

in collaboration with

Norma M. Agras
Miami Dade Community College

Robert F. Almgren
University of Chicago

Linda Falstein
University of Massachusetts, Boston

Meg Hickey
Massachusetts College of Art

John A. Lutts
University of Massachusetts, Boston

Beverly K. Michael
University of Pittsburgh

Jeremiah V. Russell
University of Massachusetts, Boston
Boston Public Schools

Software Developed by
Hubert Hohn
Massachusetts College of Art

Funded by a National Science Foundation Grant

This material is based upon work supported by the National Science foundation under Grant No. USE- 9254117

**This project was supported, in part,
by the**
National Science Foundation
Opinions expressed are those of the authors
and not necessarily those of the Foundation

ISBN: 0471-10699-2

Printed in the United States of America
10 9 8 7 6 5 4 3 2 1

Dedicated to:

Theresa Mortimer, Associate Provost at UMass/Boston, who initially brought us together and whose generous and unwavering support has been critical throughout this project

and

to all of our students who have been thoughtful critics and enthusiastic participants in the development of these materials.. Their sincere and candid comments on all phases of the course have indelibly shaped its content.

Preface

This book has been developed by a consortium of schools from an experimental course, Explorations in College Algebra, funded in part by a grant from the National Science Foundation, and initially taught at the University of Massachusetts, Boston. The course materials have now been tested in fifteen different schools. This text and the accompanying software are the result of this ongoing collaboration.

The course was designed in the spirit of the calculus reform movement and we intend the course to be effective preparation for other reformed college mathematics courses. Our goal is to give students sufficient mathematical confidence and skills so they can continue on in more advanced mathematics or other quantitatively based courses. Equally important, we hope students come to understand and appreciate the basic uses of mathematics in our society.

The content and approach was shaped by the reform guidelines that have been laid out by the various mathematical professional societies (including the MAA and AMATYC). The materials are designed to shift the focus from learning a set of discrete mechanical rules to exploring how algebra is used in the social, physical, and life sciences.

Our approach

The materials are designed around the following general principles:

- the development of algebra from data based applications
  We start with quantitative questions from the social, physical and life sciences and learn how mathematics can provide powerful strategies to help answer them. Real problems and real data motivate students to learn how to use algebraic and technological tools.

- the active involvement of students
  Our course advocates actively engaging students in class discussions and team work. The explorations included at the end of each Chapter are designed for use by small groups of students working together in or out of class. The explorations often pose open-ended problems with no single "correct" solution. Students work collaboratively to synthesize information from class lectures, the text and readings, and most importantly from their own discoveries.

- the use of multiple representations
  Graphs, tables and symbolic expressions are used to explore patterns in data, test conjectures and make generalizations about the properties of functions. Students construct their own meaning of algebraic rules and procedures by making connections among these representations.

- the communication of ideas
  Communication of ideas is stressed through use of class discussions, small group work, student presentations and written reports. Suggestions for writing strategies are included in Chapter 1, and many of the assignments require students to

describe their conclusions in writing. Students are encouraged to read the writings of others as represented in the variety of essays, articles, and reports in the Anthology of Readings.

- the integration of technology

  While we encourage the use of technology and have built many explorations and exercises around graphing calculators and computers, there are no specific technology requirements. Some schools use graphing calculators, others use computers and some a combination of both. The course was originally designed to be taught in a computer lab (Mac or IBM) equipped with a generic spreadsheet and function graphing program. The materials have now been expanded for use with a graphing calculator.

  In addition, the book comes with easy to use interactive software (for both Mac and IBM) that helps students visualize particular mathematical concepts. The software can be used by students in a lab or at home or by the instructor for in-class demonstrations.

Our process of development

The text has been developed and refined over the last three years by experts from many different disciplines. We were able to write materials and then test them out the next day in class. Our students graciously allowed us to experiment, and sometimes fail, as we tried out new materials and new approaches. Their input, more than any other, shaped our ideas.

John Wiley & Sons conducted an extensive survey of faculty members teaching college algebra courses. The results helped us decide which topics to include in the text, but also convinced us that there is no such thing as a generic college algebra course.

The materials are still under development and we hope that those of you who are using this preliminary edition will help us to make the materials stronger. Revisions will be based on your comments and suggestions.

Content

The course is motivated by the need to answer practical questions arising in science and society: Is there a relationship between income and years of education? How fast do bacteria grow? How can we describe to ourselves and our children the size and age of our universe ?

The text includes more material than college algebra courses usually include in one semester. Instructors will need to make choices and may wish to cover the chapters in a different order. It is recommended that students become familiar with the material covered in Chapters 1, 2, 3 & 4, before attempting to do the other chapters. The rest of the chapters may be covered in any order, with the exception that the material on exponents in Chapter 7 should be covered prior to material in the other chapters in Part II. We assume that students have had elementary high school algebra and are familiar with the use of symbols to represent unknown numerical quantities.

Part I focuses on algebraic applications in the social sciences. Chapter 1 introduces the basic issues of working with and writing about data. This chapter is the least like a traditional algebra chapter, but perhaps the most relevant to students' everyday lives. We have found that plunging students the first day of class into collecting data about themselves, is a successful strategy for immediately setting the tone of active student participation throughout the course. Students learn the basics of using a graphing calculator or computer. The amount of time spent on this chapter (from 2 days to 3 weeks) will vary a great deal from instructor to instructor. The time spent is probably dictated by what mathematics courses most of the students would take next and how long it takes for students to become familiar with a particular technology.

The fundamental concept of functions and their representations in tables, graphs and equations are introduced in Chapter 2 . In Chapter 3, students encounter the notion of the *average rate of change*, which becomes a central thread throughout the course.

Chapter 4 is the first of three chapters dedicated to *linear functions* and their applications. It is motivated by looking at phenomena in which the rate of change is constant and focuses on constructing, graphing and interpreting linear functions. Chapter 5 introduces a major tool of the social sciences - fitting lines to data. Students use real data about 1000 families in the US. to describe and analyze the relationship between education and income. Chapter 6 deals with systems of linear equations and their uses in a number of contexts, many economic; for example, break even costs for gas, electric or solar heating and the proposed graduated vs. flat income tax plans for the state of Massachusetts.

Part II starts with Chapter 7, "Deep Time and Deep Space." Studying the age and size of objects in the known universe provides a context for learning *scientific notation* and strategies for comparing objects of widely differing sizes. Students learn to estimate answers, a critical skill in this age of calculators and computers. The basic laws of manipulating *exponents* are extended to different bases and powers. *Logarithms* are used to scale graphs and to construct measurement systems for numbers with different orders of magnitude.

The growth of *E. coli* bacteria motivates the study of exponential functions in Chapter 8 ("Growth and Decay.") The model of exponential growth and decay is applied to human populations, radioactive decay and Medicare costs. *Exponential functions* are plotted on a semi-log graph. The text currently deals with exponential functions in the form $y = C a^x$, but future editions will include functions of the type $y = C e^{rx}$. Chapter 9 introduces the family of *polynomial functions* by first studying the *power functions*. Power functions are introduced through studying the behavior of gases and are applied to a scuba diving problem, which was the inspiration for the cover of our book.

Chapters 10 and 11 introduce the family of *quadratic functions*. A classic free fall experiment in Chapter 10 gives students some insights into how Galileo uncovered the basic laws of motion. Instructors are encouraged to arrange for students to visit a physics lab or use a motion sensor attached to a graphing calculator to collect the data for themselves. If this is not possible, data collected by other students is provided on the course disk. In Chapter 11 ("Parabolic Reflections") the properties of quadratics (symmetry, vertices, x and y-intercepts) are examined.

Part III provides a collection of diverse experiments from the social and physical sciences for use by small groups of students. Different groups can work on different explorations, or the entire class could work on the same exploration. You may want to design your own exploration and share it with us. The instructor does not have to have covered all the chapters in the book, as long as the basic notion of curve fitting has been discussed. At this point students hopefully have the basic knowledge to apply algebraic techniques developed throughout the course. These explorations do not occur at the end of a chapter focused on a particular function. Students must make their own decisions about the kinds of patterns that occur. Is there a relationship between infant mortality rate and GNP? If so, of what kind? How can we describe the rate at which sucrose is converted to fructose and glucose? What affects this rate? Groups are asked to present their findings and discuss them with the class.

Special features

Our curriculum mandates an unusual book format. As other books do, our book contains explanatory text and related exercises; in addition there are:
- explorations
- an anthology of pertinent readings
- graphing calculator instructions
- course software
- data files that can be used with graphing calculators or with spreadsheets

Text supplements include: Instructor's Manual, an Instructor's Solution Manual and a Student's Solution Manual. The supplements along with the course software and data files are available upon request from your Wiley representative or by contacting John Wiley & Sons (see address below.)

Readers' comments

The authors would be delighted to receive reactions to this preliminary edition and welcome any suggestions for homework problems or explorations that could be included in the first edition. Please contact us by mail at the University of Massachusetts at Boston, 100 Morrissey Blvd, Boston, MA 02125 or by e-mail at *algebra@umbsky.cc.umb.edu*

Readers who are interested in further information about the materials or in site testing the course, should contact: John Wiley & Sons, Inc. by mail at 605 Third Avenue New York, NY 10158 or by e-mail at *math@wiley.com.*

Acknowledgments

We wish to express our appreciation to all those who have helped and supported us in this collaborative endeavor. We are grateful for the support of the National Science Foundation, whose funding made this project possible, and for the generous help from our program officers, Elizabeth Teles and Marjorie Enneking, We wish to thank the members of our Advisory Board: Deborah Hughes Hallett, Philip and Phylis Morrison, Nicholas Rubino and Ethan Bolker for providing encouragement and invaluable advice in each phase of our work.

The text could not have been produced without the generous and on-going support we received from students, faculty, staff and administrators at UMass/Boston. Our thanks especially to the administrators who provided resources for many aspects of the project: Chancellor Sherry Penney, Provost Louis Esposito, former Provost Fuad Safwat, Associate Provost Theresa Mortimer, and Deans Patricia Davidson, Christine Armett-Kibel, Ellie Kutz. and Martin Quitt. Special thanks to Paul O'Keefe, Mike Larsen, and Carol DeSouza for helping us with the endless paperwork and procedures that are an inevitable part of obtaining and running a grant.

We are especially grateful to the following colleagues at UMass and elsewhere, for their contributions to the text. Bob Seeley, George Lukas, Rachel Skvirsky, Ron Etter, Lowell Schwartz, Max West, Tony Roman, Joe Check, Suzy Groden, Joan Lukas, Bernice Auslander, Karen Callaghan, Lou Ferleger, Art MacEwan, Randy Albelda, Rosanne Donahue, Gourish Hosangady, Bob Lee, Ken Kustin, John Looney, Brenda Cherry, Peg Cronin, Bob Morris, Margaret Zaleskas, Debra Borkovitz, Paul Foster, Elizabeth Cavicchi, Eric Entemann, Bob Martin, David Hruby, and Steve Rodi. Particular thanks are due to Barry Bluestone from whom we borrowed the concept of the FAM 1000 data set used in Chapter Five "Looking for Links."

A project cannot function without an efficient and forgiving Project Administrator. In Theresa Fougere we found a woman with a divine combination of attributes: common sense, dedication, attention to detail and a sense of humor. Our special thanks to her and others who have shared administrative duties: Clare Crawford, Marie Coleman, Aurora Alamariu, David Wilson, Val Goktuk, Jonathan Rose, Wendy Sanders, Matt Gunderson, Karen Sullivan, Eric Dunlap, Jan McLeod, Matt Smith, Lynne Bowen, Patricia Dognazzi, and John Harper.

A text designed around the application of real world data would have been impossible if not for the long and selfless hours put in by Myrna Kustin, researching and acquiring copyright permissions from around the globe.

We also wish to thank our terrific student teaching assistants: John Koveos, Tony Beckwith, Irene Blach, Philip Wan, Kristen Demopoulos, Tony Horne, Cathy Briggs and Arlene Russo. They excelled as students and as teachers.

We are deeply indebted to Peter Renz, Madalyn Stone and Dick Cluster for their gracious and supportive editorial help.

Kudos to our publisher, John Wiley & Sons. We are grateful that William Pesce gave us a chance to present our ideas to the staff at Wiley. He asked good questions and we are still searching for some of the answers. Eileen Navagh, Wayne Anderson, and many others at Wiley have been most helpful. Particular thanks goes to our wonderful editor, Ruth Baruth. After our first two hour lunch in Harvard Square, we knew that we would have a pleasant long term relationship. Ruth has provided the gentle guidance necessary for first time authors to publish a text with multiple contributors. She has generous given us many hours of practical advice and is always cheerful about helping with the endless details of publishing a book.

We are especially grateful to the 14 beta sites scattered across the U.S.. One of the joys of this project has been working with so many dedicated faculty who are searching for new ways to reach out to students. These faculty all offered incredible support and encouragement, and a wealth of helpful suggestions. Our heartful thanks to: Jean Prendergast and Keith Desrosiers, Bridgewater State College; Sandi Athanssiou, Erika Kwiatkowski, and Mark Yannotta, University of Missouri-Columbia; Judy Stubblefield, Garden City Community College; Josie Hamer, Robert Hoburg, and Bruce King, Western Connecticut State University; Peggy Tibbs and John Watson, Arkansas Tech University; Peg McPartland, Golden Gate University; Russell Reich, Sierra Nevada College; Christopher Olsen, George Washington High School; Lida McDowell, Jan Davis, and Jeff Stuart, University of Southern Mississippi; Judy Jones and Beverly Taylor, Valencia Community College.

Our families couldn't help but become caught up in this time consuming endeavor. Linda's husband, Milford, cheerfully helped subsidize a sabbatical year and her son, Kristian, compiled the text's index. Judy's husband, Gerry, became our Consortium lawyer, and her daughters, Rachel, Caroline and Kristin provided support, understanding, laughter and "whatever." All our family members ran errands, made dinners, listened to our concerns and gave us the time and space to work on the text. Our love and thanks.

Finally, we wish to thank all of our students. It is for them that this book was really written.

A Note to the Students

Mathematics is a filter, a way of looking at the world. Just as musical training increases your awareness of sounds, or knowledge of history gives a deeper perspective on the present, mathematics heightens your perception of underlying order and systematic patterns in numerical information.

Numbers are adjectives, one of many ways that help us describe things. For instance, we might say an object is yellow, red and brown, is in the shape of a 12 inch circle and smells like pepperoni with extra cheese. All of these descriptions are potentially useful facts, but the 12 inch circle is the only piece of data which gives us quantitative information. From this fact we can calculate the area of the pizza, the size of a box that can contain it, and a fair price for it relative to a 9 inch pizza. Unfortunately there are many people who are afraid to make such use of quantitative facts. If you are like many of the intelligent students we've encountered, who have little confidence in their ability to do math, this course is intended to help you overcome your fears and become comfortable describing things quantitatively.

Many skills are involved in being able to generate a useful quantitative description. In this course, the focus is much broader than in a traditional skills-based college algebra course. Much of the work may seem at first unrelated to the kind of mathematics you've done in previous courses. Our hope is that it will enable you to use quantitative reasoning throughout the rest of your life. You will be asked to *read* and analyze mathematical arguments from a variety of different sources and viewpoints, from contemporary newspaper articles and essays, to publications from the federal government. You will be asked to *reason* through your own quantitative arguments, using both traditional algebraic tools and contemporary technological ones. And you will be asked to *write* about your conclusions.

The classes may not look much like the ones you're used to. The professor might lecture a lot less, and you may find yourself working with your classmates in small groups. Our exploratory approach results in a lot more questioning and discussion than is common in math classes. Sometimes you might wonder why the professor doesn't tell you straight out what you're supposed to learn, instead of allowing you to explore the issue. The answer is that sometimes there is no "straight out" solution to the problem, and that the process of exploring can be as important as memorizing the "facts."

A former student: wrote to us that

> Doing problem solving and learning about the computer and the rules of algebra all at once was a struggle - but now that we are almost done, I can say that I thoroughly enjoyed this class. I learned how to relax and to struggle with finding answers to hard questions.

We hope that you too will enjoy doing math and that mathematics will become one of the tools you use to understand and participate in the world around you. We also hope to hear from you about your ideas for exploring algebra.

With warm regards,
Linda Kime and Judy Clark
Cambridge, Massachusetts
June, 1996

HOW TO USE DIFFERENT PARTS OF THE TEXT

This book has many parts: chapters on specific topics, explorations, exercises, readings and graphing calculator instructions. We encourage you to be flexible in how you use the materials and spend some time just browsing through the text.

CHAPTERS

The text is designed in a way that allows you to be thinking and doing as you read it. Each chapter contains sections called *Algebra Aerobics* and *Something to think about*. The Algebra Aerobics provide a chance for you to practice algebraic rules and procedures. These are skill building exercises and we encourage you to do the problems as you are reading the text. We have provided answers at the bottom of the page so you can check your work. The Something *to think about* sections pose provocative questions for you to ponder alone or discuss with others in or outside of class.

EXPLORATIONS

The Explorations allow you to experiment with the ideas presented in each chapter. They are more open-ended than the exercises and are designed to be used in parallel with reading the text. They can be done in small groups inside or outside of class. Some of the explorations and exercises assume you have access to graphing calculators or computers. We have noted in the text when this is the case. You may be asked to present your results to the class. We have found that after students' initial fears are dispelled, the presentations often became the most interesting part of the course.

EXERCISES

The exercises contain many different types of problems: skill building exercises where you practice using algebraic rules and procedures and problems that require you to reason critically and exercises that ask you to express your ideas in writing.

GRAPHING CALCULATOR APPENDIX

The Graphing Calculator Appendix provides instructions for using the TI-82 and TI-83 graphing calculators along with exercises to practice specific techniques. If you are using a graphing calculator for the first time, you should do these exercises before doing the explorations or exercises in each chapter.

ANTHOLOGY OF READINGS

The Anthology of Readings are meant to supplement the ideas developed in the text and to give you a glimpse of the complexity of the issues raised in each chapter. You might want to look at the Table of Contents and read those articles of interest whether or not they are assigned by your instructor.

We hope that you explore the ideas presented in each chapter by talking, thinking and working together.

Part I: Algebra in the Social Sciences

Exploring Data

Exploring Functions

Measuring Change

Linear Functions

Part II: Algebra in the Physical and Life Sciences

Using Powers

Exponential Functions

Power Functions and Polynomials

Quadratic Functions

Part III: Exploring On Your Own

Appendices

Part I

Algebra in the Social Sciences

Introduction to Part I

Secretary of Labor Robert Reich pronounces "learning as the key to earning" in a speech on the eve of Labor Day 1994. He asserts that "The fundamental fault line running through today's workforce is based on education and skills. ... As recently as 1979 a male college graduate earned 49 percent more than a similar man with only a high school diploma... By 1992, however, the average male college graduate was earning 83 percent more than his high-school graduate counterpart, and the notion of common prospects had faded considerably."

Presidential hopefuls battle over the virtues of a flat income tax. Republican analyst Kevin Phillips claims that the public fascination is based on the misconception that "the flat tax, by closing loopholes, will make the rich pay more." Yet according to Steve Forbes "Everyone gets a tax break with the flat tax."

Life expectancy at birth for both sexes has increased dramatically in the last century for Americans. In 1900 the average life expectancy was slightly over 49 years. By 1990 the average life expectancy had risen to 75.8 years. This is due in large part to the fact that the United States is the world's richest nation and it spends far more of its income on health care than does any other. Yet according to the United Nations' 1994 Human Development Report, people in other countries live longer and get more care.

Political and personal decisions depend on understanding the benefits of education, analyzing the merits of a flat tax, or knowing the impact of gains in life expectancy. These issues are addressed by the social sciences: political science, economics, history, demography and sociology, the disciplines that deal with the functioning of our society. Unlike physical scientists who try to discover fundamental laws of the universe, social scientists do not expect to discover exact principles that govern society. But general societal trends can be uncovered by collecting numerical data and using algebraic strategies to search for patterns. The conclusions from such research often form the basis for decisions about how our society should operate.

Part I approaches the standard topics in college algebra from the perspective of the social sciences. The tools required to deal with real data are developed as needed: verbal and visual methods of data presentation; algebraic methods of single and two-variable data analysis including numerical data summaries, rate of change calculations, linear functions and linear regression models; and systems of equations. Throughout this process we raise questions about how good the data and accompanying conclusions are.

Chapter 1

Making Sense of Data

Overview

In all sciences, physical or social, answering questions requires the collection of data. A physicist or chemist generates data using a controlled laboratory experiment; an economist or sociologist collects data using society as a laboratory. As citizens, it is essential to understand the issues involved in collecting, representing, and interpreting data. We will often have to make decisions based on data presented in the news media, textbooks, or market surveys.

Chapter 1 deals with the issues encountered using real data. We introduce mathematical and technological skills needed to develop and interpret numerical answers from real data. Then you'll be asked to interpret those answers, and communicate your results clearly to others.

The explorations let you apply these tools to data collected on your class or collected by the United States Bureau of the Census. They also provide experience with any technology (graphing calculators or computers) you will be using throughout the course.

After reading this chapter you should be able to:

- read and interpret data tables
- visualize data
- understand the differences among the terms mean, median, and mode
- construct a "60 Second Summary"
- know critical questions to ask about data and conclusions

1.1   Introduction to Variables and Data

This course starts with you. How would you describe yourself to others? Are you a 5' 6," black, 26 year-old, female engineer? Or, perhaps you are a 5' 10," Chinese, 18 year-old, male flute player, sharing an apartment with two friends? In statistical terms, characteristics like height, race, age, and major that vary from person to person are called *variables*. Information collected about a variable is called *data*.[1]

Some variables that you might use to describe yourself, such as age, height, or number of people in your household, can be represented by a number and a unit of measure (like 18 years, 6 feet or 3 people). These are called *quantitative variables*. For other variables, such as gender or college major, we use categories (like male and female, or physics, English and psychology) to classify information. These are called *qualitative variables*.

Many of the controversies in the social sciences have centered on how particular variables are defined and measured. (The Anthology reading "Units and Measurements" offers a glimpse into the history of how mankind has counted and measured.) Variables such as race and the number of people in a household can be defined and categorized in many different ways. For nearly two centuries, the categories used by the U.S. Census Bureau to classify race and ethnicity have been the subject of debate. For example, Hispanic used to be considered a racial classification. It is now considered an ethnic classification, since Hispanics could be black, or white, or of any other race. Decisions related to jobs, scholarships, distribution of federal funds, and enforcement of civil rights laws (like the Voting Rights Act) are influenced by the way people are categorized.

Data can be represented using tables, graphs and numerical descriptors such as averages. We begin by seeing how tables and graphs can be used to reveal patterns in data, starting with a small data set collected by students, and then looking at a large data set of the ages of the United States population. By comparing present to future age distribution projections, we can ask how the increase in the average age will affect issues like Social Security.

The search for ways to identify, summarize, and describe patterns in the data can be thought of as a search for a "60 Second Summary." If you had to describe, in 60 seconds, the main idea you have drawn from the data, what would you say? Creating a one minute summary may take quite a bit of time. Constructing a brief, succinct description is much more difficult than constructing a lengthy one. Keep in mind that the process is as important as the product. You will need to ask and answer a series of probing questions, focus in on a key idea, and then put your conclusions into writing. The first few sections of this chapter prepare you to construct "60 Second Summaries" of current and projected U.S. age data.

Something to think about
Describe in your own words the difference between quantitative and qualitative variables. Give examples of each. Now look at the class questionnaire in Exploration 1.1 (at the end of the chapter exercises) and identify each variable as quantitative or qualitative.

---

[1] Data is the plural of the Latin word datum (meaning something given). Hence one datum, two data.

1.2   Reading and Interpreting Data Tables

A first step in organizing raw data is to create a table. Let's start with a simple data table and look for ways to summarize the numbers it contains.

Table 1.1 displays data collected in a survey of one section of college algebra at the University of Massachusetts at Boston. The first column lists values from 18 years to 46 years for the variable "age." The second column gives the number of people of each age and represents the *frequency* or *frequency count* for each age; that is, there was one 18 year-old, three 20 year-olds, etc.

> The *frequency* or *frequency count* of any value or range of values of a variable
> is the number of times that value or values in that range occur in the data.

Table 1.1

Ages of Students in UMass Class

Age years	Frequency Count
18	1
19	1
20	3
21	1
22	1
23	3
24	0
25	0
26	1
27	1
28	1
29	1
30	2
31	0
32	0
33	1
34	0
35	0
36	1
37	0
38	0
39	1
40	0
41	1
42	0
43	0
44	0
45	0
46	1
Total	**21**

This information on students' ages can be summarized by compressing the data into intervals of 2 or more years. The first step is to decide on the interval size and how you want to construct the intervals. You can then count the number of observations (here the number of people) that fall into each of the intervals. For example, if we construct 3-year intervals, and start with 18 to 20 year-olds, then there were 5 people between 18 and 20 years of age; 5 people between 21 and 23 years; 1 person between 24 and 26 years, etc. (Note that the interval from 18 to 20 years includes those people who were 18, 19, or 20 years old.)

Table 1.2

Ages of Students in Three Year Intervals

Age Interval	Frequency Count
18-20	5
21-23	5
24-26	1
27-29	3
30-32	2
33-35	1
36-38	1
39-41	2
42-44	0
45-47	1
Total	21

Converting frequency counts into percentages

In order to get a better sense of the relative numbers of students in each age interval, we can compare the frequency count in each interval to the total frequency count for the sample. Such a comparison is called a *relative frequency* and is expressed as either a decimal or a percentage.

> The *relative frequency* of any value or range of values of a variable is the fraction (often expressed as a percentage) of all the data in the sample having that value or lying in that range of values.

Tables of data often include relative frequencies since they can be used to make comparisons among categories of data. For example, if we wanted to compare this mathematics class to another group, especially if it is not the same size, we must consider the percentages of students in each particular age interval, not just the total numbers. Why? Suppose you know that 3 students in one section were age 18 and that, 1235 students in the college were age 18. This doesn't tell you a great deal because you don't know the total number of students in the class or the college. But if you know that 3 out of 35 students (or 8.5%) of the class were 18, and 1235 out of 4231 students (or about 29%) of all the students in the college were 18, then the numbers are seen relative to the size of the class and the population of the college. We say that the relative frequency of 18 year-olds in the class is 8.5%, and the relative frequency of 18 year-olds in the college as a whole is 29%. We now know that the class has a disproportionately small number of 18 year-olds compared to the college as a whole.

The relative frequency for each age interval in our data set can be found by dividing the number of people in each age interval by the total number of people in the class. For example, there were five people in the first age interval of 18 to 20 years and the total number of students in the class was 21. How do we express the relative frequency of 18 to 20 year-olds as a percentage of the students in the class?

dividing 5 by 21	$5/21$	$= 0.238$
rounding to two places		$= 0.24$
multiplying by 100 %	$(0.24)(100\%)$	$= 24\%$

Table 1.3		
Student Ages with Relative Frequency		
Age Interval	Frequency Count	Relative Frequency (%)
18-20	5	24%
21-23	5	24%
24-26	1	5%
27-29	3	14%
30-32	2	10%
33-35	1	5%
36-38	1	5%
39-41	2	10%
42-44	0	0%
45-47	1	5%
Total	21	100%

Table 1.3 is the age data table with an additional column for the relative frequency (expressed as a percentage) of students in each age interval. Since we rounded off each of the percents, the percentages actually add to 102 %. But 21 students is 100% of our class.

Summarizing data to reveal patterns

Stop and look at the following table (Table 1.4.) for a minute. It displays information collected by the U.S. Census Bureau on the age and sex of people in the United States in 1991. Note that the counts are all in thousands. For example, the total population of the U.S. is listed as 252,177 thousand or, equivalently, 252,177,000. The table presents frequency counts in different ways. There are frequency counts of the total population (and of males and females separately) in 1-year intervals and in 5-year intervals (e.g., all those under 5 years, from 5 to 9 years, from 10 to 14 years, etc.). The frequency counts for the 5-year intervals are in bold type. In a large data table like this, it can be difficult to recognize patterns in the distribution of ages in the U.S. population. But by grouping the data and condensing the details, some overall patterns start to emerge and we have taken the first step toward a "60 Second Summary."

Table 1.4
Resident US Population, by Sex and Age: 1991

(In thousands, except as indicated. As of July 1. For derivation of estimates, see text, section 1)

Age				Age			
Total	252,177	122,979	129,198				
Under 5 years. old	19,222	9,836	9,386	**45 to 49 years. old**	14,094	6,907	7,188
Under 1 yr	4,011	2,052	1,959	45 years. old	2,832	1,392	1,440
1 years. old	3,969	2,030	1,938	46 years. old	2,821	1,384	1,437
2 years. old	3,806	1,949	1,857	47 years. old	2,848	1,392	1,456
3 years. old	3,718	1,902	1,816	48 years. old	2,856	1,396	1,459
4 years. old	3,717	1,902	1,815	49 years. old	2,737	1,342	1,395
5 to 9 years. old	18,237	9,337	8,900	**50 to 54 years. old**	11,645	5,656	5,989
5 years. old	3,702	1,897	1,806	50 years. old	2,528	1,234	1,294
6 years. old	3,681	1,884	1,797	51 years. old	2,340	1,139	1,200
7 years. old	3,575	1,829	1,746	52 years. old	2,298	1,116	1,182
8 years. old	3,512	1,797	1,715	53 years. old	2,280	1,105	1,175
9 years. old	3,767	1,930	1,836	54 years. old	2,200	1,063	1,137
10 to 14 years. old	17,671	9,051	8,620	**55 to 59 years. old**	10,423	4,987	5,436
10 years. old	3,703	1,899	1,804	55 years. old	2,129	1,023	1,106
11 years. old	3,662	1,875	1,786	56 years. old	2,195	1,054	1,141
12 years. old	3,484	1,783	1,701	57 years. old	2,068	991	1,077
13 years. old	3,414	1,746	1,668	58 years. old	1,946	927	1,019
14 years. old	3,409	1,747	1,661	59 years. old	2,085	992	1,093
15 to 19 years. old	17,205	8,834	8,371	**60 to 64 years. old**	10,582	4,945	5,637
15 years. old	3,293	1,690	1,603	60 years. old	2,124	994	1,129
16 years. old	3,362	1,732	1,630	61 years. old	2,100	994	1,106
17 years. old	3,360	1,733	1,627	62 years. old	2,076	971	1,105
18 years. old	3,383	1,733	1,650	63 years. old	2,145	1,004	1,141
19 years. old	3,808	1,948	1,860	64 years. old	2,138	982	1,156
20 to 24 years. old	19,194	9,775	9,419	**65 to 69 years. old**	10,037	4,491	5,546
20 years. old	4,080	2,087	1,992	65 years. old	2,072	939	1,132
21 years. old	3,969	2,029	1,940	66 years. old	2,069	933	1,136
22 years. old	3,732	1,902	1,829	67 years. old	2,026	910	1,116
23 years. old	3,633	1,845	1,788	68 years. old	1,911	848	1,063
24 years. old	3,781	1,912	1,869	69 years. old	1,960	862	1,099
25 to 29 years. old	20,718	10,393	10,325	**70 to 74 years. old**	8,242	3,531	4,712
25 years. old	3,837	1,935	1,902	70 years. old	1,886	828	1,058
26 years. old	4,042	2,026	2,016	71 years. old	1,731	750	981
27 years. old	4,217	2,116	2,101	72 years. old	1,649	710	939
28 years. old	4,063	2,033	2,030	73 years. old	1,518	637	881
29 years. old	4,559	2,283	2,276	74 years. old	1,458	606	852
30 to 34 years. old	22,159	11,034	11,125	**75 to 79 years. old**	6,279	2,482	3,797
30 years. old	4,482	2,234	2,247	75 years. old	1,405	576	829
31 years. old	4,414	2,200	2,214	76 years. old	1,334	540	795
32 years. old	4,371	2,173	2,197	77 years. old	1,244	491	753
33 years. old	4,372	2,169	2,203	78 years. old	1,204	462	742
34 years. old	4,520	2,258	2,263	79 years. old	1,091	414	677
35 to 39 years. old	20,518	10,174	10,344	**80 to 84 years. old**	4,035	1,406	2,629
35 years. old	4,312	2,142	2,170	80 years. old	970	352	618
36 years. old	4,189	2,079	2,109	81 years. old	883	314	568
37 years. old	4,120	2,040	2,080	82 years. old	799	279	521
38 years. old	3,778	1,867	1,912	83 years. old	739	249	489
39 years. old	4,119	2,047	2,072	84 years. old	644	211	433
40 to 44 years. old	18,754	9,258	9,496	**85 to 89 years. old**	2,090	625	1,465
40 years. old	3,829	1,891	1,938	**90 to 94 years. old**	812	201	611
41 years. old	3,716	1,833	1,883	**95 to 99 years. old**	214	45	169
42 years. old	3,629	1,788	1,841	**100 years. & over**	44	10	34
43 years. old	3,585	1,761	1,824	**Median age (yr.)**	33.1	31.9	34.3
44 years. old	3,996	1,986	2,010				

Source: U.S. Bureau of the Census, Current Population Reports, P25-1095 in *The American Almanac: Statistical: Abstract of the United States, 1993.*

Table 1.5 shows only the data on 5-year age intervals for the total population from Table 1.4. (Exploration 1.4 follows an alternate path to a "60 Second Summary" and compares the distribution of ages for men versus women.) By focusing on one aspect of the data and simplifying details, some patterns become recognizable.

Table 1.5			
Ages of 1991 US Population in 5 Year Intervals			
Age	Frequency count 1000's	Age	Frequency count 1000's
Under 5	19,222	55 to 59	10,423
5 to 9	18,237	60 to 64	10,582
10 to 14	17,671	65 to 69	10,037
15 to 19	17,205	70 to 74	8,242
20 to 24	19,194	75 to 79	6,279
25 to 29	20,718	80 to 84	4,035
30 to 34	22,159	85 to 89	2,090
35 to 39	20,518	90 to 94	812
40 to 44	18,754	95 to 99	214
45 to 49	14,094	100 & over	44
50 to 54	11,645	Total	252,177

Source: U.S. Bureau of the Census, Current Population Reports, P25-1095 in *The American Almanac: Statistical Abstract of the United States, 1993.*

Note: Includes armed forces abroad

In Table 1.5 the first and third columns display age intervals of 5 years and the second and fourth columns display the frequency count for that interval. The interval size from 5 to 9 years of age includes the endpoints of 5 years and 9 years; thus the interval includes all those people who are 5, 6, 7, 8 or 9 years of age.

Examine the table. What five year interval has the highest frequency count? As we look at the age intervals from youngest to oldest, when is the frequency count increasing? When is it decreasing?

We could further summarize the data by compressing the data into 10-year intervals and calculating the relative frequencies as shown in Table 1.6.

Table 1.6		
Ages of 1991 US Population in 10 Year Intervals		
Age	Frequency count 1000's	Relative Frequency (%)*
Under 10	37,459	15%
10 to 19	34,876	14%
20 to 29	39,912	16%
30 to 39	42,677	17%
40 to 49	32,848	13%
50 to 59	22,068	9%
60 to 69	20,620	8%
70 to 79	14,521	6%
80 & up	7,195	3%
Total	**252,177**	**100%**

Based on data from: U.S. Bureau of the Census, Current Population Reports, P25-1095 in *The American Almanac: Statistical Abstract of the United States*, 1993.
Note: Includes armed forces abroad
*Relative frequency total may not be exact because of rounding

Examine Table 1.6. What patterns can be seen? What information has been lost?

Jottings for a 60 second summary

When data are summarized, information is lost. In Tables 1.5 and 1.6 we can no longer tell how many people are a certain age: for example, we can't tell from these tables how many people are 18 years old. But we have more readable tables, that reveal certain facts that were hidden before. We first ask questions and record some observations, focusing on patterns that might be useful for constructing a 60 second summary.

Questions / observations from the table of 5-year intervals:

What's the biggest number? In what age interval does it occur?
- the biggest number is about 22 million for those between 30 and 34 years.

What are the overall trends? When are the population numbers increasing? Decreasing?
- between 0 and 19, the number of people in each successive age interval decreased
- then between 15 and 34, the number of people in each successive age interval started increasing, reaching the maximum in the 30-34 year interval.
- after 34 years, there was generally a steady decline in the number of people for each age group.

Questions / observations from the table of 10-year intervals:

What's the biggest number now? In what age interval does it occur?
- the biggest number is about 43 million for those between 30 and 39 years old

Are the overall trends still the same?
- Yes. The same broad trends are evident even after doubling the interval size. The numbers decreased initially until about age 20. Then they increased, reaching a peak for those in their thirties. After age 40, the numbers started to decrease again.

Any other observations?
- the biggest drops (of about 10,000,000 each) seem to occur between those in their thirties and forties, and those in their forties and fifties.

Take a moment to jot down in your class notebook any other patterns you can find in the data tables that might be useful for a 60 second summary.

An Introduction to Algebra Aerobics

At the end of most sections in the text you will find a short "algebra aerobics" workout with answers in a footnote. They are intended to give you practice in the mechanical skills introduced in the preceding section. In general the problems shouldn't take you more than a few minutes to do. So, we recommend reading the text with pencil in hand, and working out all these practice problems (before peeking at the answers!)

Algebra Aerobics:[2]
1. In Table 1.3, what percentage of the students are less than 30 years old?
2. In Table 1.5 what percentage of the population is 50 to 64 years old?
3. Refer to Table 1.6
    a.) Suppose that, between 1991 and 2001, the number of people over the age of 69 increases by 10% of its 1991 base value. How many people would be in that age category in 2001?
    b) Now suppose that between 2001 and 2011 the number of people over 69 decreases by 10% of the new 2001 base value. In the year 2011, how many people will there be who are at least 70 years old?
4. Fill in Table 1.7 :

Table 1.7		
Age	Frequency Count	Relative Frequency (%)
1-20	52	
21-40	35	
41-60	28	
61-80	22	
Total		

---

[2] Answers
1. $24\% + 24\% + 5\% + 14\% = 67\%$
    Note: If you calculated these by adding frequencies first, you would get: $14/21 = 0.667 = 67\%$
2. $11{,}645 + 10{,}423 + 10{,}582 = 32{,}650$ $32{,}650/252{,}177 = 0.129 = 13\%$
3. a) $14{,}521 + 7{,}195 = 21{,}716$ people over age 69 10% of $21{,}716 = (0.1)(21{,}716) = 2{,}172$
    $2{,}172 + 21{,}716 = 23{,}888$ (or take 110% of $21{,}716 = 23{,}888$)
    b) 10% of $23{,}888 = (0.1)(23{,}888 = 2{,}389)$ $23{,}888 - 2{,}389 = 21{,}499$ people over age 69
4. To find the relative frequency, divide the frequency count by the total count. To convert from a decimal to a percent multiply by 100.

Age	Frequency Count	Relative Frequency (%)
1-20	52	52/137=0.38 or 38%
21-40	35	35/137=0.26 or 26%
41-60	28	28/137=0.20 or 20%
61-80	22	22/137=0.16 or 16%
Total	137	137/137=1 or 100%

1.3   Visualizing Data

Humans are visual creatures. By converting a data table into a picture, we can recognize patterns that may be hard to discern in a list of numbers. There are many different ways to display data. (See the Anthology reading "Visualization.") The kind of graph or chart chosen can influence which aspects of the data emerge and which remain hidden. In this chapter we focus on *histograms*, which show how data are distributed over a range of values. A histogram is a graph on which the horizontal axis represents a section of the number line that has been marked off, usually in equal sized intervals (in order to facilitate comparisons between intervals). The following histograms have been constructed from the previous tables of data on U.S. population. They all show relative frequencies, but each uses a different interval size on the horizontal axis. In the histogram for individual ages the data beyond 84 years is shown in 5-year intervals since the data for individual ages was not included in the original Table.

Distribution of United States Population by Age in 1991

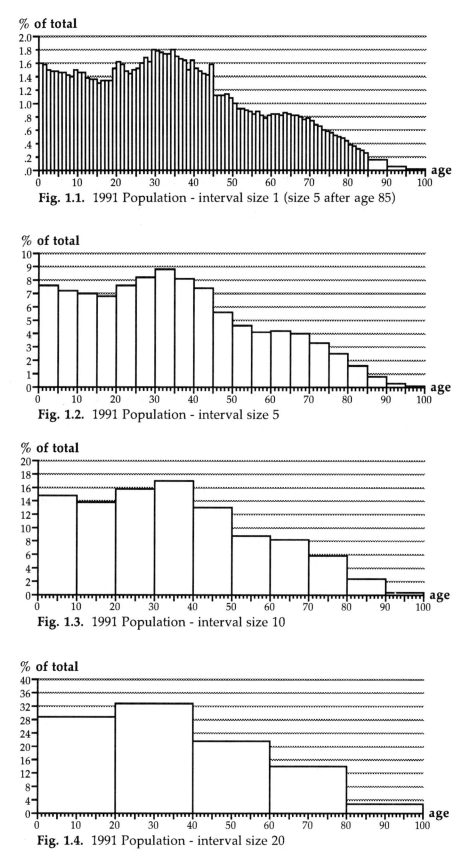

Fig. 1.1. 1991 Population - interval size 1 (size 5 after age 85)

Fig. 1.2. 1991 Population - interval size 5

Fig. 1.3. 1991 Population - interval size 10

Fig. 1.4. 1991 Population - interval size 20

Source: U.S. Bureau of the Census, Current Population Reports, P25-1095 in *The American Almanac: Statistical Abstract of the United States, 1993.*

More jottings for the 60 second summary

How do histograms help us find patterns in the data? When data are condensed by summarizing them into larger intervals, subjective decisions must be made. With a large number of small intervals, the complexity of the image can be confusing. With a small number of large intervals, meaningful information may be lost through overgeneralization. For example, the slight rise in the percentage of people in the 60-64 group compared to the 55-59 group is lost in the histogram with 10 year intervals. The decline in the percentage of people in the 10-19 group compared to the 0-9 group is lost using 20 year intervals.

Examining the graphs may lead us to useful questions about the data. To find answers to these questions it is sometimes helpful to refer back to the data table or other graphs, or to do some calculations with the data.

Some questions/observations:

What do you notice about all four histograms?
- they all show a general decline in the percentage at each age, interrupted by a large bump occurring between about 20 and 45 years

What can we say about this bump?
- the peak of this bump occurs at about age 30. Checking back in Table 1.4 we can determine that the peak actually occurs at age 29; that is, there are more people 29 years old (about 4,559,000 of them) than any other age
- on the 5-year interval histogram, the peak occurs between 30 and 35
- on the 10-year interval histogram, the peak occurs between 30 and 39

Anything else?
- there is a sharp drop at about age 45
- the histograms seem to flatten out between about 55 and 70

In your notebook jot down any other patterns you see that might be useful for the 60 second summary. For instance, you might want to estimate the percentage of the population is of working age. What percentage children? What percentage retired? How do you think this percentage breakdown might shift over the next 50 years?

Tradeoffs in representing data

Choosing the size of the interval affects what type of information is revealed in a histogram of the data. The type of graph we choose to use can also affect what aspects of the data are highlighted. The pie chart is a common graph used in newspapers and magazine articles to display relative frequencies.

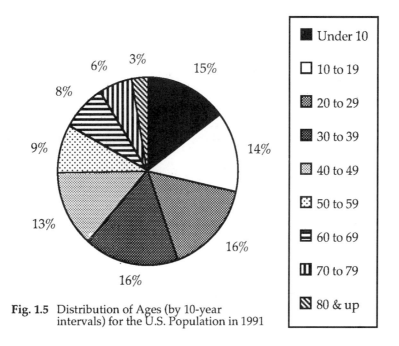

Fig. 1.5 Distribution of Ages (by 10-year intervals) for the U.S. Population in 1991

Source: U.S. Bureau of the Census, Current Population Reports, P25-1095 in *The American Almanac: Statistical Abstract of the United States, 1993.*

Something to think about

In what ways is the pie chart easier to read and interpret than the histogram for 10-year age intervals? What type of information is easier to see in the histogram?
What are the general tradeoffs in using pie charts vs. histograms?

An important aside: what a good graph should include

When you encounter a graph in a newspaper or you produce one for class, there are three elements the graph should always include:

1. An informative title that succinctly describes the graph
2. Clearly labeled axes including the units of measurement (e.g., is age being measured in years or months)
3. The source of the data, either included in the table, or identified in the text.

Algebra aerobics:[3]

1. Use the table to create a histogram and a pie chart.

Table 1.8		
Age	**Frequency Count**	**Relative Frequency (%)**
1-20	52	38%
21-40	35	26%
41-60	28	20%
61-80	22	16%
Total	**137**	**100%**

2. From the histogram in Figure 1.6, create a frequency distribution table. Assume that the total number of people represented by the histogram is 1352.

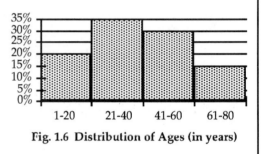

Fig. 1.6 Distribution of Ages (in years)

---

[3] Answers:

1.

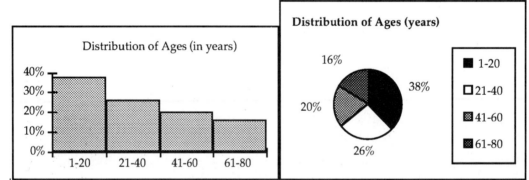

2.

Age	Relative Frequency (%)	Frequency Count	
1-20	20%	(0.20)(1352)=	270
21-40	35%	(0.35)(1352)=	473
41-60	30%	(0.30)(1352)=	406
61-80	15%	(0.15)(1352)=	203
Total	**100%**		1352

1.4   What Is Average Anyway?

In addition to graphs and tables, data are often summarized by using numerical descriptors. Three numerical descriptors, called the *mean, median,* and *mode,* offer different compact ways of describing and comparing data sets. Each descriptor can be thought of as a "one second summary" of data. Although each term is sometimes incorrectly called *average* only the mean should be referred to as the average.[4]

Mean, Median, Mode

The *mean* is the arithmetic average with which you are probably already familiar.

> The *mean* of a list of numbers is their sum divided by how many there are.

Example:

The following set of data extracted from Table 1.1 contains the ages (in years) of students in a college algebra class.

18, 19, 20, 20, 20, 21, 22, 23, 23, 23, 26, 27, 28, 29, 30, 30, 33, 36, 39, 41, 46

The mean age can be found by adding together all of the 21 observed ages and dividing the resulting sum by 21, the total number of observations.

(18+19+20+20+20+21+22+23+23+23+26+27+28+29+30+30+33+36+39+41+46) = 574

So the mean is: 574/21 = 27.33

To describe this process in general, we can represent the age of each student by a letter with subscripts; thus for a class with n students, the ages can be described as $a_1, a_2, a_3,......, a_n$

The mean of n values, $a_1, a_2, a_3,...., a_n$, is found by adding these values and dividing the total by n, the total number of students in the class.

$$\text{mean} = \frac{(a_1 + a_2 + a_3.....+a_n)}{n}$$

---

[4] The word average has an interesting derivation according to Klein's etymological dictionary. It comes originally from the Arabic word "awariyan" which means merchandise damaged by sea water. The idea being debated was that if your ships arrived with water damaged merchandise, should you have to bear all the losses yourself, or should they be spread around or "averaged" among all the other merchants? The words "averìa" in Spanish, "avaria" in Italian, "avarie" in French still mean damage.

The *median* locates the "middle" of a numerically ordered list. More formally:

> The *median* separates a numerically ordered list of numbers into two parts,
> with half the numbers at or below the median and half at or above the median.
> If the number of observations is odd, the median is the middle number in the
> ordered list. If the number of observations is even, the median is the mean
> of the two middle numbers in the ordered list.

Example:

In the previous ordered list of ages, there are 21 numbers. Since the list has an odd
number of terms, the median is the middle or 11th number which is 26 years.

$$18, 19, 20, 20, 20, 21, 22, 23, 23, 23, \mathbf{26}, 27, 28, 29, 30, 30, 33, 36, 39, 41, 46$$
$$\text{median} = 26$$

If the 22 year-old person dropped out of the class, the list would contain 20 ages. Since
the list now has an even number of terms, the median would be the mean of the
middle two numbers, in this case the 10th and 11th numbers. So the median would be
the mean of 26 and 27 which equals $(26 + 27)/2$ or 26.5

$$18, 19, 20, 20, 20, 21, 23, 23, 23, \mathbf{26, 27,} 28, 29, 30, 30, 33, 36, 39, 41, 46$$
$$\text{median} = (26+27/2=26.5$$

Neither the mean nor the median need to be one of the numbers on the original list.

> The *mode* is the number that occurs with the greatest frequency. If data are
> grouped into intervals, then the interval with the greatest frequency count
> is called the *modal interval*.

There may be more than one mode for a set of data. Of the mean, median, and mode, the
mode is the least used of the three measures.

To find the mode of a list of numbers, arrange them in ascending order and look for the
number(s) with the greatest frequency counts. In the class data on ages, there are 3 people
who are 20, and 3 people who are 23; thus both 20 and 23 are modes. This data set is said to
be *bimodal.* If the data is grouped into intervals, do the same to find the modal interval or
intervals.

Since the word *average* is sometimes used incorrectly to represent any one of these three
terms, it is often not clear what is meant by a statement like "the *average* salary for an

American is $23,000". Is $23,000 the mean– the result of adding together the salaries of all Americans and then dividing by the total number of people whose salaries were counted? Does it mean that $23,000 is the median– that half of Americans make less than $23,000 and half make more? Or is $23,000 the mode– that if you rounded off salaries to the nearest thousand, more Americans would make $23,000 than any other salary?

Although the mean, median, and mode are often confused, the meanings are distinct and the values can be quite different. Since the term *average* is used very loosely, it is important to ask what is actually meant when you encounter it.

Visualizing the mean, median and mode

Of these three measures, the mode is the only one that can be read directly from a histogram. It's the value or interval where the histogram peaks. The following data set has a mode of 7. The frequency count shows that the number 7 appeared 18 times, which was higher than the count for any other value.

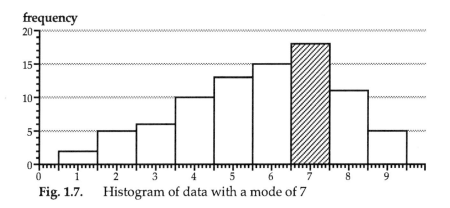

Fig. 1.7. Histogram of data with a mode of 7

If the histogram has two distinct peaks, even if they are not exactly the same height, we still call the distribution bimodal. For example, the histogram in Figure 1.8 shows the time between eruptions of probably the most famous geyser in the world, "Old Faithful" in Yellowstone National Park, Wyoming. The data represents 222 intereruption times taken during August 1978 and August 1979 (cf. *Applied Linear Regression*, 2nd ed., by S. Weisberg). The data is bimodal, with peaks occurring between 51 and 54 minutes and between 75 and 78 minutes. This suggests that the data clusters into two separate subgroups.

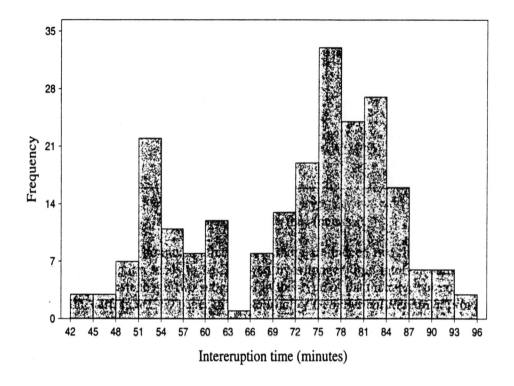

Fig. 1.8 Intereruption time (in minutes) of "Old Faithful"

Source: S. Chatterjee, M. Handcock and Jeffrey Simonoff, *A Casebook for a First Course in Statistics and Data Analysis"*, John Wiley & Sons, Inc., New York, 1995, p. 5.

Something to think about

Try sketching in your notebook what the histogram of heights of students in a coed class might look like. How would the fact that men are in general taller than women affect the shape of the histogram?

The median divides the area of the histogram in half, so that the areas to the left and right of the median are roughly equal. Table 1.4 gives the median age of the U.S. population in 1991 as 33.1 years. A vertical line drawn through the median on any of the previous age histograms of the U.S. population should divide the histogram into two approximately equal areas.

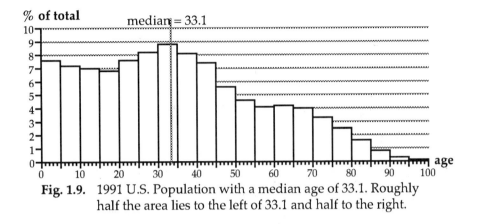

Fig. 1.9. 1991 U.S. Population with a median age of 33.1. Roughly half the area lies to the left of 33.1 and half to the right.

The precise location of the mean is difficult to visualize. If you have an intuitive feeling for center of gravity and balance, the mean is where the histogram would balance on a seesaw if the histogram were made from a solid material.

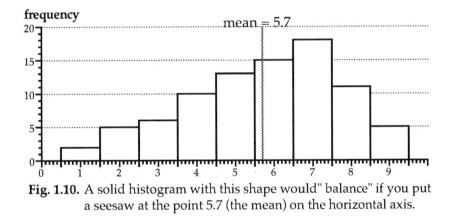

Fig. 1.10. A solid histogram with this shape would" balance" if you put a seesaw at the point 5.7 (the mean) on the horizontal axis.

The distribution of values on a histogram is *symmetric* if the right and the left halves are mirror images of each other about a center line as seen in Figure 1.11. With perfectly symmetric data the mean and the median will be the same.

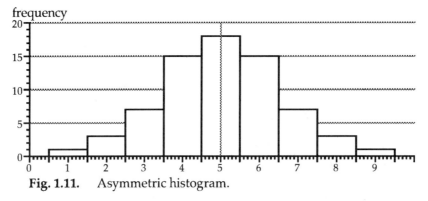

Fig. 1.11.   Asymmetric histogram.

The distribution of values is *skewed* to the right if there is a long right tail and skewed to the left is there is a long left tail. In Figure 1.12, the histogram of personal wages for 1000 people in the U.S. is skewed to the right. When there is a long right tail, you would expect the mean to be larger than the median. In this case for instance, the median personal wage is $18,000 and the mean personal wage is $21,914. A line through $18,000 would divide the area of the histogram in half. The histogram would balance on a line drawn through $21,914.

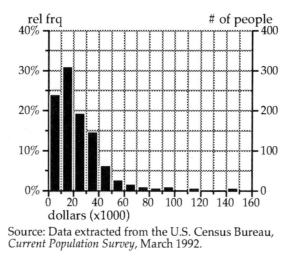

Source: Data extracted from the U.S. Census Bureau,
Current Population Survey, March 1992.
Fig. 1.12 Right skewed histogram of personal wages for 1000 individuals in U.S.

If you have a distribution skewed to the left, and hence a long left tail, you would expect the mean to be less than the median.

Something to think about

1. If the distribution is skewed to the right, why should the mean be larger than the median? If the distribution is skewed to the left, why should the mean be smaller than the median?
2. Read "The Median Isn't the Message," to find out why a right skewed distribution gives new hope to the author, a renowned Harvard scientist diagnosed with a lethal form of cancer.

Significance of the mean, median and mode

Mean, median and mode are all useful in different ways. The mean and median are ways to describe the "center" of the data and hence are called *measures of central tendency*.

The *mode* is the most common value in a set of data. The term mode is sometimes associated with the fashion industry, since mode can be thought of as the most popular data value, or the one that is the most in fashion.

We can look at the course of the mode as it moves through time like the peak of a wave. One example is the "baby boom," the explosion in numbers of babies that occurred just after World War II. College admissions officers scrutinize population histograms to plan for advancing waves of students. As a middle-aged population peak advances into old age, society needs to plan to be able to meet their housing, health care, and other needs. A moving peak also has a negative side: if a country has built additional schools for an incoming wave of high school and college students, what do they do with the facilities when the crest has passed by, and the student population falls? Understanding the mode and how it might change is important in both short and long range planning.

The special significance of *median* is that it divides the number of entries in the data set into two equal halves. If the median age in a large urban housing project is 17, then half the population is 17 or under. Hence issues of day care, recreation, supervision of minors, and kindergarten through 12th grade education should be very high priorities with the management. The other half of the population is, of course, 17 or older. If the adults in that same project included a fairly large proportion of elders, the *mean* age could be 35; by itself this number would give no indication that a majority of the residents were minors.

The major disadvantage of the median is that it is unchanged by the redistribution of values above and below the median. For example, as long as the median income is larger than the poverty level, it would remain the same if all poor people suddenly increased their incomes up to that level and everyone else's income stayed the same.

The *mean* is the most commonly cited statistic in the news media. (According to the Anthology article from the electronic magazine *CHANCE News* it is not always correctly used.) One advantage of the mean is that it can be used for calculations relating to the whole data set. Suppose a corporation wants to open a new factory similar to its other factories. If the managers know the mean cost of wages and benefits for an employee, they can make an estimate of what it will cost to employ the number of workers needed to run the new factory.

total employee cost =   (mean cost) (number of employees)

Something to think about

Why would you not want to use the median cost or mode cost of wages and benefits for an employee in this equation?

The major disadvantage of the mean is that it is affected by *outliers*, extreme values in the data set. John Schwartz comments in a *Washington Post* article ("Mean statistics: when is the average best?," December 6. 1995, p. H7.) that if Bill Gates, billionaire chief executive of Microsoft, were to move into a town with 10,000 penniless people the mean income would be more than a million dollars. This might suggest that the town is full of millionaires, which is obviously far from the truth. The median income, however, is still $0.

Adding numerical descriptors to the 60 second summary

What are the mean, median and mode for the U.S. population?
- median age is 33.1 years (given at the bottom of Table 1.4, the original U.S. population table on age).

- in the histogram with 1-year intervals (Fig. 1.1) the two most common ages seem to be 29 and 34. Checking the values in the data table we find that there were slightly more 29 year-olds than 34 year-olds. So the mode is 29. (Or we might say the data is bimodal with modes 29 and 34.)

- in the 5-year interval histogram, the modal interval is 30-34. In the 10-year histogram it is 30-39.

- no mean age is given in Census data in Table 1.4. (One of the homework problems asks you to use Table 1.6 to estimate the mean age for the U.S. population.)

Algebra aerobics:[5]
1. Calculate the mean, median and mode for the following data:
   - a.) $475, $250, $300, $450, $275, $300, $6000, $400, $300
   - b) 0.4, 0.3, 0.3, 0.7, 1.2, 0.5, 0.9, 0.4
2. Explain why the mean may be a misleading numerical summary of the data in 1a)
3. Seven members of an eight-man track team reported the following times (in minutes) for the mile run: 5.25, 4.21, 5.04, 3.58, 4.41, 4.90 and 5.14
   If the team's mean time was 4.60 minutes, what was the time of the eighth runner?

---

[5] Answers
1. a) sum = $8750   mean = $8750/9 = $972.22   median = mode = $300
   b) sum = 4.7   mean = 4.7/8 = 0.59   median = $(0.4 + 0.5)/2 = 0.45$
   The two modes are 0.3 and 0.4 (bimodal)
2. One of the values ($6000) is much higher than the others, which forces a high value for the mean.. In cases like this, the median is generally a better choice for measuring central tendency.
3. Let t represent the unknown time. The sum of the seven reported times is 32.53. Therefore,
   $(32.53 + t) / 8 = 4.6$ => $32.53 + t = 36.8$ => $t = 4.27$ minutes

Something to think about

1. If someone tells you that in his town "all of the children are above average" you might be skeptical. (This is called the "Lake Wobegon effect.") But could most (more than half) of the children be above average? Explain.

2. Why do you think most researchers use median rather than mean income when studying "typical" households?

3. Herb Caen, a Pulitzer prize winning columnist for the *San Francisco Chronicle*, remarked that a person moving from state A to state B could raise the average IQ in both states. Is he right? Explain.

1.5   Writing About Data

Whether you are constructing a 60 second summary to present to your class, writing an executive summary for a report to your boss or simply trying to understand what is going on in your business, you can help your ideas take shape by putting them down on paper. At first your notes may be mere sketches, incomplete and perhaps with errors. As you think through your ideas they will become clearer, stronger and more focused. Writing a succinct, focused summary is difficult. Though the end product may be short, the time invested in it creating it can be quite lengthy. But the skills you hone here should prove useful throughout your life. The essential steps are: to get started no matter how tentatively, and then to polish and refine your ideas. We offer here one (but certainly not the only) strategy for constructing short, clear summaries..

Constructing a 60 second summary of U.S. age data

How would you go about constructing a paragraph describing the distribution of ages in the U.S.? Many different types of summaries are possible. For a one paragraph summary, it is usually best to have a single key point and then to weave together evidence to support that point.

1)   Make jottings

Start by jotting down a list of your initial questions and observations from tables, graphs and numerical descriptors. We've included a condensed list of some of the previous jottings.

What were the general patterns in the data?
- a general steady decline in percentages for each age interval with a large bump roughly between ages 20 and 45
- between 0 and 19, the number of people in each successive 5-year age interval decreased.
- between 15 and 34, the number of people in each successive 5-year age interval increased, reaching a maximum number of about 22 million for those between 30 and 34 years.
- after 34 years, there is generally a steady decline in the number of people for each 5-year age group.

What were the values for the basic numerical descriptors: mean, median and mode?
- median age was 33.1 years
- mode was 29 years
- the mean was not available

Any other observations?
- there was a sharp drop at about age 45

2)   Pose additional questions

Exploring the data may raise additional questions such as:

- What percentage of the population are children? What percentage elderly?
- What percentage of the population is part of the work force?
- Where does the baby boom that occurred right after World War II show up in the data?
- What caused the sharp drop at age 45?

3)   Identify the key idea

The next step is to identify a key point. This forces you to reach some kind of summary statement about the data. You may want to use this statement as your *topic* or *opening* sentence.

In 1991 there was a general decline in the percentage of the population in each successive age interval which was interrupted by a large bump between 20 and 45.

4)   Provide additional supporting evidence

Construct quantitative statements that support, expand or modify your key idea. Your jottings may give enough information or as shown below you may have to go back to tables and graphs to find more information.

- about 17% of the population was under 20
- about 13% was 65 or over
- the "baby boomers" were born roughly between 1946 and 1960. Since the data is for 1991, those born between 1946 and 1960 would be between 31 and 45 years old in the data set. (So if we think of reading the individual year histogram (Figure 1.1) from right to left, we see that there is a big surge in 45 year olds (as compared to 46 year olds). This may signal the beginning of the baby boom. Continuing to read from right to left, the boom seems to peak with the 30 year olds, perhaps signaling the end of the baby boom. Thus the baby boom may be the cause of the bump, and (reading from left to right) the sharp drop at age 45.)

5)   Put ideas together

Construct a paragraph describing some aspect of the data. You may wish to start with a topic sentence containing your key idea, followed by supporting evidence that might include a graph or table. You will probably want to weave back and forth among the steps in order to organize your thoughts, then pick and choose what evidence you want to include. You could then finish with a concluding sentence.

PATTERNS IN THE 1991 U.S. POPULATION

In 1991, there was a general decline in the percentage of the population in each successive age interval which was interrupted by a large surge roughly between the ages of 20 and 45. Grouping the ages into 5-year intervals we can see in the accompanying histogram that from newborns to teenagers, the percentage of people in each successive interval decreased. Between the ages of 15 and 34, the percentage of people in each successive interval increased, reaching a maximum number of about 9 % for those between 30 and 34 years. The median age of 33.1 is located in this peak. After age 34, there is a return to a steady decline in the number of people for each age group. This overall wave pattern may be due in part to the "baby boom," the explosion of births after World War II.

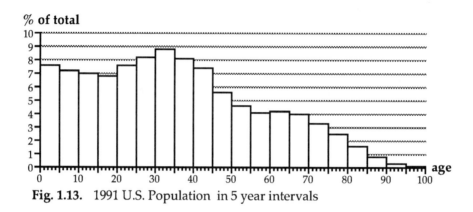

Fig. 1.13.   1991 U.S. Population in 5 year intervals

Questions that require more research

The previous paragraph offered a description of the age distribution of the U.S. population. Analysis often exposes limitations of the data and generates new questions. Some of the questions raised in the analysis of age data would require more data, such as information about death rates and birth rates, to answer them. Other questions may require research in other disciplines such as history, sociology, medicine, or political science.

- Is the baby boom really the cause of the sharp drop in the population percentage after 45 years of age?

Does the drop reflect an increase in the birth rate back in 1946? Or does it perhaps reflect a sharp increase in the death rate for people at age 46? Could it reflect some historical event (like the Vietnam War) that killed many who would have been 45 years old in 1991? Was there an influx in immigrants who are now 45 and under? The number of people at any given age is determined by a number of factors, including the birth rate, the death rate, and immigration patterns.

- What will be the impact on America's future as those in the 20 to 45 population bulge get older?

- How does the distribution of ages in the U.S. compare with those of other industrialized nations like Japan or Germany?

- How does the distribution of ages in the U.S. compare to those of developing countries like China? (There is a homework problem on this near the end of the Exercises.)

These are the kinds of questions asked by social scientists.

Starting with a question: What lies ahead?

Data are usually collected and explored for a specific purpose. Often data analysis is prompted by the need to answer a particular question, like:

How will the distribution of ages in the U.S. change over time?

In addition to collecting current demographic data, the U.S. Census Bureau makes predictions about the future. The following table and graph include the Bureau's projections for the distribution of ages in the U.S. population in the year 2050.

Table 1.9 Distribution of Ages of U.S. Population in 1991 and 2050 (projected)									
	POP. IN 1991		POP. IN 2050			POP. IN 1991		POP. IN 2050	
Age	Total (in 000's)	%	Total (in 000's)	%	Age	Total (in 000's)	%	Total (in 000's)	%
Under 5	19,222	8%	25,382	6%	60 to 64	10,582	4%	20,553	5%
5 to 9	18,237	7%	25,222	6%	65 to 69	10,037	4%	18,859	5%
10 to 14	17,671	7%	25,650	7%	70 to 74	8,242	3%	15,769	4%
15 to 19	17,205	7%	25,897	7%	75 to 79	6,279	2%	14,510	4%
20 to 24	19,194	8%	25,313	6%	80 to 84	4,035	2%	12,078	3%
25 to 29	20,718	8%	24,659	6%	85 to 89	2,090	1%	9,283	2%
30 to 34	22,159	9%	24,803	6%	90 to 94	812	0%	5,779	1%
35 to 39	20,518	8%	24,316	6%	95 to 99	214	0%	2,623	1%
40 to 44	18,754	7%	23,423	6%	100 & over	44	0%	1,208	0%
45 to 49	14,094	6%	22,266	6%	Total	252,177	100%	392,031	100%
50 to 54	11,645	5%	22,071	6%	Median Age	33		39	
55 to 59	10,423	4%	22,367	6%					

Source: Based on data from: U.S. Bureau of the Census, Current Population Reports, P25-1095 and P25-1104. Note: Includes armed forces abroad
*Relative frequency total may not be exact because of rounding

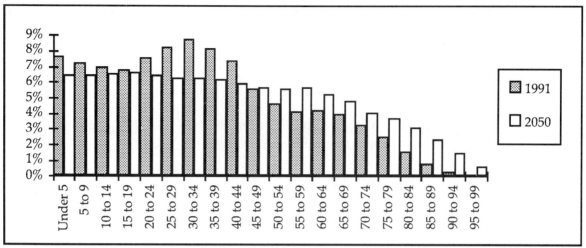

Fig. 1.14 Distribution of Ages of U.S. Population in 1991 and 2050 (Projected)

Source: Based on data from: U.S. Bureau of the Census, Current Population Reports, P25-1095 and P25-104.
Note: Includes armed forces abroad

*Relative frequency total may not be exact because of rounding

Let's continue the writing process with you taking a more active role this time. In your notebook, try going through the process of constructing your own 60 second summary comparing the two age distributions. You could work entirely on your own or flesh out the following outlines.

1)   Make jottings

      Jot down some questions and observations about the key features of the data table and graphs, expanding on what is written here and adding some observations of your own..

          What is overall change in the distribution of ages between 1991 and 2050.?
          •   there is a marked shift in ....
          •   ....

          What is the change in the mean and median ?
          •   the median is ......
          •   ....

          Other observations?
          •   there are more elderly in 2050.
          •
          •

2)   Pose additional questions

      What is the projected shift in the workforce between 1991 and 2050?

- If we assume the potential workforce consists of all those between 20 and 65, then ....

If the predictions are correct, what impact might such a dramatic shift in the distribution of age have on society? What costs would be expected to rise? to fall?

- ....
- ....

Other questions?

3)   Identify the key idea

The U.S. population is expected to age dramatically .

4)   Provide additional supporting evidence

- in 1991 the distribution of ages was very uneven. In contrast by 2050 .....
- the percentage of the population over 65 ....
- the percentage under 20 ...
- ....
- ....

5)   Put ideas together

Writing tip:

If we are to construct a paragraph comparing populations, there are two aspects to consider: organization and language. In making comparisons between two data sets, there are two different organizational strategies. You can say everything about one data set and then everything about the other data set, or you can interweave comments, comparing and contrasting as you go. There are certain vocabulary words and phrases that are particularly useful in making comparisons: *like, unlike, in contrast, although, both, whereas, on the other hand,* and *but* draw attention to similarities or differences.

Hints to get started.

Fill in the 60 second summary with supporting evidence from your notes or create your own summary. You may wish to include tables and/or graphs.

THE AGING U.S. POPULATION

Projections made by the Bureau of the Census indicate that the population of the United States will be substantially older in the year 2050. ....

While the percentage of elderly is expected to increase, the percentage of children is expected to decrease......,

(CONCLUDING SENTENCE).......

Writing tip:

Your concluding sentence could be a summary statement of the evidence such as:

The U.S. Census data clearly indicate that the U.S. population is expected to age dramatically during the next century.

Or it could suggest plausible reasons for the patterns observed:

The main cause for this shift in projected ages is probably increased life expectancy, but birth, death, and immigration rates would figure in too.

Or it could raise questions about the implications of your observations:

What impact will this dramatic aging of the U.S. population have on future spending patterns on Social Security, health care and education?

The Search for Answers

Mathematics provides the tools to describe patterns in populations. It is the job of social scientists to interpret what these patterns mean. Social scientists use these tools to form and test hypotheses, and to reach (perhaps controversial) conclusions. For example, since the U.S. Census Bureau projections indicate that there will be fewer people working, and more people retired, some policy advisers have warned that the United States is facing a social security funding crisis.[6] In a speech to the U.S. Senate, Bill Bradley produced a chart that

---

[6] The following analysis relies heavily on arguments found in a section entitled "Will You Still Feed Me?" in Mark Maier's *The Data Game: Controversies in Social Science Statistics"*, M.E.Sharpe, Armonk, New York., 1991.

showed that in 1950 one Social Security beneficiary was supported by 16.5 workers, in 1980 one beneficiary was supported by 3.3 workers, while by 2030 there may be only 2 workers supporting each beneficiary. In *Too Many Promises: The Uncertain Future of Social Security* Michael Boskin, chief economic adviser to President Bush, argues that the well-off should rely on private pension programs rather than social security benefits.

Other social scientists are not so pessimistic. While everyone seems to agree on the inevitability of an increasingly older population, some point out that the population will shrink at the bottom while it grows at the top. They argue that we should be looking at the total dependent population, which consists of the young as well as the elderly. In 1991 the sum of those under 20 and over 65 was approximately 41%. In 2050 the corresponding sum is expected to be about 46%. So the percentage of the population requiring support may only increase by about 5%. Federal expenditures for children, however, are roughly one-sixth of the amount spent on adults. Most of the financial support for children comes from their parents. To shift resources from children to the elderly would require drastic and probably unpopular political decisions. The harsh reality is that most adults are far more willing to pay the expenses for their own children than to be taxed to support an elderly population.

1.6. Does the whole process make sense?

Whether reading a quantitative argument in the newspaper or constructing your own 60 second summary, the fundamental question to keep in mind is "Does it all make sense?" Any conclusion that claims to be drawn from the analysis of data should be approached with reasoned skepticism. We need constantly to ask questions about how much we trust the data and the conclusions drawn from them.

How good are the data?

- **What definitions were used ?**

Table 1.10
Estimates of Unemployment for the Month of
November 1935 According to Five Reporting Agencies

Agency Preparing Estimates	*Estimate of Number Unemployed*
The National Industrial Conference Board	9,177,000
Government Committee on Economic Security	10,913,000
The American Federation of Labor	10,077,000
National Research League	14,173,000
Labor Research Association	17,029,000

Source: Jerome B. Cohen, "The Misuse of Statistics," *Journal of the American Statistical Association*, XXXIII, No. 204, (1938), p. 657. Reprinted with permission from *Journal of the American Statistical Associatio***ource:**.Copyright ©, 1938 by the Ameeerican Statistical Association.

During the Great Depression five reputable agencies published five very different unemployment estimates for the same month. The variance was due primarily to differences in the definitions of unemployment. For instance some estimates included people leaving school and seeking employment for the first time and some did not; some measured only blue collar unemployment, while others included unemployed professionals.

- **Was the choice of definition biased by the goal? Would another definition have been more useful?**

Different groups collecting data on the same issue may be gathering information for different purposes. The different definitions used by the above groups are tied to the group's concerns and objectives. Clearly, different arguments about the seriousness of the unemployment problem could be made depending on which estimates you choose to use for the number of unemployed people.

- **What is being measured? Is it relevant?**

In the nineteenth century, the "science" of craniology equated intelligence with brain, and hence skull size. It is possible to measure skull size with extreme accuracy, but is it reasonable to believe that head size is related to intelligence? Today we attempt to measure intelligence using the Intelligence Quotient and Scholastic Aptitude Tests. Many questions have been raised about what these tests really measure. Some people claim that they are twentieth century versions of craniology.

- **Are the measurements accurate and reliable?**

In May of 1993, *The New York Times* published an article entitled "Job Loss in Recession: Scratch Those Figures" explaining measurement errors in employment figures:

> Having declared last June that the recent recession had eliminated many more jobs than anyone had realized, the Labor Department is now canceling that view. The department says it overstated by 540,000 the number of jobs that were created in the late 1980's. And then it overstated how many jobs had disappeared in the recession.[7]

The measurements used to count jobs were faulty. Among other things, the Labor Department counted paychecks instead of the people receiving paychecks. Any person receiving both a regular check and a separate check for overtime was counted twice. Figure 1.15 from the *New York Times* article shows the original count, and what purports to be a more accurate count.

---

[7] "Job Loss in Recession: Scratch Those Figures". *The New York Times*, Friday, May 7, 1993, p. D1.

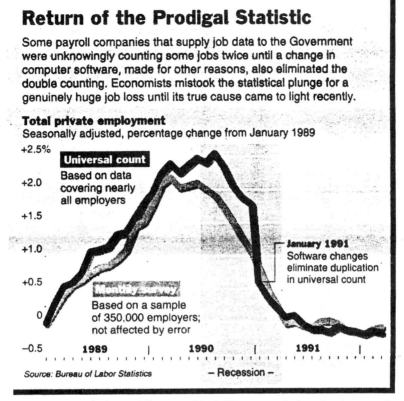

Fig. 1.15 Graph from *The New York Times* on federal miscounting of employment numbers

Source: Bureau of Labor Statistics. Copyright © 1993 by the New York Times Company. Reprinted by permission.

• **How big is the sample? How was the sample chosen? Is the sample representative of the total population?**

In the decennial census, the U.S. Census Bureau has attempted to collect data on everyone in the United States. Starting in the year 2000 however, the Bureau will collect data on 90% of the population, and by sampling the remaining 10% extrapolate information on the rest.

In his book *Flaws and Fallacies in Statistical Thinking*, Stephen Campbell cites the classic case of a poorly chosen sample. *The Literary Digest* during the 1936 Presidential election

> ... mailed out 10,000,000 ballots and had 2,300,000 returned. On the basis of this unusually large example, it confidently predicted that Alfred M. Landon would win by a comfortable margin. As it turned out, however, Franklin D. Roosevelt received 60 percent of the votes cast, a proportion representing one of the largest majorities in American presidential history.......[It turns out that the] ballots were sent primarily to upper-income types ... as a result of the magazine's having selected names from lists of its own subscribers, and telephone and automobile

owners. ...In this election, the presence of bias rendered the sample, despite its enormous size, inadequate as a representative subset of the population.[8]

How good are the conclusions?

Almost all efforts at data analysis seek, at some point, to generalize the results and extend the reach of the conclusions beyond a particular set of data. The inferential leap may be from past experiences to future ones, from a sample of a population to the whole population, or from a narrow range of a variable to a wider range. The real difficulty is in deciding when the extrapolation beyond the range of the variables is warranted and when it is merely naive. As usual, it is largely a matter of substantive judgment...."

From *Data Analysis for Politics and Policy* by Edward R. Tufte[9]

When we read conclusions drawn from data that are, as Tufte points out, often a matter of judgment we should ask:

- **Is the analysis accurate?**

"Boston researchers disclosed today [Oct. 6, 1993] that they had made a mathematical error in a major study published in July that seemed to provide reassuring news about the risk of breast cancer in women with a family history of the disease.....The study of nearly 130,000 women published in the July 21st issue of the Journal of the American Medical Association found that only 2.5 percent of all breast cancers occurred in women with a family history of the disease. When corrected, the percentage turned out to be 6 percent, much closer to previous estimates."[10]

Apparently the error occurred when a researcher carelessly transcribed numbers from two different tables and plugged them into a formula. Because the erroneous percentage obtained was so much lower than previous estimates, the report was front page news when it was published. Many women with family histories of breast cancer who had been cheered by the erroneous results were disappointed by the corrected findings.

---

[8] Campbell, Stephen, *Flaws and Fallacies in Statistical Thinking*, p. 148, © 1974, Reprinted by permission of Prentice Hall, Inc., Upper Saddle River, New Jersey.

[9] Tufte, Edward R., *Data Analysis for Politics and Policy*, Prentice-Hall, New Jersey, 1991, p. 32.

[10] "Cancer study error found. Key finding unchanged on breast disease risk", The Boston Globe, Wednesday, October 6, 1993, p. 1.

- **Could alternative conclusions better fit the data?**

When alternative plausible conclusions are derived from the same data, both conclusions may be wrong or they might both possess some truth. Stephen Campbell argues in his book, *Flaws and Fallacies in Statistical Thinking*, that:

> When equally plausible alternative conclusion can be reached from exactly the same statistical evidence, the logical link between evidence and conclusion offered is probably rather weak.
> Fact: "Fifteen individual nations have improved their infant mortality rates more than the United States since 1950."
> Conclusion offered: The state of health care in the United States is inadequate. Equally plausible alternative conclusion: The state of health care in several other countries - probably underdeveloped countries - has improved markedly since 1950, at least with respect to infant mortality.[11]

- **Are the data selectively reported?**

In his book *We're Number ONE: Where America Stands—and Falls – in the New World Order*, Andrew Shapiro challenges the assumption that America is "Number One" by showing how "statistics can be found to tell any story." One section is devoted to debunking America's Number One status on spending on education. He claims:
> *We're Number One in private spending on education.*
But ....   *We're Number 17 in public spending on education.*

> Including public and private spending, the United States spends about 7 percent of its gross domestic product (GDP) on education annually, or around $330 billion. By this measure, we do fairly well; only a few developed nations spend more overall relative to the size of their economies. But as with health care expenditures, the spending comparisons change dramatically when we distinguish between public and private sources. The United States is first in private spending on education among the nineteen major industrial nations, but only seventeenth in public spending as a percentage of GDP.[12]

In most cases you will not have access to the original data used, or have the time and energy to reconstruct the analysis done on the data. But you should always ask the kind of questions that we have raised here and critically examine the arguments made. Keep asking yourself:

Does it all make sense?

---

[11] Campbell, Stephen, *Flaws and Fallacies in Statistical Thinking*, p. 111, © 1974, Reprinted by permission of Prentice Hall, Inc., Upper Saddle River, New Jersey.

[12] Shapiro, Andrew, *We're Number ONE: Where America Stands - and Falls - In the New World Order*, Random House, New York, 1992.

Chapter 1 Summary

In statistical terms, characteristics like height, race, age, and college major that vary from person to person are called *variables*. Information collected about variables is called *data;*.

Variables can be classified as *quantitative variable* or *qualitative variables*. Quantitave variables, such as age, height, or number of people in your household can be represented by a number and a unit of measurement (for example, 18 years, 6 feet, or 3 people). Qualitative variables, such as gender or college major, require us to construct categories (like male or female; physics, English or psychology.)

The *frequency* or *frequency count* of any value or range of values of a variable is the number of times that value or values in that range occur in the data.

The *relative frequency* of any value or range of values of a variable is the fraction (often expressed as a percentage) of all the data in the sample having that value or range of values.

Data can be described using tables, graphs (such as *histograms* and *pie charts*), or numbers (like the mean).

The *mean* of a list of numbers is their sum divided by how many there are.

The *median* separates a numerically ordered list of numbers into two parts, with half the numbers at or below the median and half at or above the median. If the number of observations is odd, the median is the middle number in the ordered list. If the number of observations is even, the median is the mean of the two middle numbers in the ordered list.

The *mode* is the number that occurs with the greatest frequency. If data are grouped into intervals, then the interval with the greatest frequency count is called the *modal interval*. Modes indicate peaks in the data. If there are two distinct peaks the data is said to be *bimodal*.

Outliers are extreme, non typical values in a data set.

A good graph contains:
1.) An informative title
2.) Clearly labeled axes including the units of measurement
3.) The source of the data.

A "60 Second Summary" gives a concise focused description of patterns found in a data set.

EXERCISES

1. If you are using technology in your classroom, you need to become familiar with the features of your particular machine. This might mean completing a computer tutorial on the general features of an IBM or Mac, or working assignments on a graphing calculator. (If appropriate, you might want to refer to the basic set of instructions for the TI-82 and TI-83 contained in Appendix A. The instructions cover, chapter by chapter, the minimal TI-82 skills needed to do the chapter Explorations.) Your instructor will be able to guide you.

2. The following bar chart shows the predictions of the U.S. Census Bureau about the future racial composition of American society.

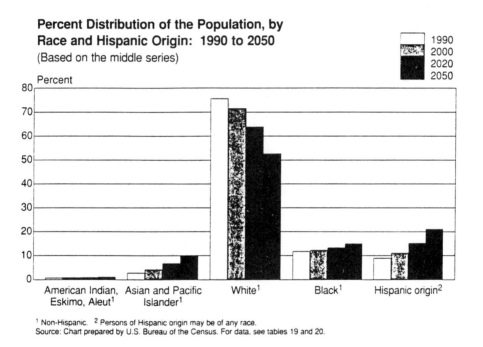

Percent Distribution of the Population, by Race and Hispanic Origin: 1990 to 2050
(Based on the middle series)

[1] Non-Hispanic. [2] Persons of Hispanic origin may be of any race.
Source: Chart prepared by U.S. Bureau of the Census. For data, see tables 19 and 20.

Source: U.S. Bureau of the Census in *The American Almanac: Statistical Abstract of the United States*, 1993.

a) Estimate the following percentages:
    American Indian, Eskimo, Aleutian in the year 2050
    Combined white and black population in the year 2020
    Non-Hispanic population in the year 1990

b) Write a topic sentence describing the overall trend.

3. This pie chart of America's spending patterns appeared in *The New York Times*, Thursday, Jan. 20th, 1994.

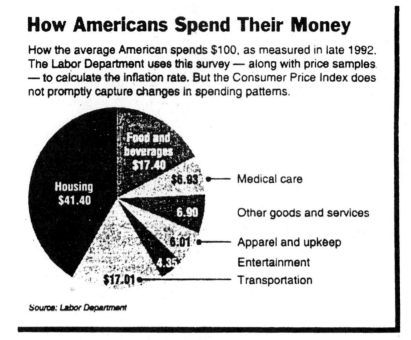

How Americans Spend Their Money

How the average American spends $100, as measured in late 1992. The Labor Department uses this survey — along with price samples — to calculate the inflation rate. But the Consumer Price Index does not promptly capture changes in spending patterns.

Food and beverages $17.40

Housing $41.40

$5.93 — Medical care

6.90 — Other goods and services

6.01 — Apparel and upkeep

4.35 — Entertainment

$17.01 — Transportation

Source: Labor Department

a) In what single category do Americans spend most of their money, and what percentage of their income does it represent?

b) If an American family had an income of $30,000 how much of it would be spent on food and beverages?

c) If you were to write a newspaper article to accompany this pie chart, what would your opening topic sentence be?

4. Below is a table of salaries taken from a survey of recent graduates (with bachelors degrees) from a well known University in Pittsburgh.

a) How many graduates were surveyed?

b) Is this quantitative or qualitative data? Explain why.

c) What is the relative frequency of people having a salary between $26,000 and $30,000?

d) Draw a histogram of the data.

Salary (in thousands)	Number
11-15	2
16-20	5
21-25	20
26-30	10
31-35	6
36-40	1

5.   Population pyramids are a form of bar chart used to depict the overall age structure of
       a society.

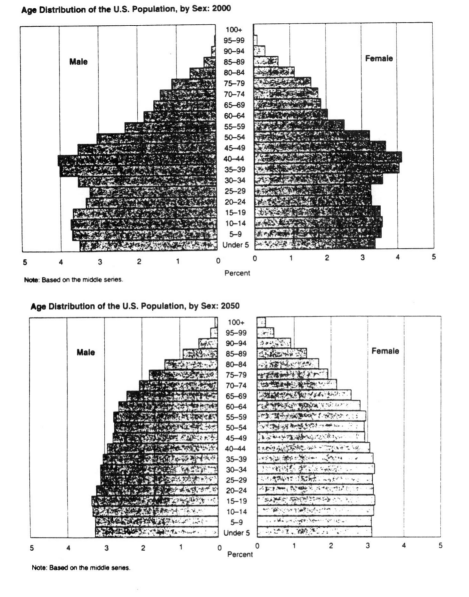

Source: "Population Projections of the United States, by Age, Sex, Race, and Hispanic Origin: 1992 to 2050" Current Population Reports, P25-1092, November 1992.

a)   Estimate the following percentages:
            Males between the ages 30-34 in 2000
            Females 85 and over in 2050
            Total of males and females between 0 and 9 in 2050

b)   The largest proportions of people are indicated by the greatest bulges in the
       pyramid. Briefly describe what these bulges suggest about trends occurring in
       American society.

6. (From Peggy Tibbs, Arkansas Tech University) This bar chart shows the distribution of tornadoes in Illinois by month for the years 1870-1910. Write a paragraph that summarizes the information in the histogram.

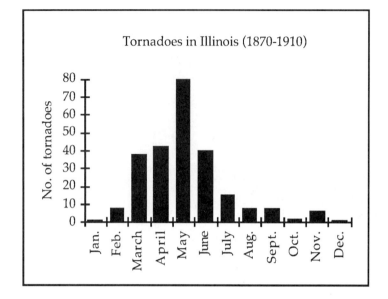

7. The diagram below shows at what point during a flight jet crashes tend to occur.

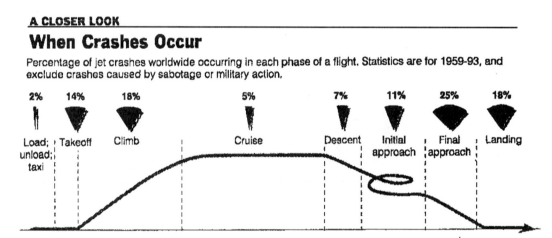

Source: Boeing; rpt. in "When Crashes Occur," *The New York Times*, 18 Dec. 1994: 36. Copyright © 1994 by the New York Times Company. Reprinted by permission.

a) Does the above chart include information about all worldwide jet crashes during 1959-1993? How do you know?
b) During what phase of a flight is a crash most likely to occur? Least likely to occur?
c) Can you tell from the diagram how many crashes occurred worldwide from 1959 to 1993? Why or why not?
d) Write an informal list of 2 or 3 features (other than those mentioned in parts b) and c) that the table does or doesn't tell you about airline crashes.

8. (From Peggy Tibbs) According to *The Arkansas Democrat-Gazette*, Dec. 2, 1995 these school districts reported the following college remediation rates (the percentage of college students from each district taking remedial courses.)

Alma: 52.9%   Eudora: 86.7%   Little Rock: 61.8%
Arkadelphis: 56.1%   Fountain Hill: 100%   Maynard: 15%
Calico Rock: 15.2%   Hamburg: 69.8%   Pea Ridge: 33.3%
Deer: 25%   Lead Hill: 50%   Russellville: 36%

Construct a histogram illustrating these percentages using intervals of 25 (i.e., 1-25%; 26-50%; 51-75%; 76-100%).

9. Try looking for some interesting graphs of data in the newspaper or in textbooks from other courses and bring them into class. Why did you find them interesting? What do the graphs tell you about the data? What questions are left unanswered by the graphs?

10. Read the Anthology article "Visualization" and come to class prepared to interpret any of the diagrams in the text. Choose a favorite format for the visual display of data and generate a similar display for a data set of your choosing.

11. Compute the mean, median, and mode for the following sets of numbers:
    a) 7, 1, -4, 7, 10, 1, 13
    b) Use numbers in part a) replacing 13 with 1014.
    c) 9.0, -2.3, 9.0, 4.5, 5.3, 4.5

12. Suppose you are waiting in the checkout line at the grocery story, and you suddenly wonder if you have enough cash. You count $15 in your pocket and 7 items in your basket. Estimate the largest "mean" item price you will be able to afford. (Sometimes, just by running your eyes over your basket, you can quickly guess whether the mean price of your selections is above or below the affordable value without adding up all the prices.)

13. a) Compute the mean, median, and mode for the list: 5, 18, 22, 46, 80, 105, 110.
    b) Now change one of the entries in the list so that the median stays the same and the mean increases.

14. Suppose that a church congregation has 100 members, each of whom tithes (donates) 10% of his or her income to the church. If the church collected $250,000 last year, what was the mean income of its members? Can you predict the median or the mode?

15. Read Stephen Jay Gould's article "The Median Isn't the Message" (in the Anthology) and explain how an understanding of statistics brought hope to a cancer victim. (P.S. Stephen Jay Gould is alive and well today.)

16. a) On the first quiz (worth 25 points) given in a section of college algebra 1 person received a score of 16, 2 people got 18, 1 got 21, 3 got 22, 1 got 23 and 1 got 25. What was the mean, median and mode score for this group of students?
    b) On the second quiz (again worth 25 points), the scores for 8 of the students were: 16, 17, 18, 20, 22, 23, 25, and 25.
       i) If the mean of the scores for the 9 students was 21, then what was the missing score?
       ii) If the median of the scores was 22, then what are possible scores for the ninth student?
       iii) If the mode was 25, then what are possible scores for the ninth student?

17. a) Write an algebraic expression for the mean of the 4 numbers: x_1, x_2, x_3, x_4
    b) Write an algebraic expression for the mean of the n numbers: $x_1, x_2, x_3, \ldots x_n$

18. Read the *CHANCE News* article and explain why the author was horrified.

19. According to the 1992 U.S. Census, the median net worth of American families was $52,200. The mean net worth was $220,300. How could there be such a wide discrepancy?

20. (From Peg McPartland, Golden Gate University)
    Consider the data given in the table below describing wolves captured and radio-collared in Montana and southern British Columbia, Canada in 1993.
    a) Look at the variables: sex, weight, and age. Are these quantitative or qualitative variables?
    b) Calculate the mean, median, and mode weights of female wolves; of male wolves; of pups.
    c) Discuss which measurement of central tendency might be more appropriate.
    d) Estimate the mean age of the females; of the males; of the total population.
    e) Write a "60 Second Summary" describing this wolf population.

PACK	CAPTURE DATE	WOLF NUMBER	EARTAG NUMBER	SEX	WEIGHT POUNDS	AGE YEARS
Spruce Creek	6/14	9318	102103	M	100	2-3
	6/16	9381	104105	F	73	1-2
North Camas	6/2	9375	none	F	77	4-5
	6/4	9378	100101	F	72	3-4
	6/4	9376	9596	F	66	2-3
	6/4	9377	9798	F	63	1-2
	6/9	9379	none	F	59	?
	6/11	9380	none	F	59	1
South Camus	5/27	9474	9091	F	61	1
	5/29	9317	9293	M	95	2-3
	5/30	8756	2627	F	77	6
Murphy Lake	6/22	1718	1718	F	73	3
	6/23	2627	2627	F	21	pup
	6/23	3637	3637	M	96	4-5
	10/15	2223	2223	F	62	pup
Sawtooth	2/26	8808	8808	M	122	5
	9/22	2829	2829	F	50	pup
	9/25	3031	3031	F	53	pup
Ninemile	8/27	4243	4243	F	73	1

Source: 1993 Annual Report of the Montana Interagency Wolf Working Group, prepared by the U.S. Fish and Wildlife Service.

21. Below is a table that shows the number of Community Hospital beds and personnel per 1,000 people for states in the Northeast and the Midwest in 1993. (Community hospitals comprise the majority of all hospitals in the U.S. They exclude long-term, psychiatric, tuberculosis, and federal hospitals.)

REGION AND STATE	Beds (per 1000)	Occupancy Rate (%)	Personnel (per 1000)
Northeast			
Maine	4.4	68.0	18.5
New Hampshire	3.4	63.7	13.8
Vermont	1.9	64.2	7.0
Massachusetts	21.1	71.5	107.8
Rhode Island	3.0	73.3	14.7
Connecticut	9.2	74.4	44.8
New York	77.4	82.8	328.7
New Jersey	31.1	77.0	121.0
Pennsylvania	53.4	72.6	230.3
mean	---	**71.9**	**98.5**
median	---	**72.3**	**71.7**
Midwest			
Ohio	41.1	60.5	176.2
Indiana	21.3	58.7	90.6
Illinois	44.1	63.5	180.0
Michigan	30.9	64.7	140.9
Wisconsin	17.7	63.4	65.5
Minnesota	18.4	66.0	55.0
Iowa	13.4	57.9	44.1
Missouri	23.6	58.9	95.9
North Dakota	4.4	64.2	12.2
South Dakota	4.3	60.6	11.4
Nebraska	8.4	55.2	25.7
Kansas	11.3	54.2	36.3
mean	---	**60.7**	**77.8**
median	---	**60.6**	**60.3**

Source: U.S. Bureau of the Census in *The American Almanac:: Statistical Abstract of the United States*, 1995-1996.

a) What is the mean and median for the number of hospital beds per 1,000 people for the Northeast and Midwest?

b) Write a summary paragraph that would describe some aspect of the information in the table.

22..   The following chart gives the ages of students in a mathematics class.

Age Interval	Frequency Count
15 - 19	2
20 - 24	8
25 - 29	4
30 - 34	3
35 - 39	2
40 - 44	1
45 - 49	1
Total	21

a) Use this information to make an estimate of the mean age of the students in the class. Show your work.

b) What is the largest value the mean could have? The smallest? Why?

23.   (Requires graphing calculator or computer)
Use the following table to generate an estimate of the mean age of the U.S. population. Show your work.

Ages of United States Population in 1991 in Ten Year Intervals	
Age	Total (thousands)
Under 10	37,459
10 to 19	34,876
20 to 29	39,912
30 to 39	42,677
40 to 49	32,848
50 to 59	22,068
60 to 69	20,620
70 to 79	14,521
80 and over	7,195
Total	252,177

Based on data from the U.S. Census Bureau, Current Population Survey, March 1992.
Note: Includes Armed Forces abroad
Total may not be exact because of rounding. Source: U.S. Census, Statistical Abstract, 1992

24.   (Requires computer with course software)
In the *Fam 100 Census Data Kit* folder open up the program called *Fam 1000* and choose "Histogram" from the menu. 1992 U.S. Census data about 1000 randomly selected U.S. individuals and their families is imbedded in this program. Experiment with creating histograms for ages, education, and different measure of income. Try using different interval sizes to see what patterns emerge. Decide on one variable (say education) and then compare the histograms of this variable for different groups of people. For example, you could compare education histograms for men and women, or for people living in two different regions of the country, or for members of two different racial groups. Pick a comparison that you think is interesting and if possible print out your histograms. Then write a 60 second summary describing your observations.

25. The following table and graphs contain information about the populations of the U.S. and China. Write a 60 second summary comparing the two populations.

Age Intervals	China Totals	China %	U.S. Totals	U.S. %
0-4	116,624	10.30	19,222	7.62
5-9	99,440	8.79	18,237	7.23
10-14	97,456	8.61	17,671	7.01
15-19	120,403	10.64	17,205	6.82
20-24	125,877	11.12	19,194	7.61
25-29	104,268	9.21	20,718	8.22
30-34	83,805	7.40	22,159	8.79
35-39	86,315	7.63	20,518	8.14
40-44	63,844	5.64	18,754	7.44
45-49	49,181	4.35	14,094	5.59
50-54	45,666	4.03	11,645	4.62
55-59	41,753	3.69	10,423	4.13
60-64	34,057	3.01	10,582	4.20
65-69	26,395	2.33	10,037	3.98
70-74	18,119	1.60	8,242	3.27
75-79	10,971	0.97	6,279	2.49
80-84	5,373	0.47	4,035	1.60
85 and up	2,338	0.21	3,160	1.25
Total	1,131,885	100.00	252,175	100.00
Median Age	25		33	
Mean Age	28		36	

Age Distribution of the 1990 Population (in 1000's) of China and the US

Source: 1990 U.S. Bureau of the Census; rpt. in *The American Almanac: Statistical Abstract of the United States*, 1993. Chinese population/statistical information reprinted with permission from the State Statistical Bureau of the People's Republic of China from *The China Statistical Yearbook*, copyright © 1991.

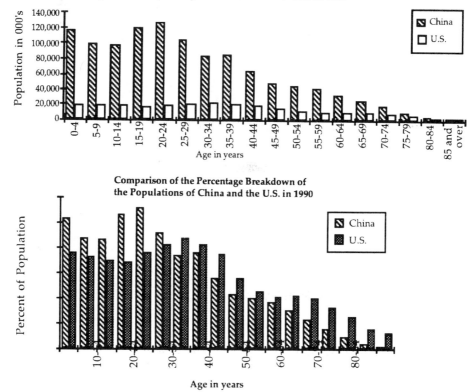

26.

A New York Times/CBS News poll asked a question about the level of spending for "welfare" and 23 % of the respondents said the nation was spending "too little" on welfare. The question was reworded with one small change: "assistance to the poor" was substituted for "welfare" and 64% of the same respondents said the nation was spending "too little".

Are these results contradictory?

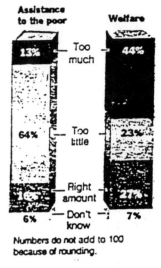

The Answer Depends On the Question

Are we spending too much, too little, or about the right amount on:

From a nationwide New York Times/CBS News Poll conducted May 6-8, 1992, results published in *The New York Times*, July 5, 1992. Copyright © 1992 by New York Times Company. Reprinted by permission.

27.   Read *The New York Times* OP-ED article," A Fragmented War on Cancer" by Hamilton Jordan, President Jimmy Carter's chief of staff (included in the Anthology of Readings.) Jordan claims that we are on the verge of a cancer epidemic. What questions are raised about the evidence he uses to support this claim?

   a)   Use what you have learned about the distribution of ages over time in the U.S. to refute his claim.

   b)   Read *The New York Times* letter to the editor by William M. London, Director of Public Health, American Council on Science and Health (In the Anthology of Readings.) London argues that Hamilton Jordan's assertions are misleading. What questions are raised by the arguments of William London? What additional data would you need to evaluate his arguments?

EXPLORATIONS

> The Explorations are intended primarily for in-class use. However most can be done outside of class if you have access to a computer or graphing calculator. The first set of Explorations can be done in sequence using the class data collected in the first Exploration or each can stand alone. You should bring a notebook to class to keep a record of your results and observations.
>
> Data that can be used with the Explorations are on graph link and spreadsheet files on the disk that accompanies the text. Basic instructions for using the
> TI-82 are included in the Appendix.

EXPLORATION 1.1
Collecting Data

Objectives
- explore some of the issues related to collecting data
- collect data about students in order to describe the class

Materials/Equipment
- Class Questionnaire
- measuring tapes in centimeters and inches
- optional measuring devices:
      eye chart
      flexibility tester constructed from cardboard box and yardstick
      measuring device for blood pressure and person who knows how to use it

Related Readings
   "Health Measurements" (includes eye chart and description of flexibility tester)
   "U.S. Government Definitions of Census Terms"

Procedure
In a small group or with a partner
1. Pick (or your instructor will assign you) one of the undefined variables on the questionnaire. Spend about 15 minutes coming up with a workable definition of that variable on the questionnaire and write the definition on the board. Be sure there is a way in which responses can be answered by a number or single letter.
2. Consult the reading "Health Measurements" if you decide to collect health data.

Discussion/Analysis
Class discussion

After all the definitions are recorded on the board, discuss your definition with the class. Is it clear? Does everyone in the class fall into one of the categories of your definition? Can anyone think of someone who might not fit into any of the categories? Modify the definition until all can agree on some wording.

Are there any other measurements or questions you would like to add to the questionnaire? Propose them to the class and, if the class agrees, pencil them in at the bottom of the questionnaire.

As a class, decide on the final version of the questionnaire. Write down the final definitions in your class notebook so you can refer to them in later classes.

Summary
In a small group or with a partner

Help each other when necessary to take measurements and fill out the entire questionnaire. Questionnaires remain anonymous, and you can leave blank any question you can't or don't want to answer.

Hand in your questionnaire to your instructor by the end of class.

Exploration Linked Homework
Read "U.S. Government Definitions of Census Terms" for a glimpse into the federal government's definitions of the variables you defined in class. How do the "class" definitions and the "official" ones differ?

Algebra Class Questionnaire

(You may leave any category blank.)

| | 1. Age (in years) |

| | 2. Sex (female = 1, male = 2) |

| | 3. Your height (inches) |

| | 4. Distance from your navel to the floor (centimeters) |

| | 5. Estimate your average travel time to school (in minutes) |

| | 6. What is your most frequent mode of transportation to school? (F = by foot, C = car, P = public transportation, B = bike O = other) |

The following variables will be defined in class. We will discuss ways of coding possible responses and then use the results to record our personal data.

| | 7. The number of people in your household |

| | 8. Your employment status |

| | 9. Your ethnic classification |

| | 10. Your attitude towards mathematics |

Health data

| | 11. Your pulse rate before jumping (beats per minute) |

| | 12. Your pulse rate after jumping for 1 minute (beats per minute) |

| | 13. Blood pressure: systolic (mm. Hg) |

| | 14. Blood pressure: diastolic (mm. Hg) |

| | 15. Flexibility (inches) |

| | 16. Vision, left eye |

| | 17. Vision, right eye |

Other Data

EXPLORATION 1.2
Organizing and Visualizing Data

Objectives
- learn techniques for organizing and graphing data using a computer (with a spreadsheet program) or a graphing calculator
- describe and analyze overall shape of single-variable data using frequency and relative frequency histograms.

Materials/Equipment
- computer with spreadsheet program or graphing calculator with list-sorting and statistical plotting features
- data from class questionnaire or other small data set in both electronic and hard copy form. If using class data, copy of questionnaire decided upon in Exploration 1.1.
- overhead projector and projection panel for computer or graphing calculator
- transparencies for printing graphs (optional). Could also use transparencies with special colored pens to draw graphs.

Related Reading
"Presentation of Mathematical Papers at Joint Mathematics Meetings and Mathfests"

Procedure
Class demonstration
1.   If you haven't used a spreadsheet or graphing calculator before you'll need a basic technical introduction. (Note if you are using a TI-82 or TI-83 graphing calculator there are basic instructions in the Appendix.) Then you'll need an electronic version of the data set from which you will choose one variable for the whole class to study (for example, age from the class data)

   If you're using a spreadsheet,
   - copy the column with the data onto a new spreadsheet.
   - graph the data. What does this graph tell you about the data?
   - sort the data and plot it again. Is this graph any better at conveying information about the data?

   If you're using a graphing calculator
   - discuss window sizes, changing interval sizes and statistical plot procedures

   Finally select an interval size and construct a frequency and then a relative frequency histogram. If possible, label one of these carefully and print it out. If you have access to a laser printer, you can print onto an overhead transparency

In a small group or with a partner

> Choose another variable from your data. Pick an interval size, and then generate both a frequency and relative frequency histogram. If possible, make copies of the histograms for both your partner and yourself.

Discussion/Analysis

> With your partner(s), analyze and jot down patterns that emerge from the data. What are some limitations of the data? What other questions are raised and how might they be resolved? In your notebook, record jottings for a 60 second summary that would describe your results.

Exploration Linked Homework

A verbal report of your results

> With your partner(s) prepare a verbal 60 second summary to give to the class. If possible use an overhead projector with a transparency of your histogram or a projector linked to your graphing calculator. If not, bring in a paper copy of your histogram. (The short pamphlet "Presentation of Mathematical Papers at Joint Mathematics Meetings and Mathfests" will show you that even mathematical professionals need very basic advice about talking before an audience.)

A written report of your results

> Construct a written 60 second summary". (See Section 1.5 of the text for some writing suggestions.)

EXPLORATION 1.3
Describing Data Using Numbers and Graphs

Objective
- describe and analyze data using numbers and graphs
- explore how to make comparisons using numbers and graphs

Materials/Equipment
- computer with spreadsheet program **or** graphing calculator
- data from class questionnaire or other small data set either as spreadsheet or graph link file
- overhead projector and projection panel for computer or graphing calculator.
- transparencies for printing or drawing graphs for overhead projector (optional)

Related Software
"Mean, median and mode" in *Statistical Graphs*

Procedure
Class demonstration
1. Calculate mean, median, and mode using software functions or graphing calculator functions.
2. Using two columns of data, do a double sort (for example, using sex and ages, sort on sex and then age.) Now compare the mean, median, and mode for different subsets (e.g., ages of men and ages of women.)

Working in small groups or with a partner
Work on a 60 second summary comparing at least two variables from your small data set.
- Decide on two variables and what approach you will take.
- Think of ways of dividing up the work. By the end of the next class, your small group should be prepared to do a short class presentation of your findings, using overhead transparencies, enlarged Xeroxes or projections from your graphing calculator.

Discussion/Analysis
Possible research topics and possible questions you might want to address are:

Pulse rates before and after exercising
What does a histogram of resting pulse rates look like? A histogram of pulse rates after exercise? What is an "average" pulse rate before exercising? After exercising? Are there any conclusions you can draw?

Heights of men vs. women
> What does a histogram of women's heights look like? A histogram of men's heights? What is an "average" height for women? For men? Are there any generalizations you can draw?

Commuting time vs. mode of transportation
> What is the "average" commuting time for the whole class? What percent of the class commutes by foot? By car? By public transportation? What is the "average" commuting time by car? By public transportation?

There are many different ways of comparing two sets of data, and there is no one right way of making comparisons. You can generate numbers such as averages, create graphs, such as histograms, or try out other ways for visualizing data. You just need to be able to describe your results.

Be creative. It may be interesting to compare blood pressure of students with different commuting times; the flexibility of males and females or the vision of students of different ages.

Exploration Linked Homework

A verbal report of your results:
> With your partner(s) prepare a 60 second summary of your results. See Sections 1.5 and 1.6 of text for guidelines. If possible use a transparency or graphing calculator projector with your talk.

A written report of your results:
> Write a 60 second summary" of your results. Include graphs and numerical descriptors in summarizing your data.

EXPLORATION 1.4
Comparing the Ages of Men and Women in the U.S.

Objective
- describe and analyze ages of men and women in the U.S. using numbers and graphs
- explore how to make comparisons using numbers and graphs

Materials/Equipment
- computer with spreadsheet program **or** graphing calculator (basic instructions for using the TI-82 with this Exploration are in the Appendix.)
- data files and hard copy of ages of men and women in U.S. (cf. Table 1.4 and its electronic versions that appear as graph link and spreadsheet files on the course disk)
- overhead projector and projection panel for computer or graphing calculator
- transparencies for printing or drawing graphs for overhead projector (optional)

Procedure
Working with a partner
1. Make predictions as to differences or similarities between the ages of men and women in the U.S. Will the number of men in relation to the number of women vary by age?

2. Using information from the enclosed table(which is also on both spreadsheet and graph link files on the course disk) find the mean and median age of men and of women in the U.S. in 1991.

3. Compress the age data using the same size interval for men and women. Find the relative frequency for each age interval for men and women. Construct frequency or relative frequency histograms showing the distribution of ages for men and women.

4. In your class notebook, jot down your observations about the age data. Are there any differences between men and women? Compare and contrast the age data using numbers such as mean and median and information from your graphs.

Discussion/analysis
Discuss your findings with your partner(s). What do the graphs and numbers reveal about the ages of men and women? What type of arguments can be made? What are the limitations of the data? What questions have been raised by your analysis?

Exploration Linked Homework
A verbal report of your results
With your partner(s) prepare a "60 Second Summary" of your results. See Sections 1.5 and 1.6 of text for guidelines. If possible use an overhead projector with a transparency or graphing calculator projector to demonstrate your results.
A written report of your results.
Write a "60 Second Summary" of your results, including graphs and numbers.

Chapter 2

An Introduction to Functions

Overview

A central theme in mathematics and its applications is the search for patterns. Mathematics offers powerful tools both for recognizing patterns and for describing them succinctly. In Chapter 1 we saw how images like histograms and pie charts, or numbers like mean and median, are used to find patterns in single variable data. In this Chapter we explore how to describe the relationship between two variables. In particular, we focus on patterns of change. Does change in one variable affect change in another? If one variable increases, when does the other variable increase, decrease, or stay the same?

If one variable depends upon another, we speak loosely of one variable being a function of another. We talk for instance of population size being a function of time. In mathematics the term *function* is used to describe a particular type of relationship between two variables. Functions can be represented by tables, graphs and algebraic expressions.

The explorations will help you develop an intuitive understanding of functions through graphs and tables.

After reading this chapter you should be able to:

- recognize from a graph when one variable (with respect to another) is increasing or decreasing and when it reaches a maximum or minimum value
- define a function
- identify independent and dependent variables, domain and range
- represent functions with tables, graphs and equations

2.1   Relationships between Two Variables

Images like histograms and pie charts, or numbers like mean and median, give important but static snapshots of patterns in single-variable data. Two-variable data contain information about how one variable changes as the other changes. We can ask questions about the effect of one variable upon another. How does the weight of a child determine the amount of medication prescribed by a pediatrician? How does median age or income change over time? Mathematics offers powerful tools both for recognizing patterns of change and describing them succinctly.

One of the most common and important cases of two variable data is the *time series*. In a time series, one of the variables is time, measured perhaps in minutes, months, years or decades. The other variable may be anything that can be measured at different times, for example median age, a child's weight, or number of voters. Some estimates claim that over 80% of the graphs in the popular press are time series.

Example 1: Table 2.1 shows a time series for median age. The first column contains the year, and the second the corresponding median age for the U.S. population.

Table 2.1
Median Age of U.S. Population 1860-1991
(data for 1995-2050 is projected)

Year	Median Age
1850	18.9
1860	19.4
1870	20.2
1880	20.9
1890	22.0
1900	22.9
1910	24.1
1920	25.3
1930	26.4
1940	29.0
1950	30.2
1960	29.5
1970	28.0
1980	30.0
1990	32.8
1991	33.1
1995	34.0
2000	35.5
2025	38.1
2050	39.0

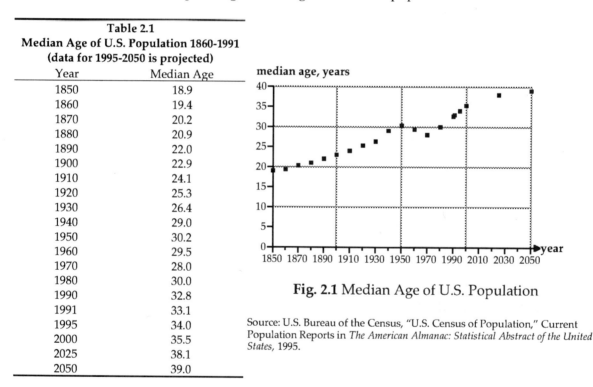

Fig. 2.1 Median Age of U.S. Population

Source: U.S. Bureau of the Census, "U.S. Census of Population," Current Population Reports in *The American Almanac: Statistical Abstract of the United States*, 1995.

We can think of the rows as ordered pairs. The first row, for example, corresponds to the ordered pair (1850, 18.9). The second row corresponds to (1860, 19.4). We can graph these ordered pairs, using the standard convention of plotting the first coordinate (year) on the horizontal axis, and the second (median age) on the vertical axis. The resulting graph is called a *scatter plot*.

It is apparent from both Table 2.1 and Figure 2.1, that as the years increase, in general the median age increases as well. This is not surprising, since we saw in Chapter 1 (and repeated in Table 2.1) that the Bureau of the Census predicted the median age of 33.1 in 1991 would rise to 39.0 by the year 2050. In Table 2.1 the *minimum* median age of 18.9 occurs in 1850, and the *maximum* median age is projected to be 39.0 in 2050.

The median age however does decrease between 1950 and 1970, dropping from 30.2 in 1950, to 29.5 in 1960, down to 28.0 in 1970. The corresponding dip is easy to spot in Figure 2.1.

Something to think about

1. What are the trade-offs in using the median age instead of the mean age?
2. What might have caused a lowering of the median age between 1950 and 1970? More babies? Fewer old people? An increase in youthful immigrants?

Example 2: Table 2.2 and Figure 2.2 show the annual federal budget surplus (+) or deficit (-) since World War II.

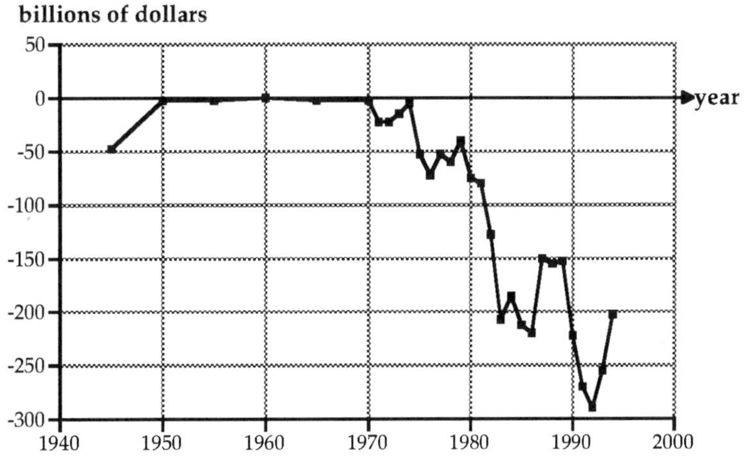

Table 2.2 Federal Budget Surplus or deficit (-) (in billions of dollars)	
Year	
1945	-47.5
1950	-3.1
1955	-3.0
1960	0.3
1965	-1.4
1970	-2.8
1971	-23.0
1972	-23.4
1973	-14.9
1974	-6.1
1975	-53.2
1976 *	-73.7
1977	-53.7
1978	-59.2
1979	-40.7
1980	-73.8
1981	-79.0
1982	-128.0
1983	-207.8
1984	-185.4
1985	-212.3
1986	-221.2
1987	-149.8
1988	-155.2
1989	-152.5
1990	-221.4
1991	-269.2
1992	-290.4
1993	-255.1
1994	-203.2

Fig.2.2. Annual federal budget surplus or deficit, billions of dollars

Source: U.S. Office of Management and Budget, *Budget of the United States Government*, annual,in *The American Almanac: Statistical Abstract of the United States*, 1995.

* The Balanced Budget and Emergency Deficit Control Act of 1985 put all the previously off-budget Federal entities into the budget and moved Social Security off-budget.

Figure 2.2 indicates that the federal budget was fairly balanced (with little debt or surplus) until about 1970. Since then the federalbudget has been running an annual deficit that has been generally increasing. If you look at the values in Table 2.2, you can see that the graph reaches a maximum in 1960 (when there was actually a surplus of 301 million) and a minimum in 1992, (when the annual deficit was almost 300 <u>billion dollars</u>). Since Table 2.2 shows the annual surplus or deficit, to calculate the total federal debt you need to take into account the cumulative effect of all the deficits and surpluses for each year together with any interest or payback of principal. (There's an exercise about this in the problem set at the end of the chapter.) It is a matter of intense national debate what the size of the annual deficit (and total debt) will be in future years. Will the graph continue its broad downwards path, or will it continue the upward trend started in 1992, and eventually return to near zero?

Something to think about

If the annual deficit returns to zero, does that mean that the federal debt is zero?

Example 3: We can describe mathematical relationships between two abstract variables x and y using equations like:

$$y = x^2 + 2x - 3$$

Think of the equation $y = x^2 + 2x - 3$ as a mathematical sentence. Some values of x and y make the sentence true, such as $x = 0$ and $y = -3$. Other values make it false, for example $x = 0$ and $y = 5$. Values for x and y that make the sentence true are called *solutions* to the equation. We can express these solutions as ordered pairs of the form (x, y).

The *solutions* to an equation in two variables x and y are the set of ordered pairs (x, y), that make the equation a true statement.

By substituting in various values for x, and finding the associated value for y we can generate a table of solutions. There are infinitely many possible solutions (since we could substitute any number for x, and find a corresponding y), so the table can only show a finite number of them. We can plot the points by hand or use technology to generate a graph of the function. Each of the points on the graph corresponds to an ordered pair of the form (x,y) which makes the equation a true statement.

Table 2.3

x	y
-4	5
-3	0
-2	-3
-1	-4
0	-3
1	0
2	5
3	12

Fig.2.3 Graph of $y = x^2 + 2x - 3$

> The *graph* of an equation in two variables is the set of points corresponding to the ordered pairs of numbers which make the equation a true statement.

We can "read" the graph as we did the time series graphs from left to right, asking what effect changes in x have on y. Initially as x increases, y decreases until y reaches an estimated minimum value of -4. (In later chapters we'll learn how to use algebra to determine this exactly.) As x continues to increase, y now increases. There is no maximum value for y.

Algebra Aerobics[1]

The following table and scatter plot show median family income over time.

Table 2.4	
Year	Median Family Income in constant 1990 $[2]
1973	35,474
1974	34,205
1975	33,328
1976	34,359
1977	34,528
1978	35,361
1979	35,262
1980	33,346
1981	32,190
1982	31,738
1983	32,378
1984	33,251
1985	33,689
1986	35,129
1987	35,632
1988	35,565
1989	36,062
1990	35,353

a) What is the maximum value for the median family income during this period? In what year does this occur? What are the coordinates of this point in Figure 2.4?

b) What is the minimum value for the median family income? In what year does this occur? What are the coordinates of this point.

c) What is the longest time period during which the median income was increasing? Decreasing?

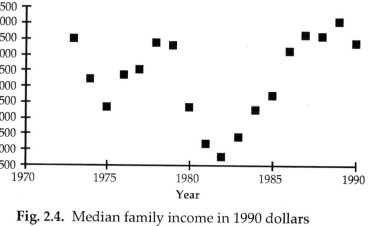

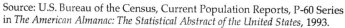

Fig. 2.4. Median family income in 1990 dollars

Source: U.S. Bureau of the Census, Current Population Reports, P-60 Series in *The American Almanac: The Statistical Abstract of the United States*, 1993.

---

[1]   a) The maximum median family income is $36,062 in 1989. The coordinates of this point: are:(1989, 36062)
     b) The minimum is $31,738 in 1982. The coordinates are: (1982, 31738)
     c) The longest period of increase was from 1982 to 1987. The longest period of decrease was from 1978 to 1982.

[2] "Constant dollars" is a measurement used by economists to compare incomes and other variables in terms of purchasing power, eliminating the effects of inflation. To say the median income in 1973 was $35,474 in "constant 1990 dollars" means that the median income in 1973 could buy an amount of goods and services that would cost $35,474 to buy in 1990. The actual median income in 1973 (measured in what economists call "current dollars") was much lower. Income corrected for inflation is sometimes called "real" income.

2.2   Functions

The definition of a function

In Section 2.1 we used ordered pairs to describe the relationships between time and real social variables (median age, the federal debt , and median family income), and the relationship between two abstract variables x and y. We think of the first variable in the ordered pair as determining the second variable; each time determining a unique value of median age, debt or income or each x determining a unique y. Equivalently, we could think of age, debt and income as being "functions" of time, or think of y as a function of x.

Informally, when we speak of something being a function of something else, we mean that one depends upon the other. How well a car runs is a function of how well it is maintained. One's political views are often a function of one's upbringing. The response of a patient to medication is a function of the dose he or she is given.

With mathematical functions, the *independent variable* is the one that we are imagining may be changed at will or that like time governs the unfolding of a process. The other variable, which only responds, is the *dependent variable*. The dependent variable is a *function* of the independent variable; if each value of the independent variable determines a <u>unique</u> value of the dependent variable.

> A variable y is a *function* of a variable x if each value of x determines a unique value of y.
> y is the *value of the function* or the *dependent variable* and x is the *independent variable*.

Examples:
Hat size is a function of how big around your head is. What you pay at the gas station is a function of the number of gallons of gas you pump into your car.

Each of the examples in Section 2.1 represented functions. For instance, in the Table and Figure 2.1 median age is a function of time, since each year corresponds to a unique value for median age. The equation $y = x^2 + 2x - 3$ describes y as a function of x, since given a value of x the equation determines a unique corresponding value for y.

Representing functions with tables, graphs, and equations

We can think of a function as defining a "rule" that allows us to determine a unique y, given an x. The rule can be described using tables, graphs, or equation;s.

- Using a table, the rule may say "find the value of x in the left column of the table and read the value of y from the right column."
- Using a graph, it may say "go along the horizontal axis a distance x, go up or down until you meet the curve, and read y from the vertical axis."

- Using an equation, it might say "take the value of x and square it, add twice the value of x, and then subtract 3, the result will be y" (or more compactly $y = x^2 + 2x - 3$.)

> The most common representations of functions are tables, graphs or equations.

Example:

The sales tax rate in Illinois is 8%; that is, for each dollar you spend in a store, the law says that you should pay a tax of 8 cents or 0.08 dollars.[3] We may write this as an equation

$$T = 0.08\,P$$

where T is the amount of sales tax you should pay and P is the price of your purchase, both measured in dollars. This equation is a rule for determining a value for T given a value for P It says "Take the given value of P and multiply it by 0.08; the result is the value of T."

P is the independent variable, and T the dependent variable; the sales tax depends upon the purchase price. The formula represents T as a function of P, since for each value of P the formula determines a unique amount of T. A function like this one used to describe a real world phenomenon is called a *mathematical model*.

Of course, one could also use this formula to make a table of values for T for many different values of P. Such tables are indeed posted next to many cash registers. We can plot the points in the table to create a graph of the function. *By convention the independent variable is graphed on the horizontal axis, and the dependent variable on the vertical axis.* The equation, table and graph are all representations of sales tax as a function of purchase price.

$T = 0.08\ P$

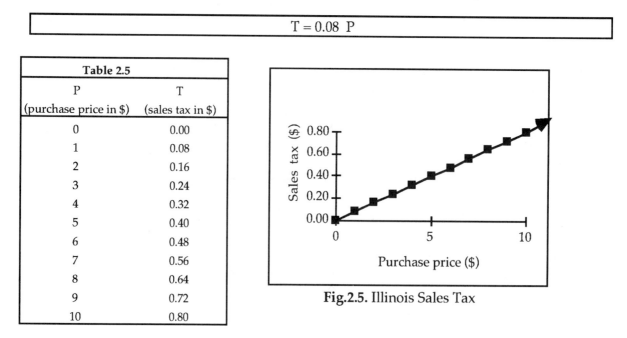

Table 2.5	
P	T
(purchase price in $)	(sales tax in $)
0	0.00
1	0.08
2	0.16
3	0.24
4	0.32
5	0.40
6	0.48
7	0.56
8	0.64
9	0.72
10	0.80

Fig.2.5. Illinois Sales Tax

---

[3] We have rounded off the sales tax to simplify calculations. The 1996 sales tax in Illinois is 7.75% and in the city of Chicago is 8.75%.

We selected a few values for P (integers between 0 and 10) in order to generate Table 2.5 and Figure 2.5. But what are all the possible values for P for which it makes sense to use the formula T = 0.08 P ? Since in our model P represents purchase price, negative values for P are meaningless. We need to restrict P to possible dollar amounts such that P≥ 0. In theory there is no upper limit on prices, so we assume P has no maximum amount. We call the set of possible values for the independent variable the *domain*.So in this example:

domain = all possible dollar amounts P greater than or equal to 0
         = all possible dollar amounts P where P ≥ 0.

What are the corresponding values for the tax T? The values for T in our model will also always be non-negative. Assuming that there is no maximum value for P, then there will be no maximum value for T. So T could assume any dollar amount such that T≥ 0. We call the set of possible values for the dependent variable the *range*. So here

range = all dollar amounts T greater than or equal to 0
       = all dollar amounts T with T ≥ 0.

The *domain* of a function is the set of possible values of the independent variable.
The *range* is the set of corresponding values of the dependent variable.

Table 2.5 only shows the sales tax for a few selected purchase prices, but we assume that we could have used any positive dollar amount for P. We have "connected the dots" on the scatter plot to suggest the many possible intermediate price values. We've placed an arrow at one end to show that the graph extends indefinitely to the upper right.

Algebra Aerobics (answers on next page):

1. a) Write the formula for computing a 15% tip in a restaurant. What are the independent and dependent variables? Is your formula a function? If so, what are reasonable choices for the domain and range?
    b) How much should you tip for an $8 meal?
    c) Use a calculator to compute a 15% tip on a total check of $26.42.
2. If we let D stand for ampicillin dosage in milligrams and W stand for a child's weight in kilograms, then the equation

    $$D = 50\ W$$

    gives a rule for finding the safe maximum daily drug dosage of ampicillin (used to treat respiratory infections) for children who weigh less than 10 kilograms (about 22 pounds).[4]
    a) What are the independent and dependent variables?
    b) Does the equation represent a function? Why?
    c) What is a reasonable domain? Range?
    d) Generate a small table and graph of the function.

---

[4] Information extracted from Anna M. Curren and Laurie D. Munday, *Math for Meds: Dosages and Solutions*, 6th ed. (San Diego: W.I. Publications, 1990): p.198.

Answers:

1. a) Tip = 15% of meal

   $T = 0.15M$

   Independent variable: M (price of meal)   Dependent variable: T (tip)

   The formula is a function since to each value of M, there corresponds a unique value of T. The domain and range are positive numbers. The highest value for M that is believed to be "reasonable" varies from person to person. The highest value of T is 15% of the highest value selected for M

   b) T=(0.15)($8)=$1.20

   c) T = (0.15)($26.42) = $3.96

   In real life we'd probably round this to the nearest nickel or dime.

2. a) Since the dosage depends upon the weight, the logical choice is W as the independent, and D as the dependent variable.

   b) In this formula, each value of W determines a unique dosage D, so D is a function of W.

   c) Values for the domain must be larger than 0 and less than 10 kilograms. So we have

   Domain = all values of W greater than 0 and less than 10

   = all values of W with $0 < W < 10$.

   As is often the case in the social sciences, the domain includes some questionable values (we don't expect a child to weigh 0 pounds), but it spans all the appropriate ones.

   Range = all values of D greater than 0 to a maximum of 500 (10•50) milligrams

   = all values of D with $0 < D < 500$

   d) The following table and graph are representations of the function. Note that (0,0) and (10,500) are calculated only to help us draw the graph and are not actually included in the model:

Daily Ampicillin Dosage by Weight

W	D
Child's Weight (kilograms)	Daily Maximum Dosage (in milligrams)
0	0
1	50
2	100
3	150
4	200
5	250
6	300
7	350
8	400
9	450
10	500

Maximum daily ampicillin dosage as a function of weight

Functional notation

There are a number of different notations used to write about functions. Functions are often named with a letter; f is a common obvious choice. We can write $f : x \longrightarrow y$ to emphasize f as a rule that sends or "maps" a value of the independent variable x to a unique value of the dependent variable y. We also write this as $y = f(x)$, which is read "y = f of x ." Sometimes x is called the input, and y the output;.

In the Illinois sales tax example, we decided that the equation $T = 0.08P$ defined T as a function of P. If we called that function rule f the following three notations are all equivalent.

$$f : P \longrightarrow 0.08P$$
$$\text{or} \qquad f(P) = 0.08P$$
$$\text{or} \qquad T = 0.08P$$

Each indicates that given a value for P we can find the corresponding value for T by multiplying the P value by 0.08.

If P = 10, the corresponding value for T can be denoted by $f(10)$. Then

$$f(10) = (0.08)(10)$$
$$= 0.8$$

Similarly

$$f(-100) = (0.08)(-100) \qquad\qquad f(0.5) = (0.08)(0.5) \qquad\qquad f(7/2) = (0.08)(7/2)$$
$$= -8 \qquad\qquad\qquad\qquad\qquad = 0.04 \qquad\qquad\qquad\qquad = 0.28$$

Example:

Let's look again at the equation

$$T = 0.08 P$$

this time divorcing it from the context of Illinois taxes. We are no longer thinking of the equation as a real world model, but as describing a relationship between two abstract variables P and T. T is still a function of P. Since P no longer represents purchase price, P doesn't have to be measured in dollars and cents. The values for P, the independent variable, can be <u>any</u> real number, positive, negative, or zero.[5] So now

domain = all real numbers

Possible values for the dependent variable T also now include any real number.

range = all real numbers

We can generate a small table of values (including negative numbers) and an associated graph. As the two arrows at the ends of the graph indicate, the line now extends indefinitely

---

[5] Recall that the real numbers are all numbers that can be written as signed decimal expressions, for example, 1, -3, 20.12, $\pi = 3.14159.....$, $1/3 = 0.33333......$ We often visualize the real numbers as points on a line that has been marked off with a zero, and divided into equally spaced unit lengths.

```
       -5      -4      -3      -2      -1       0       1       2       3       4       5
 <----------|----------|----------|----------|----------|---------|----------|----------|----------|----------|---------->
```

The real numbers include all integers, fractions and positive and negative numbers, as well as numbers such as $\sqrt{2}$ whose decimal expansions contain an infinite number of digits without a pattern that infinitely repeats, such as $2/11 = 0.1818181818.....$

in two directions. P can now assume irrational values like √2 that wouldn't have made sense as a price.

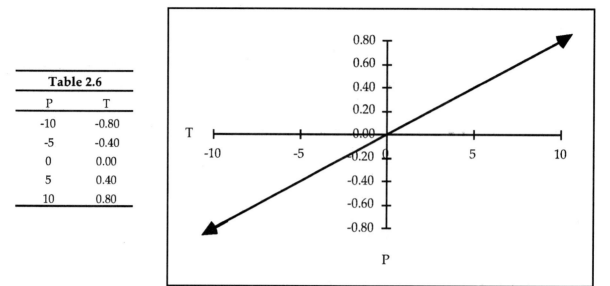

Table 2.6	
P	T
-10	-0.80
-5	-0.40
0	0.00
5	0.40
10	0.80

Fig.2.6. Graph of T = 0.08P

A note about functions and their domains

An implicit part of the definition of a function is its domain; functions and their domains go hand in hand. Hence we need to be careful when we use the same functional notation $f(P) = 0.08P$ to denote both the Illinois sales tax function (in which the domain was values of $P \geq 0$) and its abstraction (where the domain was all the real numbers). By changing the domain, we have changed the function. For example, if f represents the functional rule defining Illinois sales taxes, then $f(-1)$ is undefined. If we take f to represent an abstract relationship, then $f(-1) = (0.08)(-1)$ or -0.08.. So anytime we use a function, we must consider the function paired with its domain.

Algebra Aerobics:[6]

1. Think of your formula for generating 15% tips as an abstract mathematical equation, having nothing to do with meal prices. What is now the domain and range?

2. Think of your formula for ampicillin dosage as an abstract mathematical equation. What is now the domain? The range? How could the table and graph reflect these changes? Rewrite the function using f to represent the rule relating the abstract variables W and D. What is $f(15)$? $f(-15)$? $f(15000)$?

---

[6] Answers:

1. The domain for M and T is now all real numbers.
2. The domain and range are now all the real numbers. The table should now include negative values, and the graph will continue indefinitely in two directions.
    We could write $f(W) = 50W$. Then $f(15) = (50)(15) = 750$, $f(-15) = (50)(-15) = -750$, and $f(15000) = (50)(15000) = 750{,}000$

2.3 Studying Equations that Represent Functions

Putting equations into "function form"

Functions are easier to work with than equations. If we have an equation, like $3x^2 - 2y = 5x$, it is often difficult to find solution pairs by anything better than trial and error. Not all equations in two variables represent functions. But when an equation is a function, it is useful to solve the equation for the dependent variable in terms of the independent variable. This is called putting the equation in *function form*. Once an equation is in function form, we can easily generate many solution pairs (for tables or graphs) just by picking different values for the independent variable and using the functional rule to compute the corresponding values of the dependent variable. Many graphing calculators and graphing programs accept only equations in function form as input.

To put an equation into function form, we first need to pick our independent and dependent variables. Sometimes the choice may be obvious, other times not. There may be more than one possible correct choice. For now, in order to practice the mechanics of solving for one variable in terms of another, we adopt the standard convention that x represents the independent and y the dependent variable. So in the following examples, to put each equation into function form, we need to represent y, the dependent variable, in terms of x, the independent variable. We want:

$$y = \text{some expression involving } x.$$

We start by putting all the terms involving y on one side of the equation and everything else on the other side.

Some examples of equations that do and do not represent functions

Example 1:   Analyze the equation $4x - 3y = 6$. Can it be put in function form? Generate a table and graph of solutions to the equation. Discuss.

First, let's try to put the equation in function form.

given the equation	$4x - 3y = 6$
subtracting 4x from both sides	$-3y = 6 - 4x$
dividing both sides by -3	$(-3y)/(-3) = (6 - 4x)/(-3)$
we get	$y = -(6/3) + (4/3)x$
or equivalently	$y = (4/3)x - 2$

We now have an expression for y in terms of x. Does this expression represent y as a function of x? That is, given a particular value for x, does the equation determine a unique value for y? This is not always an easy question to answer. What do the symbols in this equation tell us to do? They say:

"To find y, take the value of x, multiply it by $(4/3)$, and then subtract 2."

Starting with a particular value for x, each step in this process determines a unique number, resulting in a unique corresponding value for y. So y is indeed a function of x. If we used f to represent the function we could write: $y = f(x)$ or $f(x) = (4/3)x - 2$.

What is the domain, the set of possible values for x? In most cases throughout this text, we implicitly assume that the domain (and range) are restricted to real numbers. Are there any other constraints on the values which x may assume? No. In this case we could use any real number for x. So

$$\text{domain} = \text{all real numbers}$$

We can generate a small table of solutions to the equation and the associated graph.

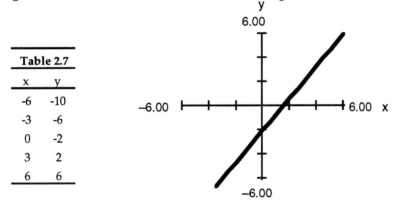

Table 2.7	
x	y
-6	-10
-3	-6
0	-2
3	2
6	6

Fig.2.7   Graph of $f(x) = (4/3)x - 2$

What is the range, the corresponding set of possible values for y? Table 2.7 and Figure 2.7 both suggest that there are no maximum or minimum values for y. This implies that

$$\text{range} = \text{all real numbers}$$

Throughout the graph, as x increases, y increases.

Example 2:   Analyze the equation $xy = 1$. Can it be put in function form? Generate a table and graph of solutions to equations. Discuss.

First, put the equation into function form.

given the equation $xy = 1$

dividing both sides by x (assumes $x \neq 0$) $y = 1/x$

Note that when we divided by x, we assumed that x was not zero, since division by zero is undefined. By doing so have we placed a constraint on x that didn't exist in the original equation, $xy = 1$? No. Since the product of x and y equals 1, neither x (nor y) could ever be zero.

Does y = 1/x describe a function? For any non-zero value for x, does the equation determine a unique value for y? Yes. The symbols say given any x, take its reciprocal. This will give us a value for y that is unique.

If we denote the function by h, then y = h(x) so h(x) = 1/x. So

h(100) = 1/100 h(-35) = 1/(-35)
                                                    = -1/35.

What is the domain of this function? We know that x cannot equal 0. Hence h(0) is undefined Are there any other limitations on x? If we choose any non-zero value for x, is 1/x a real number? Yes. Then

domain = all real numbers except 0

Table 2.8 and Figure 2.8 show some solution pairs to the equation y = 1/x. It's important to look carefully at the behavior of a graph near points or intervals that are missing from the domain. If you're using technology to graph the function, you may want to change your window size several times in order to gain a better idea of the function's shape near the y-axis where x = 0.

Table 2.8

x	h(x)
-100	-1/100
-10	-1/10
-1	-1
-1/10	-10
-1/100	-100
0	not defined
1/100	100
1/10	10
1	1
10	1/10
100	1/100

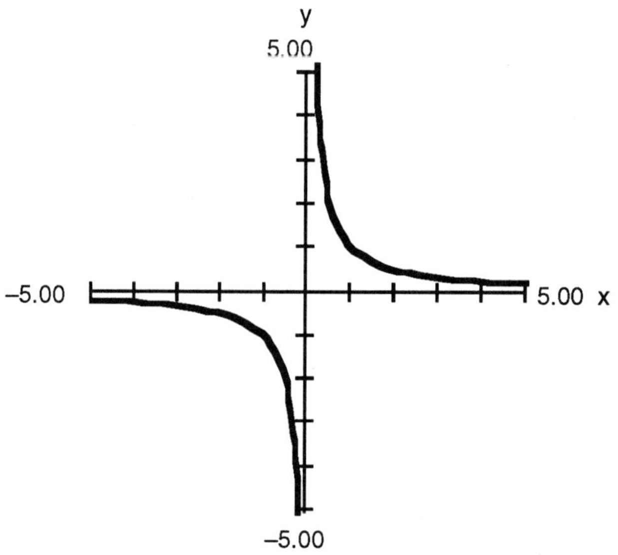

Fig.2.8 A graph of the function y = 1/x

We can see that the function appears to "blow up" when x is near zero, splitting the graph into two pieces one when x > 0 and one when x < 0.

Case 1: x > 0

When x > 0, y > 0. As x increases, the values for y decrease, approaching but never reaching zero. For example, the following points are solutions to the equation y = 1/x (see Table 2.8) and hence lie on the graph. : (1,1), (10, 1/10), (100, 1/100). Notice that as x gets larger and larger, the corresponding y gets smaller and smaller:

As x approaches 0 from the right, the values for y get arbitrarily large, as the solutions (1,1) (1/10, 10), (1/100, 100) suggest. Here as x gets closer and closer to 0, while the y values get larger and larger.

Algebra Aerobics: [7]
Can you describe what happens to the graph in Case 2, when $x < 0$?

The function $y = 1/x$ has no maximum or minimum value. The range is all real numbers except 0.

Example 3: Analyze the equation $y^2 - x = 0$. Can it be put in function form? Generate a table and graph of solutions to equations. Discuss.

Try to put the equation in function form.

given the equation $\qquad\qquad\qquad\qquad y^2 - x = 0$

add x to both sides $\qquad\qquad\qquad\qquad\quad y^2 = x$

There are now two possible solutions for y in terms of x

$$y = +\sqrt{x} \qquad (1)$$
$$\text{and} \qquad y = -\sqrt{x} \qquad\qquad (2)$$

(Remember that by definition $\sqrt{x}$ is non negative).

So for example, if $\qquad\qquad\qquad\qquad\quad y^2 = 4$

then one solution is $\qquad\qquad\qquad\qquad y = +2$

and another is $\qquad\qquad\qquad\qquad\quad y = -2$

We can combine equations (1) and (2) as: $\qquad y = \pm\sqrt{x} \qquad (3)$

We claim that (3) does <u>not</u> represent a function unless we restrict ourselves to a trivial domain of only 0. First, If we limit ourselves to values of x and y that are real, then in order to take the square root of x we must assume that x is not negative. Hence we require

$$x \geq 0$$

Second, to any positive x there are associated two different y's, namely $+\sqrt{x}$ and $-\sqrt{x}$.

---

[7] Answers:

Case 2: $x < 0$

When $x < 0$, $y < 0$. As x increases (moving from left to right but this time on the left side of the y-axis), y decreases. As x approaches 0 through negative values, the values for y move further and further down the y-axis., as the following solution points indicate: $(-1, -1)$, $(-1/10, -10)$, $(-1/100, -100)$.

Thus when x is positive or negative, as x increases, y decreases.

As x gets arbitrarily large negatively, the graph suggests that the corresponding y values approach 0. Examine for example, the solutions (-1,-1), (-10, -1/10), (-100, -1/100).

We can still generate a table and graph of solutions to the equation $y^2 - x = 0$.

Table 2.9

x	y
0	0
1	1
1	-1
2	$\sqrt{2}$
2	$-\sqrt{2}$
4	2
4	-2
9	3
9	-3

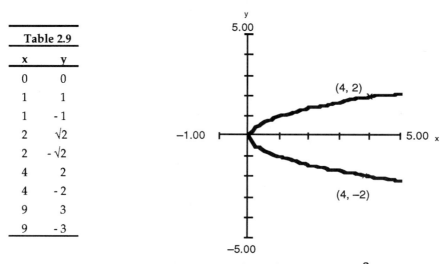

Fig.2.9 Graph of the equation $y^2 = x$

We can see on the graph that the equation $y^2 = x$ does not represent y as a function of x since in particular, when x is 4, y may be +2 or -2.

Something to think about

Can you think of an easy visual test to determine whether or not a graph is a function?

Algebra Aerobics: [8]

In problems 1 - 4 :
a)   solve for y in terms of x
b)   determine if y is a function of x
c)   if y is a function of x, determine the domain

1. $2(x-1) - 3(y+5) = 10$

2. $x^2 + 2x - 3y + 4 = 0$

3. $\dfrac{x}{2} + \dfrac{y}{3} = 1$

4. $x + \sqrt{x-2} + y = 4$

5.   Consider the function $f(x) = x^2 - 5x + 6$.
    a)   Find f(-3), f(0), and f(1)
    b)   Can you find two values of x such that f(x) = 0?

[8]   Answers

1.   a) $2x - 2 - 3y - 15 = 10 \Rightarrow y = (27-2x)/-3 = (2x - 27)/3$ or $y = \dfrac{2}{3}x - 9$

     b) y is a function of x   c)   the domain is all real numbers

2.   a) $y = \dfrac{1}{3}(x^2 + 2x + 4)$ b)   y is a function of x   c)   all real numbers

3.   a) $3x + 2y = 6 \Rightarrow y = \dfrac{-3}{2}x + 3$ b)   y is a function of x   c)   all real numbers

4.   a) $y = 4 - x - \sqrt{x-2}$ b)   y is a function of x   c) $x \geq 2$

5.   a) f(-3) = 30, f(0) = 6, f(1) = 2   b)   x = 2 or x = 3

When is a relationship <u>not</u> a function?

A two-variable table, graph or equation is not a function if it violates the definition of a function; that is, if you can find even one value for the independent variable that determines more than one corresponding value for the dependent value. That means that the relationship does not define a clear unambiguous way of getting from any value of x to a unique value for y. Graphs often offer the easiest way to recognize a non-function.

The Vertical Line Test;

Graphs do not represent functions if they fail the *vertical line; test*; that is, if you can draw any vertical line that crosses the graph two or more times. Any two points of the relationship that lie on the line would share the same first coordinate, but have two different second coordinates. So the relationship would match two different values of the dependent variable with the same value of the independent variable. In other words, for each value of x, there would not be a unique y.

For example, Figure 2.10 is a graph of the equation $y^2 = x$. We've already verified in the previous section that the equation is not a function. But the graph offers simple visual evidence. You could draw a vertical line (many in fact) which would cross the graph twice. One such line is sketched in, and it crosses the graph at (4 , 2) and (4 , –2). That means that the value x = 4 does not determine a unique value of y. It corresponds to y values of both 2 and – 2.. Figure 2.11 shows that the graph of the function $y^2 = x$ is actually a composite of two separate pieces which are functions: the upper half, $y = +\sqrt{x}$ and the lower half, $y = -\sqrt{x}$.

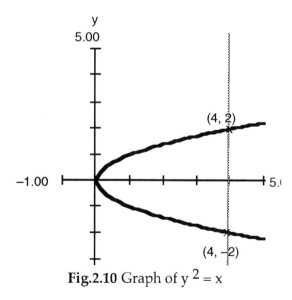

Fig.2.10 Graph of $y^2 = x$

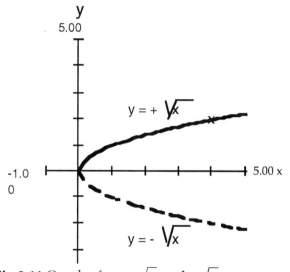
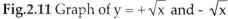

Fig.2.11 Graph of $y = +\sqrt{x}$ and $-\sqrt{x}$

Algebra Aerobics: [9]

1. Describe the graph in Figure 2.12..

   a) Is y a function of x?

   b) When is the function increasing?　　Decreasing?

   c) What seems to happen to the graph　as the value of x is near 0?

2. Consider Figure 2.13.

   a) Estimate the maximum military　sales deliveries between 1986 and　　1993.

   b) Estimate the minimum.

   c) During what years was the amount　of sales increasing?

   d) During what years was it decreasing?

y

10.00

−10.00   10.00   x

−2.00

Fig.2.12 Graph of an abstract relationship between x and y

U.S. Military Sales Deliveries to Foreign Governments

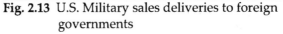

Fig. 2.13 U.S. Military sales deliveries to foreign governments

Source: U.S. Bureau of the Census, in *The American Almanac: The Statistical Abstract of the United States,* 1995

3. Which of the following graphs represent functions and which do not. Why?

   a)   b)   c)   d)

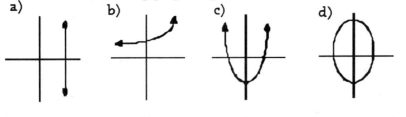

9. Answers:

1.   a) yes   b) When x < 0, the function is increasing. When x>0, the function is decreasing.
       c) The graph appears to "blow up'" meaning that the closer x gets to 0, the higher will be the value for y
2.   a) About 11 billion dollars   b) 7.5 billion dollars   c) Between 1986 and 1987 and 1989 to 1993.
       d) Between 1987 and 1989
3.   b) and c) are functions. a) and d) and not functions since they fail the vertical line test.

If y is a function of x, is x also a function of y? Answer: Not necessarily.

The Pensacola tides are a function of time.

Figure 2.14 shows the water level over a 24-hour period in Pensacola, Florida. A water level of 0 represents the mean water level.

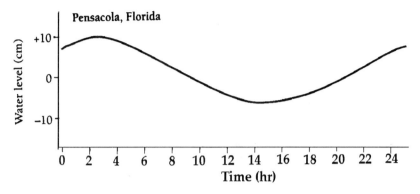

Fig. 2.14   Diurnal Tides in a 24-hour period in Pensacola, Florida

The independent variable is time, and the dependent variable is water level. The domain is from 0 to 24 hours, and the range from about -10 to +10 centimeters. From 0 to 2 hours the water level appears to be steadily increasing until it reaches a maximum of about 10 centimeters. After 2 hours the water level steadily decreases until it reaches a minimum level of about -10 centimeters at 14 hours. After 14 hours the water level starts increasing again, continuing the tidal cycle.

Is the time of day a function of the Pensacola tides?

That is, if I tell you the water level in Pensacola, can you tell me what time of day it is?

No. The same water level may be reached several times during the same day. Look at Figure 2.15 which shows the Pensacola tides, now with water level on the horizontal axis and time on the vertical axis. The graph fails the vertical line test. For instance, if you draw a vertical line corresponding to a water level of 0 centimeters (see the dotted line on the graph) the line crosses the graph in two different places (at 8 and 22 hours). That means there is not a unique hour in the 24 hour period associated with a water level of 0 centimeters. Therefore the time of day is not a function of the water level.

Something to think about

Those of you who live near the sea may notice something very unusual about Pensacola's tides. Normally there is about a 6-hour time difference between high and low tide, creating 2 high and 2 low tides a day. What is the time difference between high and low tide in Pensacola? Such a tide is called *diurnal*.

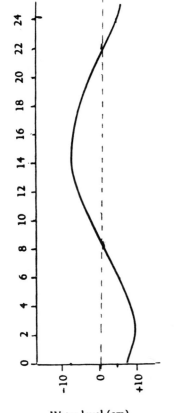

Water level (cm)

Fig. 2.15 The time of day is not a function of the tide

Algebra Aerobics: [9]

Consider the following table:

a)   Is P a function of Y?

Y	1990	1991	1992	1993	1994	1995
P	$1.4	$2.3	$0	-$0.5	$1.4	$1.2

b)   Now think of P as the independent and Y the dependent variable. Is Y a function of P?

---

[9] Answers:

   a)   P is a function of y

   b)   Y is not a function of P (one value of P yields more than one value for Y)

2.4 Finding Models to Describe Data

Deducing functional formulas from data

In the examples so far, the algebraic expression for the function was either given to us if it was known (as in the Illinois tax equation) or no algebraic formula is known (as in median income over time). Often we do not know in advance the formula giving the value of one variable in terms of the value of another. In such instances, we must first, decide whether one variable really is a function of the other, and second, determine the form of the function.

Consider the following experimental data on growth of a colony of *sordaria finicola* fungus. The area of the colony (growing in a petri dish) is measured at a series of times. We denote by t the time in hours following the introduction of the sample into the dish, and by A the area of the colony in square centimeters. Here is a table of the results.

Table 2.10	
time t (hours)	area A (sq. cm)
0	0.0
5	0.6
10	2.5
15	5.6
20	10.0
25	15.6

Our first step is to plot the data. We have drawn straight lines between successive data points to guide our eyes.

In this case, the area certainly is a function of time: there is only one measurement at each time, so there can never be two values of the area for the same time.

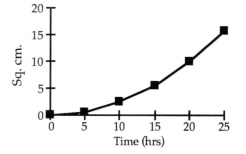

Fig. 2.16   Area of sordaria finicola colony

Let us try to guess an algebraic expression for this function.

Our first guess might be a simple proportionality as in the Illinois tax example; that is, the area might be able to be expressed as a multiplication of a simple number times the time. Since A=0 at t=0, and since substituting t=20 should give us A=10, we may guess the formula

$$A = t / 2 \quad (\text{ matches the data points } t=0, A=0 \text{ and } t=20, A=10)$$

But this formula does not fit any other data point. For example, at t=10, this formula predicts $A = 10/2 = 5$, but the measured value was A=2.4. The following table and graph show the predicted area (using the formula $A = t/2$) and the actual area.

Table 2.11		
time t (hours)	Predicted area A (sq. cm)	Actual area A (sq. cm)
0	(0)/2 = 0.0	0.0
5	(5)/2 = 2.5	0.6
10	(10)/2 = 5.0	2.5
15	(15)/2 = 7.5	5.6
20	(20)/2 = 10.0	10.0
25	(25)/40 = 12.5	15.6

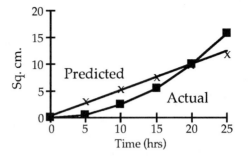

Fig. 2.17 Predicted vs. the actual area of sordaria finicola colony

The fit doesn't look very good.

As a second guess, we may try a function involving a square. If we again try to match the data point t=20, A=10, we now may write the formula
$$A = t^2 / 40$$

Carrying out the instructions on the right side, we compute $t^2/40 = 20^2/40 = 400/40 = 10$ Again, we check whether this formula correctly matches the other data points. This time the fit is very close! We have determined a single formula, a model, which summarizes all the information in the original table of data values.

Table 2.12		
time t (hours)	Predicted area A (sq. cm)	Actual area A (sq. cm)
0	$(0^2)/40 =$ 0.0	0.0
5	$(5^2)/40 =$ 0.6	0.6
10	$(10^2)/40 =$ 2.5	2.5
15	$(15^2)/40 =$ 5.6	5.6
20	$(20^2)/40 =$ 10.0	10.0
25	$(25^2)/40 =$ 15.6	15.6

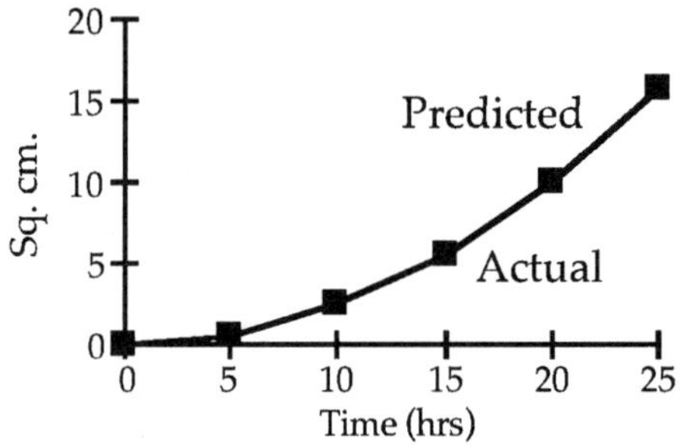

Fig. 2.18 Predicted and actual area coincide

This formula describes growth of the fungus colony for values of t between 0 and 25 hours. It may be true for times somewhat beyond 25 hours, but we do not know how far; perhaps at 26 hours the colony met the edge of the petri dish and stopped growing completely. We say that the domain of the function is the values of t from 0 to 25 hours.

We should remember that simply determining this formula does not tell us *why* the formula is true. It is simply a way of summarizing the information we have measured. However, knowing the functional expression for the data can suggest various ideas for the reason. In this case, we might guess that the colony has the shape of a circle whose radius is increasing at a constant rate.

A glimpse ahead

Scatter plots that come from real data are rarely as clean as the fungus data. Real data are usually messy. As Figure 2.19, there may be many values of the dependent variable associated with a single value of the independent variable. This particular graph shows a scatter plot of the relationship between per capita gross domestic product (GDP) measured in U.S. dollars and female life expectancy for 114 different countries. (The raw data can be found in Part III in the Nations Data Set.) The gross domestic product is a nation's total output of goods and services produced in that country and valued at market prices. "Per capita" is a Latin phrase that means per person.

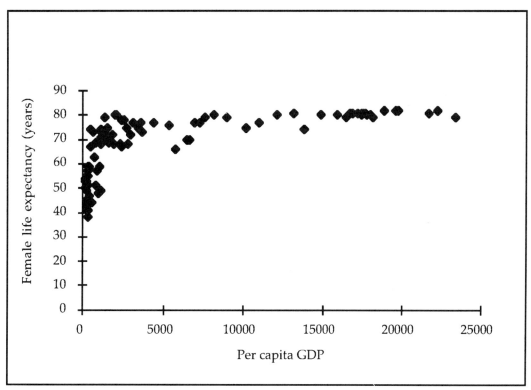

Fig.2.19 Female life expectancy vs. per capita GDP

Clearly you could not find a function that would be a perfect fit to the data. Yet the graph does seem to suggest a relationship between per capita GDP and life expectancy. If the GDP is less than $5000 per person, an increase in per capita GDP seems to be associated with an increase in female life expectancy. If the per capita GDP is greater than $5000, the scatter plot is fairly flat, suggesting an increase in per capita GDP does not increase female life expectancy. There seems to be a "ceiling" for life expectancy of about 80 years. We can imagine penciling in a function that would roughly approximate the shape of the data.

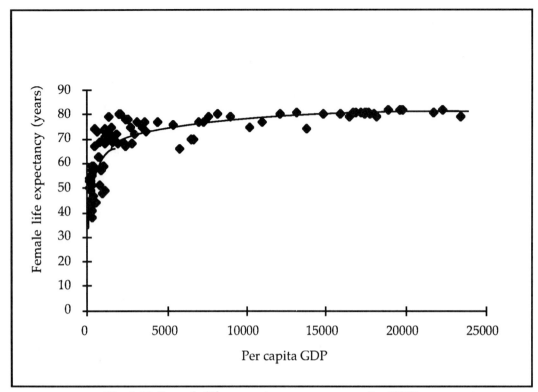

Fig.2.20 Female life expectancy vs. per capita GDP

If we could find an algebraic expression for the function, we might have a reasonable way of describing the relationship between per capita GDP and life expectancy. This function would be only an approximation of the data, a simplification of reality. But it may provide a model that can help us to describe and understand what is happening. We can then go back to the raw data and explore in more detail countries with a per capita GDP below and above $5000 or those countries which seem to fall outside of the general pattern. We can ask if the same relationship holds for male life expectancy. We could study the relationship between life expectancy and other variables like the number of doctors, or pollution levels.

Mathematical models can provide tools for understanding *what* is happening, but they do not tell us *why* it is happening. Why does female life expectancy remain fairly stable if per capita GDP is greater than $5000? Why does it increase when the per capita GDP is less than $5000? We can posit plausible explanations. We could hypothesize that greater per person national wealth implies access to better health care and hence longer life expectancy. We could refine our model to include the ceiling effect. But we should be aware that we have now taken a leap beyond the mathematical description of the data.

The rest of this text will be dedicated to building a library of functions that can serve as mathematical models for real world data sets like this one. We will study the properties of these functions to enable us to better understand how they can be used to describe the world around us.

Chapter 2 Summary

A variable y is a *function* of variable x if each value of x determines a unique value of y. y is the *value of the function* or the *dependent variable;* and x is the *independent variable;*.

A function used to describe a real world phenomenon is called a *mathematical model;*.

The *domain* of a function is the set of possible values of the independent variable. The *range* is the set of corresponding values of the dependent variable. Unless otherwise specified, the domain is assumed to be the set of all real numbers x such that the corresponding y values in the range are real.

The most common representations of functions are tables, graphs, or equations.

The *solutions* to an equation in two variables x and y are the set of ordered pairs of the form (x, y) that make the equation a true statement. The graph of an equation is the set of points corresponding to the ordered pairs solutions.

If an equation represents a function, then solving the equation for the dependent variable in terms of the independent variable is called putting the equation in *function form*. The ordered pair solutions of the function are conventionally listed with the value of the independent variable first, and the dependent second.

(independent variable,dependent variable)

Hence the graph of a function by convention would have the independent variable on the horizontal axis and the dependent on the vertical axis.

A graph does not represent a function if it fails the *vertical line test*; that is, if you can draw a vertical line that crosses the graph two or more times.

EXERCISES

1. Consider the graph of profit vs. year below.
    a) During what years did profit increase?
    b) During what years did it decrease?
    c) What was the maximum profit?
    d) What was the minimum profit?
    e) During what interval was the amount of profit stable?

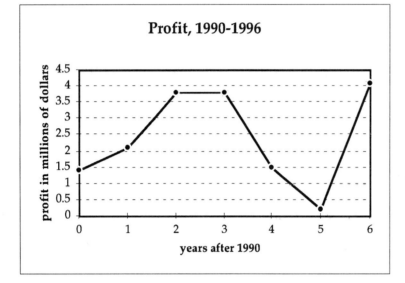

2. Sketch a plausible graph for the following and label the axes:
    a) The amount of snow on your backyard each day from December 1 to March 1
    b) The temperature during a twenty-four hour period in your home town during one day in July.
    c) The amount of water inside your fishing boat if your boat leaks a little and your fishing partner bails out water every once in a while.
    d) The amount of daylight each day of the year.
    e) The temperature of a cup of coffee left to stand.
    f) The temperature of an ice-cold drink left to stand.

3. The following graph shows the twenty-four hour temperature cycle of a normal man. The man was confined to bed in order to minimize temperature fluctuations due to activity.
   a) Estimate the man's maximum temperature.
   b) Estimate the minimum temperature.
   c) Give a short general description of the twenty-four hour temperature cycle.

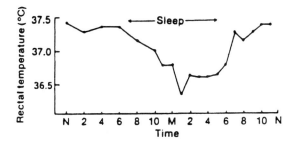

From Mountcastle V.B. *Medical Physiology*, vol. 2, ed. 14, St. Louis, Mosby-Year Book, Inc.

4. Look at the graph below showing the populations of two towns.
   a) What is the range of population size for Johnsonville? for Palm City?
   b) During what years did the population of Palm City increase?
   c) During what years did the population of Palm City decrease?
   d) When were the populations equal?

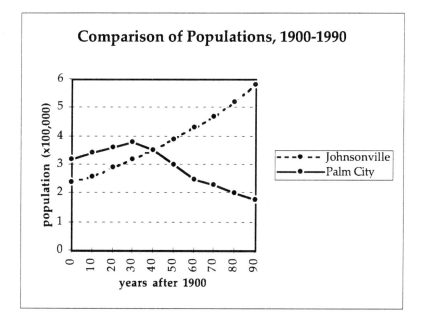

5. In each of the following three examples:
    a)   identify the independent and dependent variable
    b)   verify whether or not the example represents a function

If the example is a function then:
    c)   generate a table of values (if not already given)
    d)   generate a graph (if not already graphed)
    e)   identify the intervals on which the function is increasing or decreasing
    f)   identify any maximum or minimum values of the function

Example 1: While Table and Figure 2.2 showed the annual budget deficit, the following table and graph show the accumulated gross federal debt

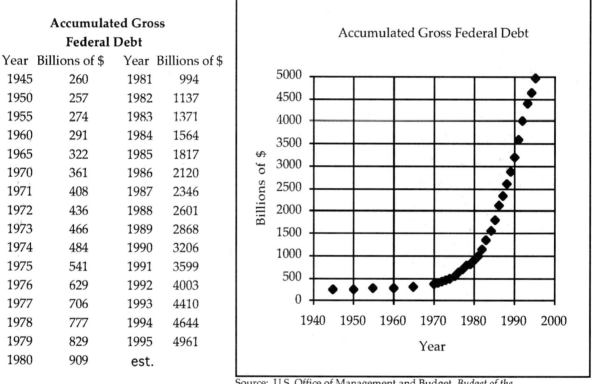

Accumulated Gross Federal Debt

Year	Billions of $	Year	Billions of $
1945	260	1981	994
1950	257	1982	1137
1955	274	1983	1371
1960	291	1984	1564
1965	322	1985	1817
1970	361	1986	2120
1971	408	1987	2346
1972	436	1988	2601
1973	466	1989	2868
1974	484	1990	3206
1975	541	1991	3599
1976	629	1992	4003
1977	706	1993	4410
1978	777	1994	4644
1979	829	1995	4961
1980	909	est.	

Source: U.S. Office of Management and Budget, *Budget of the United States Government*, annual, in *The American Almanac: Statistical Abstract of the United States*, 1995.

Example 2: The formula

$$A = 25 \, W$$

where A = ampicillin dosage in milligrams, W = child's weight in kilograms, represents the *minimum* effective pediatric daily drug dosage of ampicillin as a function of a child's weight. (You may recall that A = 50W represented the *maximum* recommended dosage.) Both formulas apply only for children weighing up to 10 kilograms.

Example 3: The following table shows the success of national regulations in controlling carbon monoxide emissions.

National Ambient Air Pollutant Concentrations: 1985 to 1994 (Air quality standard is 0 parts per million (ppm))									
	1985	1987	1988	1989	1990	1991	1992	1993	1994
Carbon monoxide (ppm)	6.97	6.69	6.38	6.34	5.87	5.55	5.18	4.88	5.57

Source: U.S. Environmental Protection Agency, National Air Quality and Emissions Trends Report annual, as taken from *The American Almanac: Statistical Abstract of the United States, 1995-1996.*

6.   Consider the chart below.
     a) During what interval did the number of arrests show a slight decrease?
     b) Estimate the ratio of the number of arrests in 1993 to the number of arrests in 1988.
     c) What can be said generally about the number of juveniles arrested for murder based on this chart?

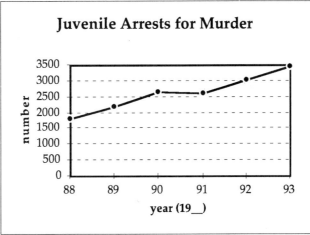

Source: Bureau of the Census, *The American Almanac: Statistical Abstract of the United States, 1995.*

7.   For persons who earn less than $20,000 a year, income tax is 16% of their income.
     a) Generate a formula that describes income tax in terms of income.
     b) What are you treating as the independent variable? The dependent variable?
     c) Does your formula represent a function? Explain.
     d) What is the domain? the range?

8.   The cost of a phone call is $1.50 for the first three minutes, then $0.32 for each additional minute.
     a)   Generate a formula that describes the price of a phone call as a function of time.
     b)   What is the independent variable? The dependent variable?
     c)   Does your formula represent a function? Explain.
     d)   What is a suitable domain? Range?
     e)   Generate a small table of values and a graph.

9. The price of gasoline is $1.24 per gallon.
    a) Generate a formula that describes the cost of buying gas as a function of the amount of gasoline purchased.
    b) What is the independent variable? The dependent variable?
    c) Does your formula represent a function? Explain.
    d) What is a suitable domain? Range?
    e) Generate a small table of values and a graph

10. The price of a bus ticket is $350. If a group of six or less travels together, the price of each ticket is reduced $20 for each person traveling in the group.
    a) Generate a formula that describes the price per ticket as a function of the number of persons traveling in a group.
    b) What is the independent variable? The dependent variable?
    c) Does your formula represent a function? Explain.
    d) What is a suitable domain? Range?

11. Consider the table below listing the weight (W) and height (H) of 5 individuals. Based on this table, is height a function of weight? Is weight a function of height?

W	120	120	125	130	135
H	54	55	58	60	56

12. Consider the table below:

Y	1990	1991	1992	1993	1994	1995
P	$1.4	$2.3	$0	-$0.5	$1.4	$1.2

    a) Is P a function of Y?
    b) What is the domain? What is the range?
    c) What is the maximum value of P? In what year did this occur?
    d) During what years was P increasing? Decreasing?
    e) Now redo the table so that the independent variable is P and the dependent variable is Y. Is Y a function of P?

13. Look at the table below.

n	-4	-3	-2	-1	0	1	2	3	4	5
p(n)	0.063	0.125	0.25	0.5	1	2	4	8	16	32

    a) Find p(-4), p(5) and p(1).
    b) For what value(s) of n does p(n) = 2?

14. Consider the function graphed below.
    a)   Find f(-3), f(0), and f(1)
    b)   Find two values of x such that f(x) = 0

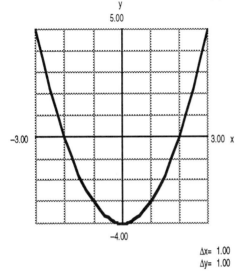

15.. Write an equivalent equation that expresses z in terms of t. Is z a function of t? Why or why not.?
    a) $3t - 5z = 10$
    b) $(2t - 3)(t + 4) - 2z = 0$
    c) $t + z^2 + 10 = 0$

16. Assuming y is a function of x, find the domain for each of the following functions:
    a) $y = 300.4 + 3.2x$

    b) $y = \sqrt{x + 2}$

    c) $y = \dfrac{5 - 2x}{2}$

    d) $y = \dfrac{1}{x + 1}$

    e) $y = 3$

17. Determine whether y is a function of x in the following examples. If the equation is not a function, find a value of x that is associated with 2 different y values.
    a) $y = x^2 + 1$
    b) $y = 3x - 2$
    c) $y = 5$
    d) $x^2 + y^2 = 25$

18.   Which of the following graphs describe functions?

a)

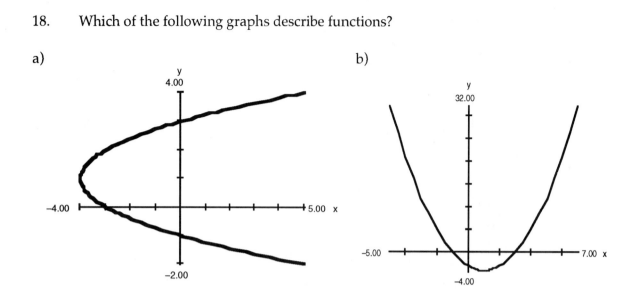

b)

c)

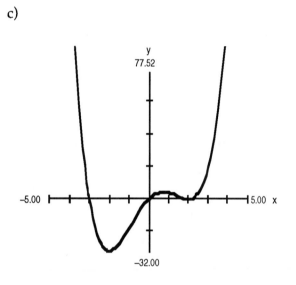

d)

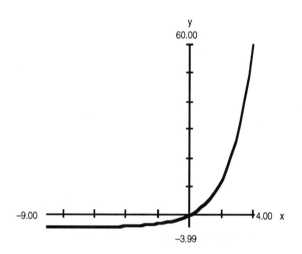

EXPLORATION 2.1
Picturing Functions

Objective
- develop an intuitive understanding of functions

Materials/Equipment
- none

Procedure
PART I

Class discussion

Bridget, the 6 year old daughter of a professor at the University of Pittsburgh, loves playing with her rubber duckie in the bath at night. Her mother drew the following graph for her math class. It shows the water level in Bridget's tub as a function of time.

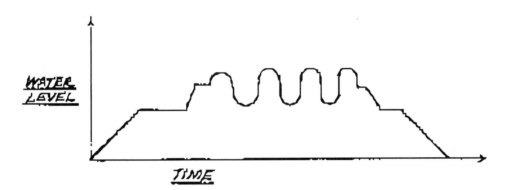

Pick out the time period during which:
- the tub is being filled
- Bridget is entering the tub
- she is playing with her rubber duckie
- she leaves the tub
- the tub is being drained ?

With a partner

Create your own graph of a function that tells a story. Be as inventive as possible. (Some students have drawn functions that: show the decibel levels during a phone conversation with a boy and girl friend; number of hours spent doing homework during one week; amount of money in their pocket during the week.)

Class discussion

Draw your graph on the blackboard and tell its story to the class.

PART II
With a partner
Generate a plausible graph for each of the following:
1. Time to drive to work as a function of the amount of snow on the road. (Note: the first inch or so may not make any difference; the domain may be only up to about a foot of snow since after that you may not be able to get to work.)
2. The length of the day as a function of the time of year.
3. The temperature of the coffee in your cup as a function of time.
4. The distance that a cannonball (or javelin or baseball) travels as a function of the angle of elevation at which is it is launched. The maximum distance is attained for elevations of around 45 degrees.
5. Your distance from home as a function of time. Assume that you leave home walking at a normal pace, realize you have forgotten your homework and run home, and then run even faster to school. You sit for a while in a classroom and then walk leisurely home.

BONUS QUESTION
Assume that water is pouring into each of these containers at a constant rate. The height of water in the container is a function of the volume of liquid. Generate a graph of this function for each container.

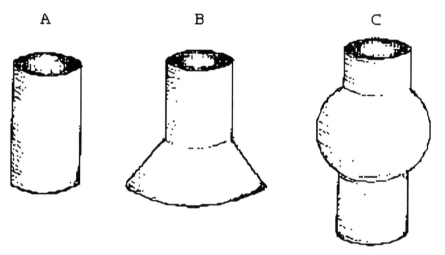

A   B   C

Discussion/Analysis
Are your graphs similar to those generated by the rest of the class? Can you agree as a class as to the basic shape of each of the graphs? Are there instances in which the graphs could look quite different?

EXPLORATION 2.2
Deducing Functional Formulas from Data

Objectives
- find and describe patterns in data
- deduce functional formulas from data tables
- extend patterns using functional formulas

Materials/Equipment
- none required

Procedure
Class discussion
- Examine the following data table. Look for a pattern in terms of how y changes when x changes. Explain in your own words how to find y in terms of x .

x	y
0	0.0
1	0.5
2	1.0
3	1.5
4	2.0

x	y
0	5
1	8
2	11
3	14
4	17

- Assume that the pattern continues indefinitely, use the rule you have found to extend the data table to include negative numbers for x.
- Check your extended data tables. Did you find a unique value for y given a particular value for x?
- Use a formula to describe the pattern that you have found. Do you think this formula describes a function? Explain.

On your own
- Use the following data tables to explain in your owns words how to find y in terms of x and extend each of the data tables using the rule you have found.
- Use a formula to describe the pattern that you have found.

x	y
0	0
1	1
2	4
3	9
4	16

x	y
0	.0
1	1
2	8
3	27
4	64

x	y
0	0
1	2
2	12
3	36
4	80

x	y
−2	0
0	10
5	35
10	60
100	510

x	y
0	-1
1	0
2	3
3	8
4	15

x	y
0	3
10	8
20	13
30	18
100	53

Discussion/Analysis

With a partner:

Compare your results. Do the formulas that you have found describe functions? Explain.

Class Discussion

Does the rest of the class agree with your results? Remember that formulas that look different may give the same results.

Exploration Linked Homework

1   a) For each of the following data tables, explain in your own words how to find y in terms of x. Using the rule you have found, extend each of the data tables to include negative numbers.

x	y
−10	10.0
0	0.0
3	0.9
8	6.4
10	10.0

x	y
0	-3
1	1
2	5
3	9
4	13

   b) Use a formula to describe the pattern you have found. Does your formula describe a function? Explain.

2.   Make up a functional formula, generate a data table and bring the data table on a separate piece of paper to class. The class will be asked to find your rule and express it as a formula.

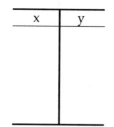

x	y

Chapter 3

Rates of Change

Overview

From graphs and tables we can see that over time the U.S. population has been steadily increasing, the percentage of people living in rural areas has been decreasing, and the number of civil disturbances has cycled up and down. Here we discuss various mathematical techniques for measuring such changes.

The *average rate of change* is one of the most useful measures for describing change. The average rate of change of the U.S. population with respect to time is calculated by dividing the change in the population by the change in time. The concept of the average rate of change can be abstracted to a calculation involving the coordinates of any two points. There are other common ways of describing change over time, including absolute or percentage change in size.

The first exploration includes an exercise in constructing the strongest possible case for your point of view. You can use all the tools developed so far: tables, graphs and numerical summaries (including mean, median, and the average rate of change). It challenges you to choose information to argue strenuously one side of an issue. Then you are asked to construct, from the same data set an equally convincing argument for the opposing side.

The second exploration will help you understand why rates of change are "averages."

After reading this chapter you should be able to:

- define the average rate of change
- calculate the slope of a line
- describe change over time in several different way

3.1 Average Rates of Change

We can think of the U.S. population as a function of time. Table 3.1 and Figure 3.1 are two representations of that function. They show the changes in the size of the U.S. population since the U.S. government conducted its first decennial census in 1790. Time, as usual, is the independent variable and population size is the dependent variable.

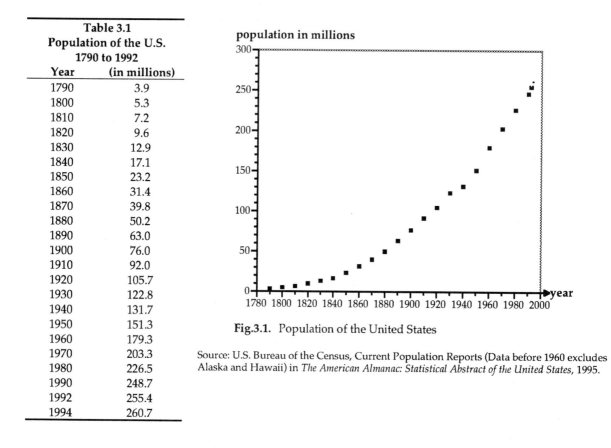

Table 3.1 Population of the U.S. 1790 to 1992	
Year	(in millions)
1790	3.9
1800	5.3
1810	7.2
1820	9.6
1830	12.9
1840	17.1
1850	23.2
1860	31.4
1870	39.8
1880	50.2
1890	63.0
1900	76.0
1910	92.0
1920	105.7
1930	122.8
1940	131.7
1950	151.3
1960	179.3
1970	203.3
1980	226.5
1990	248.7
1992	255.4
1994	260.7

Fig.3.1. Population of the United States

Source: U.S. Bureau of the Census, Current Population Reports (Data before 1960 excludes Alaska and Hawaii) in *The American Almanac: Statistical Abstract of the United States, 1995.*

Figure 3.1 clearly shows that the size of the U.S. population has been growing over the last two centuries, and growing at what looks like an increasingly rapid rate. How can we describe the change in population over time quantitatively? We could take any two points on the graph of the data and find how much the population increased during the time period between them.

Suppose we look at the change in the population between 1900 and 1992. In 1900 the population was 76.0 million; by 1992 the population had grown to 255.4 million. How much did the population increase? Since it rose from 76.0 million to 255.4 million, it increased by the difference between these two values.

change in population size = 255.4 − 76 .0 = 179.4 million

In Figure 3.2, we have drawn two parallel horizontal lines through the points (1900, 76.0) and (1992, 255.4) to the vertical population axis. The 179.4 million population change is represented by the difference in the height of the two points or the change in the variable on the vertical axis.

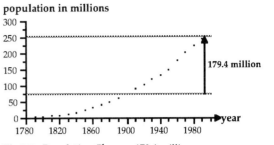

Fig.3.2. Population Change: 179.4 million

Knowing that the population changed by 179.4 million tells us nothing about how rapid the change was; this change clearly represents much more dramatic growth if it happened in 20 years than if it happened across 200 years. In this case, the length of time over which the change in population occurred is:

change in years = 1992 - 1900 = 92 years

In Figure 3.3 we have drawn two parallel vertical lines from the points (1900, 76.0) and (1992, 255.4) down to the horizontal time axis. The 92 year change is represented by the horizontal difference in the positions of the two points.

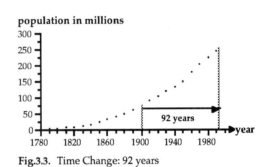

Fig.3.3. Time Change: 92 years

If a line segment is drawn connecting the two points, it forms the hypotenuse of the right triangle sketched in Fig 3.4. The length of the horizontal section of the triangle represents a change of 92 years, and the length of the vertical section of the triangle represents a change of 179.4 million in population size.

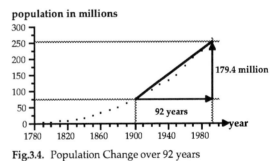

Fig.3.4. Population Change over 92 years

To find the *average rate of change* in population per year from 1900 to 1992, divide the change in the population by the change in the number of years:

average rate of change = change in population size / change in years
= 179.4 million people / 92 years
= 1.95 million people / year

This calculation shows that "on average," the population grew at a rate of 1.95 million people per year from 1900 to 1992. In the phrase "million people/year" the slash sign represents division and is read as "per," so the phrase is read as "million people per year."

Geometrically the average rate of change represents the *slope* of the line on which the two endpoints lie.

$$\text{average rate of change} = \text{slope} = \frac{\text{change in vertical variable}}{\text{change in horizontal variable}}$$

In this case the endpoints are (1900, 76.0) and (1992, 255.4), so:

$$\text{slope} = \frac{255.4 - 76.0 \text{ million people}}{1992 - 1900}$$
$$= 1.95 \text{ million people / year}$$

which is of course precisely the average rate of change.

Limitations on the use of average rates of change

Average rates of change have the same limitations as any *average*. Though the average rate of change of the U.S. population from 1900 to 1992 was 1.95 million per year, it is highly unlikely that in each year the population grew by exactly 1.95 million. The number 1.95 million per year is, as the name states, an *average*. If the arithmetic average or *mean* height of students in your class is 67 inches, you don't expect every student to be 67 inches tall.

The average rate of change depends entirely upon the end points you use to calculate the rate. If the data points do not all lie on a straight line, the average rate of change varies for different intervals. For instance, if you compare the average rates of change in population for the time intervals 1840 to 1940 and 1880 to 1980 you get two different values shown below. You can see on the graph that the slope (or average rate of change) is much steeper between 1880 to 1980 than between 1840 to 1940. Different intervals give different impressions of the rate of change in the U.S. population, so it is important to state which end points are used.

Table 3.2

Time Interval	Change in Population	Average Rate of Change
1840 to 1940	131.7 - 17.1 = 114.6 million	114.6 million/100 yrs = 1.15 million/yr
1880 to 1980	226.5 - 50.2 = 176.3 million	176.3 million/100 yrs = 1.76 million/yr

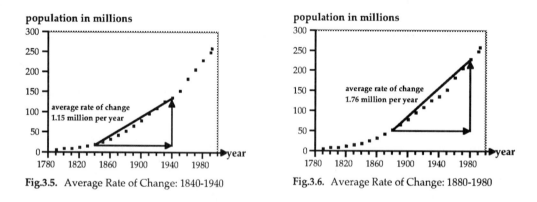

Fig.3.5. Average Rate of Change: 1840-1940   **Fig.3.6.** Average Rate of Change: 1880-1980

The average rate of change does not reflect all the fluctuations in population size that may occur between the end points. The average rate of change over a large interval gives an overall average estimate; for more specific information, the average rate of change can be calculated for smaller intervals.

Algebra Aerobics [1]

1. Suppose your weight five years ago was 135 pounds and your weight today is 143 pounds. Find the average rate of change in your weight.
2. The U.S. imported electronic equipment and accessories valued at $18.3 trillion in 1989 and valued at $27.5 trillion in 1993. Find the average rate of change in the value of the electronics imported.
3. The table below indicates the number of deaths in motor vehicle accidents in the United States.

Table 3.3 Annual Deaths in Motor Vehicle Accidents

1970	1980	1990	1992
114,638	105,718	91,983	86,777

Find the average rate of change:
   a)   from 1970 to 1980
   b)   from 1980 to 1992.

4. If you have access to the course software, open up *Linear Graphs* and select "L7: Average rates of change: $\Delta y / \Delta x$" Clicking and dragging on the mouse will fix one end point and vary a second end point. Track the corresponding average rate of change calculation to develop an intuition for how its values vary.

---

[1]
1. $(143-135)/5 = 1.6$ pounds per year

2. $(27.5-18.3)/4 = \$2.3$ trillion per year

3. a) -892 deaths per year (or a decrease in the annual deaths by 892 deaths per year)
    b) -1578.4 deaths per year or a decrease of 1,578.4 deaths per year.

3.2 A Closer Look at Average Rates of Change

The concept of the average rate of change may give a better sense of what is happening if we compare rate of change calculations over different intervals. One way to do it is to pick a fixed interval size and then calculate the average rate of change for each interval. On the U.S. population data, using an interval size of one decade, we can add a third column to the data table in which each entry represents the average population growth *per year* (i.e., the average annual rate of change) during the previous ten years.[2] If you "connected the dots" on the original graph in Figure 3.1, these numbers would represent the slopes of each of the small line segments connecting adjacent data points. A few of these calculations are shown in the last column in Table 3.4.

		Table 3.4	
	Average Annual Rates of Change of U.S. Population		
	from 1790 to 1990		
Year	Population in millions	Average Annual Rate for prior decade in millions/yr.	calculation
1790	3.9	data not available	
1800	5.3	0.14	$0.14 = (5.3 - 3.9)/10$
1810	7.2	0.19	
1820	9.6	0.24	
1830	12.9	0.33	
1840	17.1	0.42	$0.42 = (17.1 - 12.9)/10$
1850	23.2	0.61	
1860	31.4	0.82	
1870	39.8	0.84	
1880	50.2	1.04	
1890	63.0	1.28	
1900	76.0	1.30	
1910	92.0	1.60	
1920	105.7	1.37	
1930	122.8	1.71	
1940	131.7	0.89	$0.89 = (131.7-122.8)/10$
1950	151.3	1.96	
1960	179.3	2.80	
1970	203.3	2.40	
1980	226.5	2.32	
1990	248.7	2.22	

What is happening to the average rate of change over time? Start at the top of the third column and scan down the numbers. Notice that until 1910 the average rate of change increases every year. Not only is the population growing every decade, but it is growing at an increasing rate. It's like being in a car that is not only moving forward, but accelerating. A feature that was not so obvious in the original data is now evident: in the decades 1910 -1920 and 1930-1940, and in the decades following 1960 we see a decreasing rate of growth. It's like a car decelerating - it is still moving forward but it is slowing down.

---

[2] We have omitted the information about the years between 1990 and 1994 since it did not represent a full decade.

A graph with years on the horizontal axis and average rates of change on the vertical axis shows more clearly how the average rate of change fluctuates over time. The first point, corresponding to the year 1800, shows an average rate of change of 0.14 million people/year for the decade 1790-1800. The rate 1.71, corresponding to the year 1930, means that from 1920 to 1930 the population was increasing at a rate of 1.71 million people per year.

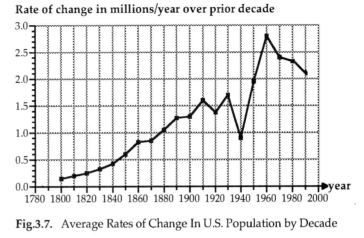

Rate of change in millions/year over prior decade

Fig.3.7. Average Rates of Change In U.S. Population by Decade

Take the time to interpret a few points, since this graph is somewhat more abstract than the previous ones. The pattern of growth was fairly steady up until about 1910. Why did it change? During the decades when the rate of growth was decreasing, did more people die, or were there fewer people born? Did the immigration rate slow down? Any one of these reasons would slow the growth rate. To form some hypotheses you may need to review history. For instance, a possible explanation for the slowdown in the decade prior to 1920 might be World War I and the 1918 flu epidemic that by 1920 had killed nearly 20,000,000 people, including about 500,000 Americans.

One obvious suspect for the big slowdown in population growth in the 1930s is the Great Depression. Look back at Fig. 3.1, the graph that shows the overall growth in the U.S. population. The decrease in the average rate of change in the 1930s is large enough to show up in our original graph as a visible slowdown in population growth.[3] Why is there a drastic decline in growth in the '30s? Perhaps during the Depression people just couldn't afford to have children.

Something to think about

What might be some reasons for the slowdown during your lifetime, from the 1960s onward? Where could you turn to find some evidence to test your hypotheses?[4] Questions arising from one set of data often prompt the search for more data.

---

[3] If you zoomed in on the original graph, the other fluctuations in growth rate would show up too.

[4] If you are really motivated you can calculate the rate of change of the rate of change!

Algebra Aerobics

Table 3.5 and Figure 3.8 show estimates for world population between 1800 and 1993.

Table 3.5 World Population

Year	Total Population (in millions)	Average Rate of Change
		n.a.
1800	910	
1850	1130	
1900	1600	
1950	2510	
1970	3702	
1980	4456	
1990	5293	
1993	5555	

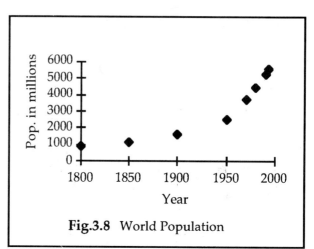

Fig.3.8 World Population

Source: U.S. Bureau of the Census, Current Population Reports (Data before 1960 excludes Alaska and Hawaii) in *The American Almanac: Statistical Abstract of the United States,* 1995.

a)   Fill in the third column on the table, calculating the average annual rate of change.

b)   Graph the average annual rate of change vs. time.

c)   During what period was the average annual rate of change the largest?

d)   Describe in general terms what happened to the average rate of change between 1880 and 1993.

Answers:

a)

Year	Total Population (in millions)	Average Annual Rate of Change (millions/yr)
1800	910	n.a.
1850	1130	4.4
1900	1600	9.4
1950	2510	18.2
1970	3702	59.6
1980	4456	75.4
1990	5293	83.7
1993	5555	87.3

b)

Average Annual Rate of Change

c)   1990-1993

d)   According to these numbers, the average annual rate of change has been increasing.

3.3 Generalizing the Concept of the Average Rate of Change

Though our examples so far described the change in one variable over time, the notion of the average rate of change can be used to describe the change in any variable with respect to another. If you have a graph that represents a plot of data points, then the average rate of change between two endpoints is always the change in the vertical variable divided by the change in the horizontal variable.

$$average\ rate\ of\ change = \frac{change\ in\ vertical\ variable}{change\ in\ horizontal\ variable}$$

If the variables represent real world quantities, which have units of measure (e.g. dollars or years), then the average rate of change has units too.

$$units\ of\ average\ rate\ of\ change = \frac{units\ of\ vertical\ variable}{units\ of\ horizontal\ variable}$$

For example, the units might be dollars/year (read as dollars per year) or pounds/person (read as pounds per person).

Geometrically the average rate of change represents a slope, an indicator of the steepness of the line connecting the two endpoints. We often refer to the change in the vertical variable as the *rise*, and the change in the horizontal variable as the *run*. If (x_1, y_1) and (x_2, y_2) are the two end points, then $(y_2 - y_1)$ gives the rise. This difference is often denoted by Δy (read as "delta y"), where Δ is the Greek letter capital D (think of D representing difference).

$$\Delta y = y_2 - y_1$$

Similarly, the run (delta x) can be represented by:

$$\Delta x = x_2 - x_1$$

So $\quad \dfrac{rise}{run} = \dfrac{y_2 - y_1}{x_2 - x_1} = \dfrac{\Delta y}{\Delta x} = \dfrac{change\ in\ y}{change\ in\ x}$

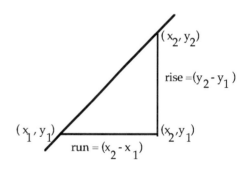

Fig.3.9. Slope = rise/run

> The average rate of change represents a *slope*.
> Given two points (x_1, y_1) and (x_2, y_2)
>
> $$average\ rate\ of\ change = slope = \frac{rise}{run} = \frac{y_2 - y_1}{x_2 - x_1} = \frac{\Delta y}{\Delta x} = \frac{change\ in\ y}{change\ in\ x}$$

The Anthology reading *Slopes* describes many of the practical applications of slopes, from cowboy boots to handicap ramps.

A note about calculating slopes: it doesn't matter which point is first

Given two points, (x_1, y_1) and (x_2, y_2), it doesn't matter which one you use as the first point when you calculate the differences; the result is the same either way. You do need to be consistent in the order in which the coordinates appear in the numerator and the denominator. If y_1 is the first term in the numerator then x_1 must be the first term in the denominator.

$$slope = \frac{y_2 - y_1}{x_2 - x_1} \qquad \text{or} \qquad slope = \frac{y_1 - y_2}{x_1 - x_2}$$

Example:
Given the two points (-2 , -6) and (7, 12) we can calculate the slope treating (-2, -6) as (x_1, y_1) and (7, 12) as (x_2, y_2). Then:

$$slope = \frac{y_2 - y_1}{x_2 - x_1} = \frac{12 - (-6)}{7 - (-2)} = \frac{18}{9} = 2$$

Or equivalently we could have used -6 and -2 as the first terms in the numerator and denominator respectively. Either way we obtain the same answer.

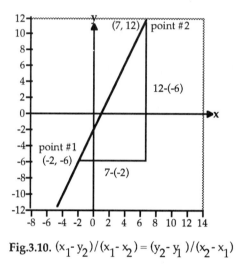

Fig.3.10. $(x_1 - y_2)/(x_1 - x_2) = (y_2 - y_1)/(x_2 - x_1)$

$$slope = \frac{y_1 - y_2}{x_1 - x_2} = \frac{-6 - 12}{-2 - 7} = \frac{-18}{-9} = 2$$

An aside: proving it doesn't matter which point is first

We saw above that the order of points didn't matter when we calculated the slope of the line connecting (-2 , -6) and (7, 12). We can show that this is true given any two points (x_1, y_1) and (x_2, y_2). Suppose we use y_2 and x_2 as the first terms in the numerator and denominator respectively. Then:

$$slope = \frac{y_2 - y_1}{x_2 - x_1}$$

multiplying by $\dfrac{-1}{-1}$
$$= \frac{-1}{-1} \times \frac{y_2 - y_1}{x_2 - x_1}$$

simplifying
$$= \frac{-y_2 + y_1}{-x_2 + x_1}$$

rearranging terms
$$= \frac{y_1 - y_2}{x_1 - x_2}$$

We end up with an equivalent expression in which y_1 and x_1 are now the first terms in the numerator and denominator respectively.

Examples of Calculating the Average Rate of Change (or Slope)

Example 1: Given Table 3.6, plot the data, calculate the average rate of change between 1850 to 1940 of the percent of the U.S. population living in rural areas, and interpret the result.

Fig.3.11. Percentage of the U.S. Population Living in Rural Areas

Solution:

Table 3.6	
% of U.S. Population Living in Rural Areas	
Year	%
1850	84.7
1860	80.2
1870	74.3
1880	71.8
1890	64.9
1900	60.3
1910	54.3
1920	48.8
1930	43.8
1940	43.5
1950	36.0
1960	30.1
1970	26.4
1980	26.3
1990	24.8

slope = -0.46 percentage points/yr

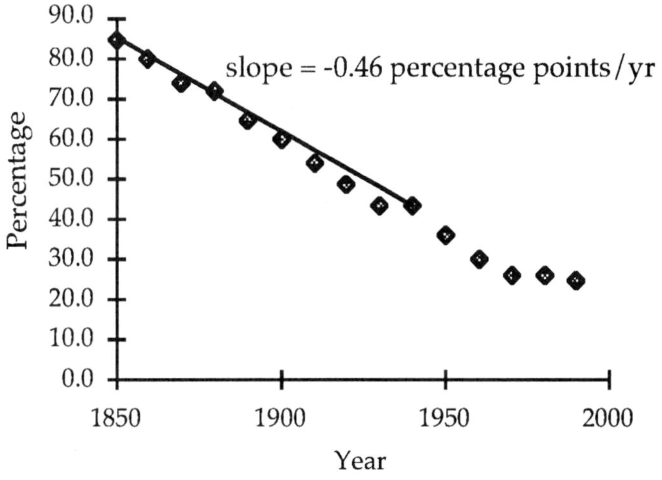

Source: Historical Statistics Colonial Times to 1970, U.S. Bureau of the Census in *The American Almanac: Statistical Abstract of the U.S., 1993*

If we connect the two endpoints (1850, 84.7%) and (1940, 43.5%) from Table 3.4 we see that the line slopes down, since the percentage of the population that is rural is *declining*. Here we have an example in which the slope or average rate of change will be negative.

$$\text{average rate of change} = \frac{\text{change in percent of rural population}}{\text{change in years}}$$

$$= \frac{43.5\% - 84.7\%}{1940 - 1850}$$

$$= \frac{-41.2\%}{90 \text{ yrs}}$$

$$= -0.46\%/\text{yr}$$

This means that the percentage of population that lived in rural areas decreased on average by 0.46 percentage points (a little under one half a percent) each year from 1850 to 1940. That may not seem like very much on an annual basis, but in less than a century the rural population went from being the large majority (84.7%) to less than half of the population (43.5%). By 1990 the rural population represented less than 25%, one quarter of the total.

Something to think about

Were 1850 and 1940 good choices for endpoints or would another time period have given an average rate of change (or slope) more typical of all the data? What seems to be happening to the rate of change between 1970 and 1990?

What kind of social and economic implications does a population shift from rural to urban of this magnitude have on society?

Example 2: We saw the following table and scatter plot for median family income during the 1970's and 1980's in an Algebra Aerobics section in Chapter 2. How could we use the information to make a case that families are better off? Worse off?

Table 3.7

Year	Median Family Income in constant 1990 $
1973	35,474
1974	34,205
1975	33,328
1976	34,359
1977	34,528
1978	35,361
1979	35,262
1980	33,346
1981	32,190
1982	31,738
1983	32,378
1984	33,251
1985	33,689
1986	35,129
1987	35,632
1988	35,565
1989	36,062
1990	35,353

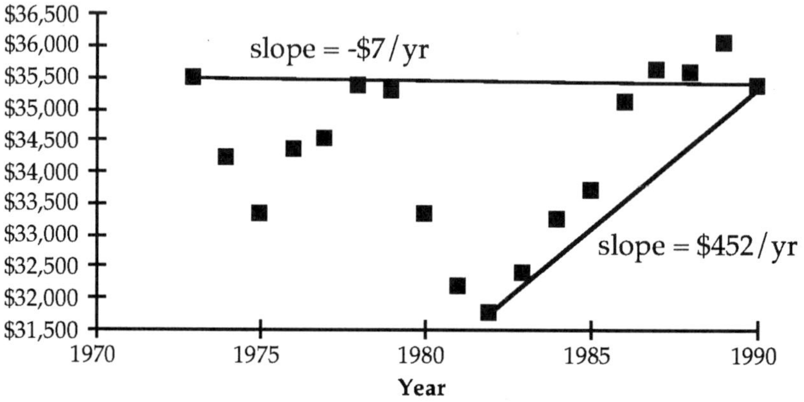

Fig. 3.12 Median family income in 1990 dollars

Source: U.S. Bureau of the Census, Current Population Reports, P-60 Series in *The American Almanac: The Statistical Abstract of the United States,* 1993.

Solution: In order to make an optimistic case that families are better off we could choose as endpoints (1982, $31738) and (1990, $35353) . Then

$$
\text{average rate of change} = \frac{\text{change in median income}}{\text{change in years}}
$$
$$
= \frac{\$35{,}353 - \$31{,}738}{1990 - 1982}
$$
$$
= \frac{\$3615}{8 \text{ yrs.}}
$$
$$
= \$452/\text{yr.}
$$

So between 1982 and 1990 the median family income *increased* by $452 per year. You can see this reflected in Figure 3.12 in the steep positive slope of the line connecting (1982, $31738) and (1990, $35353)

A gloomy interpretation of the same data could be constructed merely by changing the left hand endpoint to (1973, $35474). Then

$$\text{average rate of change} = \frac{\text{change in median income}}{\text{change in years}}$$

$$= \frac{\$35,353 - \$35,474}{1990 - 1973}$$

$$= \frac{-\$121}{17 \text{ yrs.}}$$

$$= -\$7/\text{yr.}$$

So between 1973 and 1990 the median family income *decreased* by $7 per year! A glimpse at Figure 3.12 shows the shallow negative slope of the line connecting (1973, $35474). and (1990, $35353).

Both average rates of change are correct, but they certainly give very different impressions of the well-being of American families.

Example 3: Given Table 3.8 of civil disturbances over time, plot and then connect the points, and (without doing any calculations) indicate on the graph when the average rate of change between adjacent data points is positive (+), negative (-), or zero (0).

Solution:

Table 3.8		
Civil Disturbances In US Cities		
	Period	Count
1968	Jan.-Mar.	6
	Apr-June	46
	July-Sept	25
	Oct.-Dec.	3
1969	Jan.-Mar.	5
	Apr-June	27
	July-Sept	19
	Oct.-Dec.	6
1970	Jan.-Mar.	26
	Apr-June	24
	July-Sept	20
	Oct.-Dec.	6
1971	Jan.-Mar.	12
	Apr-June	21
	July-Sept	5
	Oct.-Dec.	1
1972	Jan.-Mar.	3
	Apr-June	8
	July-Sept	5
	Oct.-Dec.	5

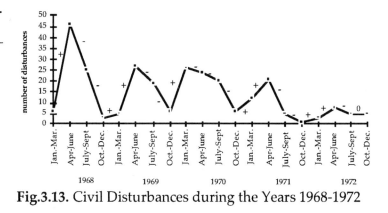

Fig.3.13. Civil Disturbances during the Years 1968-1972

Source: *Introduction to the Practice of Statistics* by Moore and McCabe. Copyright © 1989 by W.H. Freeman and Company. Used with permission.

The data are graphed in Figure 3.12. The +, -, and 0 indicate when the average rate of change between adjacent points is positive, negative, or zero, respectively. This is, of course, the same thing as indicating when the slope of the line segment is positive, negative, or zero. The largest average rate of change (or equivalently the steepest slope) seems to be between January-March and April-June in 1968. The largest negative average rate of change appears later in the same year between July-Sept and Oct-Dec in 1968.

Marking intervals on the graph with +'s and -'s is an exercise in recognizing when the slope is greater than, less than, or equal to zero, and it helps to highlight patterns in civil disturbances seen in the graph.

Civil disturbances between 1968 and 1972 occurred in cycles: the largest numbers occurred in the summer months and the smallest in the winter months. The peaks decrease over time. What was happening in America that might correlate with the peaks? This was a tumultuous period in our history. Many previously silent factions of society were finding their voices. Recall that in April 1968 Martin Luther King was assassinated and that in January 1973 the last American troops were withdrawn from Vietnam.

Example 4: Plot the points (- 3, -10) and (3, 2) and calculate the slope of the line on which they lie.

Solution:

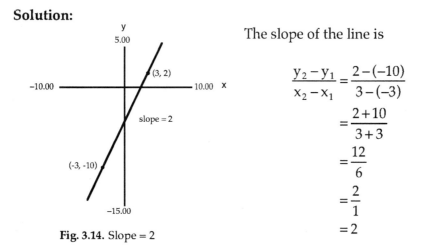

Fig. 3.14. Slope = 2

The slope of the line is

$$\frac{y_2 - y_1}{x_2 - x_1} = \frac{2 - (-10)}{3 - (-3)}$$
$$= \frac{2 + 10}{3 + 3}$$
$$= \frac{12}{6}$$
$$= \frac{2}{1}$$
$$= 2$$

Writing 2 as the ratio 2/1 indicates that if x increases by 1 unit, then y will increase by 2 units. So starting at any point on the line, if you move one unit to the right, you would have to go up 2 units to return to the line.

Example 5: Plot the points (- 6, 3) and (4, -7) and calculate the slope of the line on which they lie.

Solution:

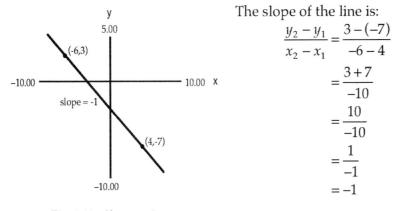

The slope of the line is:

$$\frac{y_2 - y_1}{x_2 - x_1} = \frac{3 - (-7)}{-6 - 4}$$

$$= \frac{3 + 7}{-10}$$

$$= \frac{10}{-10}$$

$$= \frac{1}{-1}$$

$$= -1$$

Fig. 3.15. Slope = -1

Writing the number -1 as the ratio -1 / 1 indicates that for an increase of 1 unit in x there is a corresponding decrease of 1 unit in y. Starting at any point on the line, if you move 1 unit to the right, you must move 1 unit down in order to stay on the line.

Algebra Aerobics [5]

1. During what intervals on the graph is the average rate of change between adjacent data points approximately zero?

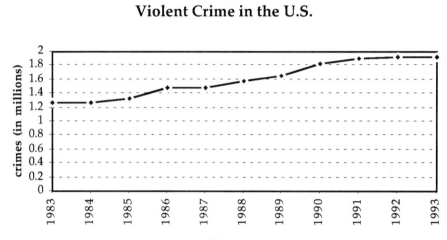

Fig.3.16.

2. Indicate on the graph when the average rate of change between adjacent data points is positive, negative, or zero.

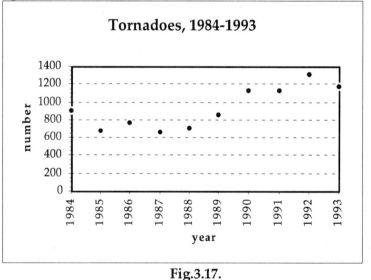

Fig.3.17.

3. Find the slope of the line that passes through the given points.
   a)   (-2,3) and (4,-5)   b)   (3,500) and (7, 500)

---

[5] Answers:

1. 1983-84; 1986-87; 1992-93 (note that the rate for 1991-92 is is just slightly greater than zero)

2. 84-85: -   85-86: +   86-87: -   87-88: +   88-89: +   89-90: +
   90-91: 0   91-92: +   92-93: -

3. a) (-5 - 3)/(4 - (-2)) = -8/6 = -4/3   b) (500 - 500) /(7 - 3) = 0/4 = 0

3.4 Other Ways to Describe Change Over Time

Describing change over time with numbers

In the popular press you may encounter several different ways of describing change over time. They are all legitimate, but may give very different impressions. For example, here are three different ways to describe the change in population between 1790 and 1800.

Table 3.9	
US Population	
Year	Population in millions
1790	3.9
1800	5.3

1. Absolute numbers
    The U.S. population increased by 1.4 million between 1790 and 1800.

    (population in 1800) – (population in 1790) = 5.3 – 3.9

    $$= 1.4 \text{ million}$$

2. Percentages
    The U.S. population grew by almost 36% between 1790 and 1800.

    $$\frac{\text{change in population}}{\text{original 1790 population}} = \frac{1.4}{3.9}$$

    $$= 0.359 \quad \text{or } 36\%$$

3. Rate of change
    The U.S. population grew at a rate of 0.14 million per year during the 1790's.

    $$\frac{\text{change in population}}{\text{change in years}} = \frac{1.4 \text{ million}}{10 \text{ years}}$$

    $$= 0.14 \text{ million per year}$$

Adding words and graphs

If you include suggestive vocabulary (the italicized words in the following examples) and a graph constructed to support your particular viewpoint, you can influence the interpretation of information. In Washington D.C. this would be referred to as "putting a spin on the data." Take a close look at the sentences below that are revisions of the sentences in 1,2, and 3 above. Note that in each case the accompanying graphs display *exactly* the same data, a plot of the two points (1790, 3.9) and (1800, 5.3) with a line connecting them.

1. The U.S. population increased by *only* 1.4 million between 1790 and 1800.

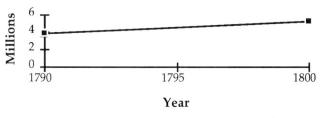

Fig.3.18. U.S. Population (in Millions)

Stretching the scale of the horizontal axis relative to the vertical axis, makes the slope of the line look shallow and hence minimizes the impression of change.

2. The U.S. population had an *explosive* growth of over 35% between 1790 and 1800.

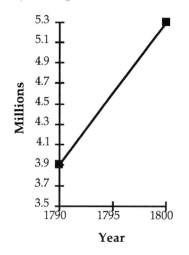

Fig.3.19. U.S. Population (in millions)

Cropping the vertical axis (which now starts at 3.5 instead of 0) and stretching the scale of the vertical axis relative to the horizontal axis makes the slope of the line look steeper and underscores the impression of change.

3. The U.S. population grew at a *reasonable* rate of 0.14 million per year during the 1790's.

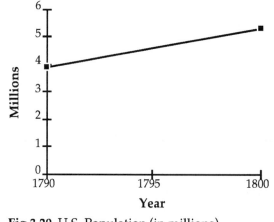

Fig.3.20. U.S. Population (in millions)

Visually the steepness of the line here seems to lie roughly half way between the previous two graphs. *In fact the slope of 0.14 million/yr is precisely the same for all three graphs.*

How could you decide upon a reasonable interpretation of the data? You might try to put the data in context by asking: How does the growth between 1790 and 1800 compare to other decades in the history of the United States? How does it compare to growth in other countries at the same time? Was this rate of growth easily accommodated or did it strain national resources and overload the infrastructure?

Though the three graphical examples are exaggerated, you need to be aware that a statistical claim is never completely free of bias. For every statistic that is quoted, others were left out. That does not mean that you should discount all statistics. But you do need to become an educated consumer, by constantly asking commonsense questions, and then coming to your own conclusions.

Something to think about

If you have the course software, open up "L4. Scaled Axes" in *Linear Graphs*. Generate a line in the upper left hand box.. The same line will appear graphed in the 3 other boxes, but with the axes scaled differently. Describe how the axes are rescaled in order to create such different impressions.

Algebra Aerobics

1. Using the given data on immigration into the U.S. between 1901 and 1990:
   a) In what decade did America have the most immigrants?
   b) Between which two consecutive decades was there the largest change (increase or decrease) in the number of immigrants?
      i) Describe the change in absolute number.
      ii) Describe the change as a percentage.

2. According to the U.S. Social Security Administration, the Social Security total annual payments have increased from $120.472 billion in 1980 to $285.980 billion in 1992.
   a) What is the dollar increase?
   b) What is the percentage increase?
   c) What is the average rate of change?

Table 3.10

Immigration: 1901-1990

Period	Number (in 000's)
1901 to 1910	8,795
1911 to 1920	5,736
1921 to 1930	4,107
1931 to 1940	528
1941 to 1950	1,035
1951 to 1960	2,515
1961 to 1970	3,322
1971 to 1980	4,493
1981 to 1990	7,338

Answers:

1.a) 1901-1910 According to Samuel Eliot Morison in *The Oxford History of the American People* "unlimited and unrestricted immigration, except for Orientals, paupers, imbeciles, and prostitutes, had been national policy down to World War I." Postwar policies limiting immigration were instituted in the 1920's.
   b) The largest change was between the 1920's and the 1930's - a decline of 3,579,000 immigrants. Morison writes that during the Great Depression immigration numbers actually became negative and "only rose after 1936 when fascism produced a new crop of refugees."
      i) -3,579,000
      ii) 3,579,000 represents (3,579,000)/(4,107,000) = 0.87 or 87% of 4,107,000. So the number of immigrants between the two decades decreased by 87%.
2.a) Social Security payments increased by $285,090,000 - $120,472,000 = $165,508,000
   b) Social Security payments increased by $165,508,000)/($120,472,000) = 1.37 or 137%
   c) The average rate of change for Social Security payments between 1980 and 1992 is
   $\frac{(\$285.090 \text{ billion}- \$120.472 \text{ billion})}{1992 - 1980} = \frac{\$165.508 \text{ billion}}{12} \approx 13.8$ billion dollars/yr.

Chapter 3 Summary

The *average rate of change* $= \dfrac{\text{change in vertical variable}}{\text{change in horizontal variable}}$

If the variables represent real world quantities, which have units of measure (e.g. dollars or years) then the average rate of change has units too.

$$\text{units of average rate of change} = \dfrac{\text{units of vertical variable}}{\text{units of horizontal variable}}$$

For example the units might be dollars/year (read as dollars per year) or pounds/person (read as pounds per person).

Geometrically the average rate of change represents a slope. Given two points (x_1, y_1) and (x_2, y_2)

$$\text{average rate of change} = slope = \dfrac{rise}{run} = \dfrac{y_2 - y_1}{x_2 - x_1} = \dfrac{\Delta y}{\Delta x}$$

where $\Delta y = y_2 - y_1$ and $\Delta x = x_2 - x_1$

Given two points, (x_1, y_1) and (x_2, y_2), it doesn't matter which one you use as the first point when you take the differences; the result is the same either way.

$$\text{slope} = \dfrac{y_2 - y_1}{x_2 - x_1} \qquad \text{or} \qquad \text{slope} = \dfrac{y_1 - y_2}{x_1 - x_2}$$

Change over time can be described using absolute numbers, percentages, or average rates of change. By using suggestive vocabulary and graphs the same data can be perceived in very different ways.

EXERCISES

1. a)   In 1980, aerospace industry profits in the U.S. were $2.59 billion, while in 1990 profits were $4.45 billion. Find the average annual rate of change.

    b)   In 1992, the aerospace industry showed a loss (negative profit) of $1.84 billion. Find the average annual rate of change from 1990 to 1992

2.   The chart below illustrates the average annual salary of professional baseball players from 1987 to 1993.

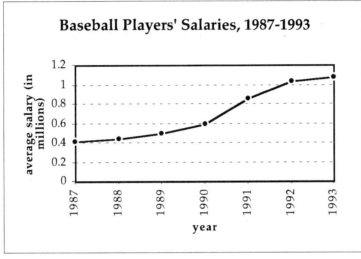

Baseball Players' Salaries, 1987-1993

a)   Fill in the table.

b)   During what one year interval was the average rate of change the lowest? Compare that part of the graph to the rest.

c)   During what one year interval was the rate of change the highest? How does that part of the graph compare to the rest?

Year	Salary ($ millions)	Rate of change over prior year
1987	0.41	N/A
1988	0.44	
1989	0.50	
1990	0.60	
1991	0.85	
1992	1.03	
1993	1.08	

3.   Using the information in the table:
    a)   Plot the data, labeling both axes and the coordinates of the points.
    b)   Calculate the average rate of change (in percentage points per year).

Year	Percent of persons 25 years old and over completing 4 or more years of college
1940	4.6%
1990	21.3%

Source: U.S. Bureau of the Census, Current Population Reports in *The American Almanac: Statistical Abstract of the United States*, 1993.

    c)   Write a topic sentence summarizing what you think is the central idea to be drawn from this data.

4. Using the information in the table:
   a) Plot the data, labeling both axes and the coordinates of the points.
   b) Calculate the slope of the line connecting the two points (in percentage points per year).

Year	Percent of persons 25 years old and over completing 4 or more years of high school
1940	24.5%
1990	77.6%

Source: U.S. Bureau of the Census, Current Population Reports in *The American Almanac: Statistical Abstract of the United States*, 1993.

   c) Write a topic sentence summarizing what you think is the central idea to be drawn from this data.

5. In each case, plot the pair of points and then calculate the slope of the line that passes through them.

   a) (3, 5) and (8, 15)
   b) (-1, 4) and (7, 0)
   c) (5, 9) and (-5, 9)

6. Read the Anthology article *Slopes* and describe two practical applications of slopes, one of which is from your own experience.

7. Use this table on life expectancy to answer the following questions.

Average Number of Years of Life Expectancy in the United States by Race and Sex, Since 1990

Life Expectancy at Birth by Year	White Males	White Females	Black Males	Black Females
1900	46.6	48.7	32.5	33.5
1950	66.5	72.2	58.9	62.7
1960	67.4	74.1	60.7	65.9
1970	68.0	75.6	60.0	68.3
1980	70.7	78.1	63.8	72.5
1987	72.1	78.9	65.2	73.6
1988	72.2	78.9	64.4	73.2
1989	72.5	79.2	64.3	73.3
1990	72.7	79.4	64.5	73.6
1991	72.9	79.6	64.6	73.8
1992	73.2	79.8	65.0	73.9
1993	73.0	79.5	64.7	73.7

Source: U.S. National Center for Health Statistics in *The American Almanac: Statistical Abstract of the US*, 1995-1996.

a) What was the average rate of change in years of life expectancy over time for white females between 1900 and 1950?

b) Which group had the largest average rate of change between 1900 and 1993?

c) Describe the gain in life expectancy for black females in at least two different ways.

8. Choose one of the following growth rate assignments, and write a statement commenting on the validity of the growth projection you calculate. (Plot time on the horizontal axis.)

    a) Look up the data for the growth of a fetus in the womb (many books on childbirth have this information). Make a chart on graph paper showing length in inches or centimeters vs. time in weeks or months up until birth. What is the average growth rate for the 9 month period? If the child grows at the same rate for the next 24 months as it did for the last month in the womb, how tall would he or she then be?

    b) Find data for growth of a human from birth to 21 years (if you know it about yourself that will do; otherwise find it in medical references on pediatrics). Plot the data on a graph in inches or centimeters vs. time in years. What is the average growth rate for the 21 year total? If humans continued to grow at the same rate as they grew in their first year, how tall would they be at age 21?

    c) Find population data for the world or some country that interests you. Plot population numbers vs. years in 10 year intervals from 1900. What is the average growth rate for the century so far? If population continues to grow at the rate it did in the last decade, what will be the total population in the year 2050? If you prefer you may use rain forest or fossil fuel data instead and calculate depletion rates.

9. Using the following information:

Persons 25 years old and over who have completed four years of high school or more		
	1940	1990
White	26.1%	79.1%
Black	7.3%	66.2%
Asian/Pacific Islander	22.6%	80.4%

Data extracted from "Educational Attainment," Population Profile of the United States 1991, US Bureau of the Census, Current Population Reports, P-23, No. 173.

    a) What has been the average rate of change (of percentage points per year) of completion of high school from 1940 to 1990 for whites? for blacks? for Asian/Pacific Islander?
    b) If these rates continue, what percentage of whites, of blacks, of Asian/Pacific Islanders will have finished high school in the year 2000?
    c) Write a paragraph describing what you see in the high school completion data, mentioning rates of change and possible projections for the near future.
    d) If these rates continue, in what year would 100% of whites have completed four years of high school or more? In what year 100% of blacks? 100% of Asian/Pacific Islanders? Do these projections make sense?

10. (Some parts of the problem require a graphing calculator or computer)
The following table and graph show the change in median age of the US population from 1850 through the present, and projected into the next century.

Median Age of U.S. Population 1860-1991 (data for 1995-2050 is projected)

Year	Median Age
1850	18.9
1860	19.4
1870	20.2
1880	20.9
1890	22.0
1900	22.9
1910	24.1
1920	25.3
1930	26.4
1940	29.0
1950	30.2
1960	29.5
1970	28.0
1980	30.0
1990	32.8
1991	33.1
1995	34.0
2000	35.5
2025	38.1
2050	39.0

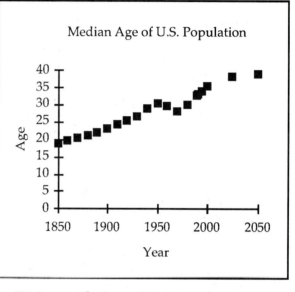

Source: U.S. Bureau of the Census, "U.S. Census of Population," Current Population Reports in *The American Almanac: Statistical Abstract of the United States*, 1995.

a) Specify the longest time period over which the median age was increasing.

b) Calculate the average rate of change between the two endpoints of this time period.

c) What is the projected average rate of change between 2000 and 2050?

d) What changes in society could make this change in median age possible? Would it necessarily mean that people are living longer?

The following questions require a graphing calculator or computer.

e) Generate a partial third column to the table, starting in 1860 and ending in 1990 that contains the average rates of change of median age over the previous decade.

f) Plot average rates of change over prior decade vs. year (columns 1 and 3 in your table.)

g) Identify periods of negative average rates of change and suggest reasons for the declining rates.

11. (Some parts of this exercise require a graphing calculator or computer.)
    The data and graphs below give a picture of the two major methods of news communication in the U.S.

Year	Newspapers (in thousands of copies printed)	Number of Newspapers published
1915	28777	2580
1920	27791	2042
1925	33739	2008
1930	39589	1942
1935	38156	1950
1940	41132	1878
1945	48384	1749
1950	53829	1772
1955	56147	1760
1960	58882	1763
1965	60358	1751
1970	62108	1748
1975	60655	1756
1980	62202	1745
1985	62766	1676
1990	62324	1611

Number of Newspapers Published vs. Year

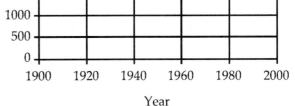

Year	Number of on air TV stations
1950	98
1955	411
1960	515
1965	569
1970	677
1975	706
1980	734
1985	883
1990	1092

Number of U.S. TV Stations vs. Year

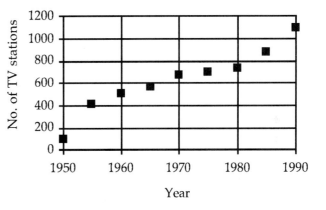

Universal Almanac © 1993 by John W. Wright.

a) Describe the trend in number of newspapers published between 1915 and 1990 in three different ways.

b) Use the U.S. population numbers from Table 3.1 to calculate and compare the number of copies of newspapers <u>per person</u> in 1920 and in 1990 .

c) What is the average annual rate of growth of TV stations for each decade since 1950? What is the average annual rate of decline of newspapers published for the same decades? Graph the results. (Use of graphing calculator or computer recommended.)

d) If TV stations continue to be started at the same rate as in the decade from 1980 to 1990, how many will there be by the year 2000? Do you think this is likely to be a reasonable projection, or might it be overly large (or small) judging from past rates of growth?

e) What trends do you see in the dissemination of news as reflected in these data? If you were seeking a job in the news reporting industry, what appear to be the greatest employment opportunities? What other data would be helpful in understanding how news is delivered in our society?

12. The data and graph below show new AIDS cases reported in Florida from 1986 to 1994. The table and graph illustrate the number of new AIDS cases reported in Florida.

Year	No. of new cases
1986	1,031
1987	1,633
1988	2,650
1989	3,448
1990	4,018
1991	5,471
1992	5,086
1993	10,958
1994	8,617

AIDS Cases Reported in Florida 1986-1994

AIDS is a relatively recent phenomenon and the rate of its growth in heavily populated urban states is of great importance in planning public health programs.

a) In a third column starting in 1987 calculate the average rate of change of AIDS cases over each preceding year. What are the units? Plot the average rate of change vs. time.

b) Find something encouraging to say about this data, using numerical facts such as percentages, average rates of change, or averages.

c) Find numerical support for something discouraging to say about the data.

d) Write a paragraph describing the status of AIDS in Florida, giving a balanced report on the number of cases and suggesting data that would be helpful for giving a more informative report. Mention percentages and average rates of change in your comments.

13. Below is an approximation of the costs of doctors' bills and Medicare between 1963 and 1979.

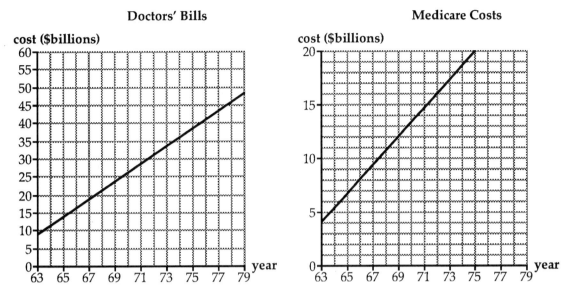

Adapted with permission from Ferleger, Lou, & Horowitz, Lucy. *Statistics for Social Change*.

  a) Which *appears* to have been growing at a faster average annual rate of change: doctors bills or Medicare costs? Why?

  b) Which actually grew at a faster average annual rate of change and how can you tell?

14. Describe the decline in air pollutant concentrations between 1983 and 1991 in three different ways.

National Ambient Air Pollutant Concentrations: 1985 to 1993								
(Air quality standard is 0 parts per million (ppm))								
	1985	1987	1988	1989	1990	1991	1992	1993	1994
Carbon monoxide (ppm)	6.97	6.69	6.38	6.34	5.87	5.55	5.18	4.88	5.57

Source: U.S. Environmental Protection Agency, National Air Quality and Emissions Trends Report annual, as taken from *The American Almanac: Statistical Abstract of the United States, 1995-1996*.

15. (Requires a graphing calculator or computer).

The following table and graph are both representations of the accumulated debt of the federal government as a function of time.

Year	Billions of $
1945	260
1950	257
1955	274
1960	291
1965	322
1970	361
1971	408
1972	436
1973	466
1974	484
1975	541
1976	629
1977	706
1978	777
1979	829
1980	909
1981	994
1982	1137
1983	1371
1984	1564
1985	1817
1986	2120
1987	2346
1988	2601
1989	2868
1990	3206
1991	3599
1992	4003
1993	4410
1994	4644
1995 (est.)	4961

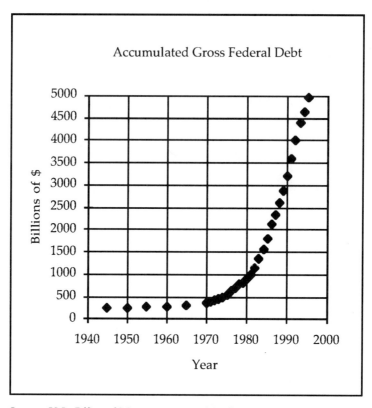

Source: U.S. Office of Management and Budget, *Budget of the United States Government*, annual, in *The American Almanac: Statistical Abstract of the United States*, 1995.

a)   Generate a third column for the table containing the average rate of change of the debt over the previous year. Describe in ordinary terms what these numbers represent.

b)   Graph the average rates of change (in your third column) vs. time. Describe what your graph and the data show.

16. Laws to regulate environmental pollution in America are a very recent phenomenon, with the first federal regulations appearing in the 1950s. The Clean Air Act, passed in 1963 and amended in 1970 established for the first time uniform national air pollution standards. The act placed national limits and a timetable on three classes of automotive pollutants: hydrocarbons (HC), carbon monoxide (CO), and nitrogen oxides (NO). The following tables show the results.

National Automobile Emission Control Standards

Model Year applicable	HC Grams/ mile	CO Grams/ mile	NO_2 * Grams/ mile
Pre-1968	8.7	87	4.4
1968	6.2	51	n.r.**
1970	4.1	34	n.r.
1972	3.0	28	n.r.
1973	3.0	28	3.1
1975	1.5	15	3.1
1975 (C)***	0.9	9	2.0
1976	1.5	15	3.1
1977	1.5	15	2.0
1978	1.5	15	2.0
1979	1.5	15	2.0
1980	0.41	7	2.0
1981 and beyond	0.41	3.4	1.0

National Automobile Emissions Estimates

(million metric tons per year)

Year	HC	CO	NO2
1970	28.3	102.6	19.9
1971	27.8	103.1	20.6
1972	28.3	104.4	21.6
1973	28.4	103.5	22.4
1974	27.1	99.6	21.8
1975	25.3	97.2	20.9
1976	27.0	102.9	22.5
1977	27.1	102.4	23.4
1978	27.8	102.1	23.3

Source: *Environmental Quality-1980: The Eleventh Annual Report of the Council on Environmental Quality,* p. 170.

* NO_2 ,nitrogen dioxide, is a form of nitrogen oxide.
** No requirement
*** California standards
Source: Portney, Paul, ed., Current *Issues in U.S. Environmental Policy,* Johns Hopkins University Press, 1978, p. 76

   a)   Argue that The Clean Air Act was a success.
   b)   Argue that The Clean Air Act was a failure. (Hint: how is it possible that the emissions per vehicle mile were down but the total amount of emissions did not improve?)

17. The graph shows the number of Nobel prizes awarded in science for various countries between 1901 and 1974. It contains accurate information, but gives the impression that the number of prize winners declined drastically in the 1970s, which was not the case. What flaw in the construction of the chart leads to this impression?

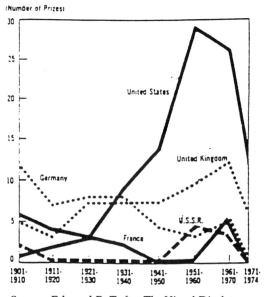

Nobel Prizes Awarded in Science, for Selected Countries, 1901-1974

Source: Edward R. Tufte, *The Visual Display of Quantitative Information*, Graphics Press, Connecticut, 1983.

18. The following four graphs show black income as a percentage of white income from 1950 to 1975. Graph (A) makes the results look more positive than graph (B), graph (C) seems to suggest change, while graph (D) seems to imply stability. How have the axes been altered in each graph in order to convey these impressions?

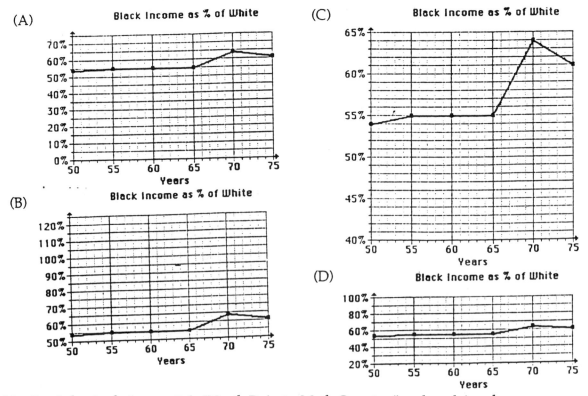

19. Read the Anthology article "North Dakota, Math Country." and explain why (according to Daniel Moynihan), it is better to live in North Dakota than Texas.

EXPLORATION 3.1
Having it your way

Objective
- to construct arguments supporting opposing points of view from the same data

Materials/Equipment
- *Student Statistical Portraits* for the University of Massachusetts/Boston and for the University of Southern Mississippi (in the Anthology of Readings) or for the student body at your institution.
- computer with spreadsheet program and printer *or* graphing calculators with projection system(optional)
- graph paper and/or overhead transparencies

Procedure
Working in small groups

Examine the data and graphs from the *Student Statistical Portrait* for either the University of Massachusetts/Boston or the University of Southern Mississippi. Explore how you would use the data to construct arguments that support at least two different points of view. For example, you could be an Admissions Officer addressing a group of prospective students or a student protesting budget cuts. Decide on the arguments you are going to make and divide up tasks among your team members.

The Rules of the Game
- Your arguments need only to be a few sentences long but you need to use graphs and numbers to support your position. You may only use legitimate numbers, but you are free to pick and choose the ones that best support your case. If you construct your own graphs, you may use whatever scaling you wish on the axes.
- For any data that represent a time series, as part of your argument, pick two appropriate end points and calculate the associated average rate of change.
- Use "loaded" vocabulary (e.g., "surged ahead", "declined drastically"). This is your chance to be outrageously biased, write absurdly flamboyant prose and commit egregious sins of omission.
- Decide as a group how to present your results to the class.

Suggested topics
Your instructor might ask your group to construct one or both sides of the arguments on one topic. If you're using data from your own institution, answer the questions provided by your instructor.

Using data from the *Student Statistical Portrait* from the University of Southern Mississippi

1   Use the data from "10 Year Enrollment Trends" by ethnic group and gender to support each of the following cases:
   a) You are a student activist trying to convince the Board of Trustees that they have not done enough to increase diversity on the USM campus.

    b) You are an administrator arguing that your office has done a good job of increasing the diversity of the student body at USM.

2   Use data on "Retention of Freshmen" and "10 Year Enrollment Trends" to support each of the following cases:

    a) You are a student trustee lobbying the State Legislature for more money for USM.

    b) You are a tax payer writing an editorial to the local paper arguing that USM does not need more money.

3.   Make the case that USM as an institution is growing and then the case that USM is shrinking. Decide on the roles you would assume in making these arguments. Use data from the "10 Year Enrollment Trends" and "Degrees Conferred" tables.

Using data from the *Student Statistical Portrait* from the University of Massachusetts/ Boston

1   You are the Dean of the College of Management. Use the data on "SAT Scores of New Freshmen by College/Program" to make the case that::

    a) the freshmen admitted to the College of Management are not as prepared as the students in the College of Arts and Sciences and therefore you need more resources to support the freshmen in your program.

    b) the freshmen admitted to the College of Management are better prepared than the students in the College of Arts and Sciences and therefore you would like to expand your program.

2.   You are an Assistant Provost for UMB lobbying the state legislature. Use the data on "Undergraduate Admissions" to present a convincing argument that:

    a) UMass is becoming less desirable as an institution for undergraduates and thus more funds are needed to strengthen the undergraduate program.

    b) UMass is becoming more desirable as an institution for undergraduates and thus more funds are needed to support the undergraduate program.

3.   Make the case that UMB as an institution is growing and then the case that UMB is shrinking. Decide on the roles you would assume in making these arguments. Use graphs and data on "Enrollment Trends" and graphs and data on "Degrees Conferred"

Exploration Linked Homework

Reporting your results verbally

    With your partner prepare a "60 Second Summary" of your arguments. If working on a computer, print out at least one transparency (if available) for your presentation. If working on a graphing calculator, enter or transfer your data to the projection panel or create a transparency by hand.

Reporting your results in writing

    Each person should write a "60 Second Summary" of the arguments. Include graphs and numerical descriptors in your summary

EXPLORATION 3.2
Why "Average" Rate of Change

Objective
- to explore two different methods for calculating average rate of change.

Materials/Equipment
- computer with spreadsheet program or graphing calculators
- U.S. Population Data 1790 to 1994 (Table 3.1) on spreadsheet file or graph link file.

Procedure
Class Demonstration using two different methods to calculate the average rate of change.
Method A
1. Enter population data into a spreadsheet or graphing calculator. Calculate the average rate of change between 1790 and 1990 using the formula,

$$\text{average rate of change} = \frac{\text{change in population}}{\text{change in years}}$$

Method B
2. Calculate the average rate of change for each ten year interval from 1790 to 1990. Sum these rates and then calculate the mean of all the average rates of change from 1790 to 1990. Compare your answer to the answer you found with Method A above. Discuss why your answers are called an *average* rate of change.

Discussion/analysis
With a partner
- Calculate the average rate of change from 1790 to 1990 using Method B and 20 year intervals. Compare your answer with your answers using 10 year intervals.
- Calculate the average rate of change from 1790 to 1990 using Method B and intervals that are not the same size. (for example include both 10 year and 20 year intervals). Did you find the same answer as the above methods? What happens when you use intervals that are not all the same size?
- Do you think Method B would work with overlapping or non-contiguous intervals? (An example of overlapping intervals would be intervals from 1900 to 1920 and 1910 to 1930. An example of non-contiguous intervals would be intervals from 1900 to 1920 and 1940 to 1960) Explain your answer.

Class Discussion
- Explain in your own words how to summarize Method B. What qualifications do you need to make if you use Method B to calculate the average rate of change over an interval. Explain why.
- What additional data would you need to use Method B to calculate the average rate of change from 1790 to 1994?

Chapter 4

When Rates of Change are Constant

Overview

For certain variables, the average rate of change remains constant. The median weight for female infants increases a fixed amount each month. Each 5 gram increase in the amount of salt dissolved in a kilogram of water lowers the freezing temperature 0.27 degrees Celsius. For every U.S. dollar you exchange, you get a fixed amount of foreign currency, depending on the daily exchange rate.

This chapter introduces the family of *linear functions*. Linear functions describe relationships in which the average rate of change (of one variable with respect to another) is constant. Their graphs are straight lines. We look at a number of different ways of finding the equation of a line and then examine the special cases of horizontal, vertical, parallel, and perpendicular lines. Finally, we study equivalent ways of describing linear relationships.

Linear functions are fundamental in mathematics and its applications to real data. Linear functions can be used to model a wide variety of relationships, such as between the number of concerts offered and average audience size per concert, between the percentage of students taking SAT tests and their scores, or between sales of current and future computer products.

The explorations provide an opportunity to discover patterns in the graphs of linear equations and make generalizations about the properties of linear equations.

After reading this chapter you should be able to:

- recognize when a rate of change is constant
- define a linear function
- generate linear functions under different conditions
- understand direct proportionality
- manipulate horizontal, vertical, parallel, and perpendicular lines
- describe linear relationships in several different ways

4.1 A First Look at Linear Relationships

What if the U.S. population had grown at a constant rate? A hypothetical example:

In Section 3.2 we calculated the average rate of change in the U.S. population between 1790 and 1800 as 0.14 million people per year. What if the average rate of change had remained constant? What if in every decade after 1790, the U.S. population had grown at exactly the same rate, namely 0.14 million per year? That would mean that starting with a population estimated at 3,900,000 in 1800, the population would grow by exactly 140,000 people each year, or equivalently 1.4 million people per decade. The slopes of all the little line segments connecting adjacent population data points would be identical- namely 0.14 million/year. Instead of the original population graph that curved upwards, we would have one absolutely straight line.

When we calculated the actual rate of change in the decades subsequent to 1790, not only was the growth rate positive each year, but the growth rate generally increased from decade to decade. Table 4.1 and Figure 4.1 compare the hypothetical and actual results.

	Table 4.1	
Year	Actual Population in millions	Population in millions IF the average rate of change remained constant at 0.14 m/yr
1790	3.9	3.9
1800	5.3	5.3
1810	7.2	6.7
1820	9.6	8.1
1830	12.9	9.5
1840	17.1	10.9
1850	23.2	12.3
1860	31.4	13.7
1870	39.8	15.1
1880	50.2	16.5
1890	63.0	17.9
1900	76.0	19.3
1910	92.0	20.7
1920	105.7	22.1
1930	122.8	23.5
1940	131.7	24.9
1950	151.3	26.3
1960	179.3	27.7
1970	203.3	29.1
1980	226.5	30.5
1990	246.7	31.9

Before 1960 the figures are for the coterminous United States. They exclude Alaska and Hawaii.
Source: U.S. Bureau of the Census, Current Population Reports in *The American Almanac: Statistical Abstract of the United States*, 1996..

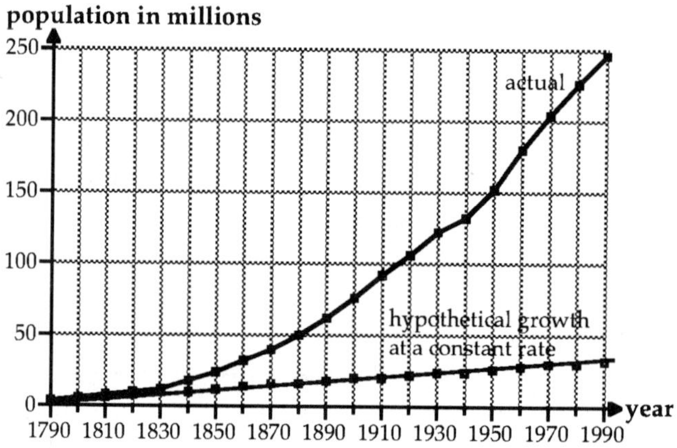

Fig. 4.1. U.S. Population: a hypothetical example

A real example of constant rate of change

Professionals in many different fields (psychology, sociology, medicine) study human growth and development. Numerous measures such as height, weight, and psychological maturity offer measures of growth for young children. By establishing a set of norms, parents and doctors can help determine whether a child's development is progressing satisfactorily.

According to the standardized growth and development charts used by many American pediatricians, the median weight for girls during their first six months of life increases at an almost constant rate.[1] Starting at 7.0 lbs. at birth, for each additional month of life the female median weight increases by 1.5 lbs. Thus 1.5 is the average rate of change in pounds per month. When the average rate of change is constant, we often drop the word *average* and just talk of the *rate of change*.

To each age there corresponds a unique median weight, so the median weight for baby girls is a function of age. Table 4.2 and Figure 4.2 show two representations of the function. Note that the domain is between 0 and 6 months.

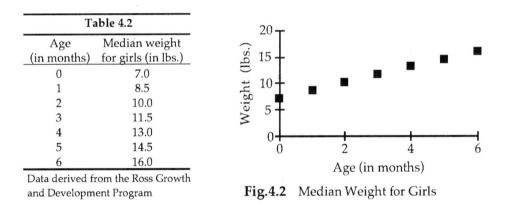

Table 4.2	
Age (in months)	Median weight for girls (in lbs.)
0	7.0
1	8.5
2	10.0
3	11.5
4	13.0
5	14.5
6	16.0

Data derived from the Ross Growth and Development Program

Fig.4.2   Median Weight for Girls

The rate of change, 1.5 lbs. per month, means that as age increases by one month, weight increases by 1.5 lbs. If we move 1 unit (1 month) to the right on the graph, then we need to move up 1.5 units (1.5 lbs.) to find the next data point. Just as in the hypothetical case of the constant U.S. population growth, the graph shows that the points lying on a straight line.

Finding a mathematical model for the relationship between female infant median weights and age

Can we find an equation, a *mathematical model*, that gives female infant weight as a function of age? Using Table 4.2, we think of the weight at birth (when age = 0) as our base value. On Figure 4.2, the base value represents the intercept on the vertical axis. In this case the base

---

[1] Adapted from Hamill PVV, Drizd TA, Johnson CL, Reed RB, Roche AF, Moore WM, Physical Growth: National Center for Health Statistics percentiles. AMJ CLIN NUTRi 32: 607-629, 1979 Data from the Fels Longitudinal Study, Wright State University School of Medicine; Yellow Springs, Ohio.

value is 7.0 lbs., that is the graph crosses the weight axis at 7. The median weight for girls starts at 7.0 lbs. at birth and increases by 1.5 lbs. every month.

Table 4.3

Age (in months)	Median weight for girls (in lbs.)	pattern	generalized expression
0	7.0	7.0 + 1.5 (0) =	7.0 + 1.5•(age in months)
1	8.5	7.0 + 1.5 (1) =	7.0 + 1.5•(age in months)
2	10.0	7.0 + 1.5 (2) =	7.0 + 1.5•(age in months)
3	11.5	7.0 + 1.5 (3) =	7.0 + 1.5•(age in months)
4	13.0	7.0 + 1.5 (4) =	7.0 + 1.5•(age in months)
5	14.5	7.0 + 1.5 (5) =	7.0 + 1.5•(age in months)
6	16.0	7.0 + 1.5 (6) =	7.0 + 1.5•(age in months)

A succinct general expression for this relationship is:

Median weight for girls = $\;$ 7.0 + 1.5 •(age in months) $\qquad\qquad$ (1)

where the median weight is in lbs. If we let W represent the median weight in lbs. and A the age in months, we can write this equation more compactly as:

$$W = \;7.0 + 1.5\,A \qquad\qquad\qquad (2)$$

This equation can be used as a *model* of the relationship between age and median weight for baby girls in their first six months of life.

Since our equation represents quantities in the real world, each term in the equation has units attached to it. W (which represents weight) and 7.0 are in lbs., 1.5 is in lbs./month and A (which represents age) is in months. The canceling of units is like canceling in fractions. So the units of the term 1.5A are

$$\left(\frac{lbs.}{\cancel{month}}\right)\cancel{month} = lbs.$$

The units of the right hand side and left hand side of our equation ought to match, and they do. That is,

$$\text{in the equation} \qquad W = \quad 7.0 + 1.5\,A$$

$$\text{the units are} \qquad lbs. = lbs. + \left(\frac{lbs.}{\cancel{month}}\right)\cancel{month}$$

$$= \text{lbs.} + \text{lbs.}$$

Our equation W = 7.0 + 1.5 A does define W as a function of A, since for each value of A (age) the equation determines a unique value of W (weight).

As a model, the equation W = 7.0 + 1.5A holds for values of A between 0 and 6; so the domain of our function is $0 \le A \le 6$. Negative values of age are meaningless and the data for the median weights for girls older than 6 months will deviate from this equation. We would certainly not expect a female to continue to gain 1.5 lbs. per month for the rest of her life!

We derived the function using only ages that were integer values (whole numbers of months), but the function can be used to *interpolate* to non-integer values of A between 0 and 6. For instance, we can use the function W = 7.0 + 1.5 A to predict the median weight for girls who are 2.5 months old.

When A = 2.5
$$\begin{aligned} W &= 7.0 + 1.5\,(2.5) \\ &= 7.0 + 3.75 \\ &= 10.75 \text{ lbs.} \end{aligned}$$

We say the point (2.5, 10.75) is a solution to or *satisfies* the equation W = 7.0 + 1.5 A

The original data set contained 7 *discrete* points, while Figure 4.3 shows the function W = 7.0 + 1.5 A, where A can be *any* real number between 0 and 6.

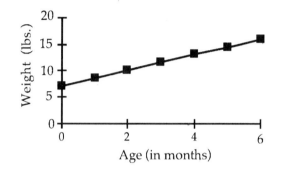

Fig.4.3   Median Weight for Girls

Something to think about

If the median birth weight for baby boys is the same as for baby girls, but they put on weight at a faster rate, which numbers in the model would change and which would stay the same?

Algebra Aerobics: [2]

1. Determine which, if any, of the following points satisfy the equation W=7.0+1.5A
    a) (2,10)   b) (5,13.5)
2. From Figure 4.3, try to estimate the weight, W, of a baby girl who is 4.5 months old. Then, use the equation (W = 7.0 + 1.5A) to calculate the corresponding value for W. How close was your estimate?
3. From the same graph, try to estimate the age of a baby girl who weighs 14 pounds. Then use the equation to calculate the value for A.

---

[2] Answers
1. a) $10 = 7.0 + 1.5(2)$ $\Rightarrow$ (2,10) satisfies the equation
    b) $13.5 \neq 7.0 + 1.5(5)$ $\Rightarrow$ (5, 13.5) does not satisfy the equation
2. From the equation, the exact weight is $W = 7.0 + 1.5(4.5) = 13.75$
3. $14 = 7.0 + 1.5A$ $\Rightarrow$ $7.0 = 1.5A$ $\Rightarrow$ $A = 7.0/1.5 = 4.67$ months

Looking at this function in the abstract

If we think of the function W = 7.0 + 1.5 A as describing a relationship between two abstract (and unitless) variables W and A, then the natural domain of the function is *all* real numbers, not just those between 0 and 6. For instance, we can let A = -6.0 and then use the equation to find the associated value of 7.0 + (1.5) (-6.0) or -2.0 for W. Table lists points that satisfy the equation. The graph of *all* the points that satisfy the equation W = 7.0 + 1.5 A is a line that extends indefinitely in both directions.

Table 4.4	
A	W = 7.0+1.5A
-6	-2.0
-5	-0.5
-4	1.0
-3	2.5
-2	4.0
-1	5.5
0	7.0
1	8.5
2	10.0
3	11.5
4	13.0
5	14.5
6	16.0
7	17.5
8	19.0

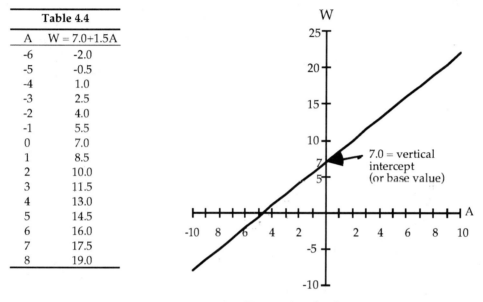

Fig.4.4 Graph of W = 7.0 + 1.5A

From Figure 4.4 and Table 4.4 we can see that the line crosses the vertical axis at the base value of 7.0. Equivalently if we let A = 0,

$$\text{then} \quad W = 7.0 + (1.5)(0)$$
$$= 7.0$$

We call 7.0 the *vertical intercept*.

Looking at the equation and Table 4.4 we can see that the rate of change of W with respect to A is 1.5. That means an increase in A by 1 unit, corresponds to an increase of W by 1.5 units.

On Figure 4.4, this means that starting anywhere on the line, if we move (or "run") 1 unit to the right, we must move up (or "rise") 1.5 units to return to the line. In particular as illustrated in Figure 4.5, if we start at the base value of 7.0 and repeatedly go over 1 and up 1.5 units, we'll stay on the line.

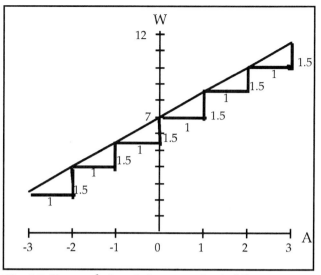

Fig.4.5 For each 1 unit increase in A, W increases by 1.5 units

Something to think about

If you run over 2 units, how many units would you have to rise up to return to the line. What if you run over 5 units? What if you run 1 unit to *the left?* (hint: think of a negative rise as a fall) 5 units to the left?

The points (–7, –3.5) and (8, 19.0) satisfy W = 7.0 + 1.5 A and hence lie on the line. The rate of change between these two points is:

$$\text{rate of change (or slope)} = \frac{(19.0 - (-3.5))}{8 - (-7)}$$

$$= \frac{22.5}{15}$$

$$= 1.5$$

Is it always true that if we pick any two points that satisfy this equation, the rate of change (or slope) between these two points is 1.5? The answer is yes. If we have two distinct points (A_1, W_1) and (A_2, W_2) that satisfy the equation $W = 7.0 + 1.5A$, then A_1 and A_2 are distinct and the following equations are true:

$$W_1 = 7.0 + 1.5A_1 \qquad (1)$$
$$W_2 = 7.0 + 1.5A_2 \qquad (2)$$

Subtracting equation (1) from equation (2) gives us:

$$W_2 - W_1 = (7.0 + 1.5A_2) - (7.0 + 1.5A_1)$$

clearing the parentheses $\qquad = 7.0 + 1.5A_2 - 7.0 - 1.5A_1$

canceling the 7.0's $\qquad = 1.5A_2 - 1.5A_1$

factoring out 1.5 $\qquad = 1.5(A_2 - A_1) \qquad (3)$

Since A_1, and A_2 are distinct, then $(A_2 - A_1) \neq 0$. Hence we can divide both sides of (3) by $(A_2 - A_1)$ to get

$$\frac{W_2 - W_1}{A_2 - A_1} = 1.5 \qquad (4)$$

The left side of this equation represents the rate of change (or slope of the line) between the two *arbitrarily chosen* points (A_1, W_1) and (A_2, W_2) that satisfied $W = 7.0 + 1.5 A$. *Hence no matter which two distinct points you choose that satisfy the equation $W = 7.0 + 1.5 A$, the rate of change (or the slope of the line) between the points is always 1.5, the coefficient of the independent variable A.*

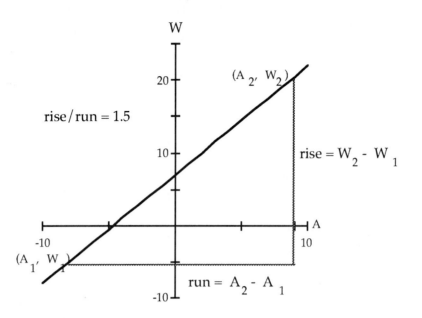

Fig.4.6 For any two points on the line W = 7.0 + 1.5A
the rise/run is always 1.5

Algebra Aerobics: [3]

1. Select any two points of the form (A,W) from Table 4.4 that satisfy the equation
    W = 7 + 1.5A. Use these points to verify that the rate of change between them is 1.5
2. Find two points that satisfy the equation W = 7 + 1.5A that are NOT in Table 4.4. First
    try using a negative value for A, then find the corresponding value for W. Then try a
    very large positive value for A, and then find the corresponding value for W.
3. If you calculate the slope by using the two points that you selected, what should your
    answer be? Now do the calculations. Were you right?

The general linear form

The function W = 7.0 + 1.5 A is an example of the general form::

$$y = \text{(y-intercept)} + \text{(rate of change)} \bullet (x)$$

All functions of this form are called *linear* since their graphs are straight lines.

---

[3] Answers
2.   For example,
         A = -20   ⇒   W = 7 + 1.5(-20) = -23
         A = 200   ⇒   W = 7 + 1.5(200) = 307
3.   slope = 1.5
      (307-(-23))/(200 - (-20)) = 330/220 = 1.5

4.2 Linear Functions

A *linear function* has the form $y = b + mx$

Its graph is a straight line where
  m is the *slope,* or *rate of change* of y with respect to x
  b is the *vertical intercept,* or the value of y when x is zero.

Alternate ways of writing a linear function

Linear functions, of course, can be written equally well in a form you've probably seen before:

$$y = mx + b$$

Social scientists tend to prefer the $y = b + mx$ format. Placing the b term first suggests starting at a base value (when x = 0) and then adding on multiples of x. We'll use both forms interchangeably; but as a rough guideline we'll use $y = mx + b$ in purely abstract examples and $y = b + mx$ in social science applications.

Though x and y are the traditional choices for the independent and dependent variables, there is nothing sacred about their use. We could just as well use A and W, as in the example in the previous section, or H_{men} and H_{women}, or Personal Income and Education, or whatever is convenient.

If *f* represents a linear function, then we could write $f(x) = mx + b$ or $f(x) = b + mx$.

Something to think about

Why are $y = mx + b$ and $y = b + mx$ equivalent statements?

Why is $y = b + mx$ a function?

Give any value for x, a linear equation gives a unique corresponding value for y.
Hence $y = b + mx$ describes y as a function of x. For example, if b = -75 and m = 25, you can find a unique value for y given that of x using the equation
$y = -75 + 25x$.

Why is b the y-intercept?

The graph of $y = b + mx$ will cross the y-axis when x = 0. When x = 0 then
$$y = b + m(0)$$
$$= b$$
For example, let $y = 4.3 - 2.7x$. Then b = 4.3 and m = -2.7. When x = 0
$$y = 4.3 - 2.7 (0)$$
$$= 4.3$$

Why is *m* the slope or rate of change?

If (x_1, y_1) and (x_2, y_2) are two distinct points that satisfy the function $y = b + mx$, then the following equations are true:

$$y_1 = b + mx_1 \qquad (1)$$
$$y_2 = b + mx_2 \qquad (2)$$

Subtracting equation (1) from equation (2) gives us:

$$y_2 - y_1 = (b + mx_2) - (b + mx_1)$$

clearing the parentheses	$= b + mx_2 - b - mx_1$
canceling b's	$= mx_2 - mx_1$
factoring out m	$= m(x_2 - x_1)$

Since y is a function and (x_1, y_1) and (x_2, y_2) are two different points, $x_2 \neq x_1$, so $x_2 - x_1 \neq 0$. So we can divide both sides by $(x_2 - x_1)$ to get:

$$\frac{y_2 - y_1}{x_2 - x_1} = m \qquad (3)$$

The left side of this equation represents the rate of change (or slope of the line) between the two arbitrarily chosen points (x_1, y_1) and (x_2, y_2) that satisfy the original equation. Equation (3) says that this rate of change is always m.

We have also just shown that we can calculate the rate of change or slope from any two points that satisfy a linear equation

> Any two points that lie on a line can be used to calculate the slope.
> That is, if (x_1, y_1) and (x_2, y_2) are two distinct points that satisfy the function $y = b + mx$, then
> $$m = \frac{y_2 - y_1}{x_2 - x_1}$$

Algebra Aerobics: [4]

1. Identify the slope, m, and the vertical intercept, b, of the line with the given equation:
   a) $y = 5x + 3$ b) $f(x) = 7.0 - 2.5x$ c) $h = -g + 10$
2. Calculate the slope of the line that passes through the given points.
   a) (4,1) and (8,11).   b)   (-3,6) and (2,6).

---

[4] Answers

1. a) $m = 5, b = 3$ b) $m = -2.5, b = 7.0$ c) $m = -1, b = 10$
2. a) $m = (11-1)/(8-4) = 10/4 = 5/2$ or 2.5
   b) $m = (6-6)/(-3-2) = 0/-5 = 0$

Another example of a linear function

For children between the ages of 2 and 12 years, the relationship between median height and age can be approximated by the linear function

$$H = 30.0 + 2.5A$$

where A = age (in years) and H = median height (in inches).[5] We can consider the function both as an abstract expression and as a model for the relationship between age and median height.

The number 2.5 is the slope or the rate of change of H with respect to A. In the model that means that the median height increases 2.5 inches per year for children between the ages of 2 and 12.

The number 30.0 is the vertical or H-intercept, the value of H when A = 0. Note that though the point
(0, 30.0) satisfies the equation, it is outside the set of points for which this model holds.

In the model H and 30.0 are in inches, 2.5 is in inches/year and A is in years. As an abstract equation the terms have no units.

Table 4.5 and Figure 4.7 show a set of points that satisfy the equation. Note that the vertical scales have been compressed for convenience.

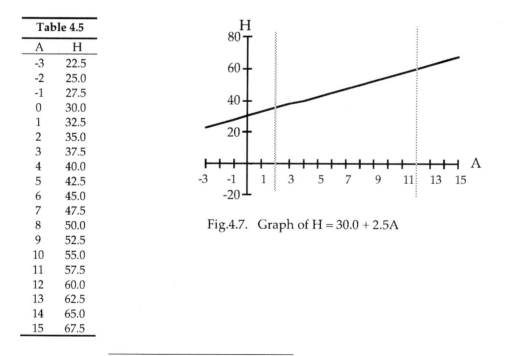

Table 4.5	
A	H
-3	22.5
-2	25.0
-1	27.5
0	30.0
1	32.5
2	35.0
3	37.5
4	40.0
5	42.5
6	45.0
7	47.5
8	50.0
9	52.5
10	55.0
11	57.5
12	60.0
13	62.5
14	65.0
15	67.5

Fig.4.7. Graph of H = 30.0 + 2.5A

[5] Information adapted from: Behrman, Richard E. and Victor C. Vaughan III, eds. *Nelson Textbook of Pediatrics*, sr. ed. Waldo E. Nelson, 12th ed. (Philadelphia: W.B Saunders, 1983), p.19.

If we think of A as representing age (in years) and H as representing median height (in inches), then only the section of the graph between the dotted lines, where $2 \leq A \leq 12$, is a model of the relationship between these two variables.

Note that since the relationship is linear, the rate of change between any two points in Table 4.5 should be 2.5 in./yr. For example, using the points (3, 37.5) and (11, 57.5), the rate of change equals:

$$\frac{\text{change in height}}{\text{change in age}} = \frac{(57.5 \text{ in.} - 37.5 \text{ in.})}{(11 \text{ years} - 3 \text{ years})}$$

$$= \frac{20 \text{ in}}{8 \text{ years}}$$

$$= 2.5 \text{ in./yr}$$

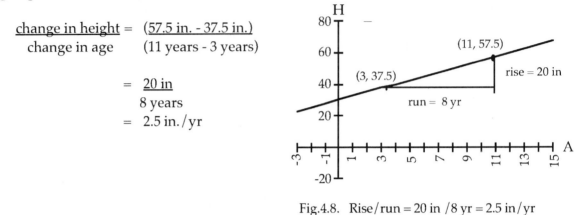

Fig.4.8. Rise/run = 20 in /8 yr = 2.5 in/yr

Algebra Aerobics: [6]

1. If $S = 20{,}000 + 1{,}000Y$ describes the annual salary, in dollars, for a person who has worked for Y years for the Acme Corporation, what is the unit of measure for 20,000 ? for 1,000?

2. In the equation $W = 7 + 1.5A$, the equation in units of measure only is
    $$\text{pounds} = \text{pounds} + (\text{pounds/month})(\text{months}).$$
    Note how month "cancels out" and you are left with:
    $$\text{pounds} = \text{pounds} + \text{pounds}$$
    Rewrite $S = 20{,}000 \; + \; 1000Y$ as an equation relating only the units of measure.
    $$\downarrow \qquad \downarrow \qquad \downarrow \; \downarrow$$
    $$\text{dollars} = \; ? \; + \; (\,?\,)(\,?\,)$$

3. If $C = 15P + 10$ describes the relation between the number of persons (P) in a dining party and the total cost in dollars (C) of their meals, what is the unit of measure for 15? for 10?
    Now rewrite this as an equation relating only units of measure.

---

[6] Answers

1. dollars (initial salary); dollars per year (annual increase)

2. dollars =   dollars + (dollars/year) (years)

3. dollars/person; dollars   dollars = (dollars/person)(persons) + dollars

4.3 Generating Linear Equations (3 Cases)

We can easily recognize equations like $y = 2000 + 1200x$ or $F = 1.5\,W - 2.7$ as linear functions. The harder problem is to generate a linear equation given different types of information about the function. Here we show three distinct ways to find an expression of the form

$$y = b + mx \qquad \text{or} \qquad y = mx + b$$

where m is the slope (or rate of change of y with respect to x) and b is the y-intercept. For each case, we generate a small table of points that satisfy the function and graph the function.

Case 1: The slope (m) and vertical intercept (b) are known

This is the easiest case, since all we need to do is substitute the values for m and b directly into the general linear equation.

Example 1: Find the equation of the line with a slope of 3/5 and a vertical intercept of -2. Generate a small table of points that satisfy the equation and then graph the equation.

Solution: The equation $y = (3/5)x - 2$ matches the general form of a linear equation $y = mx + b$, where m is the slope and b is the y-intercept. The domain is all the real numbers. Table 4.6 and Figure 4.9 show points that satisfy the equation follow.

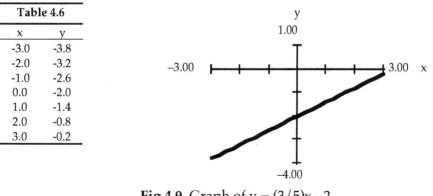

Table 4.6	
x	y
-3.0	-3.8
-2.0	-3.2
-1.0	-2.6
0.0	-2.0
1.0	-1.4
2.0	-0.8
3.0	-0.2

Fig.4.9 Graph of $y = (3/5)x - 2$

Since the slope 3/5, equals the change in y divided by the change in x, we can start at any point on the line, move 5 units to the right, and then move up 3 units to return to the line. We can also write 3/5 as 0.6 or 0.6/1. So moving 1 unit to the right, the line rises 0.6 units. This is more easily seen in Table 4.6, where y increases by 0.6 for an increase in x of 1 unit.

Example 2: Harvard Community Health Plan, an HMO in Cambridge, Massachusetts, distributes the following recommended weight formula for men. "Give yourself 106 lbs. for the first 5 feet, plus 6 lbs. for every inch over 5 ft. tall." Construct a mathematical model for the relationship. What are the recommended weights for a male 5' 8" and for a male 6' 2" tall?

Solution: We could let x represent height in inches, but since we are concerned only with heights over 5 feet let's instead let x = height in inches in *excess* of 5 feet and y = weight in lbs. Then according to the recommendations, when x = 0 inches (for a 5 foot male), y = 106 lbs. So the y-intercept is 106. For each additional inch in height, the recommendations permit 6 additional lbs. So the rate of change of height with respect to weight is a constant 6 lbs./inch. Hence the equation

$$y = \ 106 + 6x$$

gives us a mathematical model for the relationship where x = height in inches above 5 feet, and y = weight in lbs. We can think of 106 lbs. as our starting point, and 6 lbs./in as our constant rate of change. Note that y and 106 are in lbs., 6 is in lbs./inch, and x is in inches.

In our model, as in all models, we need to think about realistic values for the domain. The recommendations are for men between 5 feet (or 60 inches), and say 7 feet (or 84 inches, our guess for a reasonable maximum male height). Hence the constraints on x, the possible values for the number of inches above 5 feet, would be $0 \le x \le 24$ inches.

For a 5' 8" male, x = 8. The recommended weight would be: $y = \ 106 + 6(8)$
$$= \ 106 + 48$$
$$= \ 154 \text{ lbs.}$$

A 6' 2" male is 74" tall. Since 5' = 60", then this male would be 14" taller than 5'. So x=14". His recommended weight would be:

$$y = \ 106 + 6(14)$$
$$= \ 106 + 84$$
$$= \ 190 \text{ lbs.}$$

Table 4.7 and Figure 4.10 are alternate representations of this function.

Table 4.7	
x	y
0	106
4	130
8	154
12	178
16	202
20	226
24	250

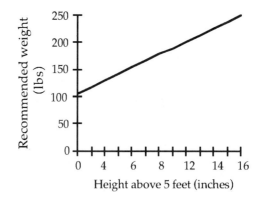

Fig. 4.10 Recommended weight for males

Example 3: Find the equation of the linear function graphed below.

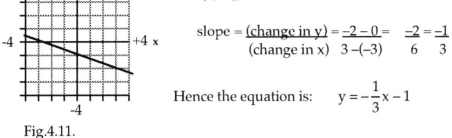

Fig.4.11.

Solution: The y-intercept is -1. We can use any two points on the line to calculate the slope. If, for example, we take (-3, 0) and (3, -2), then

$$\text{slope} = \frac{(\text{change in y})}{(\text{change in x})} = \frac{-2 - 0}{3 - (-3)} = \frac{-2}{6} = \frac{-1}{3}$$

Hence the equation is: $y = -\frac{1}{3}x - 1$

Thinking of the rise/run = -1 / 3, if we start at any point on the line and run 3 units to the right, we *drop down* 1 unit (because of the minus sign), in order to return to the line. Thinking of the rise/run in its equivalent form 1 / (-3), starting at any point on the line, if we ran 3 units *to the left* (because of the minus sign), we would have to rise 1 unit in order to return to the line.

Algebra Aerobics:

Write an equation, make an appropriate table, and sketch the graph of each of the following lines.
1.   The slope is 1.2 and the vertical intercept is -4.
2.   The slope is -400 and the vertical intercept is 300.

Answers
1. $y = 1.2x - 4$

2. $y = 300 - 400x$

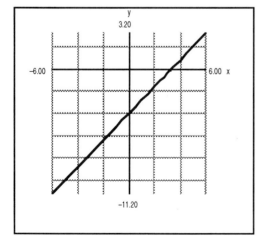

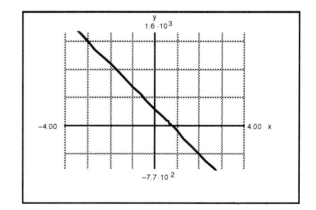

Case 2: The slope (m) and a point are known

If the slope, m, and the coordinates of a point on the line are given, we need only find the value of the vertical intercept, b, in order to write the function.

Example 1: Find the linear function with a slope of 3 and a graph that passes through the point (–1, –2).

Solution

Choose variables: Let x be the independent variable and y be the dependent variable.

Find m: We are given a slope of 3, so m = 3. The function has the form:
$$y = 3x + b$$

Find b: The point (–1, –2) satisfies the equation. So we can substitute the value –1 for x and –2 for y into the equation, and solve for b.

substitute known x and y	$-2 = 3(-1) + b$
solve for the unknown b	$-2 = -3 + b$
add 3 to both sides	$3 - 2 = b$
	$1 = b$

The function: We can substitute the values for m and b directly to get
$$y = 3x + 1$$

The domain: Since there are no constraints on x, the domain is all real numbers

Table and graph:

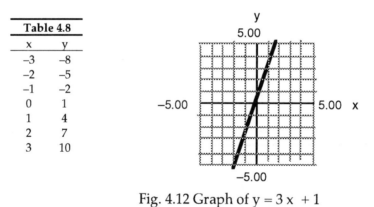

Table 4.8	
x	y
–3	–8
–2	–5
–1	–2
0	1
1	4
2	7
3	10

Fig. 4.12 Graph of y = 3 x + 1

Example 2:   A TV announcer reporting on health care says that since 1965, when the country had 1.7 million hospital beds, the number of beds has been declining by 20 thousand (or 0.02 million) a year. What equation represents this statement? When does the equation predict we will have only 1 million beds?

Solution

Choose variables: We are looking for an equation that describes the number of hospital beds, N, in terms of the independent variable time, T. That is,
$$N = b + mT$$

We could let T be the year. But since we are only interested in years from 1965 on (and to keep computations simpler), it is easier to let T = number of years since 1965.

Find m:

The statement that the number of beds is declining by 20 thousand, or equivalently 0.02 million beds a year, tells us that the average rate of change of B with respect to T is - 0.02 million beds/year. So m = -0.02 million beds/yr. The function has the form:

$$N = b - 0.02 \, T$$

Now we have to stop and think about the units for each term. T is in years, -0.02 is in millions of beds/yr. So 0.02T will be in millions of beds, and the units for b and N will be millions of beds.

Find b:

In 1965, the nation had 1.7 million beds. In terms of the variables T and N, T = 0, and N = 1.7. So the point (0, 1.7) should satisfy our equation. Substituting these values into the equation gives us:

$$1.7 = b - (0.02)(0)$$
$$\text{or} \qquad 1.7 = b$$

We might have also recognized (0, 1.7) as the vertical intercept and substituted 1.7 for b right away.

The function:

Hence our equation is:

$$N = 1.7 - 0.02T$$

where T = years since 1965, and N = number of hospital beds (in millions).

The domain:

The values for T start at 0 years (for the year 1965). If the announcement was made in 1996, the upper limit for T would be 31 years (that is, 1996 - 1965).

Table and graph:

Table 4.9	
T	N
(years since 1965)	(no. of beds in millions)
0	1.7
5	1.6
10	1.5
15	1.4
20	1.3
25	1.2
30	1.1

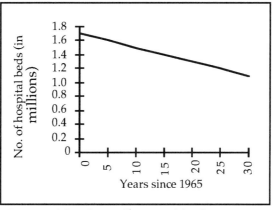

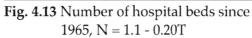

Fig. 4.13 Number of hospital beds since 1965, N = 1.1 - 0.20T

We can use our equation to predict when the number of hospital beds would reach 1 million. Since N is measured in millions of beds, we are asking when N would equal 1 million. Setting $N = 1$ we get

$$1 = 1.7 - 0.02\,T$$

subtract 1.7 from both sides	$-0.7 = -0.02T$
divide both sides by -0.02	$35 = T$

So 35 years after 1965, in the year 2000, if the rate of decline stays constant at 20,000 hospital beds a year, we will have 1 million hospital beds.

Algebra Aerobics:

For each line described below, write an equation, make an appropriate table with two or three points, and sketch the graph.

1. The slope is -4 and the line passes through (-3,1).
2. Salary increases by $1.60 per hour for every year of education, and with ten years of education, the hourly salary is $8.50.

Answers

1. $1 = (-4)(-3) + b \implies b = -11$

         $\implies y = -4x - 11$

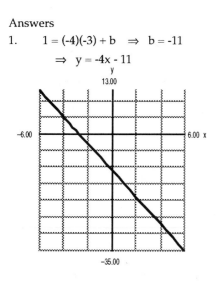

2. $8.5 = (1.6)(10) + b \implies b = -7.5$

         $\implies S = 1.6Y - 7.5$

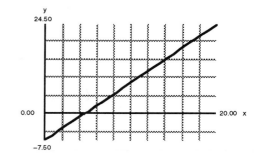

Case 3: Two points, (x_1, y_1) and (x_2, y_2) are known

In this case, both the slope (m) and the y-intercept (b) have to be found. But given two distinct points (x_1, y_1) and (x_2, y_2) on the line, we can always find the slope. Once the slope is found, we have Case 2 where the slope and a point are used to find b.

Example: Find the equation of the line passing through (–3,–5) and (–1,3).

Solution

Choose variables:	Let x be the independent variable and y be the dependent variable.
Find m:	Since both (–3,–5) and (–1,3) lie on the line:

$$m = \frac{(3 - (-5))}{(-1 - (-3))} = \frac{8}{2} = 4$$

The slope, m, is 4, and substituting the value for m, the equation is:
$$y = 4x + b$$

Find b: Since the line passes through (–1, 3), we can substitute –1 for x and 3 for y in the equation to find b:
$$3 = 4(-1) + b$$
$$3 = -4 + b$$
$$b = 7$$

The function: We can substitute these values for m and b directly into the general linear function
$$y = 4x + 7$$

The domain: x can assume any real value

Table and graph:

Table 4.10

x	y
–8	–25
–6	–17
–4	–9
–2	–1
0	7
2	15
4	23
6	31
8	39
10	47

Fig.4.14.   Graph of y = 7 + 4x

This strategy will prove useful in the next section,
where we'll talk about fitting linear models to data sets.

Algebra Aerobics: [7]
Find an equation for the line containing (-2,4) and (3,1).

[7] Answer
$m = (4 - 1)/(-2 - 3) = -0.6$ $1 = (-0.6)(3) + b$ $\Rightarrow$ $b = 2.8$ $\Rightarrow$ $y = -0.6x + 2.8$

4.4 Estimating Linear Models for Data

We've seen linear models for the median weight of baby girls and for the median height of children between the ages of 2 and 12. But where do such models come from? They are derived from observational studies of real children, like the following study. Table 4.11 and Figure 4.15 show the mean heights of a group of 161 children in Kalama, Egypt.. The children's heights were measured monthly over several years as a part of a study of nutrition in developing countries.

Table 4.11	
Mean Height of Kalama Children	
Age (months)	Height (cm)
18	76.1
19	77.0
20	78.1
21	78.2
22	78.8
23	79.7
24	79.9
25	81.1
26	81.2
27	81.8
28	82.8
29	83.5

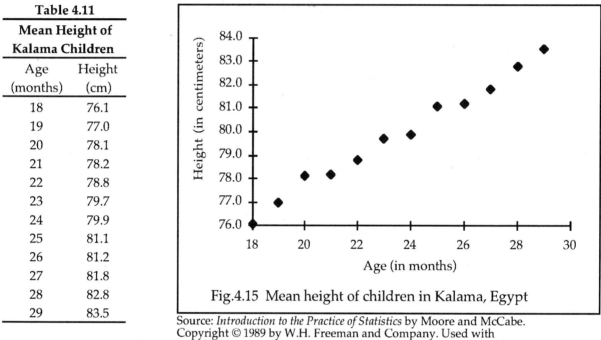

Fig.4.15 Mean height of children in Kalama, Egypt

Source: *Introduction to the Practice of Statistics* by Moore and McCabe. Copyright © 1989 by W.H. Freeman and Company. Used with permission.

While the data points do not lie exactly on a straight line, the overall pattern seems clearly linear. Now, rather than generating a line through two of the data points, let's try eyeballing a line that approximates all the data. A ruler or a piece of black thread laid down through the dots will give you a pretty accurate fit.

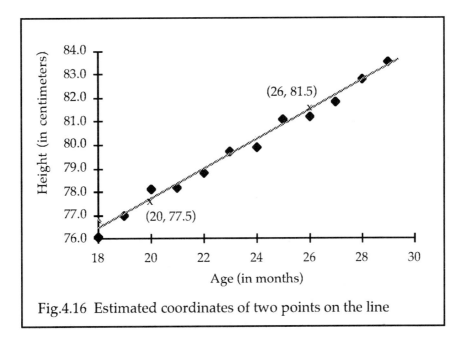

Fig.4.16 Estimated coordinates of two points on the line

What is an equation for our linear model? We can use the strategy from Case 3, when you know two points. The estimated coordinates of two points *on the line* are labeled on the graph. Note that they are NOT taken from the original data in Table 4.11. We can use these points to find the average rate of change or slope:

$$m = \frac{(81.5 - 77.5)\ \text{cm}}{(26 - 20)\ \text{months}}$$

$$= \frac{4.0\ \text{cm}}{6\ \text{months}}$$

$$= 0.67\ \text{cm}/\text{month}$$

This is the most important number in our linear model. It is a compact way of summarizing how height changes with age. *Our model predicts that for each additional month between 18 and 29 months, an "average" child will grow about 0.67 centimeters.*

Now we know our linear equation is of the form:
$$y = \quad b + 0.67x \qquad\qquad (1)$$
where x = age in months and y = mean height in centimeters.

How can we find b? We have to resist the temptation to read b directly from the graph, using the coordinates where our line crosses the height axis. Why? Isn't b always the vertical intercept? As is frequently the case in social science graphs, both the horizontal and the vertical axes are cropped. At a quick glance, what looks like the origin (the point (0,0)) actually has coordinates (18, 76). Because the horizontal axis is cropped, we can't read the vertical intercept off the graph. We'll have to calculate it.

Since the line passes through (20, 77.5) we have:

substitute (20, 77.5) in equation (1)   77.5 =   b + (0.67)(20)

simplify   77.5 =   b + 13.4

solve for b   b =   64.1 cm

Our linear equation modeling the Kalama data is:

$$y =\ \ 64.1 + 0.67\,x$$

where x = age in months and y = height in centimeters. What is the domain? The data was collected on children aged 18 to 29. We don't know of its predictive value outside these ages so:

domain =   values of x

where $18 \le x \le 29$.

Note that while the y-intercept is necessary in order to write the equation for the line, it actually lies outside the set of values for which our model applies. We have no reason to expect that growth patterns between 18 and 29 months can be projected back to give the height of a newborn baby as 64.1 cm. Compare Figure 4.17 to Figure 4.16. They both show graphs of the same equation y = 64.1 + 0.67 x. In Figure 4.16 both axes were cropped. Figure 4.17 includes the origin, (0,0) The vertical intercept is now visible and the dotted lines indicate the region which applies to our model. The moral is: when reading real life graphs, look carefully to see if the axes have been cropped.

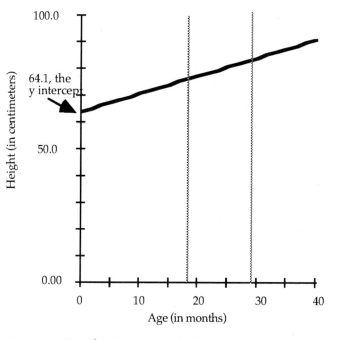

Fig.4.17   Graph of y = 64.1 + 0.67x that includes the origin (0,0). Dotted lines show region that models Kalama data

Our model offers a compact 60 second summary of the results of the Kalama height study.

A messier example:

The Kalama data set was fairly clean, partially due to the fact that we didn't have all the raw data. Table 4.11 contains one value for each month, the mean of the children's heights, not all 161 of the individual heights. For many data sets there may be many y values associated with one x value. In these cases the scatter plots look more like clouds of data.

Figure 4.18 shows mean SAT verbal scores vs. percent of seniors taking the SAT in 1989 for each state in the U.S.[8] Each point is labeled with an abbreviation of the state's name. The point labeled WA near the middle of the graph with approximate coordinates (39, 448) means that in 1989 in the state of Washington, 39% of its seniors took the SAT test and their mean verbal score was 448. There appears to be a general downward pattern. As the percentage of students taking the test increases, the mean SAT verbal score decreases. We can sketch a best fit line through the data (as shown on the scatter plot), estimate its equation and interpret its slope.

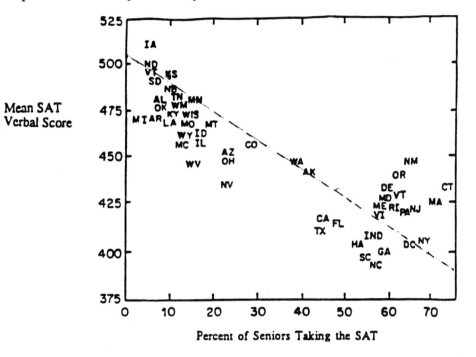

Fig. 4.18 Percent of Seniors Taking the SAT

Choose variables: Let P be the independent variable representing the percentage of seniors. Let S be the dependent variable representing the mean SAT verbal score. The equation of the line is of the form:

$$S = b + mP$$

Find m: Do a rough hand sketch of the line (for example, the dotted line in the scatter plot). In order to calculate the slope, we must estimate the coordinates of two points *on the line*, for instance (10, 490) and (70,400).

$$m = \frac{(490 - 400) \text{ SAT points}}{(10 - 70)\ \%} = \frac{90 \text{ SAT points}}{-60\%} = -1.5 \text{ SAT points } / \%$$

---

[8] From an article entitled "Verbal SAT Scores" taken from *New Liberal Arts*, edited by John Truxal, p. 30-32. The complete article appears in the anthology of readings.

$-1.5 = -1.5/1$, so on average for each additional 1% of seniors taking the SAT verbal test, our model predicts that the mean verbal score was roughly 1.5 points lower. Equivalently, when the percent of seniors taking the test increased by 10, the mean SAT verbal score dropped on average by 15 points.

Substituting -1.5 for m, the equation for the line is

$$S = b - 1.5P$$

Find b:

Since the line passes through (70, 400), we can substitute the values 70 and 400 for P and S in the equation of the line to find b.

$$400 = b - (1.5)(70)$$
$$400 = b - 105$$
$$505 = b$$

The function:

Substitute these values for m and b to get

$$S = 505 - 1.5P$$

where P = the percentage of seniors taking the SAT verbal test, and S = the mean SAT verbal score.

The domain:

Given the plotted data, a reasonable domain for P might be $5\% \leq P \leq 75\%$.

Table and graph:

The SAT scores predicted by our linear model (*not* the actual scores) are included in the Table 4.12 and Figure 4.19.

Table 4.12	
P	S
5	497.5
10	490.0
15	482.5
20	475.0
25	467.5
30	460.0
35	452.5
40	445.0
45	437.5
50	430.0
55	422.5
60	415.0
65	407.5
70	400.0
75	392.5

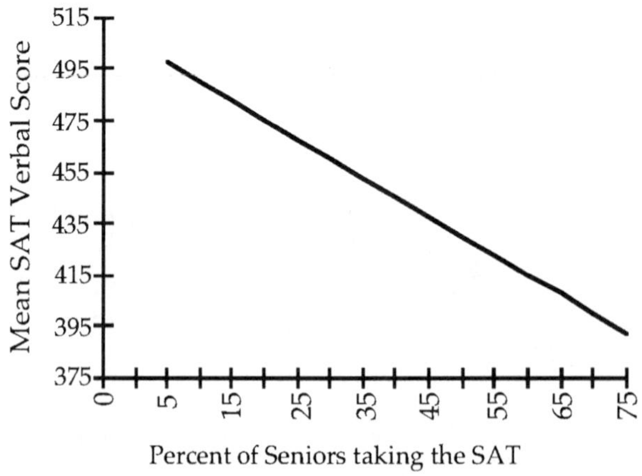

Fig. 4.19 Mathematical Model for % of Seniors taking SAT vs. Mean Verbal Score

Note that in this graph, as in the Figure 4.18, the vertical axis is cropped.

Something to think about

1. How well do the predicted scores match the real scores?

2. Look at the Anthology article "Verbal SAT Scores." What very different conclusions might you draw looking at the SAT scores as a single variable vs. looking at their relationship to the percent taking the SAT? What questions should you ask when comparisons are drawn between test scores of American students and students from other countries?

"The Paramount Convenience of Straight Lines" [9]

When analyzing relationships between two variables, linear relationships are of special importance. Not because most relationships are linear; in fact they are not. But straight lines are easily drawn and analyzed. By stretching a thread across a scatter plot, we can fit a straight line by eye almost as well as a computer. The line is usually good enough for practical purposes. Linear equations are easy to manipulate and interpret, so lines are often used as a first approximation to patterns in data.

---

[9] This title and the ideas in this subsection are from Douglas Riggs, *The Mathematical Approach to Physiological Problems*, M.I.T. Press, Cambridge, Mass.,1976. pp. 57-58.

4.5 Special Cases

Case 1:   Direct Variation

The simplest possible relationship between two variables is when one is a constant times the other. For instance, in the Illinois sales tax example T = 0.08 P; the sales tax T equals a constant (0.08 , the tax rate) times P the purchase price. We say T *varies directly with* P, or that T is *directly proportional to* P.

How can we recognize direct variation?

Linear functions in the form

$$y = mx$$

describe a relationship where x and y vary directly. Notice that we could write y = mx as y = mx + 0 which would be in the general linear form y = mx + b where b = 0. So, direct variation is a special case of the linear function where b, the y-intercept, is 0.

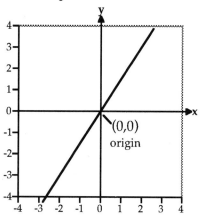

Graphs of linear equations where the vertical intercept is 0 pass through the point (0,0), which is called the *origin*. These are precisely the graphs of direct proportionalities.

Fig.4.20. Example of a relationship in which x and y vary directly, hence the line goes through the origin.

If y = mx , where m ≠ 0
the relationship between x and y is described by saying that
y varies directly with x
or that
y is directly proportional to x.

For physical situations, when quantities vary directly, m is usually positive. In an abstract case, m could be positive or negative.

Example 1: Think about situations where physical quantities vary directly. Express the relationships in the form of an equation.

Solutions

Suppose you go on a road trip, driving at a constant speed of 60 miles per hour. After one hour, you will be 60 miles away from where you started; after two hours, you will be 120 miles away; and so on. The distance d (in miles) you have traveled varies directly with the time t (in hours); in particular d= 60 t.

Suppose you are paid by the hour for performing a certain job. If your pay rate is \$7/hr, then E, the amount of money earned (in dollars), after H hours can be represented by E = 7H; that is, E is directly proportional to H.

Example 2: On July 7, 1994 you could buy \$1.345 Canadian dollars for one U.S. dollar. Find a linear function that converts American dollars into Canadian dollars at this rate.

Solution

Choose variables:	Let US \$= number of American dollars (the independent variable)
	and CD\$= number of Canadian dollars (the dependent variable)
Find b:	When US\$ = 0, the value of CD \$= 0, so b = 0.
Find m:	Each additional American dollar corresponds to 1.345 Canadian dollars, thus m = 1.345/1 = 1.345.
Construct function:	Substituting, the equation of the linear function is:
	CD\$= 1.345 * US\$
	So the number of Canadian dollars you receive is directly proportional to the number of American dollars you convert.
The domain:	US\$ ≥ 0.
Table and graph:	Table 4.13 and Figure 4.21 convert American dollars into Canadian dollars.

Table 4.13		
US\$	CD\$	
0	13.45(0)	= 0
1	1.345(1)	= 1.345
10	1.345(10)	= 13.45
100	1.345(100)	= 134.50

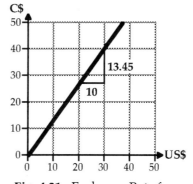

Fig. 4.21. Exchange Rate for American to Canadian dollars

Example 3: If y varies directly with x, and $y = -10$ when $x = 5$, find the value of x when $y = -8$.

Solution

Use the fact that with direct variation (0, 0) is a solution. Find m.

$$m = \frac{\text{change in y}}{\text{change in x}} \qquad = \qquad \frac{-10 - 0}{5 - 0} \qquad = -2$$

Construct function $\quad y = -2x$

Substituting -8 for y gives

$$-8 = -2x$$
$$4 = x$$

Algebra Aerobics: [10]

If y is directly proportional to x, is x directly proportional to y?

Sometimes it is more convenient to think of a linear equation of the form $y = mx$ in its equivalent form:

$$y/x = m \qquad \text{(where we assume that x is not equal to 0)}$$

We call the quotient y/x the *ratio* of y to x. When x and y are directly proportional, the ratio of y to x remains constant, and the constant is m, the slope.

> When two variables x and y are *directly proportional*
> then the *ratio* or *proportion* y/x is constant.

Example 4: Constant proportions are very common in cooking. If a good oil and vinegar dressing is 3 parts olive oil to 2 parts balsamic vinegar, or a great shandy is 12 parts beer to 32 parts ginger ale, these proportions are rates of change that can be used to make any desired quantity, for one person or a large party.

How much oil is needed for this salad dressing recipe when 7 tablespoons of vinegar are used? How much ginger ale is needed to make a great shandy when 21 pints of beer are used?

---

[10] Answer: Yes. If y is directly propotional to x, we can write $y = mx$, where $m \neq 0$. Hence $x = (1/m)y$. So $x = $ (a constant) (y). Hence x is directly proportional to y.

Solution

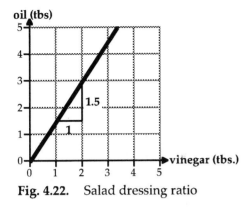

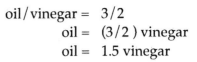

oil/vinegar =   3/2

oil =   (3/2) vinegar

oil =   1.5 vinegar

Substituting 7 tbs for vinegar then:

oil=   1.5 (7tbs)

oil=   10.5 tbs

Fig. 4.22.   Salad dressing ratio

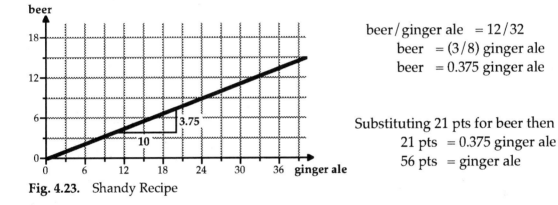

beer/ginger ale   = 12/32

beer   = (3/8) ginger ale

beer   = 0.375 ginger ale

Substituting 21 pts for beer then

21 pts   = 0.375 ginger ale

56 pts   = ginger ale

Fig. 4.23.   Shandy Recipe

Algebra Aerobics

1. Find a function that represents the relationship between distance (d) and time (t) of a moving object using the data Table 4.14. Is d directly proportional to t? Which is a more likely choice for the object:, a person walking or a moving car?

Table 4.14	
t hrs	d miles
0	0
1	5
2	10
3	15
4	20

2. For each of the following tables, determine if the variables vary directly with each other. Represent each relationship with an equation.

Table 4.15	
x	y
−2	6
−1	3
0	0
1	−3
2	−6

Table 4.16	
x	y
0	5
1	8
2	11
3	14
4	17

Table 4.17	
G Gasoline Bought # gallons	C Cost of Gasoline dollars
1	1.50
2	3.00
3	4.50
4	6.00
5	7.50

3. On the scale on a map one inch represents a geographic distance on the earth of 35 miles. What is the geographic distance on the earth between two points that are 4.5 inches apart on the map?

4. If the student/faculty ratio is 12/1, and an institution has 3000 students, how many faculty does it have?

Answers:
1. $d = 5t$, so d is directly proportional to t. The d is in miles, t is in hours, so 5 is in miles/hr. This is more likely a walking than driving pace.

2. Equation for Table 4.15 $y = -3x$ (direct variation). Equation for Table 4.16 $y = 3x + 5$ (not direct variation). Equation for Table 4.17 $C = 1.50 G$ (direct variation)

3. Let R = real geographic distance and M = distance on map. Since R is directly proportional to M, then $R = m * M$. Then $m = $ slope $ = 35$ miles/1 inch   so
    $$R = 35 * M$$
    When M = 4.5 then $R = 35$ miles/ 1 inch) * 4.5 inches = 157.5 miles

4. If S = number of students and F = number of faculty, then $S/F = 12/1$, or $S/F = 12$. If S = 3000, then $3000/F = 12$.   So $F = 3000/12 = 250$ faculty.

Special Case 2: Horizontal Lines

If a line is horizontal its slope, m, is 0. Its equation has the form y = (0)x + b or simply

$$y = \quad b.$$

This means that y does not depend upon x at all. No matter what value x assumes, there is no change in y. Hence y always has the same constant value, b.

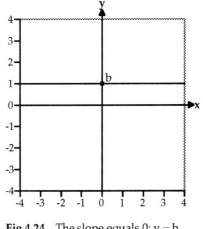

Fig.4.24. The slope equals 0: y = b

Example 1: Write the equation of a horizontal line through the point (2,3).

Solution

Choose variables:	Let x be the independent variable and y be the dependent variable.
Find m:	Since the line is horizontal, the slope, m, is 0.
Find b:	Plot the point (2,3) and sketch a horizontal line through it. Note that each point on the line is 3 units above the x–axis. So the line must cross the y–axis at (0,3) making the y–intercept, b, equal to 3 .
The function:	Substituting, the equation of the line is:

$$y = (0)x + 3$$
$$y = 3$$

The domain:	x can assume any real value
Table and graph:	

Table 4.18	
x	y
–4	3
–3	3
–2	3
–1	3
0	3
1	3
2	3
3	3
4	3

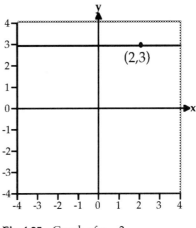

Fig.4.25. Graph of y = 3

Example 2. Table 4.19 and Figure 4.26 show the past sales of two kinds of computer disk drives: the older 5.25 inch disk drives and the newer, smaller 3.5 inch or less disk drives. (Note: a 3.5 inch disk comes with this book.) A business person would call this a "changing product mix." Describe the sales history of the 5.25 inch disk drives and construct an equation to represent these sales.

Table 4.19										
Worldwide Flexible Disk Drive Shipments–All Manufacturers (in 000's)										
	1983	1984	1985	1986	1987	1998	1989	1990	1991	1992	1993
5.25" drives	10,490	15,616	12,995	16,287	16,232	16,574	14,887	14,946	15,020	17,291	14,711
3.5" or less drives	438	1,972	3,268	6,199	12,337	18,115	23,216	29,446	32,819	41,790	51,152

Source: Disk/Trend Nov. 1994 Report. Reprinted with permission

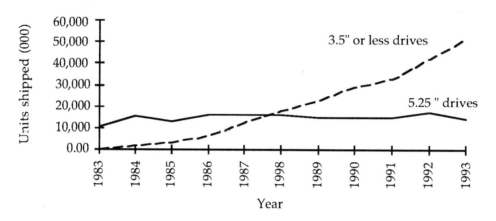

Fig. 4.26 World Wide Flexible Disk Drive Shipments

Solution:

The shipments for 5.25 inch drives remained fairly flat between 1983 and 1993 at about 15,000 drives per year. Mathematically, we could describe the horizontal line approximating 5.25 inch disk drive shipments by:

$$S_{5.25} = 15000$$

where $S_{5.25}$ = shipments for 5.25 inch drives

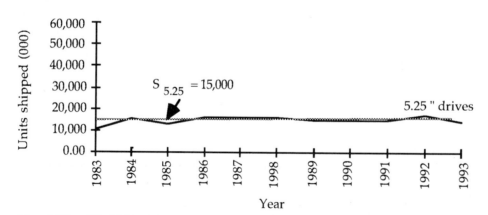

Fig. 4.27 World Wide Flexible Disk Drive Shipments
Horizontal dotted line approximates shipments for 5.25" disk drives

Algebra Aerobics: [11]

1. Find an equation for the horizontal line though (3,-5).
2. An employee for an aeronautical corporation has a starting salary of $25,000 per year. After working there for ten years and not receiving any raises, he decides to seek employment elsewhere. Write the equation and graph the employee's salary as a function of time. What is the domain for this function?

Special Case 3: Vertical Lines

If a line is vertical, its slope is undefined, so we can't use the standard form of the linear equation. However every point on a vertical line has the same x coordinate, while all values of y are allowed. When specifying a constant value c for x, with no recommendation for y, the graph of the equation is a vertical line and the equation is simply:

$$x = c$$

The points satisfying this equation are not the graph of a function, since to each x there is not associated a unique y. In particular, when x = c, there are infinitely many associated y values, so it fails the vertical line test.

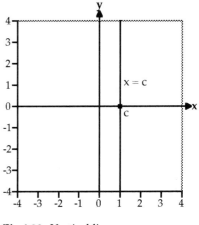

Fig.4.28. Vertical lines: $x = c$

[11] Answers

1. $y = -5$
2. $y = \$25,000$. The domain is 0 years $\leq x \leq 10$ years.

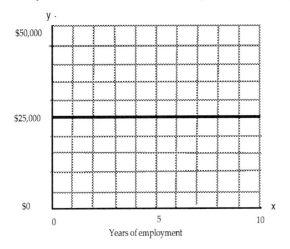

Example 1: Find the equation of the vertical line that passes through the point (–2, 1).

Solution   Let x be the independent variable and y the dependent variable. Plot the point (–2,1) and sketch a vertical line through it. Each point on the line must have an x value of –2, no matter what the y value. So the equation is:

$$x = -2.$$

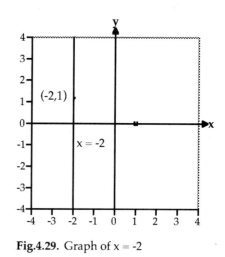

Fig.4.29. Graph of x = -2

Example 2: Vertical lines are relatively rare in the social sciences, but one theoretical example stems from economics. Economists talk about demand curves that depict the relationship between the market price and the quantity consumers demand of a commodity. (We take a closer look at demand curves in the next chapter.) You would expect that higher prices would be associated with a decreased quantity demanded by consumers. So the demand curve would normally have a negative slope. If the demand for a commodity (computer scientists or Mighty Morphin Power Rangers) is so great that an increase in price has little effect on the quantity demanded, then the demand curve becomes close to a vertical line, and is said to be *inelastic*.

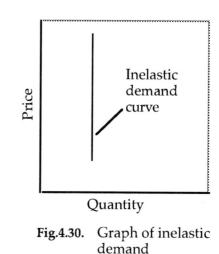

Fig.4.30.   Graph of inelastic demand

Algebra Aerobics: [12]

Write an equation for the vertical line through (3,-5).

---

[12] Answer: x = 3

Special Case 4: Parallel Lines

If two linear relationships have equal rates of change, their lines have equal slopes and the graphs are parallel lines. Geometrically this means that the graphs always maintain the same vertical distance apart, never crossing.

Example 1: The two lines y = 2.0 – 0.5x and y = –1.0 – 0.5x each have a slope of –0.5 so they are parallel.

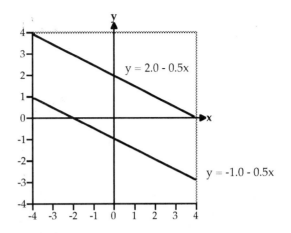

Fig.4.31. Parallel lines have equal slopes

Example 2: Between 1975 and 1985 the lines representing growth in the percentage of medical degrees awarded women doctors and women dentists in the U.S. were roughly parallel. Table 4.20 and Figure 4.32 show the percentage of medical degrees awarded to women during this time period.

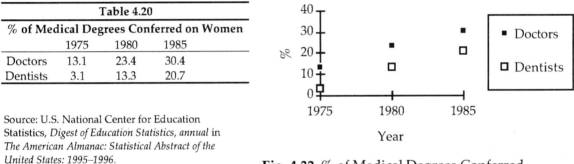

Table 4.20			
% of Medical Degrees Conferred on Women			
	1975	1980	1985
Doctors	13.1	23.4	30.4
Dentists	3.1	13.3	20.7

Source: U.S. National Center for Education Statistics, *Digest of Education Statistics, annual* in *The American Almanac: Statistical Abstract of the United States: 1995–1996.*

Fig. 4.32 % of Medical Degrees Conferred on Women

The data points for female doctors' and for dentists' degrees lie roughly on parallel lines, each with a slope of about 1.7 %/yr. That means that between 1975 and 1985 the percentage of degrees awarded to both female doctors and dentists increased by about 1.7 percentage points per year. As long as the number of women doctors and dentists is growing at the same rate, it is not possible for the number of women dentists to catch up with the number of women doctors. One plausible explanation is that similar processes may be at work causing these increases, a topic for further study by social scientists.

Example 3: In the business world, each industry has its own set of commercial "market forecasters," and each of them has developed proprietary models for forecasting industry trends.[13] A typical forecast would track 10 years of historical data and predict 2–5 years into the future. If a new product is being introduced, then there are obviously no historical data

---

[13] Dataquest and The Yankee Group are two such market forecasters in the computer industry.

available. One strategy is to look at another product previously introduced into the same market, and assume that the new product sales will follow a parallel pattern.

For example, a company in Massachusetts made the following prediction based on past sales. The company was thinking of introducing a new optical disk drive. Past disk drive sales were noted and predictions were made based on that history. The following figure shows the graph of world wide shipments of 3.5 " (and less) disk drives and the predicted shipments for the new optical disk drives. The graphs show the classic "hockey stick" pattern of growth: initial sales are fairly flat, but then as the product catches on, sales take a dramatic upward turn.

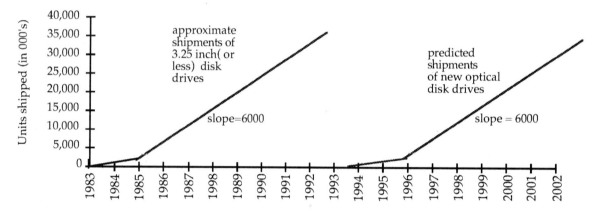

Fig. 4.33 Parallel growth for 3.25 inch disk drive and new optical disk

The rate of change during the rapid growth period of 3.25 inch (and less) disk drive shipments, and hence the prediction for the equivalent period for the new optical disk drive, is an increase of approximately 6,000 thousand (or 6 million total) units per year. The graphs of past and predicted shipments are two parallel line segments each with a slope of 6,000 thousand units per year.

Something to think about

Clearly, other factors could be considered in making predictions on sales of a new product. What other factors might be considered in the sales of new optical disk drives? How might your graph change if you considered these factors?

Algebra Aerobics: [14]
Find an equation for each line described below.
1.   It is parallel to $y = 4 - x$ and passes through the origin.
2.   It is parallel to $W = 358.9C + 2500$ and passes through (4,1000).

---

[14]   Answers

1. $y = -x$
2. $m = 358.9$ $1000 = (358.9)(4) + b$ $\Rightarrow$ $b = -435.6$ $y = 358.9x - 435.6$

Special Case 5: Perpendicular Lines

Consider a line whose slope is given by v/h. Now imagine rotating the line 90 degrees clockwise to generate a second line perpendicular to the first.

What would the slope of this new line be?

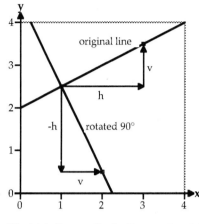

The positive vertical change, v, becomes a positive horizontal change. The horizontal change, h, becomes a negative vertical change.

The slope of the original line is v/h, and the slope of the line rotated 90 degrees clockwise is –h/v. Note that –h/v = –1/ (h/v) which is the original slope inverted and multiplied by –1.

In general, the slope of a perpendicular line is the negative reciprocal of the slope of the original line. If the slope of a line is m_1, then the slope, m_2, of a perpendicular line is $-1/m_1$.

Fig.4.34. Perpendicular lines: $m2 = -1/m1$

This is true for any pair of perpendicular lines for which slopes exist. It does not work for vertical lines since they have undefined slopes.

Example: Find the equation of a line which is perpendicular to the line $y = -2x + 3$ and that passes through the point (–2, –3).

Solution   The slope of the given line, $y = -2x + 3$, is –2 , and the slope of any line perpendicular to it will be the negative reciprocal of –2. Hence the new slope is: $-(1/-2) = 1/2$. The new line has a slope of 1/2, so its equation is
$$y = (1/2)x + b.$$

Since the point (–2,–3) lies on the new line, we can substitute –2 for x and –3 for y in order to find b.
$$-3 = (1/2)(-2) + b$$
$$-3 = -1 + b$$
$$-2 = b$$
The equation for the new line is: $y = (1/2)x - 2$.

Graph:

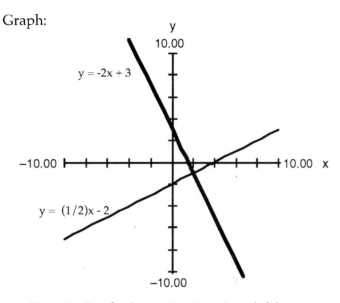

Fig. 4.35 Graph of y = - 2x + 3 and y = (1/2) x - 2

Something to think about

1. If two lines are perpendicular to each other, what do you know about the units for the two variables?

2. In this last example, is this the only line that is perpendicular to $y = -2x + 3$ and passes through the point $(-2,-3)$? Can you find the equations of other lines perpendicular to the original line $y = -2x + 3$ that do not pass through the point $(-2,-3)$?

Algebra Aerobics: [15]

1.   Find the slope of a line perpendicular to each of the following.
     a) y = 4 - 3x   b)   y = x   c)   y = 3.1x - 5.8
2.   Find an equation for the line which is perpendicular to y = 2x - 4 and passes through (3,-5). Graph both lines on the same coordinate system.

---

[15] Answers

1.   a) m = 1/3   b)   m = -1   c) m = -10/31 = -0.32
2.   m = -1/2 = -0.5   -5 = (-0.5) (3) + b   ⇒   -3.5 = b ⇒   y = -0.5x - 3.5

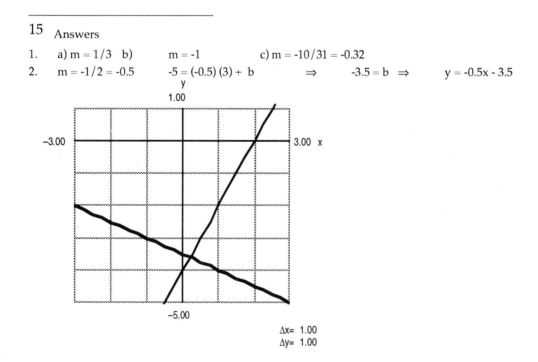

Δx= 1.00
Δy= 1.00

4.6 Equivalent Ways of Describing Linear Relationships

Most of the linear equations we've worked with so far have been in the form $y = mx + b$ or $y = b + mx$. By rearranging terms and multiplying through by constants, we could rewrite these equations in the form $Ax + By = C$ (where A, B and C are constants). Sometimes this equation is the more natural format. For example, if you have $100 to buy food for a party and liter bottles of coke cost $1.50 and bags of pretzels cost $1.00, then:

(the amount you spend on coke) + (amount you spend on pretzels) = $100.00

If x = number of bottles of coke   and y = number of bags of pretzels then
  the amount you spend on coke   =   (cost per bottle) (number of bottles) = $1.50 x
  the amount you spend on pretzels =   (cost per bag) (number of bags)   = $1.00 y

So the linear equation
$$1.50\ x + 1.00\ y = 100.00 \tag{1}$$
describes the relationship between x, the number of coke bottles, and y, the number of pretzel bags. This is in the form $Ax + By = C$ where $A = 1.50$, $B = 1.00$ and $C = 100.00$.

If we want to know how much coke we could buy, we could solve for x:
  given $1.50\ x + 1.00\ y =$ 100.00
  subtracting 1.00 y from both sides $1.50\ x=$ 100.00 - 1.00 y
  dividing both sides by 1.50 gives $x =$ (100.00 - 1.00 y) / 1.50
   using the distributive law $x =$ (100.00 / 1.50) - (1.00 / 1.50) y
  simplifying fractions $x =$ 200/3 - 2/3 y (2)

Similarly if we want to know how many bags of pretzels we could buy, we could solve for y in terms of x to get:
$$y =\ 100.00 - 1.50\ x. \tag{3}$$

So equations (1), (2) and (3) are all equivalent forms.

Equivalent ways of describing linear relationships

We've now seen a number of different ways of determining whether two variables are linearly related. These are summarized in the following table. [16]

---

[16] Note: this summary ignores the special case of vertical lines.

Two variables x and y are said to be linearly related if any *one* of the following equivalent conditions is shown to be true:

1. All of the pairs (x,y) lie on a straight line when plotted.

2. For a fixed amount of change in x there is a fixed amount of change in y.

3. The average rate of change of y with respect to x is constant.

4. There exist two constants, m (the slope) and b (the y-intercept) such that the equation $y = b + mx$ is true for all of the data pairs (x,y).

5. There exist three constants, A, B, and C such that the equation $Ax + By = C$ holds true for all pairs (x,y). (B is not equal to 0.)

How to recognize when a data set describes a linear relationship

Now we examine a data table to decide if it determines a linear relationships. Although we can check each of the five equivalent definitions, it is only necessary to prove that any ONE of the equivalent conditions is true for two variables to be linearly related.

Example 1: Adding minerals or organic compounds to water lowers its freezing point.
Antifreeze for car radiators contains glycol (an organic compound) for this purpose. Table 4.20 shows the effect of salinity (dissolved salts) on the freezing point of water.[17] Salinity is measured in the number of grams of salts dissolved in 1000 grams of water. So our units for salinity are in parts per thousand, abbreviated as *ppt*.

Table 4. 21

Salinity (ppt)	Freezing point (°C)
0	0.00
5	−0.27
10	−0.53
15	−0.80
20	−1.08
25	−1.35

[17] Data adapted from p. 522 of *Oceanography: An Introduction to the Planet Oceanus* by Paul R. Pinet, West Publishing Company (1992).

1. First graph the data. Assuming that the freezing point depends upon salinity, we choose salinity as the independent variable and graph it on the horizontal axis.

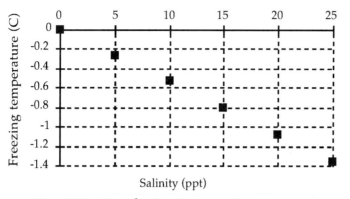

Fig. 4.36   Graph of salinity vs. freezing temperature

The points appear to lie on a straight line with a negative slope, so the relationship may be described as linear.

2. Is it true that a fixed amount of change in one variable corresponds to a fixed change in the value of the other variable? Yes. In the table we can see that each 5 ppt increase in salinity corresponds to a decrease of approximately –0.27 ° C in the freezing point. So we can consider the relationship linear.

3. Is the average rate of change constant? If we add a third column containing the average rate of change to the original table, we can see that the average rate of change remains fairly constant at about –0.054 °C/ppt. Hence we may consider the relationship as linear.

Table 4.21		
Salinity (ppt)	Freezing point (Celsius)	Average rate of change of freezing point with respect to salinity
0	0.000	n.a.
5	–0.266	–0.053
10	–0.534	–0.054
15	–0.804	–0.054
20	–1.076	–0.054
25	–1.350	–0.055

4. Can we find an m and b such that: $y = b + mx$ where $x =$ salinity (in grams of dissolved salts per 1000 gram of water) and $y =$ temperature in °C ? We can estimate a line by hand that would fit the data. Use any two points on the line, for example, (0, 0.000) and (20, –1.076), to find the equation of the line that passes through them. The slope, m is:

$$m = \text{change in temperature}/\text{change in salinity}$$
$$= (20 - 0)\ °C/(-1.076 - 0.000)\ \text{ppt}$$
$$= (20\ °C)/(-1.076\ \text{ppt})$$
$$= -0.054\ °C/\text{ppt}$$

Given our rate of change calculations in part 3, this is the value we expect. The equation of the line is then in the form:

$$y = b - 0.054\ x.$$

The line goes through the origin (0,0), so $b = 0$. (Hence our relationship represents a direct proportion.)

The equation of the line through the points (0, 0.000) and (20, –1.076) is:

$$y = -0.054\ x. \qquad (1)$$

Do the other points in the Table 4.21 satisfy this equation? Table 4.22 shows the given values for freezing point vs. the values predicted by equation (1).

	Table 4.22	
Salinity (ppt)	Actual Freezing point (°C)	Predicted Freezing point $y = -0.054\ x$
0	0.000	0.000
5	–0.266	–0.270
10	–0.534	–0.540
15	–0.804	–0.810
20	–1.076	–1.080
25	–1.350	–1.350

The results are very close, so equation (1) gives a good representation of the relationship between x and y. The relationship may be said to be linear.

5. Can we find an A, B, and C such that: $A x + B y = C$ is always true?

We can take the equation $y = -0.054x$. and write it as:

$$0.054x + y = 0.$$

This is now in the form $A x + B y = C$ with $A = 0.054$, $B = 1$ and $C = 0$.

Chapter 4 Summary

A *linear function* has the form
$$y = b + mx.$$
Its graph is a straight line where m is the *slope*, or *rate of change* of y with respect to x, and b is the *vertical intercept*, or value of y when x is 0.

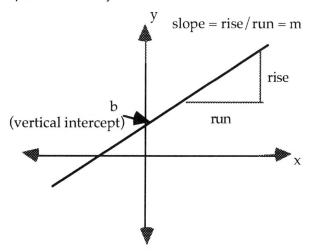

Linear functions are often written as y = mx + b. Using f to represent a linear function, we write: f(x) = mx + b.

Any two points that lie on a line can be used to calculate the slope. That is, if (x_1, y_1) and (x_2, y_2) are two distinct points that satisfy the function y = b + mx, then
$$m = \frac{rise}{run} = \frac{y_2 - y_1}{x_2 - x_1}$$
There are three general cases in which you can generate the equation of a line:
- knowing m and b
- knowing m and a point on the line
- knowing two points on the line

For a data set whose graph exhibits a linear pattern, we can eyeball a line to fit the data. The equation offers an approximate, but compact description of the data.

Lines are of special importance because they are easily drawn and manipulated. When appropriate, lines offer the simplest choice for a rough description of the patterns in the data.

Special cases of lines:
1. *Direction variation*

  y is said to *vary directly with* x or be *directly proportional to* x if is a constant m, such that:
$$y = mx$$
 This is a linear function in which the y-intercept is 0.

2. *Horizontal lines*

When the slope, m, is 0, the general linear equation becomes y = b, which is the equation of a horizontal line through the point (0, b)

3. *Vertical lines*

A vertical line is of the form x = c. Its slope is said to be undefined.

4. *Parallel lines*

Two lines are parallel if they have the same slope. For example, y = 2x + 5 and y = 2x - 7 are parallel since they both have a slope of 2.

5. *Perpendicular lines*

Two lines are perpendicular if their slopes are negative reciprocals. For example, the lines y = 2x + 5 and y = -1/2 x + 8 are perpendicular since -1/2 is the negative reciprocal of 2.

Two variables x and y are said to be linearly related if any *one* of the following equivalent conditions is shown to be true:

1. All of the pairs (x,y) lie on a straight line when plotted.
2. For a fixed amount of change in x there is a fixed amount of change in y.
3. The average rate of change of y with respect to x is constant.
4. There exist two constants, m (the slope) and b (the y-intercept) such that the equation y = b + mx is true for all of the data pairs (x,y).
5. There exist three constants, A, B, and C such that the equation Ax + By = C holds true for all pairs (x,y). (B is not equal to 0.)

EXERCISES

1. Given each set of conditions:
    a) find the equation of the line that satisfies the conditions.
    b) generate a small table of values that satisfy the equation.
    c) graph the equation.

    Conditions:
    i)   the line has a vertical intercept of -2 and a slope of 3
    ii) the line crosses the vertical axis at 3.0 and has a rate of change of -2.5
    iii) the line has a vertical intercept of 1.5 and a slope of 0

2. The equation $K = 4 F - 160$ gives the relationship between F, the temperature in degrees Fahrenheit, and K, the number of cricket chirps per minute for the snow tree cricket.

    a) Assuming F is the independent variable and K the dependent, identify the slope and vertical intercept in the above equation.
    b) Identify the units for K, 4, F and -160.
    c) Generate a small table of points that satisfy the equation. Be sure to choose realistic values for F.
    d) Calculate the slope directly from two data points. Is the value what you expected and why?
    e) Graph the equation with F on the horizontal axis and K on the vertical axis.
    f) What is a reasonable domain for this model? Indicate this domain on the graph.

3. The equation $F = 32 + (9/5) C$ gives the relationship between degrees Fahrenheit, F, and degrees Celsius, C.

    a) Assuming C is the independent variable, and F the dependent, what is the slope and the vertical intercept?
    b) Identify the units for F, 32, (9/5) and C
    c) Generate a small table of points that satisfy the equation.
    d) Calculate the slope directly from two data points. Is the value what you expected and why?
    e) Graph the equation with C on the horizontal axis and F on the vertical axis.
    f) At what temperature are the Celsius and Fahrenheit temperatures equal? Show your work.

4. Using the information in the previous 2 exercises, find an equation that describes a relationship between K, the number of cricket chirps per minute and C, the temperature in degrees Celsius.

5. The women's recommended weight formula from Harvard Community Health says: "Give yourself 100 lbs. for the first 5 ft plus 5 lbs. for every inch over 5 ft tall." Find a mathematical model for this relationship. Specify a reasonable domain for the function and then graph it.

6. For each of the following properties, construct a linear equation that satisfies those properties:

    a)   A negative slope and a positive y-intercept.

    b)   A positive slope and a vertical intercept of -10.3

    c)   A constant rate of change of $1300/yr.

(Try using some variety in your variable names.)

7. Estimate b (the y-intercept) and m (the slope) for each of the following graphs. Then for each graph write the corresponding linear function. Be sure to note the scales on the axes.

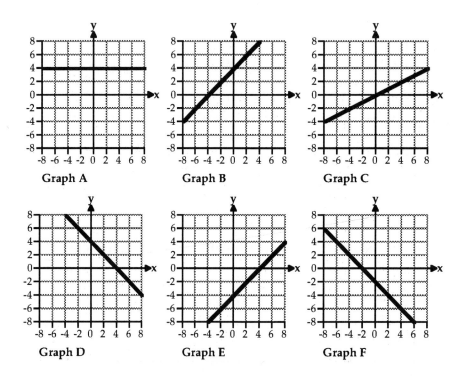

Graph A   Graph B   Graph C

Graph D   Graph E   Graph F

    A)   m =   b =   y =

    B)   m =   b =   y =

    C)   m =   b =   y =

    D)   m =   b =   y =

    E)   m =   b =   y =

    F)   m =   b =   y =

8.   (Requires course software. Instructions if needed are in the Appendix.)
     Open the program *Linear Graphs*.
     a)   Click on the button for "L2: Find constants: m & b" The program will randomly
          generate the graph of a line. Try predicting the values for m and b. Set the sliders
          to your predicted values and click on "Graph the line" to check your predictions.
          Keep generating new lines (with both integer and real coefficients) until you are
          confident in your ability to predict m and b.
     b)   Close L2 and open "L3: Two point graphs." Click on two points on the graph and
          then click on "Draw." Now calculate the values for m and b, and write down the
          equation for the line. Click on "m & b" and "Equation" to check your answers. Now
          click on the yellow box and repeat the process until you can find m and b and the
          equation accurately each time
     c)   Close L2 and open "L5: Find coordinates: (x,y)". The computer will have randomly
          generated the equation of a line that is listed at the bottom of the screen. You are
          asked to click on two points that lie on the line. Clicking on "Graph the line" will
          show you the graph of the original line and the graph and the equation of the line
          through your two points. Click on "Clear & Retry" and repeat until you are 100%
          accurate in your guesses.

9.   The following scatter plot shows the relationship between literacy rate (percentage of
     the population that can read and write) and infant mortality rate (infant deaths per 1000
     live births) for 114 countries. The raw data is in Part III, in the Nations Data Set. (You
     might wish to identify the outlier, the country with about a 20% literacy rate, and a low
     infant mortality rate of about 40 per 1000 live births.) Construct a linear model. Show all
     your work and clearly identify the variables, and units. Interpret your results.

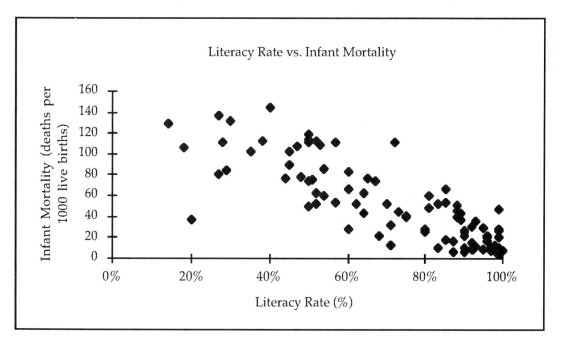

10. The following table and graph show data for the men's Olympic 16 pound shot put.

Olympic Shot Put	
year	feet thrown
1900	46
1904	49
1908	48
1912	50
1920	49
1924	49
1928	52
1932	53
1936	53
1948	56
1952	57
1956	60
1960	65
1964	67
1968	67
1972	70
1976	70
1980	70
1984	70
1988	74
1992	71

Men's Olympic 16 lb. Shot Put Results

Reprinted with permission from the Universal Press Syndicate. From *The 1996 Universal Almanac*. (The data points are missing for the early 1940's since the Olympics were not held during World War II).

a)   Could the graph be described adequately as representing a linear relationship? Draw in a line approximating the data and find its equation. Show your work and interpret your results. For ease of calculation you may wish to think of 1900 as year 0, and let your x coordinate measure the number of years since 1900. So the point (1900, 46) would become (0, 46); (1904, 48) would become (4, 48), etc.

b)   If the shot put results continued to change at the same rate, in what year would you predict that the winner will put the shot a distance of 80 feet? Does this seem like a realistic estimate? Why or why not?

11. Give specific examples of two linear equations that:
   a)   are parallel to each other
   b)   intersect at the same point on the y-axis
   c)   both go through the origin
   d)   are perpendicular to each other

12. Find the value of t if m is the slope of the line that passes through the given points.
   a) (3, t) and (-2, 1) $m = -4$
   b) (5, 6) and (t, 9) $m = 2/3$

13. For each of the following graphs compare the **m** values and then the **b** values. (You don't need to do any calculations or determine the actual equation.) Do the lines have the same slope? Are the slopes both positive, both negative, or is one negative and one positive? Do they have the same y-intercept?

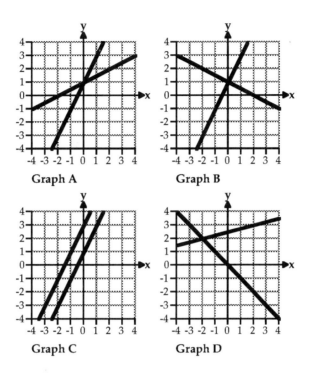

14. Find the equation of the line for each of the following set of conditions:
   a) slope = –4 and y–intercept =5.4
   b) slope = 2 and line passes through the point (5, –4)
   c) line passes through the points (3,– 4) and (6, 11)
   d) line passes through the points(1, –1) and (0, –5)
   e) line is parallel to 2 y – 7x = y +4 and passes through the point (–1, 2)

15. Find the equation of the line in the form y = mx + b. for each of the following sets of conditions. Show your work.
   a) slope is –1.62 and y–intercept is 4.20
   b) slope is $1,400/yr and line passes through the point(10 yr, $12,000)
   c) line passes through (1940, 1.3%) and (1990, 11.3%)
   d) line passes through (1971, $460) and (1979, $410)
   e) line is 1.48 x – 2.00y + 4.36 = 0
   f) line is parallel to a line y = 0.44 + 0.83x and passes through the point (0,3.89)
   g) line is horizontal and passes through (1.0 , 7.2)
   h) line is vertical and passes through (275, 1029)
   i) line is perpendicular to y = –2x + 7 and passes through (5, 2)

16 . Rewrite each of these equations in the form y = mx + b. In each case identify the slope and y-intercept and then do a quick sketch:
  a)   4 = y – 6x
  b)   2x + 3y = – 6
  c)   0.2 y – 0.04 x = 1.3
  d)   (x/2) +(y/3) = 1

17. Find the equation of the line shown on the graph below. Use this equation to create two new graphs taking care to label the scales on your new axes. For one of your graphs, choose scales that make the line appear steeper than the original graph. For your second graph, choose scales that make the line appear less steep than the original graph.

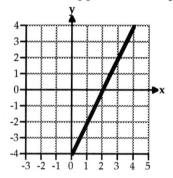

18. Match the following graphs with the equations below. Be sure to observe the scales on the axes.
  a)   y = x – 4   graph:____
  b)   –x + 4 = y   graph:____
  c)   4 = y   graph:____
  d)   y = –x – 2   graph:____
  e)   y = x + 4   graph:____
  f)   y = (1/2)x   graph:____

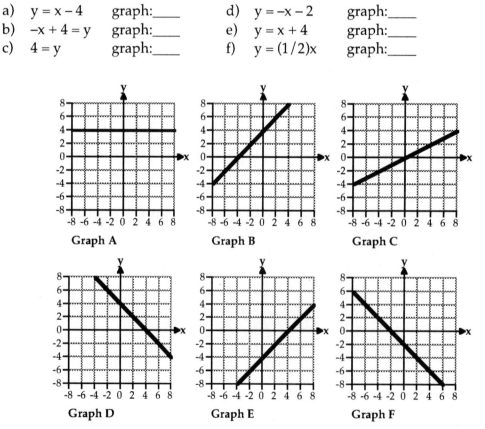

19. The y-axis, the x-axis, the line x = 6, and the line y = 12 determine the four sides of a 6 by 12 rectangle in the first quadrant of the xy plane. Imagine that this rectangle is a pool table. There are pockets at the four corners and at the points (0,6) and (6,6) in the middle of each of the longer sides. When a ball bounces off one of the sides of the table, it obeys the "pool rule," the slope of the path after the bounce is the negative of the slope before the bounce.

   a) Your pool ball is at (3,8). You hit it towards the y-axis, with slope 2.
      i) Where does it hit the y-axis?
      ii) Where does it hit next? And after that? And after that?
      iii) Does it ultimately return to (3,8)? Would it do this if the slope had been different from 2? What is special about the slope 2 for this table?
   b) A ball at (3,8) is hit toward the y-axis, and bounces off it at (0,16/3). Does it end up in one of the pockets?
   c) Your pool ball is at (2, 9). You want to shoot it into the pocket at (6, 0). Unfortunately, there is another ball at (4, 4.5) which may be in the way.
      i) Can you shoot directly into the pocket at (6,0)?
      ii) You want to get around the other ball by bouncing yours off the y-axis. If you hit the y-axis at (0,7), do you end up in the pocket? Where do you hit the line x=6?
      iii) Bouncing off the y-axis at (0,7) didn't work. However surely there is some point (0,b) on the y-axis from which the ball would bounce into the pocket at (6,0). Try to find the point.
      iv) Where, exactly, should you bounce off the y-axis to land in the pocket at (6,0)?

20. Suppose that

   for 8 years of education, the mean salary for women is approximately $6,800,
   for 12 years of education, the mean salary for women is approximately $11,600, and
   for 16 years of education, the mean salary for women is approximately $16,400.

   a) Plot this information.
   b) What sort of relationship does this information suggest between mean salary for women and education? Justify your answer.
   c) Generate an equation that could be used to model the data from the limited information given (letting E = years of education and W = personal wages.) Show your work.

21. The percentage of dentistry degrees awarded to women in the U. S. between 1970 and 1990 is shown in the data and graph below.

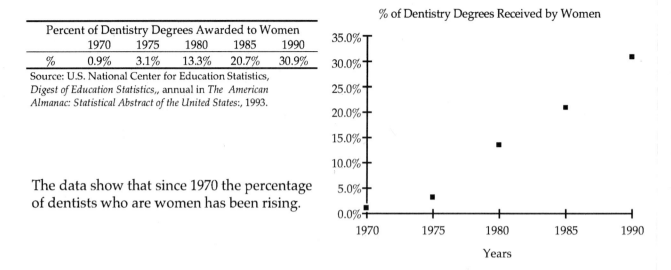

Percent of Dentistry Degrees Awarded to Women				
1970	1975	1980	1985	1990
0.9%	3.1%	13.3%	20.7%	30.9%

Source: U.S. National Center for Education Statistics, *Digest of Education Statistics,*, annual in *The American Almanac: Statistical Abstract of the United States:,* 1993.

% of Dentistry Degrees Received by Women

The data show that since 1970 the percentage of dentists who are women has been rising.

a) Rough in a line to represent the data points. What is the rate of change of percent of dentistry degrees awarded to women according to your line?

b) According to your estimate of the rate of increase, when will 100% of dentistry degrees be awarded to women?

c) Since it seems extremely unlikely that 100% of dentistry degrees will ever be granted to women, comment on what is likely to happen to the rate of growth of women's degrees in dentistry; sketch a likely graph for the continuation of the data into the next century.

d) What other data might be helpful in making an estimate for the future of women in dentistry?

EXPLORATION 4.1.a
Looking at Lines

Objective
- to find patterns in the graphs of linear equations in the form **y = mx + b**

Materials/Equipment
- computer with course software "L1. Slope/y-axis intercept graphs" and "L5. Find constants: m & b," both in *Linear Graphs.*

Procedure

Working in pairs try the following explorations and compare your findings with each other. After comparing your findings write down your observations.

Open the program *Linear Graphs* and click on the button "L1: Slope/y-axis intercept graphs." This program will allow you to construct your own lines and to explore the effects of **m** and **b** on the graph of the equation. Instructions for using this program are enclosed in the Appendix.

1. What is the effect of "m" on the graph of the equation?

    Fix a value for **b.** Construct four graphs with the same value for **b** but with different values for m. Continue to explore the effects of **m** on the graph of the equation. Jot down your observations about the effect on the line when m is positive? negative? equal to zero? When |m| (the absolute value or distance of m from zero) increases? Do you think your conclusions work for values of **m** that are not on the slider?

    Choose a new value for b and repeat your experiment, jotting down notes. Are your observations still valid? Compare your observations with your partner and describe in your notebook a 60 second summary the effect of **m** on the graph of the equation.

2. What is the effect of "b" on the graph of the equation?

    Fix a value for **m.** Construct four graphs with the same value for **m** but with different values for **b.** What is the effect on the graph of changing **b?** Jot down your observations. Would your conclusions still hold for values of **b** that are not on the slider?

    Choose a new value for **m** and repeat your experiment,. again keeping jottings. Are your observations still valid? Compare your observations with your partner. Construct a 60 second summary describing the effect of **b** on the graph of the equation.

3. Constructing lines under certain constraints

   Construct the following set of lines using "Slope/y-axis intercept graphs". Write down the equations for the lines you construct. What generalizations can you make about the lines in each case? Are the slopes or the y–intercepts of the lines related in some way?

   Construct three **parallel lines.** Record their equations in your notebook.

   Construct three lines with the **same y–intercept,** the point where the line crosses the y-axis.

   Construct a pair of lines that are **horizontal.**

   Construct a pair of lines that go **through the origin.**

   Construct a pair of lines that are **perpendicular** to each other.

   In your notebook, construct a 60 second summary of what you have learned about the equations of lines.

4. Predicting the equation of a given line.

   Close the "L1. Slope/y-axis intercept graphs" program by clicking on the square exit button in the upper–left corner of the screen. Open the program "L2. Find constants: m & b." This program randomly generates a line which you will see on your screen. Your task is to find the equation that represents that line. Predict values for **m** and **b** and use your sliders to check your predictions.

EXPLORATION 4.1.b
Looking at Lines with a Graphing Calculator

Objective
- to find patterns in the graphs of linear equations in the form $y = mx + b$

Materials/Getting started
- Set your calculator to the integer window setting. For the TI-82 or TI-83 do the following:

1. Press $\boxed{\text{ZOOM}}$, select [6:ZStandard].

2. Press $\boxed{\text{ZOOM}}$, select [8:ZInteger] $\boxed{\text{ENTER}}$.

3. Press $\boxed{\text{WINDOW}}$ to see if the settings are
    the same as Figure 1.

```
WINDOW FORMAT
 Xmin=-47
 Xmax=47
 Xscl=10
 Ymin=-31
 Ymax=31
 Yscl=10
```

Figure 1

Procedure:
- **Working in pairs try the following explorations and compare your findings with each other..**

1. Discover the effect of "m" on the graph of an equation, $y = mx$.
    a) Enter the following into your calculator and sketch the graphs to the right. To get you
        started let m=1,2,3   Try a few few other values of m such that m > 0

Y1 = x
Y2 = 2x
Y3 = 3x
Y4 =
Y5 =
Y6 =

    b) Describe the effect of multiply x by a number "m" for $y = mx$

2. Compare the effects on $y = mx$ when m < 0.
    a) Begin by comparing the lines when m = 1 and m = -1, then you pick some other negative
        values for m and graph the lines

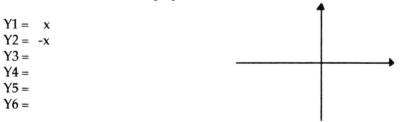

Y1 = x
Y2 = -x
Y3 =
Y4 =
Y5 =
Y6 =

    b) Alter your description in 1b above to describe the effect of multiplying x by <u>any real</u>
        <u>number</u> "m", for $y = mx$? (Remember to also explore what happens when m = 0.)

3.   Discover the effect of "b" on the graph of an equation, y = mx + b
   a)   Enter the following into your calculator and sketch the graphs to the right. To get you started let m=1 and b = 0, 20, -20 Try a few other values of b

   Y1 = x
   Y2 = x + 20
   Y3 = x - 20
   Y4 =
   Y5 =
   Y6 =

   b)   What is the effect of adding any number "b" to x for y = x + b ?
        (Hint: use trace and tell where the graph crosses the y - axis).

   c)   Choose another value for m and repeat your experiment. Are your observations still valid?

4.   Construct a 60 second summary describing the effects of both m and b on the linear equation y = mx + b.

5.   Constructing lines under certain constraints.
   a)   Construct three parallel lines and record their equations below.

   b)   Construct three lines with the same y-intercept and record their equations below.

   c)   Construct three lines that go through the origin and record their equations below.

   d)   Construct three perpendicular lines and record their equations below.
   (Hint: perpendicular lines are lines that intersect at a right angle.)

6.   On graph paper sketch the points   (2,1) and (4,5), then draw a line through the points so that the line intersects the x and y axis.
   a)   Determine the rate of change (slope of the line).
   b)   Estimate the y-intercept.
   c)   Find the equation of the line y = mx +b that goes through the points (2,1) and (4,5). What are the values of m and b? Check to see if the points satisfy your equation, if not make the necessary adjustments to your eqution.

Chapter 5

Looking For Links

Overview

Robert Reich, U.S. Secretary of Labor for the Clinton administration, believes that "Learning is the key to earning." Is this true? Is there a relationship between education and income? In this chapter, we examine how a social scientist might try to answer this question.

A case study approach is used to explore possible relationships between education and income. Using a large data set from the U.S. Census, we start by examining how certain relationships can be summarized by fitting lines to data. The resulting linear models are called *regression lines*. The next steps often involve trying to answer questions raised by the analysis. How good is the model? What other variables should be taken into consideration? Why are the variables correlated?

"Fitting lines to relationships is the major tool of data analysis." (Tufte, *Data Analysis of Politics and Policy*.) Yet we need to be aware of how to interpret what regression lines mean. A classic case of radios and insanity illustrates some of the pitfalls.

The explorations provide a hands-on opportunity to explore a large data set extracted from U.S. Census data and to conduct your own case study of education and income in the U.S.

After reading this chapter you should be able to:

- use U.S. Census data from Current Population Survey
- discuss the relationship between education and income
- use regression lines to summarize data
- understand the distinction between correlation and causation

5.1 A Case Study Using U.S. Census Data

"Does more education mean more income?" The answer may seem intuitively obvious. We could reasonably expect that having more education gives access to higher paying jobs. Is this indeed the case? Using a large set of data on education and income, we explore how a social scientist might start to answer these questions.

We use a random sample from U.S. census data. Our data set, called FAM 1000, provides information on 1000 individuals and their families. A disk accompanying the text contains: the *FAM 1000 Census Data Kit* which provides easy to use interactive software for analyzing the FAM 1000 data; FAM 1000 data stored as an Excel file; and graph link files of condensed versions of the data for use with the graphing calculator.

The Bureau of the Census, as mandated by the Constitution, conducts a nationwide census every 10 years.[1] In order to collect more frequent and up-to-date information, the Census Bureau also conducts a monthly survey of American households called the Current Population Survey or CPS.[2] The CPS is the largest survey taken between Census years. The CPS interviews members of approximately 56,000 households each month in an attempt to generate as complete and accurate a profile of the American economy as possible. Only individuals 16 or older are included in the survey. Questions are asked about race, education, housing, number of people in the house, income, and employment status.[3] The March survey is the most extensive. Our census sample data set (FAM 1000) has been extracted from the March 1992 Current Population Survey.

The following pages contain information from 57 of the 1000 individuals in our data set, along with a *data dictionary* with short definitions for each data category. Think of the data as a large rectangular array of rows and columns of facts. Each row represents all the information obtained from one particular respondent about his or her family. Each column contains the coded answers of all the respondents to one particular question. Try deciphering the information below that comes from the first row of our data array.

region	cncity	famsz	faminc	povline	marst	sex	race	eorgn	age	occup	educ	wkswrk	hrswrk	pwages	ptotinc	yrfi
1	1	3	$68,806	$10,654	1	0	1	8	61	10	8	52	48	$33,000	$33,106	1

---

[1] Extensive data on individuals is collected by institutions other than the federal government. Read the article on "Who Collects Data and Why" for an overview of the types of data collected by other institutions and the purposes to which the information is put.

[2] The article entitled "U.S. Government Data Collection" by Anthony Roman, Center for Survey Research, gives a closer look at federal data collection, and in particular the Current Population Survey.

[3] The results of the survey are used to estimate numerous economic and demographic variables like the size of the labor force, the employment rate, and income and education levels. The results are widely quoted in the popular press and are published monthly in: *The Monthly Labor Review* and *Employment and Earnings*; irregularly in the *Current Population Reports* and *Special Labor Force Reports*, and yearly in the *Statistical Abstract of the United States* and *The Economic Report of the President*.

Referring to the data dictionary, we learn that this family lives in the Northeast, outside a city, contains three people, and had an annual family income of $68,806 in 1991. The regional poverty line for a family this size is $10,654. The respondent is not married, is male, white, non-Hispanic and 61 years old. He works in a precision production craft or repair operation, has an 8th grade education, worked 52 weeks in 1991, usually 48 hours a week, earned $33,000 in personal wages and salary, had a personal total income of $33,106, and considered himself a year-round full-time worker.

Use the data dictionary in Table 5.1 and the definitions given in the readings to help you fill out a row about yourself.

Something to think about

How good are the data? Return to the questions raised about data in Chapter 1. What questions would you raise about the data collected on education and income by the U.S. Bureau of Census?

Table 5.1
Data Dictionary For March 1992 Current Population Survey [4]

Variable	Definition	Unit Of Measurement (Code And Range)
REGION	Census Region	1 = Northeast 2 = Midwest 3 = South 4 = West
CENCITY	Residence Location	0 = Central City 1 = Non-Central City (suburb or rural)
FAMSIZE	Family Size	Range is 1 to 39
FAMINC	Family Income	Range is $0 to $24,000,000
POVLINE	Poverty Line	Range is $0 to $35,000 per household
MARSTAT	Marital Status	0 = presently married 1 = presently not married
SEX	Sex	0 = male 1 = female
RACE	Race of Respondent	1 = white 2 = black 3 = American Indian, Aleut Eskimo 4 = Asian or Pacific Islander 5 = other
REORGN	Ethnic Origin of Respondent (Hispanic or other)	1 = Mexican American 2 = Chicano 3 = Mexican (Mexicano) 4 = Puerto Rican 5 = Cuban 6 = Central or South American 7 = Other Spanish 8 = All other 9 = Don't know 10= Not available
AGE	Age	Range is 16 to 90
OCCUP	Occupation Group of Respondent	1 = Executive, administrative, managerial 2 = Professional specialty

[4] Adapted from the U.S. Census Bureau, *Current Population Survey*, March 1992 by Gourish Hosangady, Computing Services, University of Massachusetts/Boston.

Variable	Definition	Unit Of Measurement (Code And Range)
OCCUP (cont)	Occupation Group of Respondent	3 = Technicians and related support 4 = Sales 5 = Administrative support, including clerical 6 = Private household service 7 = Protective service 8 = Service occupations, except protective & household 9 = Farming, forestry, and fishing 10= Precision production, craft and repair 11= Machine operators, assemblers and inspectors 12= Transportation and materials moving equipment 13= Handlers, equipment cleaners, helpers & laborers 14= Armed Forces, currently civilian
EDUC	Years of education	1 = Less than first grade 4 = First, second, third or fourth grade 6 = Fifth or sixth grade 8 = Seventh or eighth grade 9 = Ninth grade 10= Tenth grade 11= Eleventh grade 12= Twelfth grade HG= High school graduate (diploma or equivalent) SC= Some college, but no degree AO= Associate degree occupation/vocation program AP= Associate degree in academic program BD= Bachelor's degree MD= Master's degree (M.A., M.S., M.Eng., M.Ed., M.S.W., M.B.A.) PD= Professional school degree (M.D., D.D.S., D.V.M., L.L.B., J.D.) PH= Doctoral degree (Ph.D., Ed.D.)

Note: In the FAM 1000 data, substitutions of numbers for symbols were made in order to plot points. For example, we replaced HG (high school graduate),with 12, representing twelve years of school. SC (some college but no degree), we estimate as 14, or two years more than high school. Similar estimates were made for the higher education categories.

WKSWRK	Weeks worked in 1991 (even for a few hours)	Range is 0 to 52
HRSWRK	Usual hours.worked	Range is 0 to 99 per week in 1991
PWAGES	Total personal wages	Range is $0 to $200,000
PTOTINC	Personal total income	Range is $0 to $600,000
YRFT	Employed year-round, full-time	0 = not full-time all year-round 1 = full-time all year-round

Table 5.2
Information From 57 Respondents In The Fam 1000 Data Set[5]

region	cncity	famsz	faminc	povline	marst	sex	race	eorgn	age	occup	educ	wkswrk	hrswrk	pwages	ptotinc	yrft
1	1	3	$68,806	$10,654	1	0	1	8	61	10	8	52	48	$33,000	$33,106	1
1	1	2	$49,291	$9,120	1	0	1	8	54	11	HG	52	40	$26,286	$26,488	1
1	1	2	$83,509	$9,120	1	0	1	8	39	4	SC	52	45	$46,222	$54,375	1
1	1	2	$45,000	$9,120	1	0	1	8	43	10	HG	52	40	$45,000	$45,000	1
1	1	4	$60,962	$13,812	1	0	5	7	45	11	10	52	61	$22,203	$22,298	1
1	0	1	$5,213	$7,086	0	1	1	8	36	13	6	24	20	$500	$5,213	0
1	1	2	$33,986	$9,120	0	1	1	8	51	5	SC	52	40	$26,233	$27,533	1
1	1	4	$29,800	$14,278	1	0	1	8	43	12	SC	52	40	$19,000	$23,200	1
1	1	4	$55,272	$14,048	1	0	1	8	57	5	SC	52	48	$42,916	$54,644	1
1	0	4	$47,481	$14,048	0	1	1	8	19	5	SC	14	35	$4,300	$6,381	0
1	1	2	$26,866	$9,120	1	0	1	8	42	12	HG	46	40	$15,400	$17,295	0
1	0	5	$35,555	$16,662	0	0	1	8	16	13	11	16	4	$1,400	$1,400	0
1	0	5	$50,025	$17,188	0	0	1	8	18	12	12	44	26	$10,000	$10,533	0
1	0	2	$44,400	$9,120	1	1	1	8	38	3	AO	52	32	$15,000	$16,200	0
1	0	5	$37,108	$16,254	1	1	1	8	37	5	HG	52	40	$14,438	$14,875	1
1	1	3	$39,000	$10,963	1	0	1	8	29	8	HG	52	48	$32,000	$32,000	1
1	1	3	$20,212	$10,963	0	1	1	8	61	8	8	38	35	$6,600	$7,212	0
1	0	3	$20,212	$10,963	0	1	1	8	31	1	HG	48	40	$13,000	$13,000	0
1	1	3	$25,837	$10,963	0	0	1	8	43	1	BD	52	40	$22,900	$22,937	1
1	1	4	$19,080	$13,812	1	0	1	8	38	8	HG	26	40	$8,060	$10,580	0
1	0	4	$27,965	$13,812	1	1	1	8	28	1	12	52	36	$19,354	$19,499	1
1	1	1	$17,963	$7,086	0	1	1	8	40	11	HG	26	40	$13,257	$17,963	0
1	0	2	$16,842	$9,120	1	1	1	8	35	11	MD	52	40	$12,792	$12,792	1
1	1	2	$45,958	$9,120	0	1	1	8	49	5	SC	52	35	$29,544	$39,261	1
1	1	3	$38,014	$10,963	1	1	1	8	24	8	AO	52	24	$8,000	$8,007	0
1	0	3	$22,994	$10,963	1	1	1	8	32	5	HG	24	21	$4,032	$4,032	0
1	0	1	$8,500	$7,086	0	0	1	8	20	8	HG	46	40	$8,500	$8,500	0
1	0	5	$55,630	$16,662	1	1	1	8	41	1	AO	26	30	$8,500	$11,565	0
1	0	3	$41,795	$10,654	1	1	1	8	50	5	HG	52	40	$16,800	$16,800	1
1	1	2	$26,843	$8,233	1	0	1	8	68	4	BD	52	5	$9,300	$21,163	0
1	0	1	$3,860	$7,086	0	1	1	8	21	8	AP	26	15	$3,500	$3,860	0
1	0	6	$76,510	$19,160	0	1	1	8	22	3	AP	52	40	$30,500	$30,500	1
1	1	3	$84,925	$10,963	1	0	1	8	29	2	PH	52	45	$55,000	$55,175	1
1	0	2	$42,000	$8,233	1	1	1	8	61	5	BD	50	26	$9,800	$12,300	0
1	1	3	$30,352	$10,963	1	0	1	8	36	10	HG	52	40	$16,000	$16,175	1
1	0	3	$25,001	$10,963	0	0	1	8	17	13	10	45	20	$4,510	$4,510	0
1	1	4	$11,425	$13,860	0	1	1	8	35	2	HG	40	30	$6,997	$11,425	0
1	1	5	$20,000	$16,254	1	0	1	8	41	9	HG	52	99	$20,000	$20,000	1
1	1	1	$43,577	$7,086	0	1	1	8	56	2	MD	52	40	$43,000	$43,577	1
1	1	1	$9,500	$7,086	0	0	1	8	37	8	SC	52	40	$9,500	$9,500	1
1	1	1	$29,745	$7,086	0	1	1	8	50	1	MD	52	45	$25,000	$29,745	1
1	0	3	$53,160	$10,654	1	0	1	8	42	12	10	52	60	$26,870	$26,870	1
1	0	4	$40,100	$13,812	1	0	1	8	27	13	HG	52	40	$35,000	$35,050	1
1	0	4	$78,685	$14,048	0	1	1	8	22	2	BD	12	21	$480	$2,020	0
1	0	6	$146,375	$19,563	0	0	1	8	20	8	HG	52	30	$5,800	$5,810	0
1	0	3	$43,030	$10,963	1	1	1	8	30	8	SC	10	20	$1,500	$1,765	0
1	1	2	$43,300	$9,388	0	1	1	8	38	8	SC	52	40	$33,000	$43,300	1
1	0	5	$103,200	$16,941	1	0	1	8	59	2	SC	52	40	$38,000	$38,100	1
1	1	4	$79,200	$13,812	1	0	1	8	29	3	AO	51	40	$38,000	$38,100	1
1	0	2	$73,300	$9,120	1	1	1	8	30	3	BD	52	40	$31,000	$31,100	1
1	0	5	$91,388	$16,662	1	1	1	8	49	3	AO	52	26	$27,000	$27,025	0
1	0	2	$27,500	$9,120	1	1	1	8	38	5	MD	52	40	$19,000	$20,250	1
1	1	2	$29,381	$9,120	0	1	1	8	58	5	HG	52	40	$19,000	$21,381	1
1	0	5	$66,669	$16,941	1	1	1	8	48	5	AO	52	40	$16,640	$17,655	1
1	0	5	$66,669	$16,941	0	0	1	8	21	13	HG	49	40	$7,000	$7,030	0
1	0	1	$2,904	$7,086	0	1	1	8	29	6	BD	51	40	$2,900	$2,904	0
1	1	1	$33,084	$7,086	0	0	1	8	29	8	HG	52	40	$33,000	$33,084	1

[5] From the U.S. Census Bureau, *Current Population Survey*, March 1992.

5.2   Summarizing the Data: Regression Lines

What is the relationship between education and income?

In the physical sciences the relationship of variables is often quite direct; if you hang a weight on a spring it is clear, without knowing the exact relationship, that the amount the spring stretches is definitely dependent on the heaviness of the weight. Further, it is reasonably clear that the weight is the *only* important variable: the temperature, the phase of the moon, etc., can safely be neglected.

In the social and life sciences it is usually difficult to tell whether one variable truly depends on another. For example, it is certainly plausible that a person's income depends in part on how much formal education he or she has had, since we may suspect that having more education gives access to higher paying jobs, but many other factors certainly also play a role. Some of these factors, such as the person's age or type of work, are measured in the FAM 1000 data set; others, such as innate ability or good luck, may not have been measured or even be measurable. Despite this complexity, we attempt to determine as much as we can by looking at the relationship between income and education alone.

We begin to explore this question with a *scatter plot* of education and personal wages from the FAM 1000 set. If we hypothesize that income depends on education, then the convention is to graph education on the horizontal axis. Each ordered pair of data values gives a point with the coordinates (education, personal wages).[6]

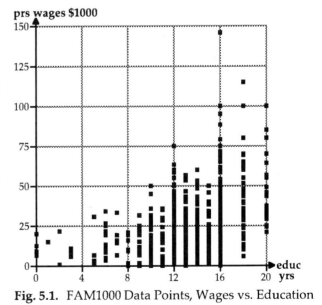

Take a moment to examine the scatter plot in Figure 5.1. There are 1000 different points, many of which are on top of other points or overlapping nearby points. Each point has two coordinates, one giving the years of education and the other giving the personal wages for a particular respondent. The other information which we have about the person is not shown. For example, the point near the top represents an individual with 16 years of education who makes $146,000 in personal wages.

Fig. 5.1. FAM1000 Data Points, Wages vs. Education

---

[6] You can form your own scatter plots using the FAM 1000 software or the Excel data file (if you are using a computer.) or.the graph link file (if using graphing calculators)

By referring back to the original data set, we may also determine that this person is a 55-year old unmarried white male part-time professional, who holds a bachelor's degree, who has 3 people in his family, and who lives in the Midwest outside a city. Some or all of this additional information may well be relevant, but it is clearly too much to display on one graph.

There appears to be an overall trend up and to the right. Everyone with less than 10 years of school earns less than $50,000, while many of those with more than 10 years of school earn more than $50,000. This suggests that there is some relationship between education and income.

How might we think of the relationship between these two variables? Clearly, personal income is not a function of education, since people who have the same amount of education earn widely different amounts. (The scatter plot obviously fails the vertical line test.)

However, we may ask the following question: Suppose that, in order to condense out a simple description of this data, we were to insist on finding a simple functional description. In fact, suppose we insisted that this simple relationship be a linear function. In Chapter 4, we fit linear functions to data informally. A formal mathematical procedure called *regression analysis* lets us determine what linear function is the "best" approximation to the data; the resulting "best-fit" line is called a *regression line*, and is similar in spirit to reporting only the mean of a set of data, rather than the entire data set. "It is often a useful and powerful method of summarizing a set of data."[7]

Intuitively, you could think of the plotted points as being small magnets attached to the paper. If you were to lay down a thin rigid iron rod in their midst, it would be pulled by all the magnets, and would stabilize in a position which was aligned as much as possible with the distribution of the magnets. Technically, for any candidate linear function, we may determine the error in how this function represents the data set by adding up the squares of the vertical distances between the line and the data points; then the regression line is the line which makes this error as small as possible. The calculations necessary to compute this line are tedious, though not difficult, and are easily carried out by computer software and graphing calculators.[8]

In Figure 5.2, we show the FAM 1000 data set, along with a regression line determined from the data points. The equation of the line is

$$\text{prs wages}_{\text{(all data)}} = -\$7{,}000 + \$2{,}190 \bullet \text{educ yrs}$$

---

[7] Tufte, Edward. *Data Analysis of Politics and Policy.* (Prentice-Hall, Englewood Cliffs, N.J.: 1974): p.65.
[8] *FAM 1000 Census Data Kit* can be used to find find regression lines. The various techniques for finding regression lines are beyond the scope of this course. If you are interested in a standard technique for generating regression lines, a summary of the method is provided in the reading *Linear Regression Summary*.

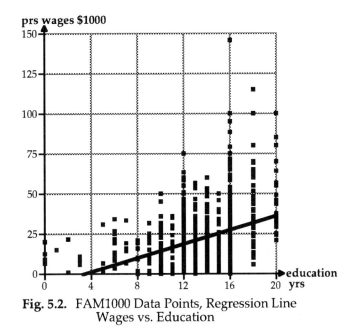

Fig. 5.2. FAM1000 Data Points, Regression Line
Wages vs. Education

This is certainly more concise that the original set of 1000 data points. Looking at the graph, you may judge with your eyes to what extent the line is a good description of the original data set. Note from the equation that the vertical intercept, – $7000, of this line is *negative*; so according to this regression line, people with three or less years of education earn *negative* income, even though all incomes in the original data set are positive. The linear model is clearly inaccurate near the left edge of the graph, so we should exclude these points from the domain of our model. $2190 represents the slope of the regression line, or the average rate of change of personal wages with respect to years of education. Thus, this model predicts that for each additional year of education, individual personal wages increases by $2190.

In the middle section of the graph, the model is perhaps not too bad. The data points are widely scattered about the line, for reasons which are clearly not captured by the linear model. Some of this scatter is simply due to randomness, and the fact that each data point is a different individual; but some of it may be due to other variables such as age which we could include by a more sophisticated analysis.

We emphasize that, although we can construct an approximate linear model for any data set, this does not mean that we really believe that the data is truly a linear relationship. In the same way, we may report the mean of a set of data, without believing that the data values are all the same number. In both cases, there are features of the original data set which we do not report, which may or may not be important.

Another way to look for structure in this data set, is to eliminate clutter by grouping together all people with the same years of education, and to plot the *mean* personal wage of each group. This sort of graph (Figure 5.3) is sometimes called the *graph of averages*.

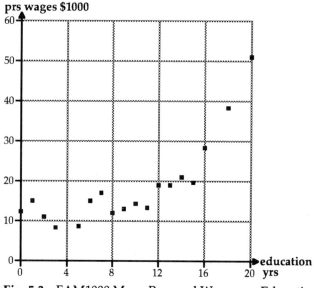

Fig. 5.3. FAM1000 Mean Personal Wages vs. Education

This plot is less cluttered. For each year of education, only the single mean income point has been graphed. For instance, the point corresponding to 12 years of education has vertical value of approximately $20,000; this means that the mean of the personal wages of everyone in the FAM 1000 data with 12 years of education, is about $20,000. The pattern is clearer: the upward trend to the right is a little more obvious.

Every time we construct a simplified representation of an original data set, we should ask ourselves what information has been suppressed. For the graph of averages, we have suppressed the spread of data in the vertical direction. We also do not see the different number of data points at each horizontal location; for example, there are only two people with two years of education, but several hundred with 16 years, and each of these sets is represented by a single point in the graph of averages.

In the graph of averages, we can see some additional features of the data which were not evident in either the full scatter plot, or in the equation of the straight-line model. For example, there are small jumps upwards at 12 years and at 16 years, representing high school and college graduation respectively. The graph is nearly flat below 12 years, and between 12 and 16 years; perhaps if you do not graduate high school or college it does not matter very much how many years you went to school. Above 16 years it slopes upwards fairly steeply, perhaps representing the payoff from graduate education. All these observations are completely suppressed by the simple linear model.

We may also compute a linear fit line to the graph of averages, again recognizing that it is not really a straight-line function. The equation of this straight line is

$$\text{prs wages}_{\text{(mean data)}} = \$5{,}000 + \$1{,}450 \bullet \text{educ yrs}$$

$5000 is the vertical intercept and $1450 is the slope or rate of change of personal wages with respect to education. Hence this model predicts that for each additional year of education the *mean* personal wage increases by $1450. Note that this linear model predicts *mean* personal wages for the *group,* not personal wages for an individual.

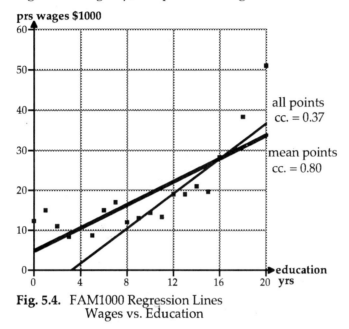

Fig. 5.4. FAM1000 Regression Lines
Wages vs. Education

Figure 5.4 shows the average data, together with the straight line representing the fit to the averages. We have also superimposed the original linear fit (the thinner line) to the full data set. Both of these straight lines are reasonable answers to the question "What straight line best describes the relationship between education and income?" and the difference between them indicates the uncertainty in answering such a question. We may argue that the benefit in income of each year of education is either $1,450, or $2,190, or anything in between.

Regression Line: How good a fit?

Once we have determined a line that approximates our data, we must ask "How good a fit is our regression line?" To help answer this, statisticians calculate a quantity called the *correlation coefficient.* This number is computed by most regression software and graphing calculators, and we have included it on our graphs, labeled "c.c." [9] The correlation coefficient is always between -1 and 1; the closer its absolute value is to 1, the better the fit and the stronger the linear association between the variables.

---

[9] We use the label *cc* for the correlation coefficient in the text and software in order to minimize confusion. In a statistics course the correlation coefficient is referred to as *Pearson's r* or just *r.*

A small correlation coefficient (close to zero) indicates that the variables do not depend linearly on each other. This may be because there is no relationship between them, or because there is a relationship which is something more complicated than linear. In future chapters we shall discuss many other possible functional relationships.

There is no absolute answer to when a correlation coefficient is "good enough" to say that the linear regression line is a good fit to the data. [10] A fit to the graph of averages always gives a higher correlation coefficient than a fit to the original data set because the scatter has been smoothed out. When in doubt, plot all the data along with the linear model, and use your best judgment. The correlation coefficient is only a tool which may help you decide among competing models or interpretations.

5.3   Correlation vs. Causation

One is tempted to conclude that increased education *causes* increased income. This may be true, but the model we used does not offer conclusive proof. This model can show how strong or weak a relationship exists between variables but we are not answering the question: "Why are the variables related?" We need to be cautious in how we interpret our findings.

Regression lines can be used to show *correlation* not *causation*. We say that two events are correlated when there is a statistical link. If we find a regression line with a correlation coefficient that is close to 1 in absolute value, a strong relationship is suggested. In our previous example, education is positively correlated with personal wages. If education increases, personal wages increase. Yet this does not prove that education causes an increase in personal wages. The reverse might be true, that is, an increase in personal wages might cause an increase in education. There may be a causal relationship, but there may also be other reasons for the association between these two variables.

Education and personal wages may be jointly caused by some other factor. Perhaps both educational opportunities and income levels are strongly affected by parental education or a history of family wealth. Thus a third variable, such as parental socio-economic status, may better account for <u>both</u> more education and higher income. We call this type of variable a *hidden variable* which may be affecting the results.

---

[10] The reading "Correlation Coefficients" will give you an intuitive sense of how the size of the correlation coefficient helps answer the question "How good a fit to the data is the regression line?"

There may also be a very strong correlation that occurs by chance between two variables that have no causal relationship at all. For example, the following graph shows a clear correlation between the number of radios and the proportion of insane people in England between 1924 and 1937.

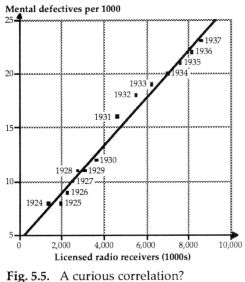

Fig. 5.5. A curious correlation?

Tufte, E.A *Data Analysis for Politics and Policy,* , © 1974, p. 90. Reprinted by permission of Prentice-Hall, Inc., Englewood Cliffs, N.J.

Are you convinced that radios cause insanity? Or are both variables just increasing with the years? We tend to accept as reasonable the argument that an increase in education causes an increase in personal wages, because the results seem intuitively possible, and they match our preconceptions. But we balk when asked to believe that an increase in radios causes an increase in insanity. Yet the arguments are based on the same sort of statistical reasoning. The flaw in the reasoning is that statistics can only show that events occur together or are correlated, but *statistics can never prove that one event causes another.* Any time you are tempted to jump to the conclusion that one event causes another because they are correlated, think about the radios in England!

5.4 Next Steps: Raising More Questions

When a strong link is found between variables, often the next step is to raise questions whose answers might provide more insight into the nature of the relationship. How can the evidence be strengthened? If we use other income measures such as total personal income or total family income, will the relationship still hold? Are there other variables that affect the relationship between education and income?

Does income depend on age?

We started our exploration by looking at how income depended on education, because it seemed natural that more education might lead to more income. But it is equally plausible that a person's income might depend on his or her age for several reasons: people generally earn more as they advance through their working career; they gain on-the-job experience in addition to formal education; and their income drops when they eventually retire. We can examine the FAM 1000 data to look for evidence to support this conclusion.

It's hard to see much when we plot all the data points. But the plot of mean personal wages versus age seems to suggest that up until about age 50, as age increases, mean personal wages increase. After age 50, as people move into middle age and retirement, mean personal wages decrease. So age does seem to affect personal wages, in a way which is roughly consistent with our intuition. But the relationship appears to be nonlinear, and so linear regression may not be a very effective tool to explore this dependence.

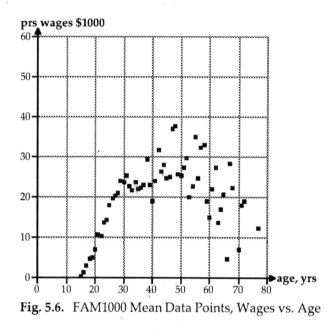

Fig. 5.6. FAM1000 Mean Data Points, Wages vs. Age

We see that the FAM 1000 data set contains internal relationships which are not obvious on a first analysis. Our example of radios vs. insanity suggests that we should also investigate the relationship between education and age.

There are a few simple ways to attempt to minimize the effect of age in analyzing the relationship between education and income. For example, we might restrict our analysis to individuals who are all roughly the same age. We could construct a scatter plot showing income vs education only for people aged between 30 and 35. Of course, this sample still would include a very diverse collection of people. More sophisticated strategies involve statistical techniques such as *multi-variable analysis,* a topic beyond the scope of this course.

Something to think about

What other variables do you think might affect income? Which measure of income would you use to test your hypothesis?

Does income depend on gender?

Despite the above cautions, we may still continue to look for simple relationships in the FAM 1000 data set. For example, we may look at whether the relationship between income and education is different for men than for women. To do this, we compute the mean personal wages for each year of education for men and women separately. By plotting the mean personal wages for men and women for each year of education, we begin to see some patterns.

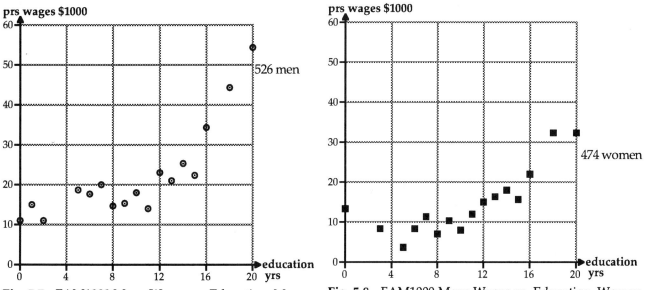

Fig. 5.7. FAM1000 Mean Wages vs. Education, Men   **Fig. 5.8.** FAM1000 Mean Wages vs. Education, Women

The graphs indicate that in general, personal wages for both men and women increase as education increases.

Something to think about

The graphs in Fig. 5.7 and 5.8 raise interesting questions. For example, in this particular data set, the mean personal wage for women is the approximately the same for both 18 years and 20 years of education. Will we find the same pattern if we look at a larger sample of data? Will this observation remain true if we examine specific professions?

If we put the data for men and women on the same graph, it is easier to make comparisons.

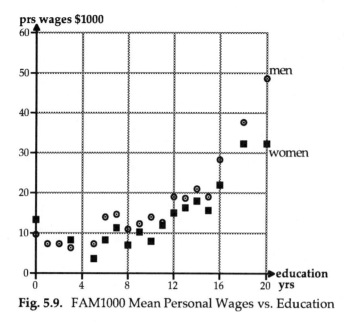

Fig. 5.9. FAM1000 Mean Personal Wages vs. Education

With the exception of zero years of education, the mean personal wages of men are consistently higher than the mean personal wages of women.

We can also examine the best fit lines for mean personal wages vs. education for men and for women shown in Figure 5.10.

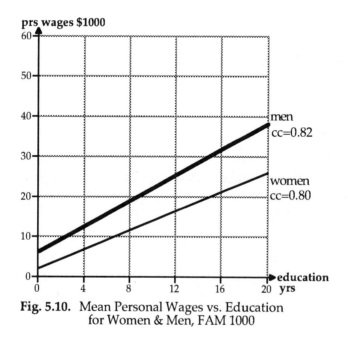

Fig. 5.10. Mean Personal Wages vs. Education
for Women & Men, FAM 1000

The linear model for mean personal wages for men is given by:

$$W_{men} = 6300 + 1600E$$

where E = years of education, W_{men} = mean personal wages for men, and the correlation co-efficient is 0.82. The rate of change of mean personal wages with respect to education is approximately \$1,600 per year. For males in this set the mean personal wage increases by roughly \$1,600 for each additional year of education.

For women the comparable linear model is:

$$W_{women} = 2100 + 1200E$$

where E = years of education, W_{women} = mean personal wages for women and the correlation coefficient is 0.80. As you might predict from the relative status of men and women in the U.S. work force, the rate of change is much lower. For women in this sample the model predicts that the mean personal wage increases by only \$1200 per year for each additional year of education. In addition, as you can see in Figures 5.9 and 5.10, the mean personal wage for any particular number of years of education is consistently lower for women than for men. The disparity seems to increase with the level of education.

Lines graphed in Figure 5.10 estimate mean personal wages for *all* the women and men in FAM 1000. By using all the data, what other variables have we ignored that we might want to consider in a more refined analysis of the impact of gender?

- Type of job
  Perhaps women and men do make the same salaries when they hold the same types of jobs, but there are more men in higher paying professions. We could compare only people within the same profession, and ask if the same level of education corresponds to the same level of personal wages for women as for men.

- Amount of work
  Typically part-time jobs pay less than full-time jobs, and more women hold part-time jobs than men. In addition there are usually more women than men who are unemployed. We could examine the personal wages of only those who are working full-time, and determine if it is still true that given equal amounts of education, women get paid less on average than men.

Something to think about

Next steps often involve hypothesizing "why?" Why do you think gender affects the relationship between education and income?

Conclusion

The real world is a messy place, and when we try to understand it, our best mathematics can be defeated by the deficiencies of actual data:

> Econometric theory is like an exquisitely balanced French recipe, spelling out precisely with how many turns to mix the sauce, how many carats of spice to add, and for how many milliseconds to bake the mixture at exactly 474 degrees of temperature. But when the statistical cook turns to raw materials, he finds that hearts of cactus fruit are unavailable, so he substitutes chunks of cantaloupe; where the recipe calls for vermicelli he uses shredded wheat; and he substitutes green garment dye for curry, ping-pong balls for turtle's eggs and, for Chalifougnac vintage 1883, a can of turpentine. [11]

In this chapter, we have illustrated some of the methods by which economists and statisticians try to make sense out of the society in which all of us live. We have also indicated some of the tremendous difficulties involved in drawing simple conclusions from data about real people in a complex society.

Questions like the ones here are actively discussed and debated every day, in Washington, in local government, in business and industry. As an educated member of society, you need to be familiar with the analytical tools that are used to identify relationships and reach conclusions, simple or not, and on which important decisions may depend. You need to be able to understand and evaluate the strengths and weaknesses of arguments presented to you, and to be able to form and defend your own judgments. Hopefully, mathematics will become one of the tools you use to understand and participate in the world around you.

---

[11] Valavanis, Stefan. *Econometrics: An Introduction to Maximum Likelihood Methods*. McGraw-Hill 1959, p. 83.

Chapter 5 Summary

Using a case study approach, this chapter examines the analytic tools and the difficulties in applying them to draw conclusions about real people in a complex society. In the physical sciences, the relationship of variables is often quite direct. In the social and life sciences it is usually difficult to tell whether one variable truly depends upon another. One of the most important tools of social scientists for studying relationships is fitting lines to data. Computers or graphing calculators can be used to calculate a best fit line, called a *regresssion line*, to a scatter plot. A number called the *correlation coefficient* indicates the strength of the linear correlation between the variables. Its value is always between -1 and 1; the closer its absolute value is to 1, the better the fit and the stronger the linear association between the variables.

Regression lines show *correlation*, not *causation*. If we find a regression line with a correlation coefficient close to 1 in absolute value, a strong relationship is suggested. For example, education is positively correlated with education. But that does not mean that increased education causes increased income. Perhaps increased income, causes more education. Maybe a third *hidden variable*, such as parental socio-economic status, may better account for both more education and higher income.

EXERCISES

1   This graph and regression line were generated from data in the FAM 1000 file on mean personal income and years of education. Round the numbers in the equation for the regression line to two significant digits and answer the following questions.

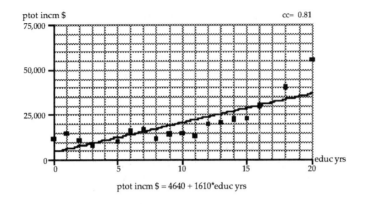

ptot incm $ = 4640 + 1610*educ yrs

   a)   What is the slope of the regression line?
   b)   Interpret the slope in this context.
   c)   By what amount does this regression line predict that mean personal income changes for 1 additional year of education? for 10 additional years of education?
   d)   What are some features of the data which are not well described by the regression line?
   e)   What are some of the limitations of the data and the model in making predictions about income?

2.   The following equation represents a best fit regression line for mean personal total income of white males vs. years of education using the FAM 1000 data set.

Personal Total Income = 6100 + 2300 • (years of education)

The correlation coefficient is 0.76 and the sample size is 202 white males.

   a)   What is the rate of change of mean personal total income with respect to years of education?
   b)   Generate a small table with three points that lie on this regression line.
   c)   Use two of these points to find the slope of the regression line.
   d)   How does this slope relate to your answer to question (a)?
   e)   Sketch the graph.

3. From the FAM 1000 data, the best fit regression line for Personal Total Income of white females vs. years of education is:

    Personal Total Income = 1300 + 1400 •(years of education)

    The correlation coefficient is 0.82 and the sample size is 414 white females.

    a) Interpret the number 1400 in this equation.
    b) Generate a small table with three points that lie on this regression line.
    c) Use two of these points to find the slope of the regression line.
    d) How does this slope relate to your answer to part (a).
    e) Sketch the graph.

4. From the FAM 1000 data, the best fit regression line for Personal Total Income for the Census category of non-white females vs. years of education is:

    Personal Total Income =4200 + 1100 • (years of education)

    The correlation coefficient is 0.76 and the sample size is 60 non-white females.

    The following figure shows the graph of averages for the mean personal total income for white and non-white females.

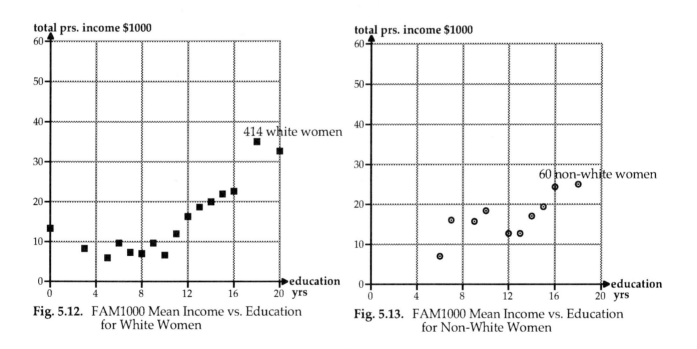

Fig. 5.12. FAM1000 Mean Income vs. Education
            for White Women

Fig. 5.13. FAM1000 Mean Income vs. Education
            for Non-White Women

Write a summary comparing the relationship between education and personal total income for white and non-white females (See Exercise 3.).

5. The term "linear regression" was first coined in 1903 by Karl Pearson as part of his efforts to understand the way physical characteristics are passed from generation to generation. He assembled measurements of the heights of fathers and their sons, from more than a thousand families, into a graph. Fathers' heights were assigned to the independent variable, F. The dependent variable, S, was the mean height of the sons with fathers of the same height. The line passing through the data points had a slope of 0.516, which is much less than 1. If on average, the sons grew to the same height as their fathers, the slope would equal 1. Tall fathers would engender tall sons and short fathers would have equally short sons. Instead, the graph shows that while the sons of tall fathers are still tall, they are not (on average) as tall as their fathers. Similarly the sons of short father are not as short as their fathers. Pearson termed this *regression;* the heights of sons *regress* back toward the height which is the mean for that population.

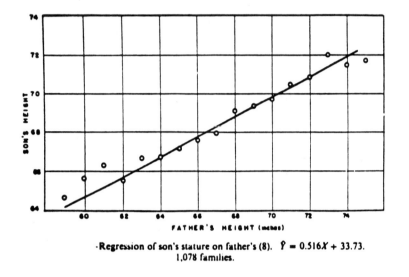

-Regression of son's stature on father's (8). $\hat{Y} = 0.516X + 33.73$.
1,078 families.

From: *Statistical Methods* by Snedecor and Cochran, 8th ed; by permission of the Iowa State University Press, © Copyright 1967.

The equation of the regression line is S = 33.73 + 0.516 F where F = father's height in inches and S = mean height of sons in inches

a) Interpret the number 0.516 in this context.
b) Use the regression line to predict the mean height of sons whose fathers are 64 inches tall and of those whose fathers are 73 inches tall.
c) If there were over 1000 families, why are there only 17 data points on this graph?

6. Read the excerpt in the Anthology from *Performing Arts - The Economic Dilemma.* Describe what the regression line tells you about the relationship between attendance per concert and the number of concerts for a major orchestra.

7.   (Optional use of graphing calculator or computer)
     The data and the graph below give the mean annual cost (in current dollars) for tuition
     and fees at public and private 4 year colleges in the U.S. since 1980.

Year	Public Ed$	Private Ed$
1980	804	3617
1981	909	4113
1982	1031	4639
1983	1148	5093
1984	1228	5556
1985	1318	6121
1986	1414	6658
1987	1537	7116
1988	1646	7722
1989	1781	8446
1990	1908	9340
1991	2137	10017
1992	2334	10449
1993	2535	11007
1994	2686	11709

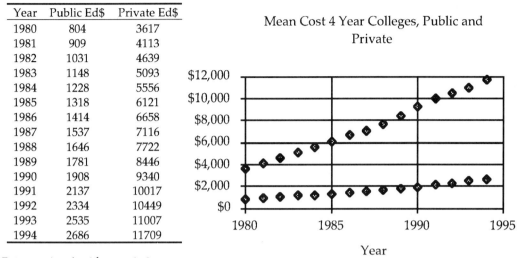

Data reprinted with permission from the Universal Press Syndicate. From the *1996 Universal Almanac.*

It is clear that the cost of education is going up, but public education is still less expensive than private. The graph suggests that costs of both public and private education vs. time can be roughly represented as straight lines.

   a) Either using technology or sketching by hand, construct two lines to represent the
       data.

   b) Using your two linear models, what is the rate of change of education cost per year
       for public and for private education? (If you sketched the lines by hand, estimate
       the coordinates of points that lie on the line, and then estimate the slope.)

   c) If the costs continue to rise at the same rates for both sorts of schools, what would
       be the respective costs for public and private education in the year 2000? Does this
       seem sensible to you? Why or why not?

8.   How did a Princeton professor's statistical analysis influence a decision by a federal
     judge in Philadelphia to give a Pennsylvania Senate seat to a losing Republican
     candidate? (Cf. Anthology reading "His stats can oust a senator or price a Bordeaux.")

9.   (Requires graphing calculator or computers)
In Chapter 4, we generated by hand a linear model for the mean heights of a group of 161 children in Kalama, Egypt.

a)   Use a graphing calculator to find the best fit regression line.

b)   Identify the variables for your model, specify the domain and interpret the slope and vertical intercept in this context. How good a fit is your line?

c)   Use your line to predict the mean height for children 26.5 months old.

Mean Height of Kalama Children	
Age (months)	Height (cm)
18	76.1
19	77.0
20	78.1
21	78.2
22	78.8
23	79.7
24	79.9
25	81.1
26	81.2
27	81.8
28	82.8
29	83.5

Source: *Introduction to the Practice of Statistics* by Moore and McCabe. Copyright © 1989 by W.H. Freeman and Company. Used with permission.

10. The following table show the calories/minute burned by a 154 lb. person at pace speeds from 2.5 to 12 miles/hour. (Note: a fast walk is about 5 miles/hr; faster than that is considered jogging (slow running). Marathons, about 26 miles, are now run in slightly over 2 hours, so that top distance runners are approaching a pace of 13 miles/hour.)

a)   Plot the data.

b)   Is the relationship between pace and calories/minute linear? If so, generate a linear model. Identify the variables, a reasonable domain for the model, and interpret the slope and vertical intercept. How good a fit is your line to the data?

c)   Describe in your own words what the model tells you about the relationship between pace and calories/minute.

Pace (miles/hr)	Calories per minute
2.5	3.0
3.0	3.7
3.5	4.2
3.8	4.9
4.0	5.5
4.5	7.0
5.0	8.3
5.5	10.1
6.0	12.0
7.0	14.0
8.0	15.6
9.0	17.5
10.0	19.6
11.0	21.7
12.0	24.5

11. (Optional use of graphing calculator or computer.)
The following table shows (for years between 1965 and 1993 and for people 18 and over): the total % of cigarette smokers; the % of all males that smoke: and the % of all females that smoke.

Year	% of smokers, 18 and older	% total male	% total female
1965	42.4	51.9	33.9
1974	37.1	43.1	32.1
1979	33.5	37.5	29.9
1983	32.1	35.1	29.5
1985	30.1	32.6	27.9
1987	28.8	31.2	26.5
1988	28.1	30.8	25.7
1990	25.5	28.4	22.8
1991	25.6	28.1	23.5
1992	26.5	28.6	24.6
1993	24.2	26.2	22.3

Source: U.S. National Center for Health Statistics, *Health, United States*, 1993 in *The American Almanac: Statistical Abstract of the United States*, 1995.

a) By hand, draw a scatter plot of the percentage of all current smokers 18 and older vs. time.

    i) Calculate the average rate of change between 1965 and 1992.

    ii) Calculate the average rate of change between 1990 and 1992.

Be sure to specify the units in each case.

b) On your graph, sketch in an approximate regression line. By estimating coordinates of points on your regression line, calculate the average rate of change of the percentage of total smokers with respect to time.

c) (optional) Using a calculator or computer, generate a regression line for the percentage of all smokers 18 and older as a function of time. (You may wish to set 1965 as year 0, and let the independent variable represent the number of years since 1965.) Record the equation and the correlation coefficient. How good a fit is this regression line to the data? Compare the rate of change for your hand generated regression line to the rate of change for the technology generated regression line.

d) Generate and record regression lines (and their associated correlation coefficients if you are using technology), for the percentage of both male and female smokers as functions of time.

e) Write a summary paragraph using the results from your graphs and calculations to describe the trends in smoking from 1965 to 1992. Would you expect this overall trend to continue? Why?

12. (Optional use of graphing calculators or computers. From Peg McPartland, Golden Gate University)

The following is a list of U.S. ski jumping records set on the 90 meter jumping hill at Howelsen Hill, Steamboat Springs, Colorado.

Year	Distance (feet)	Ski Jumper
1916	192	Ragnar Omtvedt
1917	203	Henry Hall
1950	301	Gordon Wren
1950	305	Merrill Barber
1950	307	Art Devlin
1951	316	Ansten Samuelstuen
1963	318	Gene Kotlarek
1963	322	Gene Kotlarek
1978	354	Jim Denny

Source: Tread of Pioneers Museum, Steamboat Springs, Colorado.

a) Plot the data on a scatter plot. (You may want to define 1916 to be year 0, and let your independent variable be the number of years since 1916.)

b) Do the data seem to be linear? Explain your reasoning.

c) Assuming the data are linear, construct a best fit line (by hand or using technology.) Carefully identify your variables, and the meaning of the slope and vertical intercept within the context of this problem. How good a fit is your line?

d) Use your linear model to predict the record ski jump in the year 2000. Does your estimate seem reasonable?

13. If you have access to the Nations Data Set (described in Part III and recorded as both an Excel spreadsheet and a graph link file on the disk accompanying the course), what pairs of variables would you suspect be linearly related? Which ones actually do? Hypothesize as to why. What hidden factors might be influencing the relationship?

14. (Requires graphing calculator or computer)

In the exercises in Chapter 4, we saw the following table and graph for the men's Olympic 16 pound shot put.

Olympic Shot Put	
year	feet thrown
1900	46
1904	49
1908	48
1912	50
1920	49
1924	49
1928	52
1932	53
1936	53
1948	56
1952	57
1956	60
1960	65
1964	67
1968	67
1972	70
1976	70
1980	70
1984	70
1988	74
1992	71

Men's Olympic 16 lb. Shot Put Results

Reprinted with permission from the Universal Press Syndicate. From *The 1996 Universal Almanac.*. (The data points are missing for the early 1940's since the Olympics were not held during World War II).

a) Using a graphing calculator or computer, find the equation of the best fit line and its correlation coefficient (if available)? (When you enter the data, you may want to set 1900 as year 0, and let the independent variable be the number of years since 1900.) Is the line a good fit?

b) Describe in your own words your linear model for the data, interpreting each of the terms in your equation (be sure to specify the units.)

15. (NEEDED: Computer and access to the Internet.

U.S. census data can be downloaded from the Internet. Here are a few projects that have "Homepages" that describe how to use the internet to access census data.

- http://www.census.gov/ gives access to census data and projects that have "Homepages"
- Integrated Public Use Microdata Series (IPUMS) at http://www.hist.umn.edu/~ipums/ Minnesota Historical Census Projects, University of Minnesota. IPUMS is the "largest publicly accessible computerized database on a human population. " It includes U.S. population census data from 1850 to 1990.
- Social Science Data Analysis Network (SSDAN) http://www.psc.lsa.umich.edu/SSDAN/ Population Studies Center, University of Michigan SSDAN provides "tailor-made" data sets from U.S. census data from 1950 to 1990.

Think of a question that you would like to answer about the population of your city or state. Use census data available from one of these projects to find data that can help you answer your question.

16. (Optional use of graphing calculator or computer)
    The following table and graph show the winning running time in minutes for the
    women in the Boston marathon.

Women's Boston Marathon Winning Times	
Year	Time (min.)
1972	190
1973	186
1974	167
1975	162
1976	167
1977	166
1978	165
1979	155
1980	154
1981	147
1982	150
1983	143
1984	149
1985	154
1986	145
1987	145
1988	145
1989	145
1990	145
1991	144
1992	144
1993	145
1994	142
1995	145

Boston Marathon Times for Women

Reprinted with permission from the Universal Press Syndicate.
From *The 1993 Universal Almanac*.

a) Generate a line that approximates the data (by hand or with a graphing calculator
   or function graphing program). You may wish to set 1972 as year 0. If you are
   using technology, specify the correlation coefficient. Interpret the slope of your line
   in this context.

b) If the Marathon times continued to change at the rate of your linear model, predict
   the winning running time for the women's marathon in 2010? Does that seem
   reasonable? If not, why not?

c) The graph seems to flatten out after about 1981. Construct a second regression line
   for the data from 1981 on. If you're using technology, what is the correlation
   coefficient for this line? What would this line predict for the winning running time
   for the women's marathon in 2010? Does this estimate seem more realistic than
   your previous estimate?

d) Write a short paragraph summarizing the trends in the Boston Marathon Times
   for women.

EXPLORATION 5.1
A Case Study on Education and Income in the U.S.: Part I

Objective
- to find possible relationships between educational attainment and income using the FAM 1000 data.

Materials/Equipment
If using a computer

Course software: "Education/Wage Regression" and "Regression with Multiple Subsets" in *FAM 1000 program in the *FAM 1000 Census Data Kit* folder.
Printer
Overhead transparencies (optional)

If using graphing calculators

Graphing calculators with graph link capabilities
Graph link files with FAM 1000 data on course data disk
Printout of the FAM 1000 graph link files

Related Readings
"Money Income", *Population Profile of the United States, 1995*, Bureau of the Census, Current Population Reports, Special Studies Series P23-189.

"Wealth, Income, and Poverty" from *The Data Game: Controversies in Social Science Statistics* by Mark Maier.

"Education and Income" chart from *The New York Times*, Jan. 28, 1993.

Class Discussion

We will be looking at the relationship between education and income in the U.S. We start our case study by discussing the Current Population Survey, and the information provided in FAM 1000 data set. Discuss the generation of regression lines and correlation coefficients and their interpretation. If a demonstration computer is available, view the program "Education/Wage Regression" in the *FAM 1000 program in the *FAM 1000 Census Data Kit* folder, and talk about the results.

Working with partners

Work in pairs using two computers or two calculators if possible, so you can compare two regression lines more easily. You will study the nature of the relationship between education and different income variable for various groups of people. You'll be asked to write a summary of your findings and to present your conclusions to the class.

1. Finding regression lines
   If using a computer:
   Open "Regression with Multiple Subsets" in *FAM 1000 program in the *FAM 1000 Census Data Kit*. This program allows you to find regression lines for education vs. income for different income variables and for different groups of people. Select (by clicking on the appropriate box) one of the four income variables: personal wages, personal wages per hour, personal total income or family income. Then select at least

two regression lines that it would make sense to compare (e.g., men vs. women, white vs. non-white, two or more regions of the country). You should do some browsing through the various regression line options to pick ones that are the most interesting. Print out your regression lines (on overhead transparencies if possible).

If using graphing calculators and graph link files:
Discuss the nature of summary data sets in graph link files and possible comparisons that can be made using single and double sorts of the data. Make a decision on the type of comparison you want to make. Download data from graph link files and find regression lines. Your instructor may give you hard copy of data in graph link files. Instructions for using graph link and graph link files with the TI-82 are in the Appendix.

3.   For each of the regression lines you choose, work together with your partner to record the following information in your notebooks.

Equation of the regression line
    the x variable represents:
    the y variable represents:
    the subset of the data the line represents (i.e., men? non-whites?)
    cc. =
    Whether or not the line is a good fit and why:
    The slope =
    Interpretation of the slope: (e.g., for each additional year of education, average to-tal personal wages rise by such and such an amount.)

If you have additional regression lines, supply the same information for them.

Discussion/analysis
With your partner, explore ways of comparing the two regression lines. How do the two slopes compare? Can you say that one group is better off? Are they better off no matter how many years of education? What factors are hidden or not taken into account?

Exploration Linked Homework

1.   Prepare a 60 second summary of your results. Discuss with your partner how to present your findings. What are the limitations of the data? What are the strengths and weaknesses of your analysis? What factors are hidden or not taken into account? What questions are raised? You may wish to consult the related readings or find additional sources to add to your analysis.

2.   If you have access to a computer, use "Regression with Multiple Subsets" in *FAM 1000* program in the *FAM 1000 Census Data Kit*" to look for possible relationships between income and another variable in the FAM 1000 data. (e.g. age, family size or number of work weeks).

EXPLORATION 5.2
A Case Study on Education and Income in the U.S.: Part II

Objectives
- to study educational attainment and income as single variables using data from FAM 1000 .
- to explore how to make comparison using graphs and numerical descriptors.

Materials/equipment
- computer with course software Histogram in *FAM 1000* program in the *FAM 1000 Census Data Kit* folder.

Related Readings
"Money Income" and "Educational Attainment", *Population Profile of the United States, 1995*, Bureau of the Census, Current Population Reports, Special Studies Series P23-189.

Procedure
We continue our case study by looking at income and education as single variables. For your analysis, you can use the a spreadsheet or the FAM 1000 data (on the course disk), a software program that will generate histograms of the FAM 1000 data, and the related readings. You will be asked to write a summary of your findings and to present your results to the class.

Working with partners

Working with a classmate you will compare the income distribution of two groups and explore whether there are differences between the groups in their level of income. Do men make more than women? How do income levels compare in various regions of the country or among various racial groups?

1. Open "Histogram" in the *FAM 1000* program located in the *FAM 1000 Census Data Kit* folder. It can create the histograms of your choice from FAM 1000 data. You can change the size of the interval by clicking on the arrow on the bottom of the screen and then clicking on "histogram.") Experiment by creating histograms for different measures of income, such as Personal Wages, Total Personal Income and Wages per hour for different groups of people. You can choose different interval sizes to see what patterns emerge.

2. Choose two sets of people you wish to compare, for instance men vs. women or people living in two different regions of the country or members of two different racial groups. See whether there are differences between the two groups in one of the income measures. (Click on the income measure and the category by you would like the data sorted and then click "histogram.") One partner can do one group and the other the second group and then discuss your observations. Pick a comparison that you think is interesting and print out your histograms. When printing out your histograms, put both of your names on the histogram and print two copies so you each can have one.

3.   Make jottings on the differences and similarities between the two groups for the income variable you have chosen. Include numerical descriptors such as mean and median.

Discussion/analysis

Discuss with your partner what type of arguments you can make, the limitations of the data and the questions that are raised from the data and your analysis. What is the age range of the people in your sample? Are they working full or part time? What about types of jobs, gender, race or region of the country? How might these factors affect your argument.

Exploration Linked Homework

1.   Browse through the related readings to find data and arguments that relate to the variables you were analyzing in the class exploration. The U.S. Census Current Population Reports profile American income and education levels using the data from the Current Population Surveys (our FAM 1000 data is extracted from one of these surveys).

2.   Write a 60 second summary describing your findings. Start with a topic sentence and use numerical descriptors and information from your graphs to support your topic sentence. Conclude your paragraph by summarizing your findings or suggesting a plausible explanation for your findings. Attach the histograms you used.

3.   Write a paragraph on the limitations of the data you used or on the limitations of your analysis.

4.   Conduct a similar exploration to study whether there are differences in educational attainment of two different groups of people.

Chapter 6

When Lines Meet

Overview

A collection of two or more equations relating the same variables is called a *system of equations*. A system might represent two different income tax schemes using income and tax variables, or a model for supply and demand with quantity and price variables. The place where two lines meet often has particular significance; it could represent the point at which supply equals demand, or income taxes are the same under two different tax plans. Such a point of intersection is a *solution* to the system of equations.

This chapter explores how to construct systems of linear equations, and how to solve them using graphs or algebraic methods. We discuss what the solutions mean in situations where each line is used to represent some social or physical reality.

In the exploration we investigate a system of equations modeling two different state income tax plans presented to the Massachusetts voters in 1994.

After reading this chapter you should be able to:

- find the intersection points (if any) of two linear equations
- interpret intersection points

6.1 An Economic Comparison of Solar vs. Conventional Heating Systems

On a planet with limited fuel resources, heating decisions involve both monetary and ecological considerations. Owners, builders, and architects confront this issue in choosing heating systems for new construction and renovation. Solar heating is appealing because the fuel, solar rays, is free and is easily delivered to your home without pipes or trucks. So why is solar heating rarely used?

Solar heating can work well in places that receive a fair amount of sunlight during the heating season. Areas that are usually cloudy are not good sites for solar collection. However there are many northern locations in the U.S. that have sufficient sunlight to run solar heating systems effectively. The problem is that the installation price is so high that even many years of low cost operation do not pay back the initial heavy investment. The reasons for the high installation cost are: a backup heating system, such as a wood burning stove or an electric heating system, is needed for any long periods of cloudy weather; the active capture and storage of the sun's heat requires a lot of space for heat storage in water drums, rocks, or special thick floors and walls; and the building must have a particular orientation with a series of solar collection panels and associated piping maintained on the roof.

Typical costs for three different kinds of heating systems for a three bedroom housing unit are given below.

Table 6.1		
type of system	installation cost	operation cost/year
electric	5000	1100
gas	12000	700
solar	30000	150

If only the financial cost is considered, solar heating is clearly the most costly to install and the least expensive to run. Electric heating, conversely, is the cheapest to install and the most expensive to run. Converting this information to equation form, we can find out when the solar system begins to pay back the initially higher cost. If no allowance is made for inflation or changes in fuel price, the equation for the total cost, C, is:

C = installation cost + (annual operating cost)(years of operation).

If we let Y equal the number of years of operation and use the data from Table 6.1, we can construct the following equations:

$$C_{electric} = 5000 + 1100 \, Y$$
$$C_{gas} = 12000 + 700 \, Y$$
$$C_{solar} = 30000 + 150 \, Y$$

These are recognizable as linear equations. Together they form a *system of linear equations*.

Table 6.2 gives the cost data at 5 year intervals, and Figure 6.1 shows the costs over a 40 year period for the three heating systems.

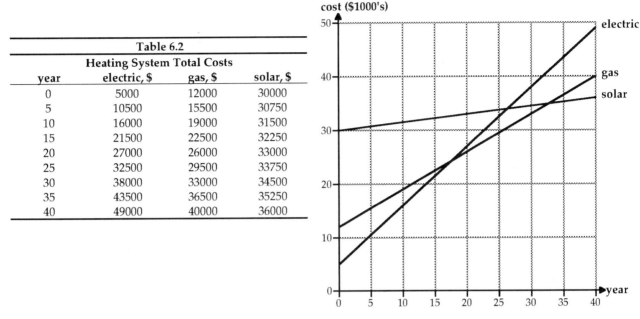

Table 6.2			
Heating System Total Costs			
year	electric, $	gas, $	solar, $
0	5000	12000	30000
5	10500	15500	30750
10	16000	19000	31500
15	21500	22500	32250
20	27000	26000	33000
25	32500	29500	33750
30	38000	33000	34500
35	43500	36500	35250
40	49000	40000	36000

Fig.6.1. Comparison of Home Heating Costs

Let's compare the costs for gas and electric heat graphed against years of operation in Figure 6.2. The point of intersection shows where the lines predict the same cost for both gas and electricity, given a certain number of years of operation. From the graph, the total cost of operation is about $24,000 for each type of energy at about year 17.

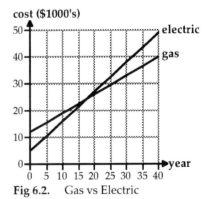

Fig 6.2.   Gas vs Electric

Compare the relative costs to the left and right of the point of intersection. How many years of operation does it take for the total cost of gas to become less than the total cost of electricity? When would the total cost of gas be higher?

Figure 6.2 shows that gas is less expensive than electricity to the right of the intersection point, after approximately 17 years of operation. To find the break-even point exactly, we find the coordinates of the intersection point. At the year Y when the gas and electric equations intersect, the coordinates satisfy both equations. At that value of Y the cost of electric heat, $C_{electric}$, equals the cost of gas heat, C_{gas}. Since the cost is the same in both cases, the equations can simply be set equal to each other to find the coordinates of the intersection point.

$$C_{electric} = 5000 + 1100Y \qquad (1)$$
$$C_{gas} = 12000 + 700Y \qquad (2)$$

If we set equation (1) equal to equation (2) $C_{electric} = C_{gas}$
substitute $5000 + 1100Y = 12000 + 700Y$
collect terms $400Y = 7000$
divide through by 400, we get $Y = 17.5 \text{ years}$

When $Y = 17.5$ years, the cost for electric or gas heating is the same, so the cost can be found by substituting this value for Y in equation (1) or (2).

substitute 17.5 for Y in equation (1) $C_{electric} = 5000 + 1100(17.5)$ (1)
                                                                           $C_{electric} = \$24{,}250$

Since we claim that the values (17.5, $24,250) solve both equations, we now check these values in equation (2):

substitute 17.5 for Y in equation (2) $C_{gas} = 12000 + 700(17.5)$
                                                                      $= 12000 + 12250$
so $C_{gas} = \$24{,}250$

The coordinates (17.5, $24,250) satisfy both equations, when $Y = 17.5$ years then $C_{electric} = \$24{,}250 = C_{gas}$. The point (17.5, $24,250) is called a *solution* to the system. Using algebra we were able to find accurate values for the coordinates of the point of intersection. We can predict from the equations that 17.5 years after installation, $24,250 has been spent on heat for either an electric or gas heating system, and that gas becomes more economical than electricity as a fuel supply thereafter.

The intersection points for electric vs. solar heating and gas vs. solar heating can be estimated using the graphs in Figures 6.3 and 6.4. The actual solutions to these systems will be found later in the text and in the homework exercises.

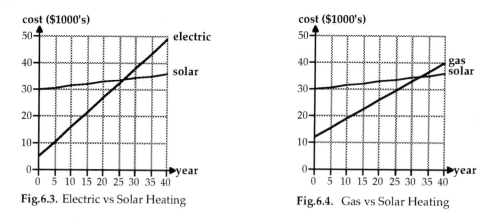

Fig.6.3. Electric vs Solar Heating   **Fig.6.4.** Gas vs Solar Heating

Home buyers in their mid thirties would probably be retired by the time the initial heavy investment in solar power began to pay back; older buyers might not live to see a return on their solar power investment. Most people don't expect to live in the same house for thirty years, and they choose a system that is most immediately economical for them. Heating systems also typically need repairs after 15 or 20 years of operation, so solar panels might fail and have to be replaced before they have paid for themselves in operational cost savings.

We made many simplifying assumptions when constructing this model. We ignored inflation and rising fuel prices, and we did not consider inevitable repair costs. We did not take into account what economists call *opportunity costs*. Opportunity costs include, for instance, the 40 years worth of interest that we could have collected if we had invested the dollars (in stocks or bonds or savings accounts) that we used to pay for startup costs. More sophisticated mathematical models might also consider the cost of depleting fuel resources and the risks of generating nuclear power. But these would simply be refinements of the sort of model we have used to give us a general idea of the financial trade-offs in using one heating system versus another.

Something to think about

Before 17.5 years our model tells us that electric systems are cheaper than gas. Using Figure 6.1, estimate the interval over which gas is the cheapest system. When does solar heating become the cheapest?

6.2 Finding Solutions to Systems of Linear Equations

We can generalize the following terminology from the previous economic model of heating systems.

A collection of linear equations that involve the same variables is called a *system of linear equations* or a set of *simultaneous linear equations;*.

A pair of real numbers is a *solution* to a system of linear equations in two variables if and only if the pair of numbers is a solution to each equation. On a graph of a system of linear equations, a solution will appear as an *intersection point*.

In this chapter we use systems of linear equations to make comparisons or to determine the combined impact of opposing forces on an economic or biological system. The key idea is to find and interpret the points where graphs of different options or interacting forces intersect.

Visualizing solutions to systems of linear equations

For a single linear equation, the graph of its solutions is a line. In a system of two linear equations, the common solutions to the two equations are where the two lines intersect. You can easily visualize the three possible outcomes. If the lines intersect once, there is one solution. If the lines are parallel they never intersect and there are no solutions. If the two lines are identical they are juxtaposed- every point on the line is a solution - so there is an infinite number of solutions.

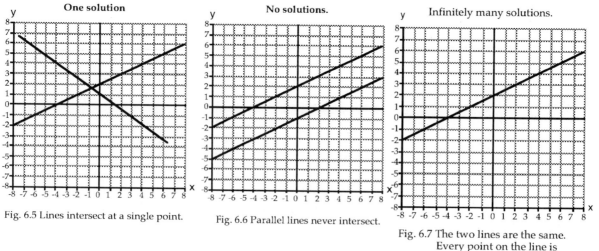

Fig. 6.5 Lines intersect at a single point.

Fig. 6.6 Parallel lines never intersect.

Fig. 6.7 The two lines are the same. Every point on the line is an intersection point.

If the lines are not parallel, the coordinates of the point of intersection can be estimated by inspecting the graph. For example, in Figure 6.5, the two lines appear to cross near the point (-0.7, 1.7). For better estimates for the coordinates of the intersection point, use a zoom function. Most function graphing programs can generate values accurate to several decimal places by this means. For exact answers, we use algebra.

Algebraic solutions to systems of linear equations

There are several possible ways to solve a system of two linear equations. Which is the most efficient method depends mainly on the form of the equations. There is no one right way or wrong way; the choice depends upon convenience. All methods give the same final answer.

Case 1: Both equations are in function form, $y = mx + b$
Strategy: Set the y's equal

We want to find the pair of values (x,y) that satisfies two equations, both of which have y written in terms of x. The steps are:

1. Set the two expressions for y equal and then solve the resulting equation to find the value for x .
2. Substitute this x value in one of the two original equations to find the corresponding y value.
3. Check your results in the other original equation.

It is a good idea to graph the equations to verify that the solution gives the coordinates of the intersection point.

Example: Find the intersection point of the two equations
$$y = 2x + 8 \qquad\qquad (1)$$
$$y = -3x - 7 \qquad\qquad (2)$$

Solution:
In both equations y is written in terms of x. At the point of intersection the two y's have the same value, so we set

$$2x + 8 = -3x - 7$$
Then solve for x
$$5x = -15$$
$$x = -3$$

The two lines cross when x = -3. In order to find the y value at the intersection, we can substitute -3 for x in either of the two original equations. Substituting into equation (1) we get:

substitute -3 for x
$$y = 2(-3) + 8$$
$$= -6 + 8$$
so
$$y = 2$$

So x = -3 and y = 2 is a solution to the system.

We can check that the solution point (-3, 2) does indeed work in *both* equations by substituting -3 for x in equation (2).

substitute -3 for x in equation (2)	$y = -3(-3) - 7$
simplify	$= 9 - 7$
so we get again that	$y = 2$

So the coordinates (-3,2) satisfy equation (2) as well as equation (1). In Figure 6.8 we get visual confirmation that (-3,2) represents the intersection point of the two lines, and is a solution to the system.

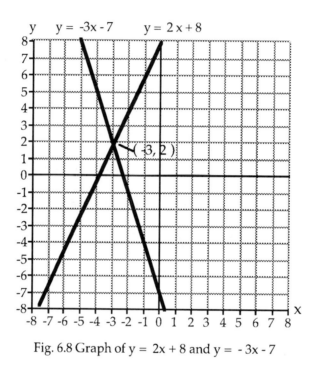

Fig. 6.8 Graph of y = 2x + 8 and y = - 3x - 7

Algebra Aerobics: [1]

Try solving the following three systems:

a) $y = x + 4$
        $y = -2x + 7$

b) $y = -1700 + 2100x$
        $y = 4700 + 1300x$

c) $F = C$
        $F = 32 + (9/5)C$

---

[1] Aswers
a) $x = 1, y = 5$ b) $x = 8, y = 15,100$ c)   both C and F = -40

Case 2: Only one equation is in function form, y = mx + b
Strategy: Substitution;

1. One equation gives y in terms of x. Substitute that expression for y in the other equation.
2. Solve the resulting equation to find a value for x.
3. Substitute this value for x into either original equation to find the corresponding y value.
4. Check your answer in the other original equation.

Example:
$$6x + 7y = 25 \qquad (1)$$
$$y = 15 + 2x \qquad (2)$$

Here equation (2) is in function form, but equation (1) isn't. We could put (1) into function form and apply the strategy in Case 1 by setting the two expressions for y equal. But *substitution* is more direct. Since at the intersection point the value for y in both equations is the same, we can substitute the expression for y from equation (2) into equation (1).

Given	$6x + 7y = 25$	(1)
substitute for y:	$6x + 7(15 + 2x) = 25$	
simplify	$6x + 105 + 14x = 25$	
	$20x = -80$	
	$x = -4$	

find a value for y when x = -4	$y = 15 + 2x$	(2)
in one of the equations	$y = 15 + 2(-4)$	
	$y = 15 - 8$	
	$y = 7$	

check that (-4, 7) satisfies		
the other equation:	$6x + 7y = 25$	(1)
	$6(-4) + 7(7) = 25$	
	$-24 + 49 = 25$	
	$25 = 25$	

Figure 6.10 shows a graph of the two equations and the intersection point at (-4, 7).

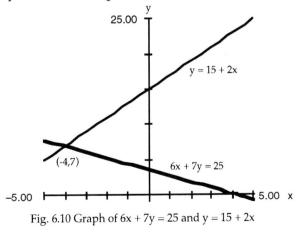

Fig. 6.10 Graph of 6x + 7y = 25 and y = 15 + 2x

Algebra Aerobics: [2]

Solve the following systems

a) $y = x + 3$ b) $z = 3w + 1$ c) $x = 2y-5$
  $5y - 2x = 21$ $9w + 4z = 11$ $4y -3x = 9$

Case 3: Neither equation is in function form.
Strategy: Elimination;

Another method called *elimination* is particularly convenient if the coefficients of the variables are small integers. The strategy is to:

1. Multiply one or both of the equations by appropriate constants, so that by adding or subtracting the equations, one variable cancels (is "eliminated").
2. Solve the new equation for the remaining variable.
3. Substitute the value found in step 2 into one of the original equations to determine a value for the "eliminated" variable.
4. Check your answer in the other original equation.

This strategy is the basis for constructing computer algorithms that are frequently used to solve linear systems.

Example:

$$x - 3y = \ \ 5 \qquad (1)$$
$$2x + y = \ -11 \qquad (2)$$

This system can be solved by multiplying each side of equation (1) by the number 2 to form an equivalent equation.

multiply equation (1) $2x - 6y = \ 10$ (3)
by 2 to find an equivalent
equation where x term has
coefficient of 2:

subtract equation (2) from $2x - 6y = \ 10$ (Remember to change
equation (3) to eliminate $-(2x + y = \ -11)$ the sign of each term
the x terms: _____   in the equation when
          $-7y = \ 21$ subtracting.)

divide by -7 $y = \ 21/-7$
          $y = \ -3$
use equation (2) to find $2x + (-3) = \ -11$ (2)
the value for x when y = -3 $2x = \ -8$
          $x = \ -4$

---

[2] Answers: a) x = 2, y = 5   b) w = 1/3, z = 2   c) x = 1, y = 3

use equation (1) to double
check if (-4, -3) is a solution
to both equations

$$x - 3y = 5 \qquad (1)$$
$$-4 - 3(-3) = 5$$
$$-4 + 9 = 5$$
$$5 = 5$$

Figure 6.11 shows the graph of the two
equations and the intersection point at (-4, -3).

Fig. 6.11 Graph of x - 3y = 5 and 2x + y = - 11

Algebra Aerobics: [3]

More practice in solving systems of equations.

a) $2y - 5x = -1$
    $3y + 5x = 11$

b) $r - 2s = 5$
    $3r - 10s = 13$

c) $3x + 2y = 16$
    $2x - 3y = -11$

[3] Answers::   a) $x = 1; y = 2$ b) $r = 6. s = 1/2$ c) $x = 2, y = 5$

Special Cases: How can you tell if there is no unique intersection point?

Example 1: No solution

Suppose you are given the following system of two linear equations.

$$y = 10000 + 1500x \qquad (1)$$
$$2y - 3000x = 50000 \qquad (2)$$

If we assume the two lines intersect, then the y's from equation (1) and (2) have the same value at some point. We can try to find this value for y by using substitution as in Case 2. We substitute the expression for y from equation (1) into equation (2).

$$2(10000 + 1500x) - 3000x = 50000$$

multiply through $\qquad 20000 + 3000x - 3000x = 50000$

simplify $\qquad\qquad\qquad\qquad 20000 = 50000 \text{ ???}$

What could this possibly mean? Where did we go wrong? We now return to the original set of equations and solve equation (2) for y in terms of x.

equation (2) $\qquad\qquad\qquad\qquad 2y - 3000x = 50000$

add 3000x to both sides: $\qquad\qquad\qquad 2y = 50000 + 3000x$

divide by 2: $\qquad\qquad\qquad\qquad\qquad y = 25000 + 1500x \qquad (3)$

Now look at the rewritten system of equations. $\qquad y = 10000 + 1500x \qquad (1)$
$$y = 25000 + 1500x \qquad (3)$$

We can see that what we really have are two parallel lines, both with the same slope of 1500, but different y-intercepts (10000 and 25000 respectively). There is no intersection point. There is no value of x that yields the same value of y in both equations, because the value of y in equation (3) must always be 15,000 greater than the corresponding y value in equation (1). Our initial premise, that the two y's were equal at some point, allowed us to substitute one expression for y for another. Our premise however was incorrect. But mathematics allows us to discover our error.

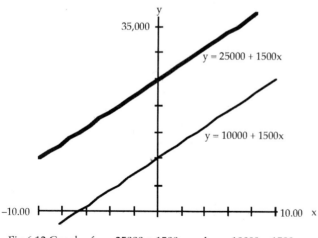

The two parallel lines are graphed in Figure 6.12. Note that the scales on the two axes are quite different.

Fig.6.12 Graph of y = 25000 + 1500x and y = 10000 + 1500x

Example 2: Infinitely many solutions

Assume we are given the following two linear equations:

$$2y + 90x = 66 \qquad\qquad (1)$$
$$45x = -y + 33 \qquad\qquad (2)$$

As always, there are multiple ways of solving the system. We could put one or both in function form and use the methods in Case 1 or 2. Instead, we try using elimination.

rewrite equation (2)	$y + 45x = 33$	(3)
multiply equation (3) by 2:	$2(y + 45x = 33)$	
to get:	$2y + 90x = 66$	(4)
So the original system becomes:	$2y + 90x = 66$	(1)
	$2y + 90x = 66$	(4)

In other words, the original equations were equivalent, and both have the same graph. Every point on the line determined by (1) is also on the line determined by (2). So there are infinitely many solutions. If we rewrote the original equations in function form, we would get $y = -45x + 33$ in both cases. Any point that lies on the line $2y + 90x = 66$ (or equivalently $y = -45x + 33$) is a solution to this system.

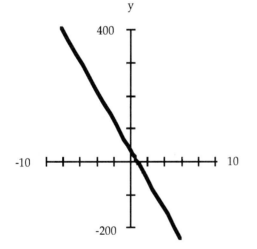

The graphs of the two equations are identical so every point on the line is a solution to the system.

Fig. 6.13 Graph of $y = -45x + 33$

Algebra Aerobics: [4]
Try these and then check your answers.

    a) $5y + 30x = 20$ b) $y = 1500 + 350x$ c) $10u + 7v = 11$
          $y = -6x + 4$ $2y = 700x + 3500$ $14v = 22 - 20u$

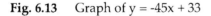

---

[4] Answers:: a) Infinitely many solutions   b) No solutions   c) Infinitely many solutions
                Equations describe same line   Parallel lines.   Equations describe same line.

6.3 Intersection Points Representing Equilibrium

When systems of equations are used to model phenomena, an intersection point often represents a state of equilibrium.

Example 1:

 In economics a graph of two lines or curves is often used to model the related concepts of supply and demand. The horizontal axis represents quantity (measured perhaps in thousands of units per month) and the vertical axis represents price (perhaps in dollars per unit).

The demand curve represents the point of view of the consumer. It typically has a negative slope, representing the notion that when the price of a product is high, consumers are reluctant to buy it, so the quantity demanded is low. As the price of the product comes down consumers are encouraged to buy, so the quantity demanded increases. Note that this discussion may be a little confusing, because economists talk about price as being the independent variable that determines demand, but they graph price on the horizontal axis.

The supply curve represents the producer's point of view. It typically has a positive slope, reflecting the idea that when the price is low, the profit is small, and producers will supply a small quantity of the product to the market. If the prices rise, a better profit can be made and producers will make larger quantities available for sale. The point where the two curves intersect represents the point where supply equals demand. At the price indicated by the intersection point, the quantity demanded by consumers equals the quantity supplied by producers. The intersection point is called the *equilibrium point*. and the price at that point is called the *equilibrium price*. At any lower price, demand would exceed supply; at any higher price, supply would exceed demand. [5]

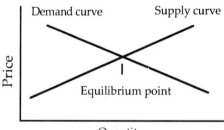

Fig. 6.14 Supply and Demand Curves

---

[5] Note that economists use the terms supply and demand *curves* even though the graphs are frequently drawn as straight lines.

Questions:

What if the overall demand increases? What if consumers want more of the product and are willing to pay a higher price to get it? What happens to the equilibrium point?

Discussion:

Pick a particular price, P, and trace that price horizontally to the original demand curve to locate the quantity consumers previously demanded (Q_{old}) at that price. If demand has increased, then that same price P is now associated with a higher quantity (Q_{new}). Since this reasoning holds for *any* price P, all points on the original demand curve shift to the right, forming a new demand curve as shown in the following graph.

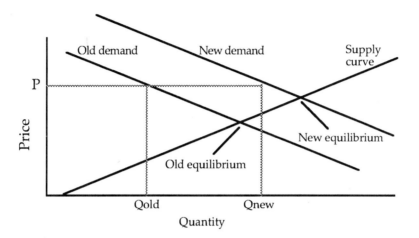

Fig. 6.15   An increase in demand shifts the demand curve to the right

The equilibrium point for the new demand curve has shifted up and to the right along the supply curve. The new equilibrium point represents a larger quantity and a higher price. In other words, an increase in consumer demand causes producers to make a larger quantity that sells at a higher price.

Example 2:

When studying populations (human or otherwise), the two primary factors affecting population size are the birth rate and the death rate. There is abundant evidence that, other things being equal, as the population density (measured in number of individuals per unit area) increases, the birth rate tends to decrease and the death rate tends to increase.[6] If we assume a fixed area such as a country, and plot population sizes against their corresponding birth rates or death rates (usually measured in births or deaths per thousand people each year), we get the following type of graph.

---

[6] Cf. Edward O. Wilson & William H. Bossert, *A Primer of Population Biolog,*" Sinauer Associates, Inc., Sunderland, Mass., 1971, p. 104.

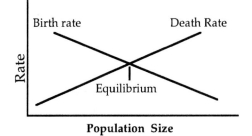

Fig. 6.16 Birth Rate and Death Rate

The intersection of the two lines represents the point at which the birth rate equals the death rate. At this point (ignoring all other factors like immigration), the growth of the population is zero. The population is said to have stabilized, and the size of the population at this point is called the *equilibrium number*.

Question:
What happens to the equilibrium point if the death rate decreases?

Answer:
A decrease in death rate means that for any population size P, the corresponding death rate D is lower. So the whole death rate curve shifts down.

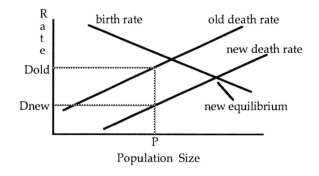

Fig. 6.17 A decrease in the death rate
lowers the death rate curve

If the death rate drops, the new equilibrium point will occur at a larger population size and a lower birth rate. On the graph, the equilibrium point moves down and to the right.

6.4 Graduated vs. Flat Income Tax: Using Piecewise Linear Functions

A Simple Model

Some taxation models are also linear systems. Income taxes may be based on either a flat or a graduated tax rate. With a flat tax rate, no matter what the income level, everyone is taxed at the same percentage. Flat taxes are often said to be unfair to those with lower incomes. They are frequently labeled *regressive* because a fixed tax percentage is considered more burdensome to someone with a low income than to someone with a high income. A graduated tax rate means that people with higher incomes pay a higher tax rate. Such a tax is called *progressive,* and is generally less popular with those who have high incomes. Whenever the issue appears on the ballot, the pros and cons of the graduated vs. the flat tax rate are hotly debated in the news media and paid political broadcasting. Of the 42 states with an income tax, 35 had a graduated income tax in 1994.

Something to think about

Not everyone agrees that flat taxes are regressive. What are some arguments for and against flat taxes being regressive?

For the tax payer there are two primary questions in comparing the effect of flat and graduated tax schemes. For what income level will the taxes be the same under both plans? And, given a certain income level, how will taxes differ under the two plans?

Taxes are influenced by many factors. Exemptions and deductions may be subtracted from income. Married people may file as couples, singles, or heads of household. For our initial comparison of flat and graduated income tax plans, we assume there is one filing status, and that exemptions and deductions have already been subtracted from income.

Under a flat tax, every additional dollar is taxed at the same percentage. So the rate of change of tax with respect to income is constant. When the rate of change is constant, the relationship is linear. Since taxes are dependent on income, we use income, I, as the independent variable and write taxes, T_F, as a function of I.. If the flat tax rate is 5%, then taxes can be represented by the following function, data table, and graph:

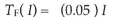

$$T_F(I) = (0.05)I$$

Table 6.3	
Income after deductions	Taxes using flat tax rate 0.05
$0	$0
$10,000	$500
$20,000	$1,000
$30,000	$1,500
$40,000	$2,000
$50,000	$2,500

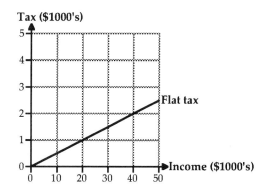

Fig.6.18. Flat Tax at a rate of 0.05

Under a graduated income tax, the tax rate changes for different portions of the income. One graduated tax scheme could include no tax on the first $10,000 of income and 10% on any income in excess of $10,000. This tax plan is represented in the following data table and graph.

Table 6.4	
Income after Deductions	Graduated tax Plan
$0	$0
$10,000	$0
$20,000	$1,000
$30,000	$2,000
$40,000	$3,000
$50,000	$4,000

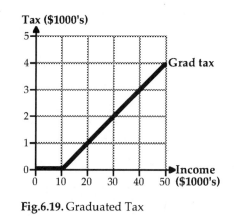

Fig.6.19. Graduated Tax

The graph of the graduated tax is pieced together with two line segments. The short horizontal segment, represents tax when income is between $0 and $10,000 (no tax required). The longer sloping segment represents taxes for incomes greater than $10,000.

The algebraic expression for the graduated tax is pieced together from two different linear formulas, each representing one of the two line segments. The value of the independent variable income determines which formula is used to evaluate the function. Such a function is a *piecewise linear function;*. The following notation describes a piecewise linear function T_G that represents the graduated tax plan.

$$T_G(I) = \begin{cases} 0 & for \ \ I \leq \$10,000 \\ 0.10(I - 10000) & for \ \ I > \$10,000 \end{cases}$$

To find T_G ($8,000), since $I \leq \$10,000$, we use the upper line of the definition where T_G (I)= 0. Hence T_G ($8,000) = 0.

To find T_G ($18,000), since $I > \$10,000$, we use the lower line of the definition where T_G (I)= 0.10(I-10000).

Substituting $18,000 for I: T_G ($18,000) = 0.10 (18000-10000)
                                        = 0.10 (8000)
So T_G ($18,000) = $800

We can compare the different tax plans by plotting the flat tax and graduated tax equations on the same graph (see Figure 6.20)

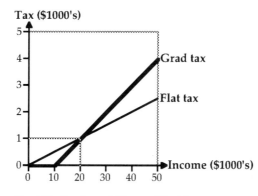

Fig.6.20. Graduated Tax vs. Flat Tax

The intersection points indicate incomes where the amount of tax is the same under both plans. From the graph, we can estimate the coordinates of the two points as ($0,$0) and ($20,000, $1,000). That is, with zero income you would pay zero taxes under both plans and with $20,000 in income you would pay what looks like $1,000 in taxes judging by the graph under both plans.

To find the actual intersection we need to set $T_F(I) = T_G(I)$. We know that $T_F(I) = 0.05\ I$. Which of the two expressions do we use for $T_G(I)$? The answer depends upon what values of I we consider. For $I \le \$10,000$, we have $T_G(I) = 0$.

If $$T_F(I) = T_G(I)$$
and $I \le \$10,000$, then $$0.05I = 0$$
So $$I = 0$$

If $I = 0$ both $T_F(I) = T_G(I)$ equal 0, so one intersection point is indeed ($0,$0).

For $I > \$10,000$, we use $T_G(I) = 0.10(I - 10000)$ and again set $T_F(I) = T_G(I)$.

Set $T_F(I) = T_G(I)$ so $$0.05I = 0.10(I - 10000)$$
multiply through $$0.05I = (0.10)\ I - (0.10)(10000)$$
simplify $$0.05I = (0.10)\ I - 1000$$
subtract $0.10I$ $$-0.05I = -1000$$
divide by -0.05 $$I = -1000/(-0.05)$$
so $$I = \$20,000$$

Each plan requires the same tax for an income of $20,000. How much tax is required? We can substitute $20,000 into either the flat tax formula or the graduated tax formula for incomes over $10,000 and solve for the tax. Using $T_F(I) = 0.05I$, substitute and solve for $T_F(\$20,000)$.

substitute $20,000 for I $$T_F(\$20,000) = (0.05)(20000)$$
$$T_F(\$20,000) = \$1000$$
The second intersection point is, as we estimated, ($20,000, $1000).

Individual voters want to know the impact on their taxes. To the left of the intersection point at ($20,000, $1,000), the flat tax is *greater* than the graduated tax for the same income. To the right of this intersection point, the flat tax is *less* than the graduated tax for the same income. This means for $I < \$20,000$, taxes are *greater* under the flat tax plan and for $I > \$20,000$, taxes will be *less* under the flat tax plan. In other words, people with incomes under $20,000 will be paying more in taxes under this flat tax plan than under this graduated income tax plan.

The Case of Massachusetts

The state of Massachusetts currently has a flat tax of 5.95% on earned income, but there has been an ongoing debate as to whether or not to change to a graduated income tax. We can apply the same sort of analysis used in the previous model to decide who would benefit under the proposed new tax.

The Massachusetts flat tax rate is represented by the following function, in which the variable I represents earned income after deductions and exemptions.

$$T_F(I) = 0.0595\ I$$

In 1994 Massachusetts voters considered a proposal, called 'Proposition 7.' Proposition 7 would have replaced the flat tax rate with graduated income tax rates (called marginal rates) as follows:

Table 6.5			
Massachusetts Graduated Income Tax Proposal [7]			
	(Marginal Rates)		
Filing Status	5.50%	8.80%	9.80%
Married/Joint	<$81,000	$81,000-$150,000	$150,000+
Married/Separate	<$40,500	$40,500 - $75,000	$75,000+
Single	<$50,200	$50,200 - $90,000	$90,000+
Head of Household	<$60,100	$60,100-$120,000	$120,000+

Questions:
For what income and filing status would the taxes be equal under both plans? Who will pay less tax and who will pay more under the graduated income tax plan? We analyze the tax for a single person, and leave the analyses of the other filing categories for you to do in an Exploration.

Discussion:
The proposed graduated income tax is designed to tax at higher rates only that portion of the individual's income which exceeds a certain threshold. For example, for those who file as single people, the graduated tax rate means that earned income under $50,200 would be

---

[7] From the Office of the Secretary of State: Michael J. Connolly, Boston, Massachusetts.

taxed at a rate of 5.5%. Any income between $50,200 and $90,000 would be taxed at 8.8%, and any income over $90,000 would be taxed at 9.8%.

The taxes for a single person lucky enough to earn $100,000 would be the sum of three different dollar amounts:

$$5.5\% \text{ on the first } \$50,200 = (0.055)(50200)$$
$$= \$2,761$$

$$8.8\% \text{ on the next } \$39,800 \text{ (the portion of income between \$50,200 and \$90,000)}$$
$$= (0.088)(39800)$$
$$\approx \$3,502 \qquad (\approx \text{ means approximately } =)$$

$$9.8\% \text{ on the remaining } \$10,000 = (0.098)(10000)$$
$$= \$980$$

The total graduated tax would be: $2,761 + $3,502 + $980 = $7,243. Under the flat tax rate the same individual pays 0.0595($100,000) or $5,950.

Table 6.6 shows the differences between the flat tax and the proposed graduated tax plan for several different income levels.

Table 6.6		
Massachusetts Taxes		
Flat Rate vs. Graduated Rate For Single People		
Income after exemptions and deductions	Current Mass income tax	Income tax under Prop. 7
$0	$0	$0
$25,000	$1,488	$1,375
$50,000	$2,975	$2,750
$75,000	$4,463	$4,943
$100,000	$5,950	$7,243

The graduated tax rate, $T_G(I)$ for single people can be represented with the following piecewise linear function.

$$T_G(I) = \begin{cases} 0.055\,I & \text{for } 0 \le I < \$50,200 \\ \$2761 + 0.088(I - \$50,200) & \text{for } \$50,200 \le I \le \$90,000 \\ \$6263 + 0.098(I - \$90,000) & \text{for } I > \$90,000 \end{cases}$$

Note that $2,761 in the second line of the definition is the tax on the first $50,200 of income (5.5% of $50,200), and $6,263 in the third line of the definition is the sum of the taxes on the first $50,200 and the next $39,800 of income (5.5% of $50,200)+ (8.8% of $39,800).

The flat and the graduated income tax for single people are compared in the following graph.

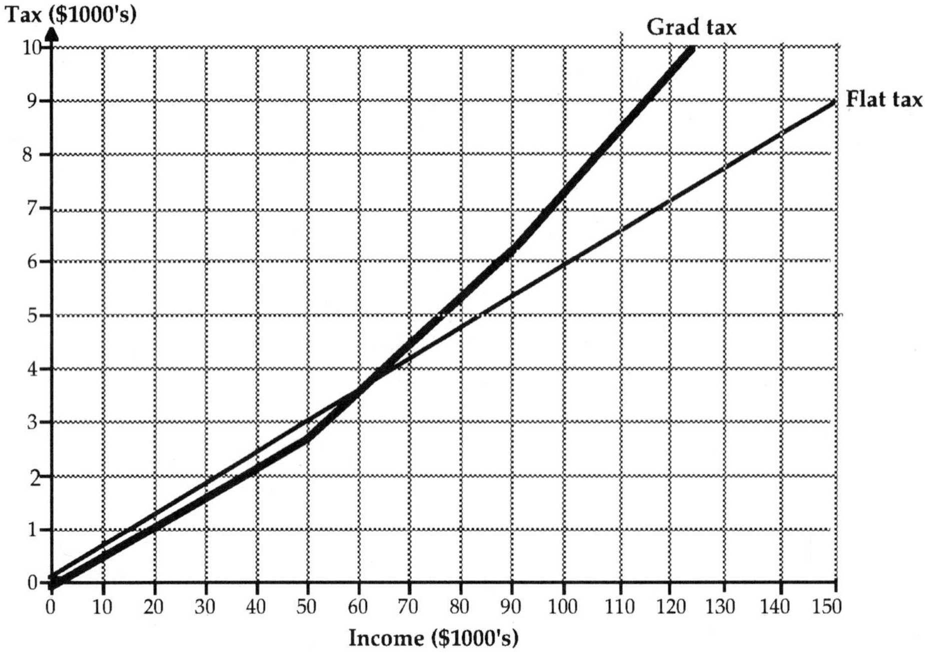

Fig.6.21. Massachusetts: Graduated Tax vs. Flat Tax

For what income would single people pay the same tax under both plans? An intersection point in the graph indicates when taxes are equal. One intersection point occurs at ($0,$0). That makes sense since under either plan if you have zero income, you pay zero taxes. The second intersection point is at approximately ($58000, $3500). That means that at an income of approximately $58,000, the taxes are the same (about $3,500). We can use our algebraic tax models to find an exact value for the intersection.

To find the intersection point we can set $T_F(I) = T_G(I)$. Since they appear to intersect at $58,000 (i.e., at an income I such that $50,200 < I \le \$90,000$, we can set the following functions equal.

$$T_F(I) = 0.0595\,I \qquad\qquad \text{flat rate plan}$$
$$T_G(I) = \$2761 + 0.088(I - \$50,200) \qquad \text{graduated rate for } \$50,200 < I < \$90,000$$

Set $T_F(I) = T_G(I)$

$$
\begin{aligned}
0.0595I &= \$2761 + 0.088(I - \$50,200) \\
0.0595I &= \$2761 + 0.088I - \$4418 \\
0.0595I &= -\$1657 + 0.088I \\
\$1657 &= 0.0285\,I
\end{aligned}
$$

So
$$I \approx \$58,140$$

Note that $50,200 < \$58,140 \le \$90,000$, so the value for I falls within the constraints on I.

Having found a value for I, we now need to find the corresponding value for taxes when $I = \$58,140$. We can use either of the original functions and substitute the value found for I.

Since	$T_F(I) =$	$0.0595\,I$
substituting $58,140 for I	$T_F(\$58,140) =$	$(0.0595)(58140)$
gives us	$T_F(\$58,140) \approx$	$\$3,460$

We can check to see if the coordinates ($58,140, $3,460) also satisfy the function T_G.

Since	$T_G(I) =$	$\$2761 + 0.088(I - \$50,200)$
substituting $58,140 for I	$T_G(\$58,140) =$	$\$2761 + 0.088(\$58140 - \$50,200)$
	$=$	$\$2761 + 0088(\$7940)$
	$\approx$	$\$2761 + \$699$
gives us	$T_G(\$58,140) \approx$	$\$3,460$

The coordinates ($58,140 , $3,460) satisfy both equations.

Single people in Massachusetts with incomes below about $58,140 would probably be happier with a graduated income tax, while those with incomes above $58,140 would probably prefer the flat income tax.

Chapter 6 Summary

A collection of linear equations that involve the same variables is called a *system of linear equations* or a set of *simultaneous linear equations*. A pair of real numbers is a *solution* to a system of linear equations in two variables if and only if the pair of numbers is a solution to each equation.

On a graph of a system of linear equations, a solution appears as an *intersection point*. In a system of two linear equations there are three possible outcomes. If the lines intersect once, there is one solution. If the lines are parallel they never intersect and there are no solutions. If the two lines are identical they are juxtaposed- every point on the line is a solution, so there are an infinite number of solutions.

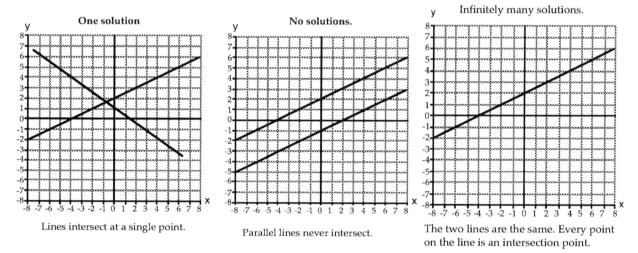

Lines intersect at a single point.

Parallel lines never intersect.

The two lines are the same. Every point on the line is an intersection point.

There are three basic strategies for algebraically finding solutions to a system of two linear equations.
1. Setting the y terms equal (when both equations are in function form)
2. Substitution (when one equation is in function form)
3. Elimination (when neither equation is in function form)

In supply and demand models in economics and in population models in biology, intersection points often represent an equilibrium, a balance point between two forces.

Some functions, like the graduated income tax, can be constructed out of pieces of several different linear functions. They are called *piecewise linear* functions.

EXERCISES

1.   Predict the *number* of solutions to each of the following systems. You don't need to find any actual solutions, but do give reasons for your answer.

a) $y = 20000 + 700x$ b) $y = 20000 + 700x$ c) $y = 20000 + 700x$
     $y = 15000 + 800x$ $y = 15000 + 700x$ $y = 20000 + 800x$

2.   Algebra Aerobics

Calculate the solution(s), if any, to each the following systems of equations. Use any method you like. Then graph each system (taking care to label both axes) and indicate on the graph where any solutions to the system lie.

I.   These are pretty straight forward:

a) $y = 12 + 2x$ d) $t = -3 + 4w$
     $y = 7 + x$ $-12w + 3t + 9 = 0$

b) $y = -11 - 2x$ e) $y = 2200x - 700$
     $y = 13 - 2x$ $y = 1300x + 4700$

c) $z = 3x - 4$ f) $3x = 5y$
     $4z + 6 = 7x$ $4y - 3x = -3$

II.   These are a little trickier (In some cases you may wish to round off answers):

a) $y = 2200x - 1800$ d) $y = 4.2 - 1.62x$
     $y = 1300x + 4700$ $1.48x - 2y + 4.36 = 0$

b) $4r + 5s = 10$ e) $xy = 1$
     $2r - 4s = -3$ $x^2y + 3x = 2$ (A nonlinear system! Give it a try.)

c) $11x + 26y = 11$
     $21x + 52y = 38$

3.   In the text the following cost equations were given for gas and solar heating.

$$C_{gas} = 12000 + 700Y$$
$$C_{solar} = 30000 + 150Y$$

where the C's stand for total cost in dollars and Y represents years since installation.

a)   Sketch the graph of this system of equations.

b)   What do the coefficients 700 and 150 represent on the graph and what do they represent in term of heating costs?

c)   What do the constant terms 12,000 and 30,000 represent on the graph and in terms of heating costs?

d)   Label the point on the graph where gas and solar heating costs are equal. Make a visual estimate of the coordinates, and interpret what the coordinates mean in terms of years and cost.

e)   Use the equations to find a better estimate for the intersection point. (You may round values to whole numbers to simplify the computations.) Show your work.

f)   During what years after installation is the total cost of solar heating more expensive? During what years is gas heating more expensive?

4.   Answer the questions in Exercise 3 for the cost equations for electric and solar heating.
$$C_{electric} = 5000 + 1100Y$$
$$C_{solar} = 30000 + 150Y$$

5.   The following are formulas predicting future raises for 4 different groups of union employees, measuring years from year 0, the date of contract:

group A: salary = 30,000 + 1500 • years
group B: salary = 30,000 + 1800 • years
group C: salary = 27,000 + 1500 • years
group D: salary = 21,000 + 2100 • years

a)   Will group A ever earn more than group B? Explain.

b)   Will group C ever catch up to group A? Explain.

c)   Which group will be making the most money in 5 years? How much will that be?

d)   Will group D ever catch up to group C? If so, after how many years and at what salary?

6.   Assume that you are given the graph of a standard model for supply and demand.

a)   What if the overall demand decreases, that is, at each price level consumers purchase a smaller quantity? Sketch a graph showing the supply curve and both the original and the changed demand curve. (Be sure to label the graph carefully.) How has the demand curve changed? Describe the shift in the equilibrium point in terms of price and quantity.

b) What if the supply decreases, that is, at each price level manufacturers produce lower quantities? Sketch a graph showing the demand curve and both the original and the changed supply curve. (Be sure to label the graph carefully.) How has the supply curve changed? Describe the shift in the equilibrium point in terms of price and quantity.

7. Assume that you are given a graph of a standard model graphing birth rate and death rate as a function of population size.

a) What if the death rate increases, that is at each population level the death rate is higher? Sketch a graph showing the birth rate and both the original and the changed death rate. (Be sure to label the graph carefully.) Describe the shift in the equilibrium point.

b) What if the birth rate increases, that is, as each population level the birth rate is higher? Sketch a graph showing the death rate and both the original and the changed birth rate. (Be sure to label the graph carefully.) Describe the shift in the equilibrium point.

8. The following table taken from a pediatric text provides a set of formulas for the approximate "average" height and weight of normal infants and children

Weight		Kilograms	Pounds
(a)	at birth	3.25	7
(b)	3-12 months	(age(mo) + 9)/2	(age(mo) + 11)
(c)	1-5 years	age(yr) x 2 + 8	(age(yr) x 5 + 17)
(d)	6-12 years	(age(yr) x 7 - 5)/2	(age(yr) x 7 + 5)

Height		Centimeters	Inches
(a)	at birth	50	20
(b)	at 1 year	75	30
(c)	2-12 years	age(yr) x 6 + 77	(age(yr) x 2.5 + 30)

Source: Behrman, R.E., & Vaughan, V.C. (editors). *Nelson Textbook of Pediatrics*, 12th edition, W.B> Saunders, Philadelphia, 1983, p. 19.

For children from birth to age 12, construct and graph a piecewise linear function for:
a) weight in kilograms
b) weight in pounds (how does this model compare to the model in Section 4.1?)
c) height in centimeters
d) height in inches
Note that at certain end points (for example at 1 year = 12 months), two formulas in the table may give you conflicting values. In these cases, decide which formula you wish to use, and be sure your function notation reflects your choice.

9.   Find out what kind of income tax your state has. Is it flat or graduated? Construct and graph a function that describes your state's income tax for one filing status. Identify the income tax for various income levels.

10. a)   (From Lida McDowell, University of Southern Mississippi)
         Construct and graph a piecewise linear function for Mississippi's graduated income tax for a single person as shown in the table.

Filing status	3.00%	4.00%	5.00%
Single	≤ $5,000	$5,001 - $10,000	$10,000 +

   b)   (From Sandi Athanassiou, University of Missouri-Columbia)
        Compare Missouri's graduated income tax for a single person as shown below in the table, with that of Mississippi shown above. All other things being equal, for what income levels would a single person be better off living in Mississippi? For what income levels would a single person be better off in Missouri?

Income	Marginal tax rate
≤$1,000	1.50%
$1,001-$2,000	2.00%
$2,001-$3,000	2.50%
$3,001-$4,000	3.00%
$4,001-$5,000	3.50%
$5,001-$6,000	4.00%
$6,001-$7,000	4.50%
$7,001-$8,000	5.00%
$8,001-$9,000	5.50%
> $9,000	6.00%

   c) What other financial considerations should be taken into consideration in comparing Mississippi and Missouri?

11. A graduated income tax is proposed in Borduria to replace an existing flat rate of 8% on all income. The new proposal is that persons will pay no tax on their first $20,000 income, 5% on their income over $20,000 to $100,000, and 10% on their income over $100,000.

   a)   Construct a table of values that shows how much tax persons will pay under both the existing 8% flat tax and the proposed new tax for each of the following incomes: $0, $20,000; $50,000; $100,000; $150,000; $200,000.

   b)   Placing income on the horizontal axis, construct a graph of tax dollars vs. income for the 8% flat tax.

c) On the same graph plot tax dollars vs. income for the proposed new graduated tax.

d) Construct a function that describes tax dollars under the existing 8% tax as a function of income.

e) Construct a piecewise function that describes tax dollars under the proposed new tax rates as a function of income.

f) Use your *graph* to estimate income level for which the taxes are the same under both plans. What plan is best for those below this income? For those above this income?

g) Use your *equations* to find the coordinates that represent the point at which the taxes are the same for both plans. Label this point on your graph.

h) If the median income in the state is $27,000 and the mean income is $35,000, do you think the new graduated tax would be voted in by the people?

Note: Borduria is a fictional totalitarian state in the Balkans that figures in the adventures of TinTin.)

12. (Computer required.) This exercise links together the analysis in the previous exercise with the FAM 1000 data.

a) It is interesting to see how many people benefit from one tax model as compared to another.

How many of the respondents in the FAM 1000 survey pay less tax with the graduated income tax than with the flat tax. (You can answer this exactly using a SORT command in the Excel sheet, or approximately, using the histogram for income distribution.)

b) In setting tax rates, the government has to calculate how much tax will be raised with a given model. To illustrate this calculation, calculate how much tax would be raised from all of the FAM 1000 respondents combined, under each of the tax plans.

i) How much tax is raised from these respondents with the flat tax? (You can calculate this using the mean income.)

ii) How much tax is raised from these respondents under the proposed new tax? (For this, you cannot use the mean income. You can calculate it exactly using a spreadsheet with all the FAM 1000 data, or you can make a good estimate using the histogram for distribution of income.)

c)   Whether a given law passes depends partly on how many people it favors, but also (to an unfortunate degree) on the resources available to those whom it benefits, or penalizes.

Estimate the total income available to those who pay less tax with the 8% flat tax than with the proposed graduated tax. Compare this to the total income of the other taxpayers.

13. Heart health is a prime concern, because heart disease is the leading cause of death in the U.S. Aerobic activities like walking, jogging, and running are recommended for cardiovascular fitness, because they increase the heart's strength and stamina.

a)   A typical training recommendation for a beginner is to walk at a moderate pace of about 3.5 miles per hour for 20 minutes. Construct a function that describes the distance traveled ($D_{beginner}$), as a function of time, T, *in minutes,* for someone maintaining this pace. (Hint you need to convert the pace into miles per minute.) Construct a small table of solutions and graph the function, using a reasonable domain.

b)   A more advanced training routine is to walk at a pace of 3.75 miles per hour for 10 minutes, and then jog at 5.25 miles per hour for 10 minutes. Construct a piecewise linear function that gives the total distance ($D_{advanced}$), as a function of time, T, *in minutes.* Generate a small table of solution and add the graph of this function to your graph in part a).

c)   Do these two graphs intersect? If so, what does the intersection point represent?

EXPLORATION 6.1.
Flat vs. Graduated Income Tax: Who Benefits?

Objective
- to compare the effect of different tax plans on individuals in different income brackets.
- to interpret intersection points

Materials/Equipment
- spreadsheet or graphing calculator (optional)
- worksheet on systems of equations in graphing calculator appendix
- graph paper
- enclosed background reading and worksheet on tax rate plans for Massachusetts

Procedure
(If you need practice on your calculator graphing systems of equations and determining intersection points, do the worksheet in the graphing calculator Appendix, Sec. 6.1)

The questions we explore are:
> For what income will taxes be equal under both plans?
> Who will benefit under the graduated income tax plan as compared to the flat tax plan?
> Who will pay more taxes under the graduated plan as compared to the flat tax plan?

In a small group or with a partner
1. Review the enclosed background information extracted from Section 7.4 "The Case of Massachusetts" regarding the flat rate vs. graduated tax plans in Massachusetts. Choose a category of people (other than single people which is used in text) on which you wish to focus.

2. Using the attached worksheet, generate a data table for taxes under the flat rate plan and the graduated plan for the group you have chosen to study.

3. Model each of these plans using taxes as a function of income. Note that the graduated income tax function will be a piecewise linear function.

Results
1. Graph the functions representing the two tax plans on the same grid. You can do this by hand and check your results using a spreadsheet or graphing calculator. Verify that the points in the data table lie on the graph. Instructions for graphing piecewise linear functions on the graphing calculator are in the Appendix.

2. Estimate from your graph any intersection points for the two functions. Use your equations to calculate exact values for the points of intersection.

3.   (Extra credit): Use your results to make changes in the tax plans.
     Decide on a different income for which taxes will be equal under both plans. You can
     use what you know about the distribution of income in the U.S. from the FAM 1000
     data to make your decision.
     Alter one or both of the original functions such that both tax plans will generate the
     same tax given the income you have chosen.

Analysis of your results

- Interpret your findings. What do the intersection points tell you about the differences
  between the tax plans?
- What information is useful in deciding on the merits of each of the plans?

Exploration Linked Homework
Reporting your results

   Take a stance for or against a graduated income tax. Using supportive quantitative
   evidence, write a 60 second summary for a voter's pamphlet advocating your position.
   Present your arguments to the class.

Background Information (from Section 7.4 in the text)
Flat rate vs. Graduated rate: The *Case of Massachusetts*

The state of Massachusetts currently has a flat tax rate of 5.95%. There has been an ongoing
debate as to whether or not to change to a graduated income tax plan. The questions we
explore are: For what income will the taxes be equal? Who will benefit under the proposed
graduated tax plan?

In Massachusetts a graduated income tax proposal, called "Proposition 7", was on the ballet
in 1994. "Proposition 7" proposed that the state's flat tax plan be replaced with a graduated
income tax (called marginal rates) as follows:

Filing Status	5.5%	8.80%	9.80%
Married/Joint	<$81,000	$81,000-$150,000	$150,000+
Married/Separate	<$40,500	$40,500-$75,000	$75,000+
Single	<$50,200	$50,200-$90,000	$90,000+
Head of Household	<$60,100	$60,100-$120,000	$120,000+

Graduated Tax Plan for Various Income Levels

This tax rate structure is designed to tax at the higher rate only that portion of the individual's income that exceeds the threshold. For example, for those who file as single people,
the graduated tax rate means that earned income under $50,200 would be taxed at a rate of
5.5% while any income over $50,200 would be taxed at higher rates. If a single person
earned $100,000, then he or she would be taxed at 5.5% on the first $50,200, 8.8% on the next
$39,800 (the portion of income between $50,200 and $90,000) and 9.8% on the final $10,000.

Worksheet for Exploration 6.1. Flat vs. Graduated Income Tax: Who Benefits?

Selected filing status:_____

Data Table for Massachusetts Taxes
Flat vs. Graduated Rate

Income after exemptions and deductions	Current Flat Tax	Graduated Tax Under Proposition 7
$0		
$25,000		
$50,000		
$75,000		
$100,000		
$150,000		
$200,000		

Function for Current Flat Tax Plan

$$T_{flat}(I) =$$

Function for Proposition 7 Tax Plan

$$T_{grad}(I) = \begin{cases} \end{cases}$$

Include a graph of the two functions

Estimation of intersection point or points from graphs_____

Use your equations to find intersection point or points. Show your work.

EXPLORATION 6.2
Flat vs. Graduated Income Tax: What Are The Assumptions? [8]

Objective

- analyze the effect of different deductions on federal income taxes
- examine how to interpret intersecting line segments
- become familiar with standard 1040 federal income tax form

Materials

- *The New York Times* article How a Flat Tax Would Work, For You and for Them"; (In Anthology of Readings) This article should be read before doing the exploration.
- spreadsheet or graphing calculator (optional)
- graph paper
- enclosed forms for four case studies, tax rate schedules, 1040 tax return form, flat tax hypothetical form

Procedure

The questions we will be exploring are:
  What effect will different deductions have on the tax that is owed?
  When will taxes be equal under the current and proposed flat tax?
  What predictions would you make about the effect of a flat tax on the revenue generated by federal taxes?

In a small group:
1. Each person should choose one of the case studies and fill out two tax forms, the U.S. 1040 federal tax form and the flat tax hypothetical form for the person described. Notice that all of the case studies are on single people.
2. As a group, analyze how the deductions allowed for single people influence the amount of taxes that are owed under each of the tax plans. Jot down your observations and any questions you may have.
3. Construct a function representing the flat tax proposed (on the hypothetical form) for single people who are not heads of household and do not have dependent children. Use taxes as a function of total income before deductions. (Note that in all the previous examples, we have used adjusted income after deductions).
4. Make predictions as to how deductions will affect taxes given the current tax plan. Based on the case studies which your group has analyzed and on your own knowledge of deductions, each person should form a different estimate for the average total deduction for single people, who are not heads of household and do not have dependents. Each person should use his or her estimate to construct a piecewise linear function that

---

[8] Adapted from an Exploration by Peggy Tibbs, Arkansas Technical University

represents the current tax plan, (treating taxes as a function of total income). One person in the group should use the minimum deduction allowed for single people •
Each person should have a different piecewise function, representing different assumptions about deductions.

Results

1. Graph your two functions (one representing the current tax plan and the other representing the proposed flat tax) on the same grid. You can do this by hand and check your results using a spreadsheet or graphing calculator. Make sure to label the graph in terms of which average deduction you have assumed. Instructions for graphing piecewise functions on the graphing calculator are in the Appendix.
2. Estimate from your graph any intersection points for the two functions. Use your equations to calculate exact values for the points of intersection.
3. Now compare your results to those of the rest of your group.

Analysis of Results

- Interpret your group's findings. What do the intersection points tell you about the differences between the tax plans? Be sure to include the horizontal intersection points in your analysis. Jot down your observations and any questions you may have.
- When will a single person, who is not head of household and does not have dependents, pay more federal taxes under the proposed flat tax plan than under the current tax plan?
- How will the tax plans affect how much income the federal government collects? Be sure to give reasons for your answers.
- What information is useful in deciding on the merits of each of the plans?

Exploration Linked Homework

Reporting your results
Prepare a balanced article reporting on the hypothetical flat tax and the current federal tax plan. Each small group may want to prepare one report, with individual members assigned to work on different aspects.

CASE STUDY #1: Jane Jackson

Jane Jackson SSN# 444-80-2221, age 61, lives with her married son, Marvin, and Marvin's wife, Laura. They live on a farm in the state of Kansas. Jane has an eighth grade education, and works as a carpenter. She earned $33,000 last year. She has a small savings account, on which she earned $106. interest. Marvin earned $35,700 last year. Laura stayed home and ran the farm, which belongs to Marvin and Laura. They had a bad year on the farm; their losses were $5,450. Jane had medical expenses of $3,000 that were unreimbursed by her insurance company. She contributed $1,200 to her church. Jane spent $500 on tools, and claimed $300 for travel expenses related to her work. Jane's husband, Joe, died in December. 1994. Jane moved in with her son the following February. She paid $350 to a local moving company.

CASE STUDY #2: Benjamin Smith

Benjamin Smith, age 43, SSN#222-77-9999, lives in a suburb of Newark, New Jersey. He attended a community college for 1 year, but did not complete any degree or course of study. He works as a bus driver, and earned $19,000 last year. He and his wife, Louise, were divorced 3 years ago. Benjamin has custody of their three children (Bertha, age 9, Billy, age 10, and Louie, age 12). Louise paid alimony of $4,200 last year. Benjamin won $6,600 gambling at Atlantic City. He had $500 in unreimbursed medical expenses. He bought a house and paid $2,317 in interest on his mortgage and $1,200 in property taxes. He donated $50 to charity.

CASE STUDY #3: Roberta Jenkins

Roberta Jenkins, SSN# 333-00-8989, is a logger in Oregon. She is a high school graduate, 41 years old and widowed. She has four children, all living with her. They are Rachel, age 3, Roger , age 7, and the twins Rhonda and Rupert, age 12. Her mother, who lives next door, serves as a babysitter. Roberta earned $20,000 last year. She has no savings. Roberta is renting a small house, at $300 per month. She has no health insurance, and had $1,500 in medical expenses last year. She contributed $250 to charity. She had to borrow $500 from the bank to fix her truck and paid $60 in interest on the loan.

CASE STUDY #4: Aaron Anderson

Aaron Anderson, SSN# 333-00-999, is 29 years old and a high school graduate. He is divorced, and has custody of his 2 boys, Chad, age 8, and Charles, age 9. He lives in rural Mississippi where he earned $32,000 last year working as a garbage collector. His ex-wife paid him $7,000 in alimony last year. He rents an apartment at $450 per month. He had $500 in unreimbursed medical expenses. He did not contribute to any church or charity. He lost $450 gambling. Last year, his apartment caught fire. He did not have home-owner's insurance, and the apartment owner's insurance only covered the apartment itself, not the contents. His cost to replace clothes, furniture, etc. was $3,000. (He received some items of furniture and clothing at no cost from the Red Cross.)

1995
Tax Rate
Schedules

Caution: *Use **only** if your taxable income (Form 1040, line 37) is $100,000 or more. If less, use the **Tax Table.** Even though you cannot use the tax rate schedules below if your taxable income is less than $100,000, all levels of taxable income are shown so taxpayers can see the tax rate that applies to each level.*

Schedule X—Use if your filing status is **Single**

If the amount on Form 1040, line 37, is: Over—	But not over—	Enter on Form 1040, line 38	of the amount over—
$0	$23,350	.... 15%	$0
23,350	56,550	$3,502.50 + 28%	23,350
56,550	117,950	12,798.50 + 31%	56,550
117,950	256,500	31,832.50 + 36%	117,950
256,500	....	81,710.50 + 39.6%	256,500

Schedule Y-1—Use if your filing status is **Married filing jointly** or **Qualifying widow(er)**

If the amount on Form 1040, line 37, is: Over—	But not over—	Enter on Form 1040, line 38	of the amount over—
$0	$39,000	.... 15%	$0
39,000	94,250	$5,850.00 + 28%	39,000
94,250	143,600	21,320.00 + 31%	94,250
143,600	256,500	36,618.50 + 36%	143,600
256,500	....	77,262.50 + 39.6%	256,500

Schedule Y-2—Use if your filing status is **Married filing separately**

If the amount on Form 1040, line 37, is: Over—	But not over—	Enter on Form 1040, line 38	of the amount over—
$0	$19,500	.... 15%	$0
19,500	47,125	$2,925.00 + 28%	19,500
47,125	71,800	10,660.00 + 31%	47,125
71,800	128,250	18,309.25 + 36%	71,800
128,250	....	38,631.25 + 39.6%	128,250

Schedule Z—Use if your filing status is **Head of household**

If the amount on Form 1040, line 37, is: Over—	But not over—	Enter on Form 1040, line 38	of the amount over—
$0	$31,250	.... 15%	$0
31,250	80,750	$4,687.50 + 28%	31,250
80,750	130,800	18,547.50 + 31%	80,750
130,800	256,500	34,063.00 + 36%	130,800
256,500	....	79,315.00 + 39.6%	256,500

Form **1040**

Department of the Treasury—Internal Revenue Service

U.S. Individual Income Tax Return **1995** (5) IRS Use Only—Do not write or staple in this space.

For the year Jan. 1–Dec. 31, 1995, or other tax year beginning , 1995, ending , 19   OMB No. 15

Label

(See instructions on page 11.)

Use the IRS label. Otherwise, please print or type.

L A B E L   H E R E

Your first name and initial	Last name
If a joint return, spouse's first name and initial	Last name
Home address (number and street). If you have a P.O. box, see page 11.	Apt. no.
City, town or post office, state, and ZIP code. If you have a foreign address, see page 11.	

Your social security nu

Spouse's social security

For Privacy Act a Paperwork Reduc Act Notice, see p

Presidential Election Campaign
(See page 11.)

Do you want $3 to go to this fund? . . . .

If a joint return, does your spouse want $3 to go to this fund? . . . .

Yes | No | Note: *Check will not chan tax or reduce refund.*

Filing Status

(See page 11.)

Check only one box.

1. [ ] Single
2. [ ] Married filing joint return (even if only one had income)
3. [ ] Married filing separate return. Enter spouse's social security no. above and full name here. ▶ _____
4. [ ] Head of household (with qualifying person). (See page 12.) If the qualifying person is a child but not your dep enter this child's name here. ▶ _____
5. [ ] Qualifying widow(er) with dependent child (year spouse died ▶ 19). (See page 12.)

Exemptions

(See page 12.)

If more than six dependents, see page 13.

6a [ ] **Yourself.** If your parent (or someone else) can claim you as a dependent on his or her tax return, **do not** check box 6a. But be sure to check the box on line 33b on page 2 .

b [ ] **Spouse** . . . .

c **Dependents:**

(1) First name Last name	(2) Dependent's social security number. If born in 1995, see page 13.	(3) Dependent's relationship to you	(4) No. of months lived in your home in 1995

No. of boxes checked on 6a and 6b

No. of your children on 6c who:

• lived with you

• didn't live with you due to divorce or separation (see page 14)

Dependents on 6c not entered above

d If your child didn't live with you but is claimed as your dependent under a pre-1985 agreement, check here ▶ [ ]

e Total number of exemptions claimed . . . .

Add numbers entered on lines above ▶

Income

Attach Copy B of your Forms W-2, W-2G, and 1099-R here.

If you did not get a W-2, see page 14.

Enclose, but do not attach, your payment and payment voucher. See page 33.

7	Wages, salaries, tips, etc. Attach Form(s) W-2	7
8a	**Taxable** interest income (see page 15). Attach Schedule B if over $400	8a
b	**Tax-exempt** interest (see page 15). DON'T include on line 8a **8b**	
9	Dividend income. Attach Schedule B if over $400	9
10	Taxable refunds, credits, or offsets of state and local income taxes (see page 15) . .	10
11	Alimony received	11
12	Business income or (loss). Attach Schedule C or C-EZ	12
13	Capital gain or (loss). If required, attach Schedule D (see page 16)	13
14	Other gains or (losses). Attach Form 4797	14
15a	Total IRA distributions . **15a** b Taxable amount (see page 16)	15b
16a	Total pensions and annuities **16a** b Taxable amount (see page 16)	16b
17	Rental real estate, royalties, partnerships, S corporations, trusts, etc. Attach Schedule E	17
18	Farm income or (loss). Attach Schedule F	18
19	Unemployment compensation (see page 17)	19
20a	Social security benefits **20a** b Taxable amount (see page 18)	20b
21	Other income. List type and amount—see page 18	21
22	Add the amounts in the far right column for lines 7 through 21. This is your **total income** ▶	22

Adjustments to Income

23a	Your IRA deduction (see page 19)	**23a**
b	Spouse's IRA deduction (see page 19)	**23b**
24	Moving expenses. Attach Form 3903 or 3903-F . . .	**24**
25	One-half of self-employment tax	**25**
26	Self-employed health insurance deduction (see page 21)	**26**
27	Keogh & self-employed SEP plans. If SEP, check ▶ [ ]	**27**
28	Penalty on early withdrawal of savings	**28**
29	Alimony paid. Recipient's SSN ▶ _____	**29**
30	Add lines 23a through 29. These are your **total adjustments** . . . . ▶	30

Adjusted Gross Income

31 Subtract line 30 from line 22. This is your **adjusted gross income**. If less than $26,673 and a child lived with you (less than $9,230 if a child didn't live with you), see "Earned Income Credit" on page 27 ▶ | 31 |

Cat. No. 11320B

Form **104**

Tax Compu-tation

(See page 23.)

32 Amount from line 31 (adjusted gross income) . . . . | **32** |

33a Check if: ☐ **You** were 65 or older, ☐ Blind; ☐ **Spouse** was 65 or older, ☐ Blind.
Add the number of boxes checked above and enter the total here . . . . ▶ **33a** |

b If your parent (or someone else) can claim you as a dependent, check here . ▶ **33b** ☐

c If you are married filing separately and your spouse itemizes deductions or you are a dual-status alien, see page 23 and check here . . . . ▶ **33c** ☐

34 Enter the larger of your: { **Itemized deductions** from Schedule A, line 28, **OR**
Standard deduction shown below for your filing status. **But if you checked any box on line 33a or b,** go to page 23 to find your standard deduction. If you checked box 33c, your standard deduction is zero.
• Single—$3,900   • Married filing jointly or Qualifying widow(er)—$6,550
• Head of household—$5,750   • Married filing separately—$3,275 } | **34** |

35 Subtract line 34 from line 32 . . . . | **35** |

36 If line 32 is $86,025 or less, multiply $2,500 by the total number of exemptions claimed on line 6e. If line 32 is over $86,025, see the worksheet on page 23 for the amount to enter . | **36** |

If you want the IRS to figure your tax, see page 35.

37 **Taxable income.** Subtract line 36 from line 35. If line 36 is more than line 35, enter -0- | **37** |

38 Tax. Check if from **a** ☐ Tax Table, **b** ☐ Tax Rate Schedules, **c** ☐ Capital Gain Tax Worksheet, or **d** ☐ Form 8615 (see page 24). Amount from Form(s) 8814 ▶ **e** _____ | **38** |

39 Additional taxes. Check if from **a** ☐ Form 4970 **b** ☐ Form 4972 . . . . | **39** |

40 Add lines 38 and 39 . . . . ▶ | **40** |

Credits

(See page 24.)

41 Credit for child and dependent care expenses. Attach Form 2441 | **41** |
42 Credit for the elderly or the disabled. Attach Schedule R . . | **42** |
43 Foreign tax credit. Attach Form 1116 . . . . . . | **43** |
44 Other credits (see page 25). Check if from **a** ☐ Form 3800
b ☐ Form 8396 **c** ☐ Form 8801 **d** ☐ Form (specify) _____ | **44** |
45 Add lines 41 through 44 . . . . | **45** |
46 Subtract line 45 from line 40. If line 45 is more than line 40, enter -0- . . . . ▶ | **46** |

Other Taxes

(See page 25.)

47 Self-employment tax. Attach Schedule SE . . . . | **47** |
48 Alternative minimum tax. Attach Form 6251 . . . . | **48** |
49 Recapture taxes. Check if from **a** ☐ Form 4255 **b** ☐ Form 8611 **c** ☐ Form 8828 . | **49** |
50 Social security and Medicare tax on tip income not reported to employer. Attach Form 4137 | **50** |
51 Tax on qualified retirement plans, including IRAs. If required, attach Form 5329 . . . | **51** |
52 Advance earned income credit payments from Form W-2 . . . . . . | **52** |
53 Household employment taxes. Attach Schedule H . . . . | **53** |
54 Add lines 46 through 53. This is your **total tax** . . . . ▶ | **54** |

Payments

Attach Forms W-2, W-2G, and 1099-R on the front.

55 Federal income tax withheld. If any is from Form(s) 1099, check ▶ ☐ | **55** |
56 1995 estimated tax payments and amount applied from 1994 return . | **56** |
57 **Earned income credit.** Attach Schedule EIC if you have a qualifying child. Nontaxable earned income: amount ▶ _____
and type ▶ .... | **57** |
58 Amount paid with Form 4868 (extension request) . . . . | **58** |
59 Excess social security and RRTA tax withheld (see page 32) | **59** |
60 Other payments. Check if from **a** ☐ Form 2439 **b** ☐ Form 4136 | **60** |
61 Add lines 55 through 60. These are your **total payments** . . . . ▶ | **61** |

Refund or Amount You Owe

62 If line 61 is more than line 54, subtract line 54 from line 61. This is the amount you **OVERPAID**. . | **62** |
63 Amount of line 62 you want **REFUNDED TO YOU**. . . . . ▶ | **63** |
64 Amount of line 62 you want **APPLIED TO YOUR 1996 ESTIMATED TAX** ▶ | **64** |
65 If line 54 is more than line 61, subtract line 61 from line 54. This is the **AMOUNT YOU OWE**. For details on how to pay and use **Form 1040-V**, Payment Voucher, see page 33 . . ▶ | **65** |
66 Estimated tax penalty (see page 33). Also include on line 65 | **66** |

Sign Here

Keep a copy of this return for your records.

Under penalties of perjury, I declare that I have examined this return and accompanying schedules and statements, and to the best of my knowledge and belief, they are true, correct, and complete. Declaration of preparer (other than taxpayer) is based on all information of which preparer has any knowledge.

▶ Your signature	Date	Your occupation
▶ Spouse's signature. If a joint return, BOTH must sign.	Date	Spouse's occupation

Paid Preparer's Use Only

Preparer's signature ▶	Date	Check if self-employed ☐	Preparer's social security no.
Firm's name (or yours if self-employed) and address ▶		EIN	
		ZIP code	

Schedule A—Itemized Deductions

(Schedule B is on back)

▶ Attach to Form 1040. ▶ See Instructions for Schedules A and B (Form 1040).

OMB No. 1545-0

19**95**

Attachment
Sequence No.

Name(s) shown on Form 1040 | Your social security n

Medical and Dental Expenses		*Caution: Do not include expenses reimbursed or paid by others.*	
	1	Medical and dental expenses (see page A-1)	1
	2	Enter amount from Form 1040, line 32. [2]	
	3	Multiply line 2 above by 7.5% (.075)	3
	4	Subtract line 3 from line 1. If line 3 is more than line 1, enter -0-	4
Taxes You Paid (See page A-1.)	5	State and local income taxes	5
	6	Real estate taxes (see page A-2)	6
	7	Personal property taxes	7
	8	Other taxes. List type and amount ▶	8
	9	Add lines 5 through 8	9
Interest You Paid (See page A-2.)	10	Home mortgage interest and points reported to you on Form 1098	10
	11	Home mortgage interest not reported to you on Form 1098. If paid to the person from whom you bought the home, see page A-3 and show that person's name, identifying no., and address ▶	
Note: Personal interest is not deductible.	12	Points not reported to you on Form 1098. See page A-3 for special rules	11
			12
	13	Investment interest. If required, attach Form 4952. (See page A-3.)	13
	14	Add lines 10 through 13	14
Gifts to Charity If you made a gift and got a benefit for it, see page A-3.	15	Gifts by cash or check. If you made any gift of $250 or more, see page A-3	15
	16	Other than by cash or check. If any gift of $250 or more, see page A-3. If over $500, you **MUST** attach Form 8283	16
	17	Carryover from prior year	17
	18	Add lines 15 through 17	18
Casualty and Theft Losses	19	Casualty or theft loss(es). Attach Form 4684. (See page A-4.)	19
Job Expenses and Most Other Miscellaneous Deductions (See page A-5 for expenses to deduct here.)	20	Unreimbursed employee expenses—job travel, union dues, job education, etc. If required, you **MUST** attach Form 2106 or 2106-EZ. (See page A-5.) ▶	20
	21	Tax preparation fees	21
	22	Other expenses—investment, safe deposit box, etc. List type and amount ▶	22
	23	Add lines 20 through 22	23
	24	Enter amount from Form 1040, line 32. [24]	
	25	Multiply line 24 above by 2% (.02)	25
	26	Subtract line 25 from line 23. If line 25 is more than line 23, enter -0-	26
Other Miscellaneous Deductions	27	Other—from list on page A-5. List type and amount ▶	27
Total Itemized Deductions	28	Is Form 1040, line 32, over $114,700 (over $57,350 if married filing separately)? **NO.** Your deduction is not limited. Add the amounts in the far right column for lines 4 through 27. Also, enter on Form 1040, line 34, the **larger** of this amount or your standard deduction. **YES.** Your deduction may be limited. See page A-5 for the amount to enter.	28

For Paperwork Reduction Act Notice, see Form 1040 instructions.　　　　Cat. No. 11330X　　　　Schedule A (Form 1040

FLAT TAX HYPOTHETICAL FORM

NAME (IF JOINT RETURN, ALSO GIVE SPOUSE'S NAME)	YOUR SOCIAL SECURITY NUMBER
	SPOUSE'S SOC. SEC. NUMBER
PRESENT HOME ADDRESS	
CITY, TOWN, POST OFFICE, STATE, AND ZIP CODE	YOUR OCCUPATION
	SPOUSE'S OCCUPATION

1. Wages and salary   1._____

2. Pension and retirement benefits   2._____

3. Total compensation (line 1 plus line 2)   3._____

4. Personal allowances
    a) $16,500 for married filing jointly   4a._____
    b) $9,500 for single   4b._____
    c) $14,000 for single head of household   4c._____

5. Number of dependents, not including spouse   5._____

6. Personal allowances for dependents (line 5 multiplied by $4,500)   6._____

7. Total personal allowances (line 4 plus line 6)   7._____

8. Taxable compensation
    (line 3 less line 7, is positive; otherwise zero)   8._____

9. Tax (19% of line 8)   9._____

Part II

Algebra in the Physical and Life Sciences

Overview

Part II will give you experience with algebraic tools often used in modeling physical phenomena. Here we introduce several non-linear families of functions: (exponential, power and polynomial) and show their many uses, primarily in the physical sciences, but also in some other disciplines.

Astronomy must deal with numbers from the incredibly large to the extremely small. It provides a meaningful context for the development of a number notation that can help us understand the amazing range of sizes found in the universe. Scientific notation is the international method for compactly expressing numbers of any size. Logarithmic scales provide a convenient way to display numbers of widely differing sizes.

Biology provides the setting for a discussion of modeling growth and decay. We analyze the growth of bacteria, and see how the same model can be applied to applications in nuclear physics, music, epidemiology and finance.

Examples from chemistry motivate the study of power and polynomial functions. Laws for the behavior of gases tell us when helium balloons rise and why scuba divers don't hold their breath.

We turn next to physics for an understanding of how scientists can uncover the "rules of the universe," such as the basic physical laws that govern gravity. We record the motion of freely falling bodies and construct mathematical models to describe their behavior.

Chapter 7

Deep Time And Deep Space

Overview

We inhabit a minuscule portion of a vast universe that has been evolving for billions of years. Understanding where we fit in the history of the known universe and where we are relative to other planets and galaxies, was traditionally the province of philosophers and religious leaders. Scientists now tackle these questions, and we have come to some understanding of the age and size of our universe.

The scale of the universe is truly "astronomical." *Scientific notation* allows scientists easily to specify and compare magnitudes from the sub-atomic to the cosmic. Scientific notation, like our number system in general, is based on *powers of ten*. The laws of exponents governing the manipulation of expressions written in base 10 can be extended to other bases. *Logarithms* are used to scale graphs or to construct measurement systems that deal with numbers of widely varying size.

The first exploration helps you to develop an understanding of the relative size and age of various objects in the universe, while offering practice in the laws of exponents and unit conversions. The second exploration is based on Kepler's discoveries about the laws governing planetary motion.

After reading this chapter you should be able to:

- use powers of ten and the metric system
- write expressions in scientific notation
- convert from one unit of measure to another
- perform operations on expressions of the form a^n
- perform operations on expressions of the form $a^{n/m}$
- compare numbers of widely differing sizes
- calculate logarithms base 10

7.1 Powers of Ten

Measuring Time and Space

On a daily basis we encounter quantities measured in tenths, tens, hundreds or perhaps thousands. Finance or politics may bring us news of "1.2 billion people living in China" or "the federal debt of over $4 trillion dollars." In the physical sciences the range of numbers encountered is much larger. A more convenient system is needed. *Scientific notation* was developed, not only to write, but to help understand and compare the sizes found in our universe, from the largest object we know, the observable universe, to the tiniest, the minuscule quarks oscillating inside the nucleus of an atom. We use examples from deep space and time to demonstrate the power of scientific notation.

The Big Bang

In 1929 the American astronomer Edwin Hubble published an astounding paper that established the basic framework for measuring time and space in our universe. He claimed that the universe was expanding. From his observations of galaxies Hubble predicted that, depending on its total mass, the universe would either expand forever or collapse back upon itself.[1] Since then most astronomers and cosmologists have become convinced that Hubble's controversial theory was correct, and they have started to investigate its consequences. If the universe is expanding, they asked, when did it start to expand, and how big was it at that time? The commonly accepted answer to this question is that somewhere between 8 and 15 billion years ago the universe began expanding rapidly from an infinitesimally small point. This event is referred to as the "Big Bang" and the universe has been expanding ever since. If this is correct, then everything that we have ever seen or measured has occurred within the last 15 billion or so years.[2]

Scientists picture the universe since the Big Bang as an expanding sphere. They believe that the rim of that sphere still contains background radiation left over from its beginning, and they estimate that about 1% of the static on a TV set (when it's not hooked up to cable) is due to the radiation residue of the Big Bang. The size of this expanding sphere determines the outer limits for the universe as we know it.

---

[1] Cosmologists are unable to estimate the total mass of the universe, since they are in the embarrassing position of not being able to find about 90% of it. Scientists call this missing mass *dark matter*, which describes not only its invisibility but scientists' own mystification. Some recent experiments at Los Alamos National Laboratory in New Mexico have produced evidence that some particles called *neutrinos*, previously thought to be weightless, may actually have mass and may constitute a major component of dark matter.

[2] The actual age of the universe is still a hotly debated topic. Observations made by the repaired Hubble space telescope give estimates for the age of the universe somewhere between 8 and 12 billion years old. Yet cosmologists are fairly certain that the oldest stars in the Milky Way are at least 14 billion years old: a troubling contradiction to resolve. Keep reading the newspaper for the latest figures.

Deep Space [3]

Measuring the Universe: The Metric System

The international scientific community (and most of the rest of the world) uses the metric system, a system of measurements based on the meter (which is about 39.37 inches, a little over three feet). In daily life Americans have resisted converting to the metric system, and still use the English system of inches, feet and yards. So in order to understand scientific measurements we need to know how to convert between metric and English units. [4] The following table shows the conversions for three standard units of length: the meter, the kilometer, and the centimeter.

Table 7.1				
Metric Unit	**Abbreviation**	**In meters**	**Equivalence in English Units**	**Informal conversion**
meter	m		3.28 feet	the width of a twin bed a little more than a yard
kilometer	km	1000 meters	0.62 miles	a casual 12 minute walk a little over a half a mile
centimeter	cm	0.01 meters	0.39 inches	the width of a finger nail a little under a half an inch

All of the measurements in the following examples and most of those throughout Part II will be in metric units.

The Observable Universe

Current measurements with the most advanced scientific instruments generate a best guess for the radius of the observable universe at about 100,000,000,000,000,000,000,000,000 meters, or "one hundred trillion trillion meters." Obviously we need a more convenient way to read, write, and say this number.

100,000,000,000,000,000,000,000,000 is a 1 with 26 zeroes after it. You could also think of it as 26 tens multiplied together. In order to avoid writing a large number of zeroes, exponents can be used as a shorthand.

- 10^{26} means: a 1 with 26 zeroes after it
- 10^{26} means: $10 \bullet 10 \bullet 10 \bullet \ldots \bullet 10$, the product twenty six 10's.
- 10^{26} is spoken as "ten to the twenty–sixth" or "ten to the twenty–sixth power."

---

[3] For a real appreciation of the size of things in the Universe, we highly recommend the video and related book by Philip and Phylis Morrison and the Office of Charles and Ray Eames entitled *Powers of Ten*. A room in the Smithsonian Museum in Washington, D.C. is dedicated just to running this remarkable video.

[4] Eventual conversion to a metric system seems inevitable. There is once again a renewed push for a "Metric America."

> When n is a positive integer,
> 10^n is defined as the product of n 10's, or
> 1 followed by n zeroes.

So the estimated size of the radius of the observable universe is 10^{26} meters.

The Milky Way

Through a telescope we can see some galaxies and clusters of galaxies, most of them elliptical in shape. With the naked eye, we can see about four galaxy clusters, members of what is called the Local Group. Most of the objects we see in the night sky are the luminous balls of gas we call stars, that are in our own galaxy, the Milky Way. There are approximately 100 billion (100,000,000,000 or 10^{11}) stars in the Milky Way. Most of these are thought to have planets circling them, though the first confirmation of a planet outside our solar system came only in 1993. We can also see a few nebulae, large clouds of dust and gas. In North America we can observe the Horsehead Nebula, the Great Nebula in Orion, and the Nebulae in Pleiades. The Milky Way has a relatively unusual spiral shape, and the Earth is near the edge of one of the long spiral arms. The radius of the Milky Way is approximately 1,000,000,000,000,000,000,000 or 10^{21} meters.

Our Solar System

The few other sky objects we can see regularly with the naked eye (outside of the moon, airplanes, and man–made satellites) are other planets in our own solar system. Unlike stars, planets do not glow; like the moon, we see them by light they reflect from the sun.

The approximate radius of our solar system is 1,000,000,000,000 or 10^{12} meters. In the metric system the prefix *tera* means 10^{12}, so this distance is one *terameter*.

Our Sun

The radius of our own sun, a modest star about half way through its life cycle, is approximately 1,000,000,000 or 10^9 meters (one billion meters). In the metric system, *giga* is the prefix for 10^9, so the distance is one *gigameter*.

The Earth

Our home, the third planet orbiting our sun, has a radius of somewhat less than 10,000,000 or 10^7 meters.

Us

Human beings are roughly in the middle of the scale of measurable objects in the universe. Human heights (including children) vary from about one third of a meter to two meters so a rough average is 1 meter tall. In order to continue the system of writing all sizes using powers of ten, we need a way to express 1 as a power of 10.

Since $10^3 = 1000$, $10^2 = 100$, and $10^1 = 10$, a logical way to continue the system would be to say that $10^0 = 1$. If you think of reducing a power of ten by 1 as equivalent to dividing by 10, the following table gives justification for defining 10^0 as equal to 1.

$$10^2 = \frac{10^3}{10} = \frac{1000}{10} = 100$$

$$10^1 = \frac{10^2}{10} = \frac{100}{10} = 10$$

$$10^0 = \frac{10^1}{10} = \frac{10}{10} = 1$$

Hence the convention:

10^0 is defined to be 1.

Humans average approximately 1 or 10^0 meter tall.

DNA Molecules

A DNA strand provides genetic information for a human being. It is made up of a chain of building blocks called nucleotides. The chain is tightly coiled into a double helix, but stretched out it would measure about 0.01 meters in length. How does this DNA length translate into a power of ten? 0.01 (or one hundredth) $= \frac{1}{100} = \frac{1}{10^2}$. When 10 to a power occurs in the denominator of a fraction, it can be brought up into the numerator if the sign of the power is reversed. This means that $0.01 = \frac{1}{10^2}$ is written as 10^{-2}. So a DNA strand, uncoiled and measured lengthwise, is approximately 10^{-2} meters or one *centimeter*.

By using negative exponents, we can logically continue the powers of ten series to smaller numbers. It is still true that reducing the power by 1 is equivalent to dividing by 10:

$10^3 = 1000$, $10^2 = 100$, $10^1 = 10$, $10^0 = 1$, $10^{-1} = 0.1$, $10^{-2} = 0.01$

Since
$$10^1 = \frac{10^2}{10} = \frac{100}{10} = 10$$

$$10^0 = \frac{10^1}{10} = \frac{10}{10} = 1$$

$$10^{-1} = \frac{10^0}{10} = \frac{1}{10} = 0.1$$

$$10^{-2} = \frac{10^{-1}}{10} = \frac{0.1}{10} = 0.01$$

When n is a positive integer, 10^{-n} is defined to be $\dfrac{1}{10^n}$, or equivalently, a decimal point followed by n-1 zeroes, followed by a 1.

Living Cells

The approximate size of the radius of a living cell is $0.00001 = \dfrac{1}{100,000} = \dfrac{1}{10^5} = 10^{-5}$ meters

Atoms

Atoms have an approximate average radius of $0.0000000001 = \dfrac{1}{10,000,000,000} =$

$\dfrac{1}{10^{10}} = 10^{-10}$ meters, (a unit commonly called an *Angstrom*).

Hydrogen Atoms

The hydrogen atom, the smallest of the atoms, has an approximate radius of

$0.00000000001 = \dfrac{1}{100,000,000,000} = \dfrac{1}{10^{11}} = 10^{-11}$ meters.

Protons

A proton, a positively charged particle usually located in the nucleus of the atom, has an

approximate radius of $0.000000000000001 = \dfrac{1}{1,000,000,000,000,000} = \dfrac{1}{10^{15}} = 10^{-15}$ meters

or one *femtometer*.

Quarks

Quarks are nature's strange elementary building blocks. Three particular quarks form a proton, three other sorts form a neutron. Different quark combinations make more transitory particles. A quark's size is approximately $0.0000000000000001 =$

$= \dfrac{1}{10,000,000,000,000,000} = \dfrac{1}{10^{16}} = 10^{-16}$ meters

Summary Of Powers Of Ten[5]

The following table summarizes the powers of ten notation.

Positive Powers of 10	$10^9 =$	1,000,000,000
	$10^6 =$	1,000,000
	$10^3 =$	1,000
	$10^2 =$	100
	$10^1 =$	10
special case: Zero Power of 10	$10^0 =$	1
Negative Powers of 10	$10^{-1} =$	.1
	$10^{-2} =$	.01
	$10^{-3} =$	.001
	$10^{-6} =$	.000001
	$10^{-9} =$	.000000001

In general,

When n is a positive integer:

$10^n = 10 \bullet 10 \bullet 10 \bullet \ldots \bullet 10$ (the product of n tens) or 1 followed by n zeroes

$10^{-n} = \dfrac{1}{10^n} = 1$ divided by 10 n times or

a decimal point followed by (n–1) zeroes and a 1

$10^0 = 1$

Something to think about

1. Scientists don't know what existed before the Big Bang, or what will happen to the universe if it does eventually collapse back upon itself. One hypothesis is that our universe oscillates: endlessly cycling through a Big Bang, an outward expansion, followed by an eventual collapse back into an infinitely small point, only to be reborn with another Big Bang.
2. Scientists also speculate about what might exist outside our universe. Are there perhaps other multiple universes, which we currently cannot detect?

---

[5] The metric prefixes for multiples or subdivisions by powers of ten are:

atto-	a	10^{-18}	centi-		10^{-2}	kilo-	k	10^3
femto-	f	10^{-15}	deci-		10^{-1}	mega-	M	10^6
pico-	p	10^{-12}	unit		10^0	giga-	G	10^9
nano-	n	10^{-9}	deka-		10^1	tera-	T	10^{12}
picto-	μ	10^{-6}	hecto-		10^2	peta-	P	10^{15}
milli-	m	10^{-3}				exa-	E	10^{18}

Thus millimeter (mm) or kilometer (km).

Algebra Aerobics: [6]
1. Express as a power of 10:
   a)   10,000,000,000   b)   0.000 000 000 000 01
2. Express in standard notation (without exponents):
   a) 10^{-9} b) 10^{13}

---

[6] Answers:
1. a) 10^{10} b) 10^{-14}
2. a)   0.000 000 001   b)   10,000,000,000,000

7.2 Scientific Notation

The previous examples used sizes estimated to the nearest power of ten, without worrying about the exact numbers, because getting the power of ten (or the *order of magnitude*) correct is the first consideration when dealing with physical objects. You have to be in the ball park before you can find your exact seat. The first step is to know whether something is closer to a thousand or to ten billion meters in size. Once the order of magnitude is known the number can be refined to more decimal places of accuracy.

For example, a more accurate measure of the radius of a hydrogen atom is actually 0.0000000000529 meters across. This can be written more compactly by using *scientific notation*; that is, we can rewrite it as a number equal to or greater than 1 and less than 10, multiplied by ten to some power. In order to convert a number into scientific notation:

- Write down the first non–zero digit and all the digits following it. Put a decimal point right after the first digit. This new number is called the *coefficient*. In this case the coefficient is 5.29.

- Figure out what power of ten is needed to convert the coefficient back to the original number. In this case the decimal place has to move eleven places to the left to get 0.<u>0000000000</u>529. This is equivalent to dividing 5.29 by 10 eleven times, or dividing it by 10^{11}.

$$0.0000000000529 = \frac{5.29}{10^{11}}$$

$$= 5.29\left(\frac{1}{10^{11}}\right)$$

But $\dfrac{1}{10^{11}} = 10^{-11}$ so we get: $= 5.29 \bullet 10^{-11}$

This number is now said to be in scientific notation. [7]

What about negative numbers? −0.0000000000529 would be written as $-5.29 \bullet 10^{-11}$. So the coefficient is −5.29.

Whether the coefficient, which we call N, is positive or negative, we insist that the coefficient stripped of its sign, (called its *absolute value* and written $|N|$), be such that $1 \le |N| < 10$. Note that if $N \ge 0$, then $|N| = N$. For example: $|2| = 2$, $|11.57| = 11.57$, and $|0| = 0$. But if $N < 0$, then $|N| = -N$ (here since N is negative, −N is actually positive). For example $|-2| = -(-2) = 2$ and $|-11.57| = -(-11.57) = 11.57$. The important fact is that $|N| \ge 0$.

---

[7] Most calculators or computers automatically translate a number into scientific notation when it is too large or small to fit into the display. The notation is often slightly modified by using the letter E (short for "exponent") to replace the "times 10 to some power" part, so $3.0 \bullet 10^{26}$ might appear as 3.0 E+26. The number after the E tells how many places and which direction to move the decimal point.

2,000,000 or 2 million or $2 \cdot 10^6$ are all correct representations of the same number. The one you choose to use depends upon the context. In general scientific notation is convenient for computations, but in presentations you should use the form that your audience will find easier. For example, most people would find it easier to think of a gross (a dozen dozen) as 144 rather than $1.44 \cdot 10^2$.

The expression $|N|$ (where N is a real number) is read "the absolute value of N."
  If $N \geq 0$, then $|N| = N$.
  If $N < 0$, then $|N| = -N$.

Any number, positive or negative, can be written in scientific notation; that is, written as the product of a coefficient N multiplied by ten to some power, where $1 \leq |N| < 10$.

A number is in *scientific notation* if it is in the form
  $N \cdot 10^n$
where $1 \leq |N| < 10$ and n is an integer (positive, negative, or zero).

Deep Time

Powers of ten and scientific notation can be used to record the progress of the universe through time.

Age of the Universe

The current best estimates place the age of our universe at about 15,000,000,000 years or about 15 billion years. In scientific notation the universe is $1.5 \cdot 10^{10}$ years old.

Age of the Earth

The Earth is believed to have been formed during the last third of the Universe's existence – about 4,600,000,000 or 4.6 billion years ago. In scientific notation the number is $4.6 \cdot 10^9$ years.

Age of Pangaea

About 200,000,000 or 200 million years ago, all the earth's continents collided to form one giant land mass now referred to as Pangaea. Pangaea existed $2.0 \cdot 10^8$ years ago.

Age of Human Life

Homo sapiens first walked on earth about 100,000 or $1.0 \bullet 10^5$ years ago. In the life of the universe, this is almost nothing. If all of time since the Big Bang were scaled down into a single year, with the Big Bang on January 1st, our early human ancestors would not appear until about 10:30 p.m. on Dec. 31st, New Year's Eve.[8]

Algebra Aerobics: [9]

1. Avogadro's number, $6.02 \bullet 10^{23}$, gives the number of molecules in one mole of any gas. Express this number in standard notation.

2. The distance between the earth and its moon is 384,000,000 meters. Express this in scientific notation.

3. An *Angstrom* (denoted by Å), a unit commonly used to measure atoms, is 0.000 000 01 cm. Express its size using scientific notation.

4. The width of a DNA double helix is approximately 2 nanometers or $2 \bullet 10^{-9}$ meters. Express the width in standard notation.

---

8 For an interesting version of this famous metaphor played out, watch Carl Sagan's video *Cosmos* or read his book *Dragons of Eden*.

9 Answers:

1. 602,000,000,000,000,000,000,000
2. $3.84 \bullet 10^8$ meters
3. 10^{-8} centimeters.
4. 0.000000002 meters

7.3 Expressions of the Form a ^{n}

In order to answer questions about how far away a star is, how fast light travels, or how much bigger the sun is than the earth, we need to be able to manipulate expressions of the form a ^{n}.

Expressions of the form a^n

Humans have a particular fondness for the number 10, perhaps because we have ten fingers and ten toes. Whether we count money or stars, people or planets, we represent the answer in a number built up out of powers of ten.

Apart from the fact that we are accustomed to a number system based on 10, there is nothing particularly special about 10. We can write and manipulate expressions of the general form a^n (like $(-5.1)^3$ or x^{-2}), just as well as expressions of the form 10^n .

In the expression a^n , where a is any real number,
the number a is called the *base* and n is called the *exponent*.

If a is any non-zero real number and n is a positive integer:
  $a^n =$ $a \bullet a \bullet a \bullet ... \bullet a$ (n factors of a)
  $a^{-n} =$ $1/a^n = 1$ divided by n factors of a
  $a^0 =$ 1

Multiplication

When multiplying very large or very small numbers, there is an advantage to working in scientific notation. Calculations are simplified and answers are more compact. For example, suppose we calculate the distance between the earth and the star Deneb, one of the brightest stars in our sky, located in the constellation Cygnus, the Swan. The distance that light travels in one year (called a *light year*) is 5.88 trillion miles, and Deneb is 1,600 light years from the earth. Then the distance from the earth to Deneb is:

$$1{,}600 \text{ light years} = (1600) \bullet (5{,}880{,}000{,}000{,}000 \text{ miles})$$
$$= (1.6 \bullet 10^3) \bullet (5.88 \bullet 10^{12} \text{ miles})$$
$$= (1.6 \bullet 5.88) \bullet (10^3 \bullet 10^{12}) \text{ miles}$$
$$\approx 9.4 \bullet (10^3 \bullet 10^{12}) \text{ miles}$$

The product $10^3 \bullet 10^{12}$ means $(10 \bullet 10 \bullet 10) \bullet (10 \bullet 10 \bullet 10 \bullet 10 \bullet 10 \bullet 10 \bullet 10 \bullet 10 \bullet 10 \bullet 10 \bullet 10 \bullet 10)$.
The number of times 10 is multiplied together is $3 + 12 = 15$
The answer is written more compactly as $10^3 \bullet 10^{12} = 10^{15}$

The exponent 15 is the sum of the exponents 3 and 12. This example illustrates a general rule based on the definition of exponents: to *multiply* two exponential terms with the same base (in this case 10), *add* the exponents. If you forget the rule and need to perform calculations, think about what the exponent does to the base.

We get $\quad\quad\quad\quad 9.4 \bullet (10^3 \bullet 10^{12})$ miles $\approx \quad 9.4 \bullet 10^{15}$ miles.

So Deneb is approximately $9.4 \bullet 10^{15}$ or 9.4 quadrillion miles away.

The general rule for multiplying two exponential terms with the same base is:

> If a is any non-zero real number and m and n are integers,
> $$a^m \bullet a^n = a^{(m+n)}$$

Example 1:   The rule works no matter what the base:
$$7^3 \bullet 7^2 = (7 \bullet 7 \bullet 7) \bullet (7 \bullet 7) = 7^{3+2} = 7^5.$$
$$w^3 \bullet w^5 = (w \bullet w \bullet w) \bullet (w \bullet w \bullet w \bullet w \bullet w) = w^{3+5} = w^8$$
$$\text{(where w is a real number)}$$

Example 2:   The rule still holds when one or more power is negative.

$$x^2 \bullet x^{-5} = x^2 \bullet (1/x^5) \quad \text{(where x is any non-zero real number)}$$
$$= x^2/x^5$$
$$= \frac{x \bullet x}{x \bullet x \bullet x \bullet x \bullet x}$$

Canceling like terms:
$$= \frac{1}{x \bullet x \bullet x}$$
$$= \frac{1}{x^3}$$
$$= x^{-3}$$

Note that $-3 = 2 + (-5)$.

A few more examples: $10^{-11} \bullet 10^{-2} = 10^{-13}$

$$(-1)^3 \bullet (-1)^5 = (-1)^8 = +1$$

$$(4.2)^2 \bullet (4.2)^{-7} = 4.2^{-5} = \frac{1}{4.2^5} = \frac{1}{1306.9} \approx 0.000765 \approx 7.65 \bullet 10^{-4}$$

$$\left(\frac{1}{2}\right)^{-11}\left(\frac{1}{2}\right)^{-2} = \left(\frac{1}{2}\right)^{-13} = \left(\frac{1}{\left(\frac{1}{2}\right)}\right)^{13} = 2^{13} = 8192 \approx 8.2 \bullet 10^3$$

$$x^{-2}(x^5 + x^{-6}) = x^{-2}x^5 + x^{-2}x^{-6}$$
$$= x^{(-2+5)} + x^{(-2-6)}$$
$$= x^3 + x^{-8}$$

Common Errors

1) Confusing a term like $-x^4$ with $(-x)^4$. For example, $-2^4 = -(2^4) = -16$. But $(-2)^4 = (-2)(-2)(-2)(-2) = +16$. You have to remember what is being raised to the power. In the expression -2^4, order of operations says to compute the power first, before applying the negation sign. In the expression $(-2)^4$, everything inside the parentheses is raised to the fourth power. So the 2 is negated first, and then (-2) is raised to the 4th power.

Example: $(-3a)^2$ means that everything in the parentheses is squared. So $(-3a)^2 = (-3a)(-3a) = (-3)(-3)a^2 = 9a^2$.

2) It is important to remember that in order to combine the exponents of two multiplied terms, the terms <u>must have the same base</u>. For example:
$$(9300)^2 \bullet (9300)^6 = (9300)^8$$
But an expression like $81^2 \bullet 47^6$, which means: $(81 \bullet 81) \bullet (47 \bullet 47 \bullet 47 \bullet 47 \bullet 47 \bullet 47)$, cannot be simplified by adding exponents since *the bases are not equal*.

3) An expression consisting of terms with different exponents which are *added*, such as $10^2 + 10^3$, <u>cannot</u> be simplified by adding exponents. A common error is to add the exponents. But $10^2 + 10^3 \neq 10^5$. If we do the calculations, we can see that these expressions are not equal. $100 + 1000 \neq 10,000$

A sum of terms with the <u>same</u> exponents <u>and</u> bases can be simplified, for example $10^3 + 10^3 = 2 \bullet 10^3$. Why can these terms be combined?

Use the distributive property		
to rewrite $10^3 + 10^3$	$10^3 + 10^3 \quad =$	$10^3(1+1)$
and then add and substitute	$=$	$10^3 \bullet 2$
and use commutative property	$=$	$2 \bullet 10^3$

Check	$10^3 + 10^3 \quad =$	$2 \bullet 10^3$
Apply exponents	$1000 + 1000 \quad =$	$2 \bullet 1000$
and perform operations	$2000 \quad =$	2000

Examples:
$7^3 \bullet 6^5$ cannot be simplified since the bases 7 and 6 are different
$6^5 \bullet 6^5 \bullet 6^5 = 6^{15}$, but $6^5 + 6^5 + 6^5 = 3 \bullet 6^5$

Something to think about

Using the distributive property, rewrite the expression $10^3 + 10^4$ by pulling out a common factor.

Algebra Aerobics: [10]

1. Simplify: a) $10^5 \cdot 10^7$ b) $8^6 \cdot 8^{-4}$ c) $z^{-5} \cdot z^{-4}$
                 d) $5^5 \cdot 6^7$ e) $7^3 + 7^3$ f) $(-2)^2$ and -2^2

2. A typical TV signal, traveling at the speed of light, takes $3.3 \cdot 10^{-6}$ seconds to travel one kilometer. Estimate long would it take the signal to travel across the U.S. (a distance of approximately 4,300 km.)

Division

In comparing two objects of about the same size it is common to subtract one size from the other and say, for instance, that one person is six inches taller than another. This method of comparison is not effective for objects that have vastly different sizes. To say that the difference between the estimated radius of our solar system (1 terameter or 1,000,000,000,000 meters) and the approximate size of a human ($10^0 = 1$ meter) is $1,000,000,000,0000 - 1 = 999,999,999,999$ meters is not particularly useful. In fact, since our measure of the solar system certainly isn't accurate to within 1 meter this difference is meaningless.

So instead of subtracting one size from another, a more useful method for comparing objects of wildly different size is to calculate the ratio of the two sizes. For example, the radius of the sun is approximately 10^9 meters and the radius of the earth is about 10^7 meters. One way to answer the question "How many <u>times</u> larger is the sun than the earth?" is to form the ratio of the two radii:

$$\frac{\text{radius of the sun}}{\text{radius of the earth}} = \frac{10^9 \text{ meters}}{10^7 \text{ meters}}$$

We can use the rule for multiplying terms of the form a^n to simplify this division problem.

$$\frac{10^9 \text{ meters}}{10^7 \text{ meters}} = 10^9 \cdot \left(\frac{1}{10^7}\right)$$

using negative powers
$$= 10^9 \cdot 10^{-7}$$
multiplication rule for exponents
$$= 10^{9+(-7)}$$
$$= 10^2 \quad \text{(note the units cancel, so } 10^2 \text{ is unitless.)}$$

Thus, the radius of the sun is approximately 10^2 or 100 times larger than the radius of the earth.

---

[10] Answers

1.   a) 10^{12} b) 8^2 c) z^{-9} d) cannot be simplified because bases are different   e) $2 \cdot 7^3$
     f) $(-2)^2 = 4$, but $-2^2 = -(2^2) = -4$

2. $(3.3 \cdot 10^{-6}) \cdot (4.3 \cdot 10^3) = (3.3 \cdot 4.3) \cdot (10^{-6} \cdot 10^3) \approx 14 \cdot 10^{-3} = 1.4 \cdot 10^{-2}$ or 0.014 sec. So it would take less than 2 hundredths of a second for the signal to cross the U.S.

The exponent, 2, is the difference of the original exponents, 9 and 7 (that is, 9–7 = 2). To *divide* two exponential terms with the same base (in this case 10), *subtract* the exponent in the denominator from the exponent in the numerator.

In general:

> If a is any non-zero real number and m and n are integers, $\dfrac{a^m}{a^n} = a^{(m-n)}$.

Examples:

$$10^2/10^{-7} = 10^{2-(-7)} \quad = 10^9$$
$$10^2/10^6 = 10^{2-6} \quad = 10^{-4}$$
$$10^{-11}/10^{-2} = 10^{-11-(-2)} = 10^{-9}$$
$$10^8/10^3 = 10^{8-3} \quad = 10^5$$

$$6^2/6^{-7} = 6^{2-(-7)}$$
$$= 6^9$$

$$(-5)^2/(-5)^6 = (-5)^{2-6}$$
$$= (-5)^{-4}$$

$$7.2^{-11}/7.2^{-2} = 7.2^{-11-(-2)}$$
$$= 7.2^{-9}$$

$$z^8/z^3 = z^{(8-3)}$$
$$= z^5$$

Example:

According to the U.S. Bureau of the Census, in 1995 the estimated gross federal debt was 4.96 trillion dollars and the estimated U.S. population was 264 million. What was the approximate federal debt *per person*?

Solution:

$$\frac{\text{federal debt}}{\text{U.S. population}} = \frac{4.96 \bullet 10^{12}\,\text{dollars}}{2.64 \bullet 10^9\,\text{people}} = \left(\frac{4.96}{2.64}\right) \bullet \left(\frac{10^{12}}{10^9}\right) \frac{\text{dollars}}{\text{people}} \approx 1.88 \bullet 10^3 \frac{\text{dollars}}{\text{people}}$$

So the federal debt amounted to $1.88 \bullet 10^3$ or 1,880 dollars per person.

Example:

The distance from the earth to the sun is approximately 150 million kilometers. If the speed of light is $3.00 \bullet 10^5$ km/sec, how long does it take light from the sun to reach the earth?

Solution:

The formula relating distance, d , rate, r, and time, t, is: $d = r \bullet t$

If we solve the equation for t , we get: $t = d/r$

Divide distance by rate (speed)

$$t = \frac{150 \bullet 10^6 \, km}{3.00 \bullet 10^5 \, km \, / \, sec}$$

and then group terms

(note: $\dfrac{1}{\left(km\big/sec\right)} = \dfrac{sec}{km}$)

$$= \left(\frac{150}{3.00}\right)\left(\frac{10^6}{10^5}\right) km \, \frac{sec}{km}$$

simplify

$$= 50 \bullet 10^1 sec$$
$$= 500 \, sec$$

So it takes about 500 seconds or about 8.3 minutes (500/60) for light to reach us from the sun.

Algebra Aerobics: [11]

1. Calculate
   a) $10^5 / 10^7$
   b) $8^6 / 8^{-4}$
   c) $3^{-5} / 3^{-4}$
   d) $5^5 / 6^7$
   e) $7^3 / 7^3$

2. Show why $\dfrac{1}{\left(km\big/sec\right)} =$ means $1 \div \left(\dfrac{km}{sec}\right) = 1 \bullet \left(\dfrac{sec}{km}\right) = \left(\dfrac{sec}{km}\right)$

3.
   a) Japan has a population of approximately 125.5 million people (in 1995) and a size of about 145.9 thousand square miles. What is the population density, the number of people per square mile?
   b) The U.S. has a population of approximately 263.8 million people (in 1995) with a total land area of about 864.7 thousand square miles. What is the population density of the U.S.?
   c) Compare the population densities of Japan and the U.S.

---

[11] Answers:

1. a) 10^{-2} b) 8^{10} c) 3^{-1} d) cannot be simplified, different bases   e) $7^0 = 1$

2. $\dfrac{1}{\left(km\big/sec\right)} =$ means $1 \div \left(\dfrac{km}{sec}\right) = 1 \bullet \left(\dfrac{sec}{km}\right) = \left(\dfrac{sec}{km}\right)$

3. a) $\dfrac{125.5 \bullet 10^6 \text{ people}}{145.9 \bullet 10^3 \text{ sq. miles}}$ $\approx 0.86 \bullet 10^3$ people/sq. mile ≈ 860 people per square mile in Japan.

   b) $\dfrac{263.8 \bullet 10^6 \text{ people}}{864.7 \bullet 10^3 \text{ sq. miles}}$ $\approx 0.31 \bullet 10^3$ people/sq. mile ≈ 310 people per square mile in the U.S.

   c) Japan is much more densely populated, with almost 3 times as many people per square mile as the U.S.

Raising a power to a power

We previously compared the radii of the sun and earth in answer to the question: "How many times larger is the sun than the earth?" Another way to answer the question is to compare the volumes of each object. The sun and earth are both roughly spherical, and

$$\text{volume of sphere} = 4/3 \, \pi \, r^3$$

The radius of the sun is approximately 10^9 meters so

$$\text{volume of sun} \approx 4/3 \, \pi \, (10^9)^3 \, m^3$$

In words, the expression $(10^9)^3$ tells us to take the ninth power or 10 and cube it. We could do the calculations by evaluating the expression inside of the ( ) and then cubing it. Thus

$$
\begin{aligned}
(10^9)^3 &= (1{,}000{,}000{,}000)^3 \\
&= 1{,}000{,}000{,}000{,}000{,}000{,}000{,}000{,}000{,}000 \\
\text{rewriting as a power of 10} \qquad &= 10^{27}
\end{aligned}
$$

We could also rewrite $(10^9)^3$ using our knowledge about exponents:

$$
\begin{aligned}
\text{definition of exponents} \qquad (10^9)^3 &= (10^9)\bullet(10^9)\bullet(10^9) \\
\text{multiplication rule for exponents} \qquad &= 10^{(9+9+9)} \\
\text{adding} \qquad &= 10^{27} \\
\text{rewriting 27} \qquad &= 10^{(9\bullet3)}
\end{aligned}
$$

Note that 27, the power of the base 10, can be obtained by multiplying the exponents 9 and 3 in the original expression $(10^9)^3$. This makes sense if you think about the meaning of multiplying expressions with exponents.

So

$$\text{volume of sun} \approx 4/3 \, \pi \, (10^9)^3 \, m^3 = (4/3)\bullet10^{27} \, m^3$$

Since the radius of the earth is about 10^7 meters, the ratio of the two volumes is:

$$= \frac{\left(10^9\right)^3}{\left(10^7\right)^3} \qquad \text{(note that the units, } m^3, \text{ cancel)}$$

$$= \frac{10^{27}}{10^{21}} \qquad = 10^6$$

So the *volume* of the sun is approximately $10^6 = 1{,}000{,}000$ or one million times larger than the volume of the earth!

In general, to raise a power of a base a to a power, multiply the exponents.

If a is any non-zero real number and m and n are integers, $\left(a^m\right)^n = a^{(m\bullet n)}$.

Example:
$$(11^2)^4 = 11^2 \cdot 11^2 \cdot 11^2 \cdot 11^2$$
$$= 11^{2+2+2+2}$$
$$= 11^{2 \cdot 4}$$
$$= 11^8$$

More examples:

$$(10^2)^{-7} = 10^{2 \cdot (-7)} \qquad = 10^{-14}$$
$$(10^2)^6 = 10^{2 \cdot 6} \qquad = 10^{12}$$
$$(10^{-11})^{-2} = 10^{-11 \cdot (-2)} \qquad = 10^{22}$$
$$(10^{-8})^3 = 10^{-8 \cdot 3} \qquad = 10^{-24}$$

$$(3^2)^{-7} = 3^{(2 \cdot -7)} \qquad = 3^{-14}$$

$$(115^2)^6 = (115)^{2 \cdot 6} \qquad = 115^{12}$$

$$(3.7^{-11})^{-2} = 3.7^{(-11 \cdot -2)} \qquad = 3.7^{22}$$

$$(w^8)^3 = w^{(8 \cdot 3)} \qquad = w^{24}$$

Example:
Describe the dimensions of a cubic volume of space that would be large enough to contain our solar system.

Since the radius of the solar system is approximately 10^{12} meters, its diameter would be twice that size, approximately $2 \cdot 10^{12}$ m. To contain this spherical volume in an immense cube of space, one side of the cube would have to be at least $2 \cdot 10^{12}$ m long. Since all sides of a cube are the same length, and *volume = length•width•height*, the volume of the cube would be $(2 \cdot 10^{12}$ m$)^3$. Note that the dimensions (meters) must be cubed as well to get the unit of volume m^3, meters cubed.

$$(2 \cdot 10^{12} \text{ m})^3 = (2)^3 \cdot (10^{12})^3 \cdot m^3$$
$$= 8 \cdot 10^{36} \text{ meters cubed (or cubic meters)}$$

Something to think about
What can you say about $(-1)^n$ when n is an even integer? An odd integer?
For any real number a, what can you say about $(-a)^n$ when n is an even integer? An odd integer?

We can summarize the rules for exponents.

If a is any non-zero real number and m and n are integers,

$$a^m \bullet a^n = a^{(m+n)}$$

$$\frac{a^m}{a^n} = a^{(m-n)}$$

$$\left(a^m\right)^n = a^{(m \bullet n)}$$

Algebra Aerobics:[12]

1. Calculate   a) $(10^4)^5$ b) $(10^4)^{-5}$ c) $(7^{-2})^{-3}$ d) $(x^4)^5$

2. The radius of Jupiter, the largest of the planets in our solar system, is approximately 71.4 km. Estimate the surface area and the volume of Jupiter. (Note: the surface area, S, of a sphere is given by $S = 4\pi r^2$.)

Estimating Answers

By using scientific notation and the rules for calculating with powers of ten, we can often make quick estimates of answers to complicated calculations. In this age of calculators and computers, it is easy to trust an answer that looks correct but, because of an error in entering the data, is off by several orders of magnitude. Being able to roughly estimate the size of an answer in advance is a critical skill.

Example 1:

Estimate $\dfrac{382,152 \bullet 490,572,261}{32,091 \bullet 1,942}$

Solution::

Round off to one significant digit $\dfrac{400,000 \bullet 500,000,000}{30,000 \bullet 2,000}$

Rewrite in scientific notation $\dfrac{(4 \bullet 10^5) \bullet (5 \bullet 10^8)}{(3 \bullet 10^4) \bullet (2 \bullet 10^3)}$

---

[12]   Answers
1. a) 10^{20} b) 10^{-20} c) 7^6 d) x^{20}
2. Surface area of Jupiter $= 4\pi r^2 = 4\pi (71.4 \text{ km})^2 \approx 64,062.85 \text{ km}^2 \approx 6.4 \bullet 10^4 \text{ km}^2$
   Volume of Jupiter $= 4/3 \pi r^3 = 4/3 \pi (71.4 \text{ km})^3 \approx 1,524,695.9 \text{ km}^3 \approx 1.5 \bullet 10^6 \text{ km}^3$

Group the coefficients
and the powers of ten

$$\frac{(4 \bullet 5)}{(3 \bullet 2)} \bullet \frac{(10^5 \bullet 10^8)}{(10^4 \bullet 10^3)}$$

Simplify each expression

$$\frac{20}{6} \bullet \frac{10^{13}}{10^7}$$

$$3.33 \bullet 10^6$$

A calculator gives the answer
to the original problem as   3,008,199.5951

Example 2:

Without a calculator, estimate: $0.0000083492 \bullet 15697$.

Solution:

$$
\begin{aligned}
0.0000083492 \bullet 15697 &\approx 0.000008 \bullet 20000 \\
&\approx (8 \bullet 10^{-6}) \bullet (2 \bullet 10^4) \\
&\approx (8 \bullet 2) \bullet (10^{-6} \bullet 10^4) \\
&\approx 16 \bullet 10^{-2} \\
&\approx 0.16 \\
&\approx 0.2
\end{aligned}
$$

A calculator gives the answer to the original problem as 0.1310573.

Example 3:

In 1995 the world population was approximately 5.734 billion people. There are roughly 57.9 million square miles of land on the earth, of which about 22% are favorable for agriculture. Estimate how many people per square mile of farmable land there were in 1995.

Solution:

$$
\begin{aligned}
\frac{\text{size of world population}}{\text{amount of farmable land}} &= \frac{5.734 \bullet 10^9 \text{ people}}{0.22 \bullet 57.9 \bullet 10^6 \text{ sq.mile}} \\
&\approx \frac{6 \bullet 10^9 \text{ people}}{0.2 \bullet 60 \bullet 10^6 \text{ sq.mile}} \\
&= \frac{6 \bullet 10^9 \text{ people}}{12 \bullet 10^6 \text{ sq.mile}} \\
&= 0.5 \bullet 10^3 \text{ people / sq.mile} \\
&= 5 \bullet 10^2 \text{ people / sq.mile}
\end{aligned}
$$

So there are roughly $5 \bullet 10^2$ or 500 people /sq. mile of farmable land in the world.

A calculator gives 450 people/sq. mile of farmable land.

Algebra Aerobics: [13]

1. Estimate to one significant digit:

    a) $0.0002976 \cdot 43990000$

    b) $\dfrac{453,897 \cdot 2,390,702}{0.00438}$

2. Only about 3/7 of the land favorable for agriculture is now in use. Using the facts in example 3, estimate the number of people /sq. mile of farmable land that is being used. (Will your estimate be larger or smaller than 500 people/sq. mile?)

---

[13] Answers:

1. a) $\left(3 \cdot 10^{-4}\right)\left(4 \cdot 10^{7}\right) = 12 \cdot 10^{3} = 1.2 \cdot 10^{4} \approx 1 \cdot 10^{4}$

   b) $\dfrac{(5 \cdot 10^{5})(2 \cdot 10^{6})}{4 \cdot 10^{-3}} = \dfrac{5 \cdot 2}{4} \cdot \dfrac{10^{5}10^{6}}{10^{-3}} = \dfrac{10}{4} \cdot \dfrac{10^{11}}{10^{-3}} = 2.5 \cdot 10^{14} \approx 3 \cdot 10^{14}$

2. Since only 3/7 of the farmable land is in use, we need to multiply the amount of land in the denominator by 3/7. Multiplying the denominator by 3/7 is equivalent to multiplying the whole fraction by 7/3. So (7/3) (450 people/sq. mile) = 1,050 people per square mile. So the earth must sustain about 1,050 people for each square mile of farmed earth.

7.4   Unit conversions

Problems in science constantly require converting back and forth between different units of measure. In order to do so, we need to be comfortable with the laws of exponents and the basic metric and English units (see Table 7.1). The following unit conversion examples describe a strategy based upon *conversion factors*.

Unit conversion within the metric system

Example:

Light travels at a speed of approximately $3.00 \bullet 10^5$ kilometers per second (km/sec). Describe the speed of light in meters per second (m/sec).

Solution:

The prefix "kilo" means thousand. 1 kilometer (km) is equal to 1000 or 10^3 meters (m).

$$1 \text{ km} = 10^3 \text{ m}$$

We can rewrite the equation as: $1 = \dfrac{10^3 \text{ m}}{1 \text{ km}}$

The ratio 10^3 m/1 km is called a *conversion factor* because we can use it to convert from kilometers to meters.

Multiply the original quantity by the conversion factor. Since our conversion factor is equivalent to one, we are not changing the value of the original quantity. If we look just at the units when we multiply the speed of light (in km/sec) times the conversion factor (in m/km), we have:

$$\frac{\text{km}}{\text{sec}} \bullet \frac{\text{m}}{\text{km}}$$

We end up with m/sec. So:

$$3.00 \bullet 10^5 \frac{\text{km}}{\text{sec}} \bullet \frac{10^3 \text{m}}{1 \text{ km}} = 3.00 \bullet 10^5 \bullet 10^3 \frac{\text{m}}{\text{sec}}$$

$$= 3.00 \bullet 10^8 \frac{\text{m}}{\text{sec}}$$

Hence light travels at approximately $3.00 \bullet 10^8$ meters per second.

When solving unit conversion problems, the crucial question always is: "What is the right conversion factor?"

Example:

Check your answer by converting $3.00 \cdot 10^8$ m/sec back to km/sec.

Solution:

Here we use the same strategy but a different conversion factor.

$$1 \text{ km} = 10^3 \text{ m}$$

This time we need a conversion factor that will cancel out meters and give us km/sec.

Rewrite the equation as: $\dfrac{1 \text{ km}}{10^3 \text{ m}} = 1$

Multiply $3.00 \cdot 10^8$ m/sec by the conversion factor.

$$3.00 \cdot 10^8 \frac{\cancel{m}}{sec} \cdot \frac{1 \text{ km}}{10^3 \cancel{m}} = 3.00 \cdot \frac{10^8 \text{ km}}{10^3 \text{ sec}}$$

$$= 3.00 \cdot 10^5 \frac{\text{km}}{\text{sec}}$$

Unit conversion between metric and English systems

Example:

If light travels $3.00 \cdot 10^5$ km/sec., how many <u>miles</u> does light travel in a second?

Solution:

The crucial question is: "What conversion factor should be used?"
From Table 7.1 we know that
$$1 \text{ km} = 0.62 \text{ miles}$$

Rewrite the equation as: $\quad 1 = \dfrac{0.62 \text{ miles}}{1 \text{ km}}$

Multiply $3.00 \cdot 10^5$ km/sec. by the conversion factor

$$3.00 \cdot 10^5 \frac{\cancel{km}}{sec} \cdot \frac{0.62 \text{ miles}}{1 \cancel{km}} = 1.86 \cdot 10^5 \frac{\text{miles}}{\text{sec}}$$

So light travels $1.86 \cdot 10^5$ or 186,000 miles/sec.

Using multiple conversion factors

Example:

Light travels $3.00 \cdot 10^5$ km/sec. How many kilometers does in travel in one <u>year</u>?

Solution:

Here our strategy is to use more than one conversion factor. Use your calculator to perform the following calculations:

$$3.00 \cdot 10^5 \frac{km}{sec} \cdot \frac{60\ sec}{1\ min} \cdot \frac{60\ min}{1\ hr} \cdot \frac{24\ hr}{1\ day} \cdot \frac{365\ day}{1\ year} = 94,608,000 \cdot 10^5 \frac{km}{year}$$

$$\approx 9.46 \cdot 10^7 \cdot 10^5 \frac{km}{year}$$

$$= 9.46 \cdot 10^{12} \frac{km}{year}$$

So a *light year* is approximately equal to $9.46 \cdot 10^{12}$ kilometers.

Algebra Aerobics: [14]

1. We previously asserted that a light year was about $5.88 \cdot 10^{12}$ miles. Verify that $9.46 \cdot 10^{12}$ kilometers $\approx 5.88 \cdot 10^{12}$ miles.
2. 1 Angstrom $= 10^{-8}$ cm. Express 1 Angstrom in meters.
3. If a road sign says the distance to Quebec is 135 km, how many miles away is it?
4. The mean distance from our sun to Jupiter is $7.8 \cdot 10^8$ kilometers. Express this distance in meters.

Something to think about

1. Estimate the number of heart beats in a lifetime.
2. A nanosecond is 10^{-9} sec; modern computers can perform on the order of one operation every nanosecond. Approximately how many feet does light (and hence an electrical signal) travel in one nanosecond?

---

[14] Answers:
1. From Table 7.1 we have 1 km = 0.62 miles. So the conversion factor is 0.62 miles/1 km.
   Hence $9.46 \cdot 10^{12}$ km $\cdot \dfrac{0.62\ miles}{1\ km}$. $\approx 5.87 \cdot 10^{12}$ miles , which is very close to $5.88 \cdot 10^{12}$ miles

2. 1 m = 100 cm , so the conversion factor for converting from cm to m is 1 m/ 100 cm..
   So 10^{-8} cm $\cdot \dfrac{1\ m}{100\ cm}$ = $10^{-8} \cdot \dfrac{1\ m}{10^2}$ = 10^{-8-2} m = 10^{-10} m = 1 Angstrom in meters.

3. 1 km = 0.62 miles, so the conversion factor for converting from km to miles is 0.62 miles/1 km.
   Hence 135 km $\cdot \dfrac{0.62\ miles}{1\ km}$ = 83.7 miles.

4. 1 km = 1000 meters, so the conversion factor from km to meters is 1000 meters/ 1 km.
   Hence $7.8 \cdot 10^8$ km $\cdot \dfrac{1000\ meters}{1\ km}$ = $\dfrac{7.8 \cdot 10^8 \cdot 10^3\ meters}{1}$ = $7.8 \cdot 10^{11}$ meters

7.5 Expressions of the Form a$^{m/n}$

So far we have derived rules for operating with expressions of the form a ^{n}, where n is a positive or negative integer. These rules can be extended to expressions of the form **a$^{m/n}$** where the exponent is a fraction. We need first to consider what an expression like **a$^{m/n}$** means.

The expression m/n can also be written as m•(1/n) or (1/n)•m. If the laws of exponents are consistent, then

$$a^{m/n} = (a^m)^{(1/n)} = (a^{1/n})^{(m)}$$

What does a$^{(1/n)}$ mean? Raising **a** to the (1/n)th power is the opposite of raising **a** to the nth power; it is an operation called "taking a *root*."

Expressions of the form a$^{1/2}$: square roots

The expression $a^{(1/2)}$ is called the *square root* of *a*, and is often written as $\sqrt{a}$ The square root of *a* is the *non–negative* number *b* such that $b^2 = a$. Of course, both –2 squared and 2 squared are equal to 4, but the square root function is defined to give *only the positive root*. If both –2 and 2 are to be considered, we write $\pm\sqrt{4}$ which means "plus or minus the square root of 4."

$$9^{(1/2)} = \sqrt{9} = +3 \qquad \text{since } 3^2 = 9 \text{ (note } \sqrt{9} \text{ does not equal } -3 \text{ since}$$
$$\text{square roots are always non-negative}$$
$$16^{(1/2)} = \sqrt{16} = +4 \qquad \text{since } 4^2 = 16 \text{ (note } \sqrt{16} \text{ is not } -4)$$

In the real numbers, $\sqrt{a}$ is defined only when *a* is non–negative.

For example, in the real numbers, the term $(-4)^{(1/2)} = \sqrt{-4}$ is undefined, since there is no real number *b* such that $b^2 = -4$

In general

If a is any non-negative real number,

$a^{(1/2)} = \sqrt{a}$ where $\sqrt{a}$ is that non-negative number b such that $b^2 = a$.

Many calculators have a square root key, often labeled "$\sqrt{}$" or perhaps *sqrt*. You can also calculate square roots by raising a number to the 1/2 or 0.5 power using the "^" key, as in 4^.5. Try using a calculator to find $\sqrt{4}$ and $\sqrt{9}$. What happens when you calculate $\sqrt{-4}$?

In any but the simplest cases where the square root is immediately obvious, you will probably use the calculator. For example, use your calculator to find:

$$8^{(1/2)} = \sqrt{8} \approx 2.8284$$

Double check the answer by verifying that $2.8284^2 \approx 8$.

Square roots appear quite frequently in real life computations.

Example 1: We can solve for the radius, r, in the formulas for the area of a circle, $A = \pi r^2$ and the surface of a sphere, $S = 4\pi r^2$. We get $r = \sqrt{\dfrac{A}{\pi}}$ and $r = \sqrt{\dfrac{S}{4\pi}} = \dfrac{1}{2}\sqrt{\dfrac{S}{\pi}}$ respectively.

Example 2: If water is dripping at a constant rate from a leaky pipe, making a circular puddle on the floor, and the puddle depth stays the same as it spreads out, then the radius, R, of the puddle is directly proportional to the square root of the time, T. So $R = k\,T^{(1/2)}$ for some constant k.

Example 3: The function $S = \sqrt{30d}$ describes the relationship between S, the speed of a car and the square root of d, the distance a car skids after applying the brakes on a dry tar road.

If we don't know the square root of some number x and don't have a calculator handy, we can estimate the square root by bracketing it with two numbers, a and b, for which we do know the square root. If $x \geq 1$ and $a < x < b$, then $\sqrt{a} < \sqrt{x} < \sqrt{b}$. For example, to estimate $\sqrt{10}$:

we know $9 < 10 < 16$

so $\sqrt{9} < \sqrt{10} < \sqrt{16}$

and $3 < \sqrt{10} < 4$

We know that $\sqrt{10}$ lies somewhere between 3 and 4, probably closer to 3.

Example: Estimate $\sqrt{27}$.

we know $25 < 27 < 36$

hence $\sqrt{25} < \sqrt{27} < \sqrt{36}$

and $5 < \sqrt{27} < 6$

So $\sqrt{27}$ lies somewhere between 5 and 6.

Expressions of the form a$^{1/n}$: nth roots

$a^{(1/n)}$ denotes the nth root of a, often written as $\sqrt[n]{a}$. The nth root of a is that number b such that $b^n = a$.

Examples

$$8^{(1/3)} = \sqrt[3]{8}$$
$$= 2 \qquad \text{(since } 2\,^3 = 8\text{)}$$

$$16^{(1/4)} = \sqrt[4]{16}$$
$$= 2 \qquad \text{(since } 2\,^4 = 16\text{)}$$

If a is any non-negative real number and n is a positive integer, then
$a^{(1/n)}$, or $\sqrt[n]{a}$ is that number b such that $b^n = a$.

Assuming the nth root exists, you can find its value on a calculator. For example, to estimate a cube root, raise the number to the (1/3) or as an approximation, the .33 power.

$$83125^{(1/5)} = \sqrt[5]{83125}$$
$$= 5 \qquad \text{calculating 83125 to the (1/2) or .2 power}$$

Double check your answer by verifying that $5\,^5 = 83125$.

Example:

Assume you are inflating a balloon from a tank of compressed air at a constant rate of volume per time and the balloon is flimsy enough so that the air in the balloon is not significantly compressed. Then (until the balloon explodes or you turn off the compressed air) the radius, R, of the balloon is proportional to the cube root of time, t,. That is,

$$R = k\,t^{\,(1/3)}$$

for some constant k.

Expressions of the form a$^{m/n}$

What about $a^{m/n}$? Thinking of it as $(a^m)^{(1/n)}$ means that you would first raise the base, a, to the mth power and then take the nth root of that. Or you could evaluate the equivalent expression $(a^{(1/n)})^m$: find the nth root of a and then raise that result to the mth power. Some examples follow:

$$2^{(3/2)} = (2^{(1/2)})^3$$
$$\approx (1.414)^3$$
$$\approx 2.8271$$

or equivalently

$$2^{(3/2)} = (2^3)^{(1/2)}$$
$$= (8)^{(1/2)}$$

using a calculator

$$\approx 2.8271$$

We could, of course, use a calculator to compute $2^{(3/2)}$ (or $2^{1.5}$) directly by raising 2 to the 3/2 or 1.5 power: 2^1.5.

Exponents expressed as ratios, of the form m/n, are called *rational exponents.* The following familiar set of laws also holds for rational exponents:

If a is any non-negative number, and p and q are rational numbers, then

$$a^p \bullet a^q = a^{(p+q)}$$
$$a^p / a^q = a^{(p-q)} \ (a \neq 0)$$
$$(a^p)^q = a^{(p \bullet q)}$$

We have dealt with integer exponents and rational (fractional) exponents in expressions like $8^{(2/3)}$. Terms with irrational exponents, such as 2^{π}, are well defined, but are beyond the scope of this course. (You may recall that irrational real numbers are those that cannot be expressed as a ratio of integers.) However the same rules for exponents apply.

Algebra Aerobics: [15]

1.   Evaluate each of the following without a calculator.

   a) $81^{\frac{1}{2}}$ b) $8^{-\frac{1}{3}}$

2.   Find the product by first expressing in exponent form:

   a) $\sqrt{2}\sqrt[3]{2}$ b) $\sqrt{3}\sqrt[3]{9}$ c) $\sqrt{5}\sqrt[4]{5}$

3.   Find the quotient by first expressing in exponent form:

   a) $\dfrac{\sqrt{2}}{\sqrt[3]{2}}$ b) $\dfrac{2}{\sqrt[4]{2}}$ c) $\dfrac{\sqrt[4]{5}}{\sqrt[3]{5}}$

4.   Without a calculator, determine between what two integers the given number lies.

   a) $\sqrt{29}$ b) $\sqrt{92}$

5.   Simplify: a) $\dfrac{t^{-3}t^{0}}{\left(t^{-4}\right)^{3}}$ b) $\dfrac{v^{-3}w^{7}}{\left(v^{-2}\right)^{3}w^{-10}}$

---

[15] Answers

1.   a)   9   b)   1/2

2.   a) $2^{\frac{1}{2}}2^{\frac{1}{3}}=2^{\frac{5}{6}}$ or $\sqrt[6]{32}$ b) $3^{\frac{1}{2}}9^{\frac{1}{3}}=3^{\frac{1}{2}}3^{\frac{2}{3}}=3^{\frac{7}{6}}$ or $\sqrt[6]{3^{7}}=3\sqrt[6]{3}$

   c) $5^{\frac{1}{2}}5^{\frac{1}{4}}=5^{\frac{3}{4}}$ or $\sqrt[4]{5^{3}}=\sqrt[4]{125}$

3.   a) $\dfrac{2^{\frac{1}{2}}}{2^{\frac{1}{3}}}=2^{\frac{1}{2}-\frac{1}{3}}=2^{\frac{1}{6}}$ or $\sqrt[6]{2}$ b) $\dfrac{2^{1}}{2^{\frac{1}{4}}}=2^{1-\frac{1}{4}}=2^{\frac{3}{4}}$ or $\sqrt[4]{2^{3}}=\sqrt[4]{8}$

   c) $5^{\frac{1}{4}-\frac{1}{3}}=5^{-\frac{1}{12}}$

4.   a)   5 and 6   b)   9 and 10

5.   a) $t^{-3-(-12)}=t^{9}$ b) $v^{-3-(-6)}w^{7-(-10)}=v^{3}w^{17}$

7.6 Orders Of Magnitude

Comparing Numbers Of Widely Differing Size

We have seen that a useful method of comparing objects of wildly different sizes is to measure the ratio rather than the difference of the sizes. The ratio can be estimated by computing *orders of magnitude*; that is, by how many times would we have to multiply (or divide) by ten to convert one size into the other. Each power of ten represents one order of magnitude.

For example, the radius of the observable universe is approx. 10^{26} meters and the radius of our solar system is approx. 10^{12} meters. To compare the radius of the observable universe to the radius of our solar system, calculate the ratio:

$$\frac{\text{radius of the universe}}{\text{radius of the solar system}} = \frac{10^{26}\,\text{meters}}{10^{12}\,\text{meters}}$$
$$= 10^{26-12}$$
$$= 10^{14}$$

The radius of the universe is 10^{14} times larger than the radius of the solar system; that is, we would have to multiply the radius of the solar system by 10 fourteen times in order to obtain the radius of the universe. Each factor of 10 is counted as a single order or magnitude, so the radius of the universe is *14 orders of magnitude greater* than the radius of the solar system. Equivalently we could say that the radius of the solar system is *14 orders of magnitude smaller* than the radius of the universe.

In general, when something is one order of magnitude larger than a *reference object*, it is ten times larger. You would *multiply* the *reference size* by 10 to get the other size. If the object is two orders of magnitude larger, it is 100 or 10^2 times larger so you would multiply by 100. If it is an order of magnitude smaller, the object is the size of the reference object *divided* by 10. Two orders of magnitude smaller means you would *divide* the reference size by 100 or 10^2. The following examples demonstrate order of magnitude comparisons:

- We calculated previously that the radius of the sun is 100 or 10^2 times larger than the radius of the earth. This is equivalent to saying that the radius of the sun is 2 orders of magnitude larger. We found that the volume of the sun is 10^9 times or 9 orders of magnitude larger than the volume of the earth.

- The sun, at 10^9 meters, is 20 orders of magnitude larger than a hydrogen atom, 10^{-11} meters, since if you multiply 10^{-11} meters by 10 twenty times you get 10^9 meters.

$$\frac{\text{radius of sun}}{\text{radius of the hydrogen atom}} = \frac{10^9\,\text{meters}}{10^{-11}\,\text{meters}}$$
$$= 10^{9-(-11)}$$
$$= 10^{20}$$

- A human is approximately *12 orders of magnitude smaller* than the solar system, since you would have to divide 10^{12} meters, the radius of the solar system, by 10 twelve times in order to get 1 (or 10^0) meter, the height of a human.

$$\frac{\text{size of human}}{\text{size of solar system}} = \frac{10^0 \text{ meters}}{10^{12} \text{ meters}}$$

$$= 10^{0-12}$$

$$= 10^{-12}$$

- Surprisingly enough, the radius of a living cell is approximately 3 orders of magnitude *smaller* than one of the single strands of DNA it contains, if the DNA is uncoiled and measured lengthwise.

$$\frac{\text{length of DNA strand}}{\text{radius of the living cell}} = \frac{10^{-2} \text{ meters}}{10^{-5} \text{ meters}}$$

$$= 10^{-2-(-5)}$$

$$= 10^{-2+5}$$

$$= 10^3$$

An Example of a Measurement Scale Based on Orders of Magnitude Scales: The Richter Scale

The *Richter scale*, designed by the American Charles Richter in 1935, allows us to compare earthquakes throughout the world. The Richter scale measures the maximum vertical ground movement (tremors) as recorded on an instrument called a seismograph. The size of earthquakes varies widely, so Richter designed the scale to measure order of magnitude differences. The scale ranges from less than 1 to over 8. Each increase of 1 unit on the Richter scale represents an increase of 10 times in the maximum tremor size of the earthquake. So an increase from 2.5 to 3.5 indicates a tenfold increase in maximum tremor size. An increase from 2.5 to 4.5 indicates a maximum tremor 100 times larger.

Table 7.2 contains some typical values on the Richter scale along with a description of how humans near the center (called the *epicenter*) of an earthquake would perceive its effects. There is no theoretical upper limit on the Richter scale, but the biggest earthquakes measured so far have registered as 8.6 on the Richter scale (in Japan and Chile).[16]

---

[16] The following table gives values on the Richter scale and the approximate number per year (worldwide).

Richter scale magnitude	Number per year (worldwide)
2	More than 100,000
4.5	A few thousand
7	16 to 18
8	1 or 2

Source: C.C. Plummer and D. McGeary, *Physical Geology*, 5th ed., Dubuque, Ia., Wm. C. Brown, 1991, p. 352.

Table 7.2	
Richter scale magnitude	**Description**
2.5	Generally not felt, but recorded on seismographs
3.5	Felt by many people
4.5	Felt by all, some slight local damage may occur
6	Considerable damage in ordinary buildings, a destructive earthquake
7	"Major" earthquake, most masonry and frame structures destroyed, ground badly cracked
8.0 and above	"Great" earthquake, a few a decade worldwide, almost total or total destruction. Bridges collapse, major openings in ground, and tremors are visible.

Algebra Aerobics: [17]

1. In 1987 Los Angeles had an earthquake that measured 5.9 on the Richter scale. In 1988 Armenia had an earthquake that measured 6.9 on the Richter scale. Compare the size of the two earthquakes.

2. In 1983 Hawaii had an earthquake that measured 6.6 on the Richter scale. Compare the size of this earthquake to the largest ever recorded (8.6 on the Richter scale).

---

[17] Answers:

1. The Armenian earthquake had tremors 10 times larger than those in Los Angeles.
2. The maximum tremor size of the Hawaiian earthquake of 1983 was 100 times smaller than the maximum tremor size of the largest earthquake.

Graphing Numbers Of Widely Differing Sizes

If the sizes of the various objects in our solar system are plotted on a standard evenly divided axis, we get the following uninformative picture:

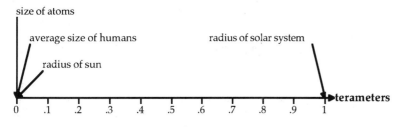

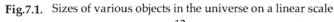

Fig.7.1. Sizes of various objects in the universe on a linear scale

(note: one terameter $= 10^{12}$ meters)

The largest value stands alone, and all the others are so small when measured in terameters that they all appear to be zero. When objects of widely different orders of magnitude are compared on an evenly divided linear scale, the effect is similar to pointing out an ant in a picture of a baseball stadium. A more effective way of plotting sizes with different orders of magnitude is to change from an axis that has integers evenly spaced along it (like the real number line) to an axis that has orders of magnitude (powers of 10) evenly spaced along it. This is usually called a *logarithmic* or *log scale*. The previous data on a logarithmic scale is more informative:

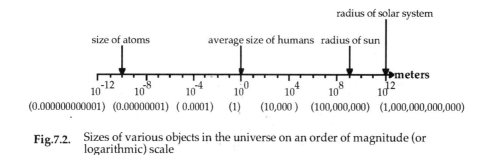

Fig.7.2. Sizes of various objects in the universe on an order of magnitude (or logarithmic) scale

Graphing sizes on a log scale is extremely useful, but since the scale on the axis is no longer linear it has to be read very carefully. If you graph on the real number line, each move of one unit to the right it is equivalent to *adding* 1 to the number, and each move of k units to the right is equivalent to *adding* k to the number.

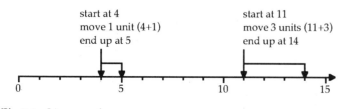

Fig.7.3. Linear scale

The log scale is not so familiar. One unit of length on a log scale is chosen to represent a change of one order of magnitude. Moving one unit to the right is equivalent to *multiplying by 10*. Moving from 10^4 to 10^5 is equivalent to multiplying 10^4 by 10. Moving over 3 units to the right is equivalent to *multiplying* the starting number by 10^3 or 1,000.

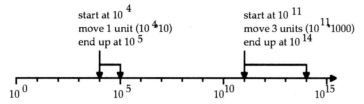

Fig.7.4. Order of magnitude (or logarithmic) scale

In effect a linear scale is an "additive" scale and a logarithmic scale is a "multiplicative" scale.

Algebra Aerobics: [18]

1. Place each of the following on the order of magnitude scale (Figure 7.2)
    a) The radius of the earth: 6,370,000 meters
    b) The radius of a virus: 0.0000007 meters
2. Now plot on Figure 7.2 and object whose radius is:
    a) 2 orders of magnitude small than that of the earth
    b) 6 orders of magnitude larger than that of a virus.

---

[18] Answers
1. a) $6{,}370{,}000 \approx 10{,}000{,}000 = 10^7 \Rightarrow$ plot it at 10^7
    b) $0.0000007 = 7 \bullet 10^{-7} \approx 10 \bullet 10^{-7} = 10^{-6} \Rightarrow$ plot it at 10^{-6}
2. a)   plot at 10^5 b)   plot at 10^0

The Electromagnetic Spectrum

Over 300 hundred years ago it was discovered that when ordinary "white" light passes through a prism, a band of color like a rainbow emerges from the other side. The colors are always spread out in a certain order: red, orange, yellow, green, blue, indigo, and violet. (You may have learned the abbreviation ROY G. BIV in science class to help you remember the color sequence.) We can understand why light contains all the different colors if we think of light as waves of radiation. These waves are all traveling at the same speed - the speed of light , $3 \cdot 10^{10}$ cm/sec (186,000 miles/sec). The distance between the crest of one wave to the next is called the *wavelength*.

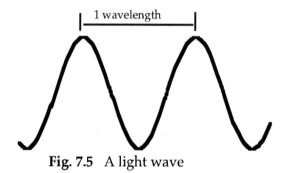

Fig. 7.5   A light wave

Lights of different wavelengths appear as different colors.

The wavelengths of light are very short, just a few hundred-thousandths of a centimeter. Scientists usually use the *Angstrom* (denoted by Å), a unit of length which is 0.000 000 01 or 10^{-8} cm. The wavelength of violet light is approximately 4000 Angstroms (4000 Å), while the wavelength of red light is approximately 6500 Angstroms (6500 Å). Light waves much shorter than 4000 Å , or much longer than 6500 Å, are outside the range of human visibility. Such radiations can however be measured by scientific instruments. At wavelengths shorter than violet, the radiation is called *ultraviolet* This is the type of radiation that sunscreen or sunglasses are used to protect against. At wavelengths longer than red, the radiation is called *infrared*. While human vision cannot detect infrared radiation, night vision scopes like those used by the military can pick up infrared signals emitted by warm bodies.

Like visible or invisible light, electric and magnetic fields (like x-rays and radio waves) can also be thought of as radiation waves moving rapidly through space. These radiations are referred to as *electromagnetic radiation*. We can draw the entire *electromagnetic spectrum* on a single order of magnitude (or logarithmic scale) ranging from wavelengths shorter than 1 Å to wavelengths many meters long.

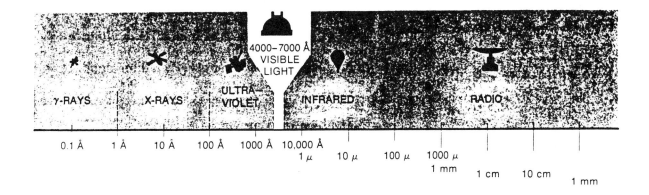

Fig. 7.6 The Electromagnetic Spectrum: An Order of Magnitude Scale
Source: Jay M Pasachoff and Marc L. Kutner, *University Astronomy*, 1978, W.B. Saunders Company, Philadelphia, p. 21.

Algebra Aerobics: [19]

1. Refer to the electromagnetic spectrum.
   a) Give a reasonable size for the wavelength of an x-ray in Angstroms. Translate this size into centimeters.
   b) Give a reasonable size for a radio wave. Translate this size into centimeters.
   c) How many orders of magnitude larger is the radio wave than the x-ray?

[19] Answers:

1 a)   10 Å or $10 \cdot 10^{-8} = (1.0 \cdot 10^1) \cdot 10^{-8} = 1.0 \cdot 10^{-7}$ cm
   b)   1 cm $= 1.0 \cdot 10^0$
   c)   7 orders of magnitude larger

7.7   Logarithms Base 10

The relationship between multiplying numbers and adding their exponents had been recognized for many years before the idea and the properties of *logarithms* were clearly worked out. Their original use was to help carry out trigonometric calculations for astronomical problems. Since then, the logarithm and the ideas that underly it have been of tremendous scientific and practical importance. A slide rule, for example, multiplies numbers by mechanically adding distances which represent their logarithms, and we use logarithmic scales to measure physical quantities from sound intensities to earthquake strengths.

Defining logarithms base 10

When handling very large or very small numbers, we have seen that it is often easier to write the number using powers of 10. For example,

$$100{,}000 \;=\; 10^{\,5}$$

Recall that we refer to 10 as the base and 5 as the power (or exponent). We say that:
   "100,000 equals the base 10 to the 5th power "

But we could also rephrase this as:
   " 5 is the power of the base 10 that is needed to produce 100,000"

The more technical way to say this is:
   " 5 is the *logarithm* of 100,000 to the base 10"

In symbols we write:
$$5 = \; \log_{10} 100{,}000$$
The two statements:

$$100{,}000 \;=\; 10^5 \qquad \text{and} \qquad 5 = \log_{10} 100{,}000$$
are two ways of saying exactly the same thing.

In general,

The *logarithm base 10 of* x is the power of 10 you need to produce x:

$$\log_{10} x \;=\; c \quad \text{means that} \quad 10^{\,c} = x$$

Example: Since
$$1,000,000,000 = 10^9$$
then
$$\log_{10} 1,000,000,000 = 9$$

and we say that the logarithm of 1,000,000,000 base 10 is 9. When we write 1,000,000,000 as a power of 10 the exponent is 9.

Example: Since
$$1 = 10^0$$
then:
$$\log_{10} 1 = 0$$
and we say that the logarithm of 1 base 10 is 0.

Example: What about the logarithm base 10 of decimals, like 0.00001 ?

Since
$$0.00001 = 10^{-5}$$
then
$$\log_{10} 0.00001 = -5$$
and we say that the logarithm of 0.00001 base 10 is –5.

$\log_{10} x$ is not defined when x is negative or x is zero. Why? If $\log_{10} x = c$ where $x \leq 0$, then $10^c = x$ (a number ≤ 0). But 10 to any power will never produce a number that is negative or zero, so $\log_{10} x$ is not defined if x is a negative number or zero.

$\log_{10} x$ is not defined when x is 0 or negative.

Table 7.3 gives a sample set of values for x and their associated logarithms base 10. Each time, in order to find the logarithm base 10, we write the number as a power of 10 , and the logarithm is just the exponent.

Table 7.3	
x	$\text{Log}_{10} x$
0.0001 $(=10^{-4})$	–4
0.001 $(=10^{-3})$	–3
0.01 $(=10^{-2})$	–2
0.1 $(=10^{-1})$	–1
1 $(=10^0)$	0
10 $(=10^1)$	1
100 $(=10^2)$	2
1000 $(=10^3)$	3
10000 $(=10^4)$	4

Logarithms base 10 are used frequently in our base 10 number system, and are called *common logarithms*. We frequently drop the 10 in the notation $\log_{10} x$ and just write log x.

Logarithms base 10 are called *common logarithms*.
$\log_{10} x$ is frequently abbreviated as log **x**.

Algebra Aerobics: [20]
Find the logarithm base 10 of:

      a)   10,000,000   b) 0.0000001

Finding the logarithm base 10 of numbers between 1 and 10

When a number (like 100 or 0.001) can be written as an integer power of 10, it's easy to find its logarithm base 10. We just write the number as a power of ten, and the logarithm is the resulting exponent. But what about the logarithm of other numbers, like 2?

Example: Estimate $\log_{10} 2$.

$$\log 2 = c \quad \text{is true if and only if} \quad 10^c = 2$$

We need to find a number c such that 10 raised to that power will give us 2. Since

$$1 \quad < \quad 2 \quad < \quad 10$$

and

$$10^0 = 1, \quad 10^c = 2, \quad 10^1 = 10$$

Then

$$10^0 \quad < \quad 10^c \quad < \quad 10^1$$

So we might suspect that $\quad\quad 0 \quad < \quad c \quad < \quad 1$

We can estimate values of 10^c where c is between 0 and 1, using a calculator. (Remember that $10^{0.1}$ (or $10^{1/10}$) means the 10th root of 10.)

$$10^{0.1} \approx 1.258925$$
$$10^{0.2} \approx 1.584893$$
$$10^{0.3} \approx 1.995263$$
$$10^{0.4} \approx 2.511886$$

So $10^{0.3}$ is very close to 2, while $10^{0.4}$ is larger than 2. Hence a good estimate is:

$$\log 2 \approx 0.3$$

---

[20] a) Since $10,000,000 = 10^7$, then log 10,000,000 = 7
   b) Since $0.0000001 = 10^{-7}$, then log 0.0000001 = -7

By trial and error we could come even closer to the actual value.

$$10^{0.31} \approx 2.041738$$
$$10^{0.301} \approx 1.999862$$
$$10^{0.3015} \approx 2.002166 \qquad (1)$$

Giving us, an even better estimate, namely:

$$\log 2 \approx 0.3015$$

The appendices of many textbooks contain "lookup tables" that contain values for log x when x is between 1 and 10.

Algebra Aerobics: [21]
Try using your calculator to estimate:
a)   log 3   b)   log 6

Finding the logarithm base 10 of positive numbers *not* between 1 and 10

How can we find the logarithm of numbers that are not between 1 and 10 ? Our knowledge of scientific notation can help.

Example: What is log 2000 ?

Write 2000 in scientific notation $2000 = 2 \cdot 10^3$
and substitute $10^{0.3015}$ for 2 [see (1) above] $2000 \approx 10^{0.3015} \cdot 10^3$
then combine powers $2000 \approx 10^{3.3015}$
and rewrite as a logarithm $\log 2000 \approx 3.3015$

So by writing any positive number in scientific notation, finding the logarithm boils down to finding the logarithm of a number between 1 and 10.

Algebra Aerobics: (answers are on the bottom of the next page)
Use the answers from the previous algebraic aerobics to estimate values for:
        a) log 3,000,000   b)   log 0.006

---

[21] a)   Since $10^{0.4} \approx 2.511886$ and $10^{0.5} \approx 3.162277$ then log 3 $\approx$ 0.5. An even better estimate is
          log 3 $\approx$ 0.48 since $10^{0.48} \approx 3.019952$.
    b)   Since $10^{0.7} \approx 5.011872$ and $10^{0.8} \approx 6.30957$ then log 6 $\approx$ 0.8. An even better estimate is
          log 6 $\approx$ 0.78 since $10^{0.78} \approx 6.025596$

Using a scientific calculator to find logarithms base 10

If you own a scientific calculator, you may have noticed a key marked **LOG.** [22] This key will generate log x for you for any x > 0. Once you understand what logarithms represent, the actual calculation can be done mechanically.

Example: Use your calculator to check your previous estimates of 0.3015 for log 2 and 3.3015 for log 2 000.

The calculator gives: $\log 2 \approx .301299957$
and $\log 2\,000 \approx 3.301029996$
So our estimates were pretty close.

Another example of a logarithmic scale: the Decibel Scale

We have seen intuitive descriptions of scales that measure orders of magnitude for earthquakes and electromagnetic waves. Now that we have developed a working knowledge of base 10 logarithms we can define such scales more precisely, and see why they are called logarithmic.

The Decibel Scale

We measure noise levels in units called *decibels*, abbreviated as dB. The name is in honor of the inventor of the telephone, Alexander Graham Bell. Small changes in noise levels (especially if it is already pretty noisy) may be difficult for humans to detect. So the decibel scale (like the Richter scale) is a logarithmic scale, one that measures orders of magnitude changes.

If we designate by I_0 the intensity of a sound at the threshold of human hearing (10^{-16} watts/cm^2), and we let I represent the intensity of an arbitrary sound (measured in watts/cm^2), then the noise level N of that sound measured in decibels is defined to be:

$$N = 10 \log (I/I_0)$$

Answers to Algebra Aerobics from previous page

a)   Write 3,000,000 in scientific notation $3,000,000 = 3 \cdot 10^6$

and substitute $10^{0.48}$ for 3 $3,000,000 \approx 10^{0.48} \cdot 10^6$

then combine powers $3,000,000 \approx 10^{6.48}$

and rewrite as a logarithm $\log 3,000,000 \approx 6.48$

b)   Write 0.006 in scientific notation $0.006 = 6 \cdot 10^{-3}$

and substitute $10^{0.78}$ for 6 $0.006 \approx 10^{0.78} \cdot 10^{-3}$

then combine powers $0.006 \approx 10^{0.78-3} \approx 10^{-2.22}$

and rewrite as a logarithm $\log 0.006 \approx -2.22$

---

[22] If you are working with spreadsheet on a computer, it will have a LOG function.

The expression (I/I_0) gives the *relative intensity* of a sound compared to the reference value of I_0.

The decibel scale is called a logarithmic scale since N is a constant times the logarithm of relative intensity.

Example 1: What is the decibel level of a sound just at the edge of human hearing (i.e., where $I = I_0$) ?

If $I = I_0$, then	$N = 10 \log (I_0/I_0)$
$I_0/I_0 = 1$	$= 10 \log 1$
$\log 1 = 0$	$= 10 (0)$
	$= 0$ decibels

Example 2: What is the decibel level of a sound whose relative intensity is 100 times the threshold? (that is, $I = 100\ I_0$)?

If $I = 100\ I_0$, then	$N = 10 \log(100\ I_0/I_0)$
$I_0/I_0 = 1$	$= 10 \log 100$
$\log 100 = 2$	$= 10 \bullet 2$
	$= 20$ decibels

Here is the decibel scale showing how much the relative intensity (the ratio I / I_0) of a sound source must increase for people to perceive differences. Note that each time we <u>add</u> 10 units on the decibel scale, we must <u>multiply</u> the relative intensity by 10 (or one order of magnitude).

Table 7.4		
Decibels **($10 \log (I/I_0)$)**	**Relative Intensity** **(I/I_0)**	**Average Perception**
0	1	Threshold of hearing
10	10	Sound proof room, very faint
20	100	Whisper, rustle of leaves
30	1,000	Quiet conversation, faint
40	10,000	Quiet home, private office
50	100,000	Average conversation, moderate
60	1,000,000	Noisy home, average office
70	10,000,000	Average radio, average factory, loud
80	100,000,000	Noisy office, average street noise
90	1,000,000,000	Loud truck, police whistle, very loud
100	10,000,000,000	Loud street noise, noisy factory
110	100,000,000,000	Elevated train, deafening
120	1,000,000,000,000	Thunder of artillery, nearby jack hammer
130	10,000,000,000,000	Threshold of pain, ears hurt

Example 3: What is the decibel level of a typical rock band playing with an intensity of 10^{-5} watts/cm^2)? How much more intense is the band than an average conversation?

Let $I = 10^{-5}$ watts/cm^2 and since $I_0 = 10^{-16}$ watts/cm^2

	$N =$	$10 \log (I/I_0)$
substitute for I and I_0	$N =$	$10 \log(10^{-5}/10^{-16})$
perform operations on powers	$=$	$10 \log (10^{11})$
simplify logs	$=$	$10 \cdot 11$
	$=$	110 decibels

So the noise level of a typical rock band is about 110 decibels.

According to the previous table, an average conversation measures about 50 decibels. So the noise level of the rock band is 60 decibels higher. Each 10 decibels correspond to a 1 order of magnitude increase in the relative intensity. So the rock band is about 6 orders of magnitude (10^6 or 1,000,000 times) more intense than an average conversation.

Algebra Aerobics: [23]
Use a calculator to find values for:
a)   log 3000   b) log 6   c)   log (-257)
Confirm each of your estimates by verifying that 10 raised to that power gives the original number.

---

[23] a) $\log 3000 \approx 3.47712155$
We can double check our answer by calculating $10^{3.47712155} = 3000.00204 \approx 3000$

b) $\log 6 \approx .7781512504$
We can double check our answer by calculating $10^{.7781512504} = 6$

c)   log (-257) generates an error message, since log x is not defined for negative values for x (or when x =0).

Chapter 7 Summary

Scientists believe that somewhere between 8 and 20 billion years ago the universe began expanding rapidly from an infinitesimally small point. This event is referred to as the "Big Bang," and the universe has been expanding ever since.

In describing the size of objects in the universe, the scientific community (and most other countries) use the metric system, a measurement system based on the meter. The following table shows the conversion into English units for three of the most common metric length measurements.

Metric Unit	Abbreviation	In meters	Equivalence in English Units	Informal conversion
meter	m		3.28 feet	the width of a twin bed a little more than a yard
kilometer	km	1000 meters	0.62 miles	a casual 12 minute walk a little over a half a mile
centimeter	cm	0.01 meters	0.39 inches	the width of a finger nail a little under a half an inch

Powers of ten are useful in describing the scale of objects. When n is a positive integer we define:

$10^n = 10 \bullet 10 \bullet 10 \bullet \ldots \bullet 10$ (the product of n tens) or 1 followed by n zeroes

$10^{-n} = \dfrac{1}{10^n} = 1$ divided by 10 n times or a decimal point followed by (n–1) zeroes and a 1.

$10^0 = 1$

In order to write very small or large numbers compactly, we use scientific notation. A number is in *scientific notation* if it is in the form

$N \bullet 10^n$

where $1 \leq |N| < 10$ and n is an integer (positive, negative, or zero).

Note that the expression $|N|$ (where N is a real number) denotes "the absolute value of N." If $N \geq 0$, then $|N| = N$. If $N < 0$, then $|N| = -N$.

The laws of exponents:

If **a** is any non-zero real number and **m** and **n** are real numbers then,

$$a^m \bullet a^n = a^{(m+n)}$$

$$\frac{a^m}{a^n} = a^{(m-n)}$$

$$\left(a^m\right)^n = a^{(m \bullet n)}$$

Fractional powers:

If a is any positive real number,
$a^{(1/2)} = \sqrt{a}$ where $\sqrt{a}$ is that positive number b such that $b^2 = a$.
If a is any non-negative real number and n is a positive integer, then
$a^{(1/n)}$, or $\sqrt[n]{a}$ is that number b such that $b^n = a$.

When comparing two objects of widely different sizes, it is common to use *orders of magnitude*. For example, to compare the radius of the universe to the radius of the solar system, first calculate the ratio of the two sizes.

$$\frac{\text{radius of the universe}}{\text{radius of the solar system}} = \frac{10^{26}\,\text{meters}}{10^{12}\,\text{meters}}$$

$$= 10^{\,26-12}$$
$$= 10^{\,14}$$

The radius of the universe is 10^{14} times larger, or 14 *orders of magnitude* larger than the radius of the solar system. Each factor of 10 is counted as one order of magnitude. When something is one order of magnitude larger than a *reference object,* it is ten times larger. Two order of magnitude larger, means it is 100 or 10^2 time larger. If it is an order of magnitude smaller, the object is the size of the reference object *divided* by 10. Two orders of magnitude smaller means you would *divide* the reference size by 100 or 10^2.

Orders of magnitude (or logarithmic) scales are used to graph objects of widely differing sizes.

The *logarithm base 10 of x* is the power of 10 you need to produce x. So

$$\log_{10} = c \quad \text{means that} \quad 10^{\,c} = x$$

Logarithms base 10 are called *common logarithms* and $\log_{10} x$ is frequently abbreviated as log x.

log x is not defined when x is 0 or negative.

EXERCISES

1.   Write each expression as a power of ten.

   a) $10 \cdot 10 \cdot 10 \cdot 10 \cdot 10 \cdot 10$ c)   one billion

   b) $\dfrac{1}{10 \cdot 10 \cdot 10 \cdot 10 \cdot 10}$ e)   one thousandth

2.   Write each of the following in scientific notation

   a)   0.00029   c)   720,000

   b)   654.456   d)   0.00000000001

3.   Why are each of the following expression *not* in scientific notation? Rewrite each in scientific notation.

   a) $25 \cdot 10^{4}$ b) $0.56 \cdot 10^{-3}$

4.   Write each of the following in decimal form

   a) $7.23 \cdot 10^{5}$ c) $1.0 \cdot 10^{-3}$

   b) $5.26 \cdot 10^{-4}$ d) $1.5 \cdot 10^{6}$

5.   Evaluate and write the results using scientific notation

   a) $(2.3 \cdot 10^{4})(2.0 \cdot 10^{6})$ d) $\dfrac{3.25 \cdot 10^{8}}{6.29 \cdot 10^{15}}$

   b) $(3.7 \cdot 10^{-5})(1.1 \cdot 10^{8})$ e) $(6.2 \cdot 10^{52})^{3}$

   c) $\dfrac{8.19 \cdot 10^{23}}{5.37 \cdot 10^{12}}$ f) $(5.1 \cdot 10^{-11})^{2}$

6.   Write each of the following in scientific notation.

   a) $725 \cdot 10^{23}$ d) $-725 \cdot 10^{23}$

   b) $725 \cdot 10^{-23}$ e) $-725 \cdot 10^{-23}$

   c) $\dfrac{1}{725 \cdot 10^{23}}$

7. Estimate without a calculator. Show your work, writing your answers in scientific notation. Use a calculator to verify your answers.

a) $0.000359 \bullet 0.00000243$

b) $2,968,001,000 \bullet 189,000$

c) $0.000079 \bullet 31,140,284,788$

d) $\dfrac{4083692 \bullet 49312}{213 \bullet 1945}$

8. Using the sizes from the text, determine the size difference in orders of magnitude for the following:

a) Pangaea compared to the earth
b) Protons compared to the Milky Way
c) Atoms compared to quarks

9. The Republic of China was estimated in 1995 to have about 1,260,000,000 people, and Monaco about 32,000. Monaco has an area of 0.75 sq. miles, and China has an area of 3,704,000 sq. miles.

a) Express the populations and geographic areas in scientific notation.

b) By what order of magnitude is China's population bigger than Monaco's?

c) What is the population density (people per square mile) for each country?

d) Write a paragraph comparing and contrasting the population size and density for these two nations, one very large and the other very small.

10. The Heavenly G

This is a mind exercise to bring home the notion of the linkage between time and space. Everything that you see is history, in the sense that there is always a time lag between when something happens and when you see it happening. This is due to the time needed for light to travel from the event to you. When you look in the mirror, you see yourself not as you are, but as you were nanoseconds ago. But even more disturbingly, what you see as simultaneous events, did not occur simultaneously. To illustrate this, consider the large constellation called the "Heavenly G" pictured on the accompanying star chart. It is a G–shaped group of 9 bright stars, 7 of them are what astronomers call of first magnitude, which means they are among the brightest stars. In order, they are Capella, Castor, Pollux, Procyon, Sirius, Rigel, Aldebaran, and Betelgeuse.

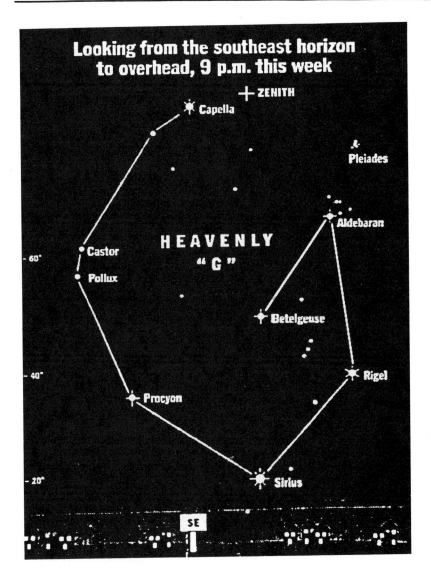

Source: Alan M. MacRobert, "The Heavenly 'G,'" *Boston Globe*, 10 Jan. 1994: 42.
Copyright © Alan M. MacRobert. Printed with permission.

The table below, gives their approximate distances in light years.

Name	Distance in light years
Capella	46
Castor	47
Pollux	33
Procyon	11
Sirius	9
Rigel	500
Aldebaran	64
Betelgeuse	300

Now imagine being in an observatory, looking through the telescope at these stars. You are observing light received simultaneously on earth from all these nine different sources. Answer the following questions about the Heavenly G.

a)   From which star are you getting the most recent information? How old is the information?

b)   From which star are you getting the oldest information? How old is the information?

c)   Suppose Betelgeuse blew up 250 years ago. Would we know that yet? Why or why not? Could it have disintegrated 350 years ago? Why or why not?

11.   Hubble's Law

Hubble's Law states that galaxies are receding from one another at velocities directly proportional to the distances separating them. The following graph illustrates that Hubble's Law holds true across the known universe.

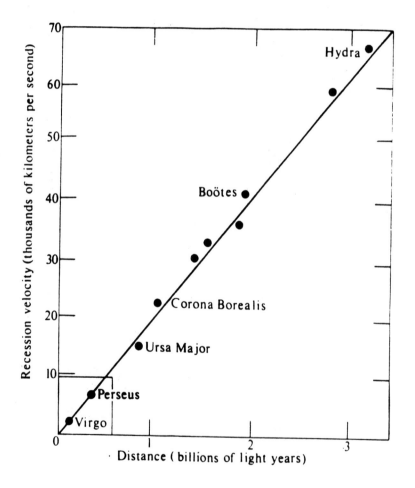

Source: "Coming of Age in the Milky Way" by Timothy Ferris; copyright © by Timothy Ferris; by permission of William Morrow & Company, Inc.

The plot includes ten major clusters of galaxies. The boxed area at the lower left represents the galaxies observed by Hubble when he discovered the law. The easiest way to understand this graph is to think of the earth as being at the center of the Universe (i.e., at 0 distance) and not moving (i.e., at 0 velocity). In other words, imagine the earth at the origin of the graph. (A favorite fantasy of humans.) Think of the horizontal axis as measuring the distance of a galaxy cluster from the earth, and the vertical axis as measuring the velocity at which a galaxy cluster is moving away from the earth. Then answer the following questions:

a)   Identify the coordinates of two data points that lie on the regression line that is drawn in on the graph. (Hint: for the horizontal coordinate it is easier to use numbers like 1 or 2 with the units being billions, and for the vertical coordinate use numbers like 10 or 20 with the units being in thousand of kilometers per second.)

b)   Use the points in item a) to calculate the slope of the line. The slope is called the Hubble constant.

c)   What does the slope mean in terms of distance from earth and recession velocity?

d)   Calculate an equation for your line in the form: y = mx + b. Show your work.

12.   Estimating the age of the earth

Below is an amusing graph that shows a roughly linear relationship between the "scientifically" calculated age of the earth and the year the calculation was published. For instance in about 1935 Ellsworth calculated that the earth was about 2 billion years old. The triangle on the horizontal axis represents the presently accepted age of the earth. The age is plotted as a function of the year the calculation was published. The triangle on the horizontal coordinate represents the presently accepted age of the earth.

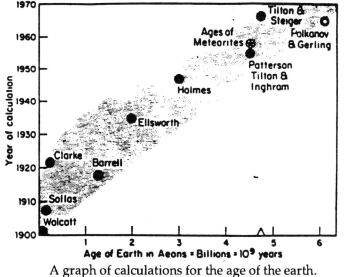

A graph of calculations for the age of the earth.

From *American Scientist*, © 1980, Research Triangle Park, North Carolina.

a)   Who calculated that the earth was less than 1 billion years old? Give the co-ordinates of the points that give this information.

b)   In about what year did scientists start putting the age of the earth at over a billion years old? Give the coordinates that represent this point.

c)   On your graph sketch an approximation of a "best fit" line for these points. Use two points on the line to calculate the slope of the line.

d)   Interpret the slope of your line.

13.   Simplify when possible, writing your answer as an expression with exponents.

   a) $10^4 \cdot 10^3$ f) $10^7 / 10^2$

   b) $10^4 + 10^3$ f) z^7 / z^2

   c) $10^3 + 10^3$ g) 256^0

   d) $x^5 \cdot x^{10}$ h) $\dfrac{3^5 \cdot 3^2}{3^8}$

   e) $(x^5)^{10}$ i) $4^5 \cdot (4^2)^3$

14.   Evaluate when $x = 2$
   a) x^3 d) $-x^2$

   b) $x^{1/2}$ e) $-3x^2$

   c) $(-x)^2$ f) x^0

15.   In order to compare the sizes of different objects, we need to use the same unit of measure.

   a)   Convert each of these to a common unit.
        The radius of the moon is approximately 1,922,400 yards.
        The radius of the earth is approximately 6,300 km.
        The radius of the sun is approximately 432,000 miles.

   b)   Determine the order of magnitude difference between:
        i)   the surface areas of the moon and the earth.
        ii)   the volume of the sun and the moon.

16. The distance that light travels in 1 year (a light year) is $5.88 \bullet 10^{12}$ miles. If a star is $2.4 \bullet 10^{8}$ light years from Earth, what is this distance in miles?

17. A TV signal traveling at the speed of light takes about $8 \bullet 10^{-5}$ seconds to travel 15 miles. How long would it take the signal to travel a distance of 3,000 miles?

18. Convert the following to feet and express your answers in scientific notation.

    a) The radius of the solar system is approximately 10^{12} meters.

    b) The radius of a proton is approximately 10^{-15} meters.

19. The speed of light is approximately $1.86 \bullet 10^{5}$ miles/sec.

    a) Write this number in decimal form and express your answer in words.

    b) Convert the speed of light into meters/year.

20. Water boils (changes phase from liquid to gas) at 100 degrees Celsius. The temperature of the core of the sun is 20 million degrees Celsius. By what order of magnitude is the sun's core hotter than the boiling temperature of water?

21. The average distance from the earth to the sun is about 150,000,000 km. and the average distance from the planet Venus to the sun is about 108,000,000 km.

    a) Convert these distances to scientific notation.

    b) Divide the distance from Venus to the sun by the distance from the earth to the Sun and express your answer in scientific notation.

    c) The distance from the earth to the sun is called 1 astronomical unit (1 A.U.) How many A.U. is Venus from the sun?

    d) Pluto is 5,900,000,000 km from the sun. How many A.U.'s is it from the sun?

22. Calculate the following:

    a) $4^{1/2}$ e) $8^{2/3}$

    b) $-4^{1/2}$ f) $-(8^{2/3})$

    c) $27^{1/3}$ g) $16^{1/4}$

    d) $-27^{1/3}$ h) $16^{3/4}$

23. Graphing exponents

Constellation I
Reduce each of the following expressions to the form: $r^a \cdot n^b$, then plot the exponents as points with coordinates (a,b) on graph paper. Do you recognize the constellation?

a) $\dfrac{(r^5)^2}{n^6 \cdot r^{-1}}$

b) $\dfrac{r^{10}}{r^{21} \cdot (n^{-2})^3}$

c) $\dfrac{(r \cdot n^5)^0 \cdot r \cdot n^3}{r^{-.4} \cdot n^5}$

d) $\dfrac{n}{r^{.8}}$

e) $\dfrac{r^{-6} \cdot r^{-13} \cdot r^7}{(r \cdot n^2)^3}$

f) $(r^3 \cdot n^2)^3$

g) $\dfrac{1}{r^{-.3} \cdot n^{.6}}$

Constellation II
Reduce each of the following expressions to the form: $u^a \cdot m^b$, then plot the exponents as points with coordinates (a,b) on graph paper. Do you recognize the constellation?

a) $\dfrac{(u^2)^2 \cdot m}{u^2 \cdot m^{-4}}$

b) $\dfrac{u^{\frac{-9}{5}} \cdot m^3}{(umu^2)^1 \cdot m^{-1}}$

c) $\dfrac{u^2 \cdot u^{-4}}{u^3 \cdot (m^{-2})^3}$

d) $\dfrac{(um^2)^3 \cdot u^2}{(um)^4}$

e) $\dfrac{u^{\frac{-3}{2}} \cdot u^{\frac{-7}{2}} \cdot m^1 \cdot (m^3)^3}{(um)^2}$

f) $\dfrac{1}{u^{12} \cdot m^{-9}}$

g) $\dfrac{(mu)^0 \cdot (u^{10})^{-1} \cdot m^{\frac{1}{4}}}{(m^{-3} \cdot u^{\frac{-1}{3}})^3}$

Constellation III
Reduce each of the following expressions to the form: $p^a \cdot c^b$, then plot the exponents as points with coordinates (a,b) on graph paper. What constellation appears?

a) $\dfrac{p^7 \cdot p^{-3} \cdot (c^7)^0}{p^4}$

b) $\dfrac{(p^{-2})^{-3} \cdot c^4}{p^{-5} \cdot c^{7.5}}$

c) $\dfrac{p^{-5}}{p^{2.5} \cdot p^2 \cdot c^{-2}}$

d) $\dfrac{(p^2 \cdot c^3)^2}{c^5}$

e) $\dfrac{1}{(p^{0.5})^3 \cdot (c^{-13/9})^3}$

24. Without using a calculator, find two *adjacent* integers such that one is smaller and one larger than each of the following: (for example $3 < \pi < 4$). Show your reasoning.

   a) $\sqrt{13}$ b) $\sqrt{22}$ c) $\sqrt{40}$

   Now verity your answers using a calculator.

25. a) Read the chapter entitled "The Cosmic Calendar" from Carl Sagan's book *The Dragons of Eden*.

   b) Carl Sagan tried to give meaning to the cosmic chronology by imagining the 15 billion year lifetime of the universe compressed into the span of one calendar year. To get a more personal perspective, consider your date of birth as the time the Big Bang took place. Map the following five cosmic events onto your own life span.

      The Big Bang
      Creation of the earth
      First life on earth
      First Homo sapiens
      American Revolution

   Once you have done the necessary mathematical calculations and placed your results on either a chart or a timeline, then form a topic sentence and write a playful paragraph about what you were supposedly doing when these cosmic events took place. Hand in your calculations along with your writing.

26. (Refer to Figure 7.6 in the text.)

   a) How many orders of magnitude larger is a radio wave with a wavelength of 10 centimeters than an x-ray with wave length of 10 Angstroms?

   b) How many orders of magnitude smaller is the maximum wave length x-ray compared to the minimum wave length radio wave?

27. Rewrite the following statements using logs.

   a) $10^2 = 100$ b) $10^7 = 10,000,000$ c) $10^{-2} = 0.01$

   Rewrite the following statements using exponents

   d) $\log 10 = 1$ e) $\log 10,000 = 4$ f) $\log 0.0001 = -4$

28. Do the following exercise without a calculator:

         a) Find the following values:

        log 100

        log 1,000

        log 10,000,000

What is happening to the values of log x as x gets larger?

        b) Find the following values:

        log 0.1

        log 0.001

        log 0.00001

What is happening to the values of log x as x gets closer to 0?

        c) What is log 0 ?

        d) Find the following values:

        log (–0.1)

        log (–10)

        log (–100)

What do you know about log x when x is a negative number?

29. Without using a calculator for each number in the form log x, find some numbers a and b such that $a < \log x < b$. Justify your answer. Then verify your answers with a calculator.

    a)   log 11   b)   log 12,000   c)   log 0.125

30. Use a calculator to determine the following logs. Double check each answer by writing down the equivalent statement using exponents, and then verify your statement using a calculator.

    a)   log 11   b)   log 12,000   c)   log 0.125

31   (Refer to Table 7.4)

    a) How much more intense is a nearby jack hammer than a quiet conversation?

    b) How much more intense is a police whistle than an average radio?

    c) How much less intense is a whisper than the average office?

EXPLORATION 7.1
The Scale and the Tale of the Universe

Objective
- to gain an understanding of the relative size and relative age of objects in the universe using scientific notation and unit conversions.

Materials/Equipment
- tape, pins, paper and string to generate a large wall graph (optional)
- enclosed worksheet and conversion table

Related readings
Excerpts from: "Powers of Ten" and "Cosmic Calendar" in the Anthology

Related videos
Powers of Ten and *Cosmic Calendar*

Related Software
Powers of Ten in Exponential Graphs

Procedure

Work in small groups. Each group should work on a separate subset of objects on the enclosed worksheet.

1. Convert the age and size of objects so they can be compared. You can refer to the enclosed tables that describe what the metric prefixes mean (like kilo-, giga- and tera-) and how to convert between English and metric units.

2. Generate on the blackboard or on the wall a blank graph whose axes are marked off in orders of magnitude, with the units on the vertical axis representing age of object, ranging from 10^0 to 10^{10} years, and the units on the horizontal axis representing size of object, ranging from 10^{-12} to 10^{27} years.

3. Each small group should plot the coordinates of their selected objects (age in years, size in meters) on the graph. You might want to draw and label a small picture of your object.

Discussion/Analysis

- Scan the plotted objects from left to right, looking only at relative size. Does your graph make sense in terms of what you know about the size of these objects?

- Now scan the plotted objects from top to bottom, only considering relative age. Does your graph makes sense in terms of what you know about the relative age of these objects?

- Compare the largest and smallest objects in the table. By how many orders of magnitude do they differ? Compare the sizes of other pairs of objects in the table. Compare the ages of the oldest and youngest objects in the table. By how many orders of magnitude do they differ?

Conversion Table			
Metric Unit	Abbreviation	In meters	Equivalence in English Units
meter	m		3.28 feet
kilometer	km	1000 meters	0.62 miles
centimeter	cm	0.01 meters	0.39 inches
light year		$9.46 \cdot 10^{15}$ meters	$5.88 \cdot 10^{12}$ miles

The metric prefixes for multiples or subdivisions by powers of ten are:

atto-	a	10^{-18}	centi-	10^{-2}	kilo-	k	10^{3}
femto-	f	10^{-15}	deci-	10^{-1}	mega-	M	10^{6}
pico-	p	10^{-12}	unit	10^{0}	giga-	G	10^{9}
nano-	n	10^{-9}	deka-	10^{1}	tera-	T	10^{12}
picto-	μ	10^{-6}	hecto-	10^{2}	peta-	P	10^{15}
milli-	m	10^{-3}			exa-	E	10^{18}

OBJECT	AGE (in years)	SIZE (of radius)	In Scientific Notation	
			AGE (in years)	SIZE (in meters)
Observable Universe	15 billion (?)	10^{26} meters		
Surtsey (Earth's newest land mass)	30 years	0.5 miles		
Pleiades (a galactic cluster)	100 million	32.6 light years		
First living organisms on earth	3.5 billion	.00005 meters		
Pangaea (Earth's prehistoric supercontinent)	200 million	4500 miles		
First Homo sapiens sapiens	100 thousand	100 centimeters		
First Tyrannosaurus Rex	200 million	20 feet		
Eukaryotes (first cells with nuclei)	2 billion	.00005 meters		
Earth	5 billion	6,400 kilometers		
Beginnings of Milky Way Galaxy	14 billion	50,000 light years		
First atoms	15 billion	.0000000001 meters		
Our sun	5 billion	1 gigameter		
Our solar system	5 billion	1 terameter		

EXPLORATION 7.2
Patterns in the Position and Motion of the Planets

Objective

- to explore patterns in the positions and motions of the planets and discover aspects of Kepler's Law.

Introduction

Johannes Kepler discovered 400 years ago (before Newton's mechanics) a law which relates the period of planets with their average distance from the sun. His strong belief, that the Solar system was governed by harmonious laws, drove him to try to discover hidden patterns and correlations in the positions and motions of the planets. He used the trial and error method and continued his search for years. Below are some hints to discovering one of Kepler's Laws.

At the time of his work, Kepler did not know the distance from the sun to each planet in terms of measures of distance such as the kilometer. But he was able to determine the distance from each planet to the sun in terms of the distance from the earth to the sun, now called the astronomical unit or **A.U.** for short. The **A.U.** is a unit in which the distance from the earth to the sun is equal to 1. The first column in the table below gives the average distance from the sun to each of the planets in astronomical units or **A.U. 's.**

Patterns in the Position and Motion of the Planets: Kepler's Discovery

KEPLER'S THIRD LAW

THE FIRST PLANET TABLE (INNER PLANETARY SYSTEM)

PLANET	AVERAGE DISTANCE FROM SUN (A. U.)	CUBE OF THE DISTANCE (A. U.^{3})	ORBITAL PERIOD (YEARS)	SQUARE OF THE ORBITAL PERIOD (YEARS ^{2})
MERCURY	0.3870		0.2408	
VENUS	0.7232		0.6151	
EARTH	1.0000		1.0000	
MARS	1.5233		1.8807	
JUPITER	5.2025		11.8619	
SATURN	9.5387		29.4557	

Data from: *Encyclopedia of Astronomy*, *Second Edition*. S. Parker & J. Pasachoff.
Table 1. Elements of Planetary Orbits. copyright © 1993. McGraw -Hill, Inc.
Reprinted with permission.

The planets, Uranus, Neptune, and Pluto were discovered after Kepler made his discovery. Check to see if the relationship you found above holds true for these three planets.

THE SECOND PLANET TABLE (OUTER PLANETARY SYSTEM)

PLANET	AVERAGE DISTANCE FROM SUN (A. U.)	CUBE OF THE DISTANCE (A. U. ^{3})	ORBITAL PERIOD (YEARS)	SQUARE OF THE ORBITAL PERIOD (YEARS ^{2})
URANUS	19.1911		84.0086	
NEPTUNE	30.0601		164.7839	
PLUTO	39.5254		248.5900	

1 A.U. = 149.6 •10 ^{6} km ;   1 Year = 365.26 days.

Data from: *Encyclopedia of Astronomy , Second Edition.* S. Parker & J. Pasachoff. Table 1. Elements of Planetary Orbits. copyright © 1993. McGraw -Hill, Inc. Reprinted with permission.

Summarizing your results

- Express your results in words.
- Use an equation to summarize your results.
- Do your conclusions hold for all of the planets?

Chapter 8

Growth and Decay

Overview

To a biologist it is important to understand how populations grow. To determine the best strategy for managing lobster fisheries off the coast of New England requires a solid understanding of the growth or decline of the lobster population. To control insects pests such as mosquitoes, ticks, and gypsy moths, or to maximize the yield of pennicillin from the fungus *Penicillium notatum*, requires a knowledge of population dynamics.

So far we have examined the linear model which can be used to describe phenomena that grow (or decline) at a constant rate of change. This chapter introduces the family of *exponential functions* that can be used to describe phenomena that grow (or decay) faster and faster over time. Exponential functions are used to calculate the size of a bacterial population that doubles every 20 minutes, or the remaining amount of a radioactive element that decays by 50% every 3000 years. They can also be used to model compound interest rates, musical pitch and inflation.

The explorations help you understand the mathematical properties of exponential functions and give you some experience fitting a model to actual laboratory data on the growth of *E. coli* bacteria under differing conditions.

After reading this chapter you should be able to:

- recognize the characteristics of exponential functions
- understand the differences between exponential and linear growth
- construct and interpret graphs of exponential functions
- use exponential models of growth and decay
- plot exponential functions on a semi-log graph
- simulate population dynamics

8.1 Modeling Growth

The growth of *E. coli* bacteria

Measuring and predicting growth is of concern to population biologists, ecologists, demographers, economists, and politicians alike. By studying the growth of a population of the bacteria *E. coli*, we can construct a simple model that can be used to describe the growth of cells, countries, or money.

Bacteria are very tiny, single-celled organisms that are by far the most numerous organisms on earth. One of the most frequently studied bacteria is *E. coli*, a rod-shaped bacterium approximately 10^{-6} meters (or 1 micron) long that inhabits the intestinal tracts of humans.[1] The cells of *E. coli* reproduce by a process called fission: the cell splits in half, forming two "daughter cells."

The rate at which fission occurs depends on several conditions; in particular, available nutrients and temperature. Under ideal conditions *E. coli* can divide every 20 minutes. If we start with an initial population of 100 *E. coli* bacteria that doubles every 20 minute time period, we generate the data in Table 8.1. The initial 100 bacteria double to become 200 bacteria at the end of the first time period, double again to become 400 at the end of the second time period, and keep on doubling to become 800, 1600, 3200, etc., at the end of successive time periods. At the end of the 24th time period (= 24•20 minutes = 480 minutes = 8 hours), the initial 100 bacteria in our model have grown to over 1.6 billion bacteria.

Because the numbers become astronomical so quickly, we run into the problems we saw in Chapter 7 when graphing numbers of widely different sizes.

Table 8.1	
20 minute time periods	No. of *E. coli* bacteria
0	100
1	200
2	400
3	800
4	1,600
5	3,200
6	6,400
7	12,800
8	25,600
9	51,200
10	102,400
11	204,800
12	409,600
13	819,200
14	1,638,400
15	3,276,800
16	6,553,600
17	13,107,200
18	26,214,400
19	52,428,800
20	104,857,600
21	209,715,200
22	419,430,400
23	838,860,800
24	1,677,721,600

---

[1] Most types of *E. coli* are beneficial to humans, aiding in human digestion. A few types are lethal. You may have read about deaths resulting from people eating certain deadly strains of *E. coli* bacteria in undercooked hamburgers.

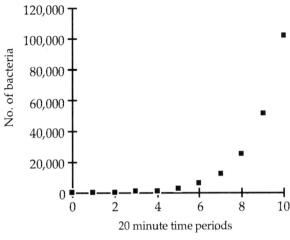

Figure 8.1 shows a graph of the data in Table 8.1 for only the first 10 time periods. We can see from the graph that the relationship between number of bacteria and time is not linear. The number of bacteria seems to be increasing more and more rapidly over time.

Fig. 8.1   Growth of *E. coli*   bacteria

A mathematical model for *E. coli* growth

If we examine the values in Table 8.1, we see that the initial number of 100 bacteria is repeatedly doubled or multiplied by 2. If we record in a third column the number of times we multiply 2 times the original value of 100, we can begin to see a pattern emerge.

time period	*E. coli* bacteria		generalized expression
0	100	100	$=100 \cdot 2^0$
1	200	$100 \cdot 2$	$=100 \cdot 2^1$
2	400	$100 \cdot 2 \cdot 2$	$=100 \cdot 2^2$
3	800	$100 \cdot 2 \cdot 2 \cdot 2$	$=100 \cdot 2^3$
4	1,600	$100 \cdot 2 \cdot 2 \cdot 2 \cdot 2$	$=100 \cdot 2^4$
5	3,200	$100 \cdot 2 \cdot 2 \cdot 2 \cdot 2 \cdot 2$	$=100 \cdot 2^5$
6	6,400	$100 \cdot 2 \cdot 2 \cdot 2 \cdot 2 \cdot 2 \cdot 2$	$=100 \cdot 2^6$
7	12,800	$100 \cdot 2 \cdot 2 \cdot 2 \cdot 2 \cdot 2 \cdot 2 \cdot 2$	$=100 \cdot 2^7$
8	25,600	$100 \cdot 2 \cdot 2 \cdot 2 \cdot 2 \cdot 2 \cdot 2 \cdot 2 \cdot 2$	$=100 \cdot 2^8$
9	51,200	$100 \cdot 2 \cdot 2 \cdot 2 \cdot 2 \cdot 2 \cdot 2 \cdot 2 \cdot 2 \cdot 2$	$=100 \cdot 2^9$
10	102,400	$100 \cdot 2 \cdot 2 \cdot 2 \cdot 2 \cdot 2 \cdot 2 \cdot 2 \cdot 2 \cdot 2 \cdot 2$	$=100 \cdot 2^{10}$

Table 8.2

Remembering that 2^0 is equal to 1 by definition, we can describe the relationship by:

$$\text{number of } E.\ coli \text{ bacteria} = \ 100 \cdot 2^{\text{time period}}$$

If we let N = number of bacteria, and t = time, we can write the equation more compactly as:

$$N = \ 100 \cdot 2^t$$

The dependent variable **N** is a function of the independent variable **t**, since each value of **t** determines a unique value for **N**. The domain of our function, as a model of bacteria growth, includes values of t between 0 and some unspecified positive number. This function is called *exponential* since the independent variable, **t**, occurs in the exponent of the *base* 2. The number 100 is the *initial bacteria population*. We also call 2 the *growth factor* or the multiple by which the population grows each time period. The population is doubling, or stated another way, is growing by 100% each time period.

Linear vs. exponential growth

The bacteria population grows faster and faster as time goes on, in a behavior typical of exponential functions. Exponential growth is *multiplicative*, which means that for each unit increase in the independent variable, we *multiply* the value of the dependent variable by the growth factor. Contrast this with linear functions, which are *additive*, which means that for each unit increase in the independent variable, we must *add* a fixed amount to the dependent variable. In the exponential function, $N = 100 \cdot 2^t$, for each increase of 1 unit in time, t, we *multiply* N by 2; but in the linear function, $N = 100 + 2t$, for each increase of 1 unit in time, t, we *add* 2 to N.

	Table 8.3		
	Linear Function		
t	N = 100+ 2t		generalized expression
0	100	100	=100+2•0
1	102	100+2	=100+2•1
2	104	100+2+2	=100+2•2
3	106	100+2+2+2	=100+2•3
4	108	100+2+2+2+2	=100+2•4
5	110	100+2+2+2+2+2	=100+2•5
6	112	100+2+2+2+2+2+2	=100+2•6
7	114	100+2+2+2+2+2+2+2	=100+2•7
8	116	100+2+2+2+2+2+2+2+2	=100+2•8
9	118	100+2+2+2+2+2+2+2+2+2	=100+2•9
10	120	100+2+2+2+2+2+2+2+2+2+2	=100+2•10

Table 8.3 shows the additive nature of the growth of the linear function $N = 100 + 2t$.

	Table 8.4	
	Linear Function	Exponential Function
time	$N = 100+ 2t$	$N = 100 \cdot 2^t$
t	pattern	pattern
0	100+2•0 = 100	100 • 2^0 = 100
1	100+2•1 = 102	100 • 2^1 = 200
2	100+2•2 = 104	100 • 2^2 = 400
3	100+2•3 = 106	100 • 2^3 = 800
4	100+2•4 = 108	100 • 2^4 = 1600
5	100+2•5 = 110	100 • 2^5 = 3200
6	100+2•6 = 112	100 • 2^6 = 6400
7	100+2•7 = 114	100 • 2^7 = 12,800
8	100+2•8 = 116	100 • 2^8 = 25,600
9	100+2•9 = 118	100 • 2^9 = 51,200
10	100+2•10 =120	100• 2^{10} =102,400

Exponential growth is more rapid as time goes on, as can be seen in Table 8.4. In both functions when t = 0, N = 100. After 10 time periods the values for N are strikingly different: 102,400 for the exponential function versus 120 for the linear function.

Every exponential growth function will eventually dominate every linear growth function.

Algebra Aerobics:

1.   Fill in the following table.

t	N = 10 + 3t	N = 10•3^t
0		
0.5		
1		
1.5		
2		
4		

2.   Sketch a graph of the linear function for $0 \leq t \leq 10$. What is the slope? The vertical intercept?

3.   Sketch the graph of the exponential function for $0 \leq t \leq 4$. What is the vertical intercept?

4.   On the same coordinate plane, sketch both functions for $0 \leq t \leq 2$. Compare the graphs.

Answers

1.1.

t	N = 10 + 3t	N = 10•3^t
0	10	10
0.5	11.5	17.32
1	13	30
1.5	14.5	51.96
2	16	90
4	22	810

2.

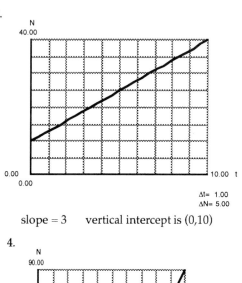

slope = 3   vertical intercept is (0,10)

3.

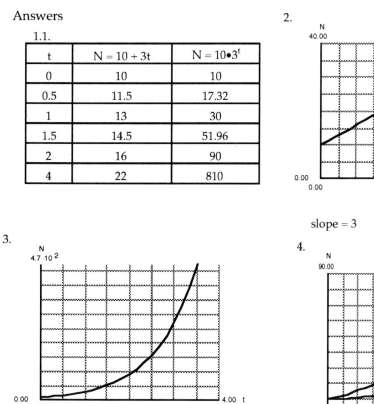

vertical intercept is (0,10)

4.

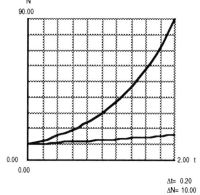

The exponential function rises faster than the linear function, and the rate at which it rises also increases.

The average rate of change of linear vs. exponential functions

Another way to compare linear and exponential functions is to examine their average rates of change. We can ask how each function changes over time. Recall that

$$\text{average rate of change} = \frac{\text{change in N}}{\text{change in t}} = \frac{\Delta N}{\Delta t}$$

Examine Table 8.5 which contains the average rate of change for both the linear and exponential function, where $\Delta t = 1$.

	Table 8.5				
	Linear Function		**Exponential Function**		
	$N = 100 + 2t$	Average rate	$N = 100 \cdot 2^t$	Average rate	
t	**N**	**of change**	**N**	**of change**	
0	100		100		
1	102	2	200	100	
2	104	2	400	200	
3	106	2	800	400	
4	108	2	1,600	800	
5	110	2	3,200	1,600	
6	112	2	6,400	3,200	
7	114	2	12,800	6,400	
8	116	2	25,600	12,800	
9	118	2	51,200	25,600	
10	120	2	102,400	51,200	

For all linear functions, the average rate of change is constant. For our linear function, $N = 100 + 2t$, we can tell from the equation, Table 8.5, and Figure 8.2, that the rate of change is constant at 2 units. For our exponential function, $N = 100 \cdot 2^t$, the average rate of change calculations in Table 8.5 and their graph in Figure 8.3 indicate that the average rate of change is *exponential*.

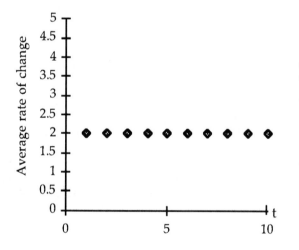

Fig. 8.2: The average rate of change of the linear function $N = 100 + 2t$

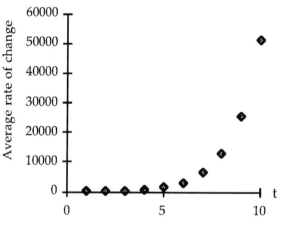

Fig. 8.3: The average rate of change of an exponential function

The exponential function in the abstract

We can detach the equation $N = 100 \cdot 2^t$ from any physical context and treat it as describing a relationship between two abstract variables t and N. The domain (the set of possible values for t) is then all the real numbers. Table 8.6 shows some values for N when $-10 < t < +10$.

Remember that $\qquad 2^{-2} = \dfrac{1}{2^2} = \dfrac{1}{4} = 0.25, \qquad$ so $\qquad 100\,(2^{-2}) = 25$

Similarly $\qquad 2^{-4} = \dfrac{1}{2^4} = \dfrac{1}{16} = 0.0625, \qquad$ so $\qquad 100(2^{-4}) = 6.25$

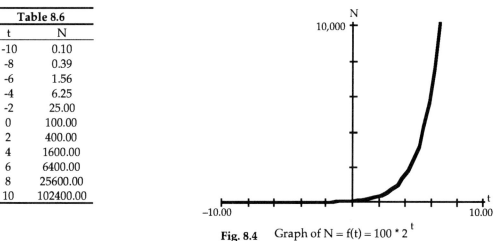

Table 8.6	
t	N
-10	0.10
-8	0.39
-6	1.56
-4	6.25
-2	25.00
0	100.00
2	400.00
4	1600.00
6	6400.00
8	25600.00
10	102400.00

Fig. 8.4   Graph of $N = f(t) = 100 * 2^t$

As values of t move from 0 to -10 to -100, values of 2^t and therefore $100 \cdot 2^t$ remain positive, but come closer and closer to zero. Hence the graph always lies above the t-axis, coming extremely close to it, but never touching it.

The value of N when t=0 is often referred to as the *initial quantity*. From Table 8.6 we can see that in this case the initial quantity is 100. The base 2 represents the *growth factor*. Each time t increases by 1, the value of N doubles (is multiplied by 2 or increases by 100%).

Algebra Aerobics:
Make a table and sketch the graph for $y = 3^x$ for $-3 \le x \le 3$

Answers:

x	$y = 3^x$
-3	$3^{-3} = 1/27 \approx 0.037$
-1	$3^{-1} = 1/3 \approx 0.333$
0	$3^0 = 1$
1	$3^1 = 3$
3	$3^3 = 27$

Looking at real growth data for *E. coli* bacteria

The idealized model for *E. coli* bacteria (in Table 8.1) showed phenomenal growth. It is not surprising that when we actually go to a biology lab and watch real bacteria grow, the growth is substantially slower. Conditions are rarely ideal; bacteria die, nutrients are used up, and temperatures may not be optimal. Table 8.7 and Figure 8.5 show some real data collected by a graduate student in biology. A solution was inoculated with an indeterminate amount of bacteria at time t = 0, and then the number of *E coli* cells per milliliter was measured at regular time periods (of 20 minutes each). For the first two time periods, the volume of bacteria was so small that it could not be measured.

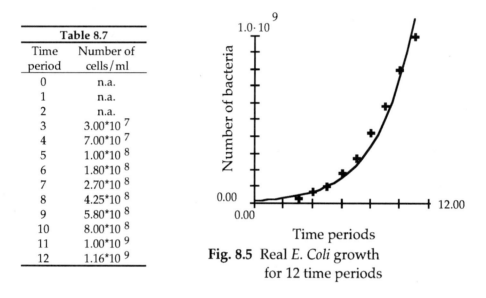

Table 8.7	
Time period	Number of cells/ml
0	n.a.
1	n.a.
2	n.a.
3	$3.00*10^7$
4	$7.00*10^7$
5	$1.00*10^8$
6	$1.80*10^8$
7	$2.70*10^8$
8	$4.25*10^8$
9	$5.80*10^8$
10	$8.00*10^8$
11	$1.00*10^9$
12	$1.16*10^9$

Fig. 8.5 Real *E. Coli* growth for 12 time periods

Using a curve fitting program to find a best fit exponential function to model the *E. coli* growth, we get:

$$N = (1.37 \bullet 10^7) \bullet 1.5^t$$

where t represents the number of 20 minute time periods and $1.37 \bullet 10^7$ (or about 13.7 million) is the estimated number of *E. coli* cells at t = 0, when the bacteria were initially placed in the broth. The base of the exponential function, 1.5, is the growth factor. For each increase of 1 unit in time t, N is multiplied by 1.5, which increases the value of N by 50%. So in the real experiment, the number of bacteria actually increased by 50% each 20 minute time period, instead of doubling as in our first idealized example.

Limitations of the model

Clearly our model described the potential growth, not the actual growth of *E. coli* bacteria. If *E. coli* bacteria really doubled every 20 minutes, then the offspring from a single cell could cover the entire surface of the earth with a layer a foot deep in less than 36 hours! The bacterial growth rate in the lab was half of that of the model, but even this growth rate can't be sustained for long. At the slower rate of a 50% increase every 20 minutes, the offspring of a single cell would still cover the surface of the earth with a foot deep layer in less than 3 days. Under most circumstances exponential growth is quickly restricted by environmental limits imposed by shortages of food, space, oxygen, predation, or the accumulation of waste products.

So, while the growth may be exponential at first, eventually the growth slows down. Figure 8.6 shows the growth of *E. coli* over 24 time periods (or 8 hours), instead of the original 12 time periods (or 4 hours). The first part of the curve (up until the vertical dotted line) repeats the original data points with their exponential growth pattern. Then the growth slows down and the curve flattens out as it approaches a maximum population size (called the *carrying capacity*). This S-shaped curve, typical of real population growth, is called a *sigmoid*, the Greek letter S. Our exponential model is applicable to only a part of the growth curve, for values of t roughly between 0 and 12 time periods. The arithmetic of exponentials leads us to the inevitable conclusion that in the long-term, the *growth rate* of populations– whether they consist of bacteria or mosquitoes or humans– must approach zero.

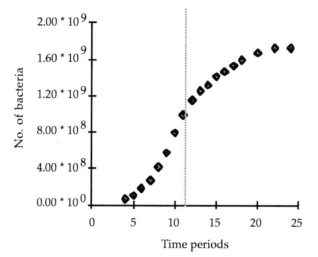

Fig. 8.6 Plot of E. coli growth for 24 time periods, showing an
initial exponentail growth that taper off.

8.2   Visualizing Exponential Functions

The bacterial growth model belongs to a class of functions called exponential functions.

An *exponential function* has the form

$$y = C \bullet a^x \qquad (a > 0 \text{ and } a \neq 1)$$

where a is the *base* of the exponent x, the dependent variable.

Predicting what the graph of an exponential function will look like [2]

Given any exponential function in the form

$$y = \ C \bullet a^x \qquad \text{where } a > 0$$

we can predict the general shape of the graph of the function. By examining values for a and C it is possible to tell the direction and relative steepness of the curve, and where the curve crosses the y-axis.

The effect of the base *a*

Case 1: *a* > 1

We compare the graphs of the three different exponential functions. Each function has $C = 1$, but different values for *a*, namely $a = 2$, $a = 3$, and $a = 4$.

$$y = 2^x$$
$$y = 3^x$$
$$y = 4^x$$

To find specific values for the function $y = 2^x$, we need to apply the rules of exponents. For example, when $y = 2^x$ and $x = 3$, then $y = 2^3 = 2 \bullet 2 \bullet 2 = 8$

When $y = 2^x$ and $x = -2$, then

$$y = 2^{-2} = \frac{1}{2^2} = \frac{1}{4}$$

Examine Table 8.8 and Figure 8.7 for the values and graphs of the functions $y = 2^x$, $y = 3^x$, and $y = 4^x$.

---

[2] We strongly suggest using a graphing calculator or "E4: Exponential Sliders" in *Quadratic Graphs* in the course software to follow along with the steps in the text.

Table 8.8			
x	$y = 2^x$	$y = 3^x$	$y = 4^x$
-2	1/4	1/9	1/16
-1	1/2	1/3	1/4
0	1	1	1
1	2	3	4
2	4	9	16
3	8	27	64
4	16	81	256

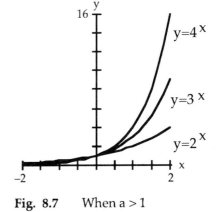

For each function, y is greater than zero for any value of x, so the graphs lie completely above the x-axis, never touching it.

Fig. 8.7   When a > 1

What happens to the values for y as x increases? For each of these curves, as the value of x increases, the value of y also increases and the curve slopes up. Thus, exponential functions where the base $a > 1$, are used to model *growth*.

Comparing curves, we see that the larger the value for a, the steeper the graph. Thus when $x > 0$, the graph of $y = 4^x$ lies above the graphs of $y = 2^x$ or $y = 3^x$. When x<0, the graph of $y = 4^x$ lies below the graphs of $y = 2^x$ and $y = 3^x$.

The graph of $y = 3^x$ lies between the graphs of $y = 2^x$ and $y = 4^x$ except at x =0. When x=0, then y =1 for all three functions. Thus the three graphs intersect at (0,1).

Something to think about

For what values of x is $2^x < 3^x < 4^x$? For what values of x is $2^x > 3^x > 4^x$?

Case 2: 0 < a < 1

Again in order to isolate the effect of a, we examine three functions in which C = 1 and vary a between 0 and 1. Examine Table 8.9 and Figure 8.8.

$$y = 0.1^x$$
$$y = 0.2^x$$
$$y = 0.5^x$$

Table 8.9			
x	$y = 0.1^x$	$y = 0.2^x$	$y = 0.5^x$
-2	100	25	4
-1	10	5	2
0	1	1	1
1	0.1	0.2	0.5
2	0 .01	0.04	0.25
3	0.001	0.008	0.125
4	0.0001	0.0016	0.625

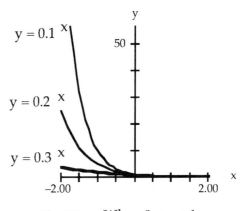
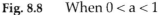

Fig. 8.8   When 0 < a < 1

Again, y is greater than zero for any value of x, so the graphs lie above the x-axis.

Notice that now the values for y *decrease* as x increases in value. Each time x increases by 1, y is multiplied by the base a, which in this case is a number less than 1, and therefore reduces the value of y. Thus, exponential functions, where $0 < a < 1$, are used to model *decay*.

Comparing curves, we can see that the smaller the value of a, the more rapidly the function declines. Thus the graph of $y = 0.1^x$ falls faster towards the horizontal axis than the graphs of $y = 0.2^x$ or $y = 0.5^x$.

The graph of $y = 0.2^x$ lies between the graphs of $y = 0.1^x$ and $y = 0.5^x$, except at x=0, y=1 where all three intersect.

Something to think about

For what values of x is it true that $0.1^x < 0.2^x < 0.5^x$?
For what values of x is it true that $0.1^x > 0.2^x > 0.5^x$?
For what value of x is $0.1^x = 0.2^x = 0.5^x$?

The effect of the coefficient C

When examining the effect of a, we compared graphs of functions with the same C value, but different values for a. Now we examine functions with the same a value and different values for C. In the following three functions the value of a is held constant at 2, and C is given different positive values:

$$y = 1 \cdot 2^x$$
$$y = 3 \cdot 2^x$$
$$y = 5 \cdot 2^x$$

Examine Table 8.10 and Figure 8.9. C has two major effects.

	Table 8.10		
x	$y = 2^x$	$y = (3)2^x$	$y = (5)2^x$
-2	1/4	3/4	5/4
-1	1/2	3/2	5/2
0	1	3	5
1	2	6	10
2	4	12	20
3	8	24	40
4	16	48	80

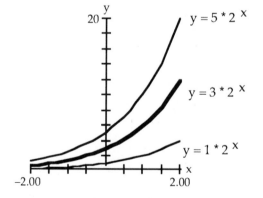

Fig. 8.9 Changing the value for C

First, changing C changes the y intercept. Given any function of the form:

$$y = Ca^x \qquad \text{where } a > 0$$

when x=0, then	$y = Ca^0$
since $a^0 = 1$	$y = C(1)$
so	$y = C$

When $x = 0$, $y = C$, so C is the y-intercept. C is often referred to as the *initial quantity*. Recall in our *E. coli* model $N = (1.37 \cdot 10^7) \cdot 1.5^t$, the value $(1.37 \cdot 10^7)$ represented an estimate of the initial bacteria, the number of bacteria at the start of the experiment when $t = 0$.

Second, an increase in positive values for C corresponds to larger function values for y. If $C > 1$, then the graph of $y = C \cdot a^x$ lies above that of $y = a^x$. If $0 < C < 1$, then the graph of $y = C \cdot a^x$ lies between that of $y = a^x$ and the x-axis. The relative heights always differ by the factor C for a given value of x.

Algebra Aerobics:[3]

Identify each of the following as a growth or decay exponential.
a) $y = 4^x$
b) $y = 0.99^x$
c) $y = 2^{-x}$
d) $y = 10^{1-t}$
e) $y = \left(\dfrac{2}{3}\right)^x$

Something to think about

Will our generalizations about the effects of C hold true if we choose different values for *a*?
Will our generalizations about the effects of *a* hold true for different values for C?
What happens if $C = 0$? What happens if $C < 0$?
What happens if $a = 1$?

---

[3] Answers

a) growth (base > 1)   b) decay (base < 1)   c) decay $\left(2^{-x} = \left(\dfrac{1}{2}\right)^x \Rightarrow \text{base} < 1\right)$

d) decay $\left(10^{1-t} = 10^1 10^{-t} = 10\left(\dfrac{1}{10}\right)^t \Rightarrow \text{base} < 1\right)$ e)   decay (base < 1)

8.3 An Expanded Definition of an Exponential Function

All exponential functions can be expressed the following form.

An *exponential function* has the form:

$$y = C \bullet a^x \qquad (a > 0)$$

where a is the *base* and C is the y-intercept or *initial quantity* (when x = 0).

If a > 1, we have *exponential growth* $\qquad$ (C $\neq$ 0)
If 0 < a < 1, we have *exponential decay* $\qquad$ (C $\neq$ 0)

Algebra Aerobics:[4]

Determine which of the following functions are exponential. Identify each exponential function as a growth or decay exponential and find the vertical intercept.

a) $A = 100(1.02)^t$ $\qquad$ b) $y = 4(3)^x$ $\qquad$ c) $y = 0.3(10)^x$

d) $y = 100x + 3$ $\qquad$ e) $y = 2^x$ $\qquad$ f) $y = x^2$

Exponential growth expressed in percentages

Exponential functions can be used to model any phenomenon that has a constant growth factor, that is, any phenomenon that increases (or decreases) by a fixed *multiple* at regular intervals. If the growth factor is 2, then the function is in the form $y = C \bullet 2^x$, so the value for y is multiplied by 2 each time x increases by 1. If the growth factor is 1.05, then the function looks like $y = C \bullet (1.05)^x$, and the value for y is multiplied by 1.05 when x increases by 1.

All exponential functions can be expressed in terms of percentages, called *growth (or decay) rates*. Multiplying by 2 is equivalent to a 100% increase. Multiplying by 1.05 is equivalent to a 5% increase. For example,

$$
\begin{aligned}
1.05 \bullet (\$1,000) &= (1.00 + 0.05)(\$1,000) \\
&= (1.00 \bullet \$1,000) \; + \; (0.05 \bullet \$1,000) \\
&= \$1,000 \; + \; (5\% \text{ of } \$1,000)
\end{aligned}
$$

---

4 $\quad$ **Answers**

a) $\quad$ growth exponential; (0,100) $\qquad$ b) $\quad$ growth exponential; (0,4)

c) $\quad$ growth exponential; (0,0.3) $\qquad$ d) $\quad$ not exponential (linear)

e) $\quad$ growth exponential; (0,1) $\qquad$ f) $\quad$ not exponential (quadratic)

The fixed *percentage increase (or decrease)* is called the *growth (or decay) rate*. In common usage, exponential growth is usually described in terms of its growth rate. How can we express the relationship between the growth rate and the growth factor mathematically? If r denotes the growth rate (in decimal form), then 1+r is the growth factor. A growth rate of 10% (or 0.10 in decimal form), corresponds to a growth factor of 1 + 0.10 or 1.10. If r denotes the decay rate (in decimal form), then 1 − r is the decay factor. A decay rate of 7% (or 0.07 in decimal form) corresponds to a decay factor of 1 − 0.07 or 0.93.

If a is the *growth factor* and r is the *growth rate* (the percentage increase in decimal form), then a = 1 + r and

$$y = C \bullet a^x = C(1 + r)^x$$

For example, if r is a growth rate equal to 0.25, then a = 1.25

If a is the *growth factor* and r is the *decay rate* (the percentage decrease in decimal form), then a = 1 − r and

$$y = C \bullet a^x = C(1 - r)^x$$

For example, if r is a decay rate equal to 0.25, then a = 0.75

Algebra Aerobics:[5]

Fill in the following table

function	initial value	growth or decay?	growth or decay factor	growth or decay rate
$A = 4(1.03)^t$				
$A = 10(0.98)^t$				
	1,000		150 %	
	30		96 %	
	$50,000	growth		7.05 %
	200 grams		51%	

---

[5] Answers

function	initial value	growth or decay?	growth or decay factor	growth or decay rate
$A = 4(1.03)^t$	4	growth	1.03 or 103 %	0.03 or 3 %
$A = 10(0.98)^t$	10	decay	0.98 or 98 %	0.02 or 2 %
$y = 1000(1.5)^x$	1,000	growth	150 %	50 % or 0.5
$y = 30(0.96)^x$	30	decay	96 %	4% or 0.04
$A = 50,000(1.0705)^x$	$50,000	growth	1.0705 or 107.05 %	7.05 %
$y = 200(0.51)^x$	200 grams	decay	51%	49% or 0.49

8.4 Examples of Exponential Growth and Decay

Medicare costs

The costs for almost every aspect of health care in America have risen dramatically over the last 30 years. One of the central issues in the ongoing debate about containing health care costs is the amount of federal dollars spent on Medicare. Medicare is a federal program that since July 1966 has provided two coordinated plans for nearly all people age 65 and over: a hospital insurance plan and a voluntary supplementary medical insurance plan. Table 8.11 and the graph in Fig. 8.10 show annual Medicare expenses (in billions of dollars) since 1970.

Table 8.11	
Year	Medicare Exp. (bil. of dollars)
1970	7.7
1971	8.5
1972	9.4
1973	10.8
1974	13.5
1975	16.4
1976	19.8
1977	23.0
1978	26.8
1979	31.0
1980	37.5
1981	44.9
1982	52.5
1983	59.8
1984	66.5
1985	72.2
1986	76.9
1987	82.3
1988	89.4
1989	102.6
1990	112.1
1991	123.3
1992	138.3
1993	154.2

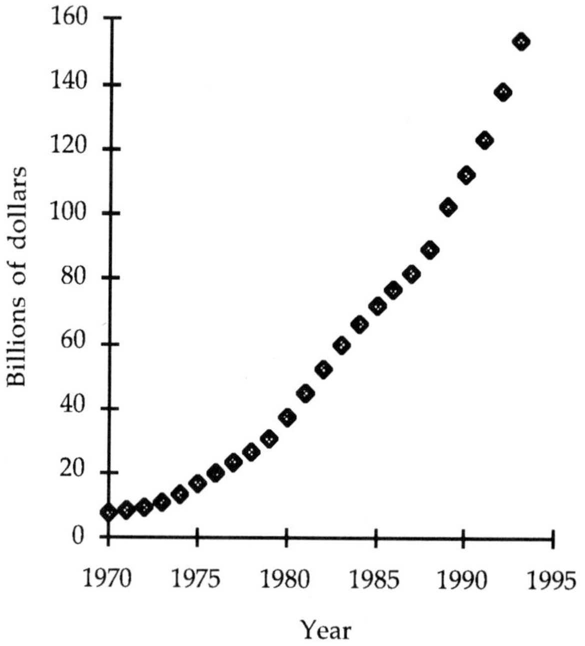

Fig. 8.10 Medicare expenditures (in billions)
Source: U.S. Bureau of the Census in *The American Almanac: Statistical Abstract of the United States, 1995-1996*

To simplify computations, it is convenient to make the base year, 1970, correspond to $x = 0$. Then 1971 would be represented by $x = 1$: 1972 by $x = 2$; 1980 by $x = 10$, etc.[6] A curve fitting program gives a best fit exponential function as:

$$y = 8.5 \cdot 1.14^x$$

where x = years since 1970 and y = Medicare expenses in billions of dollars. The function is

[6] If we did not set $x=0$ for our baseline year, and generated a best fit exponential function for the original data, we'd get an equation something like $\quad y = (3.15 \cdot 10^{-114}) \cdot 1.14^x$.
The base, or growth factor, is the same, namely 1.14. The coefficient of $3.15 \cdot 10^{-114}$ looks puzzling however. $3.15 * 10^{-114}$ is a very small number, a decimal place followed by 113 zeroes, and then 315. How could this minuscule number represent our initial costs? Remember that in the data, our x values started at 1970. The function graphing program is projecting backwards to when $x = 0$, almost 2,000 years ago to 0 A.D. The value $3.15 \cdot 10^{-114}$ is meaningless. The two graphs would appear identical except for the labels on the x-axis.

graphed in Figure 8.11 The initial quantity of 8.5 billion is the estimated value for Medicare expenses for 1970 in our *model*. It does not represent the exact value of 7.7 billion. The base of 1.14 is the growth factor. That means that according to our model, multiplying each year's Medicare expenses by 1.14 would approximate the next year's expenses. Multiplying by 1.14 is equivalent to calculating 114%. This represents an annual growth rate of 14%, quite a substantial amount.

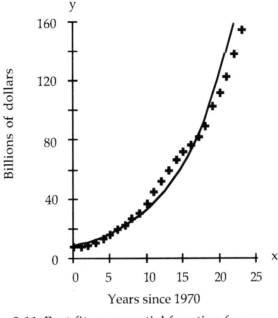

Fig. 8.11 Best fit exponential function for Medicare Expenses

U.S. population growth

Between 1960 and 1990, the U.S. grew from about 180 million to 250 million, an increase of approximately 40% in the 30 -year period. If the population continues to grow at the same rate, what would the U.S. population be in the year 2020? In 2050?

Each 30 years the population is assumed to increase by 40%; that is, it is multiplied by a factor of 1.40. Given a population of 250 million in 1990, 30 years later in 2020 the estimated population would be 1.40•250 = 350 million. 30 years after 2020 in the year 2050, the predicted population would be 1.40•350 = 490 million. If this growth rate continued, we would reach 1 billion people in about 2120. Within 11 or 12 more centuries we would have one person per square foot of land area in the United States.

Radioactive decay

One of the toxic radioactive byproducts of nuclear fission is Strontium 90. A nuclear accident, like the one in Chernobyl, can release clouds of gas containing Strontium 90. The clouds deposit the Strontium 90 onto vegetation eaten by cows. Humans ingest Strontium 90 from the cows' milk. Strontium 90 replaces calcium in bones, causing cancer. Strontium 90 is particularly insidious because it has a *half-life* of approximately 28 years. That means that every 28 years about half (or 50%) of the existing Strontium 90 has decayed into non toxic, stable Zirconium 90, but the other half is still in your bones. The following function describes the relationship between S, the amount of remaining Strontium 90, and T, the number of time periods (a time period here represents 28 years). 100 represents the initial amount of Strontium 90 in milligrams.

$$S = 100\left(\frac{1}{2}\right)^T$$

Since we usually don't measure time periods in terms of 28 years, we could let t = time measured in years. Then

$$T = \frac{t}{28}$$

So the equation becomes

$$S = 100\left(\frac{1}{2}\right)^{\left(\frac{t}{28}\right)}$$

Table 8.12 and Figure 8.12 show the decay 0f 100 mg. of Strontium 90 over time.

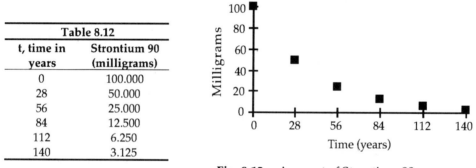

Table 8.12	
t, time in years	Strontium 90 (milligrams)
0	100.000
28	50.000
56	25.000
84	12.500
112	6.250
140	3.125

Fig. 8.12   Amount of Strontium 90

It takes about 28 years for half of the Strontium 90 to decay, but half still remains. It takes an additional 28 years for the amount to halve again, still leaving 25 grams out of the original 100.

The *half-life* of an exponentially decaying function is the time for it to halve.

The *doubling time* of an exponentially increasing function is the time for it to double.

The "rule of 70": A rule of thumb for calculating doubling times

A simple way to understand the significance of percentage growth rates is to compute the doubling time. If some quantity is growing at R percent per year, then its doubling time is approximately $\dfrac{70}{R}$ years. If the quantity is growing at R percent per *month*, then $\dfrac{70}{R}$ gives its doubling time in *months*. The same reasoning holds for any unit of time.

Example 1: Suppose that at age 25, you invest $1,000 in a retirement account, which grows at 5% per year. Since $70/5 = 14$, your investment will double approximately every 14 years. So if you retire 42 years later at age 67, then three doubling periods will have elapsed ($3 \bullet 14 = 42$). So your investment (disregarding inflation) will be increased in value 2^3 times and be worth $8,000.

Example 2: According to an article in the *Los Angeles Times*, Mexico's National Institute of Geography, Information and Statistics concluded from 1995 census data that almost half of Mexico's 91.1 million people now live in overcrowded cities and that the population nationwide will continue to grow an average of 1.8 percent each year. A private consultant claimed that the growth rate alone "means that the population will double in 40 years." Is the consultant right? Using the rule of 70, the approximate doubling time for a 1.8 growth rate would be $70/1.8 \approx 38.9$, which rounded up would give 40 years. So 40 years seems like a reasonable estimate for the doubling time.

Algebra Aerobics:[7]
1.   Find the doubling time using the "rule of 70" for:
a) $y = 100(1.02)^x$ x in years
b) $y = 10,000(1.005)^t$ t in months
c)   annual growth rate of 8.1 %
d)   annual growth factor of 106.5 %

2.   Use the rule of 70 to approximate the growth rate when the doubling time is:
a)   10 years   b)   5 minutes   c)   25 seconds

Something to think about
Using a graphing calculator or a spreadsheet, test the "Rule of 70." Can you find a value for R where this rule does not work so well?

---

[7] Answers
1.   a) $70/2 = 35$ years   b)$70/0.5 = 140$ months
      c) $70/8.1 = 8.64$ years   d)   factor $= 106.5\% \Rightarrow$ growth rate $= 6.5\%$ $70/6.5 = 10.77 \approx 11$ years
2.   a) $70/R = 10$
         $70 = 10\,R$
         $70/10 = R = 7\%$ per year   b) $R = 70/5 = 14\%$ per minute   c) $R = 70/25 = 2.8\%$ per second

Compound interest

One of the most common examples of exponential growth is compound interest. For example, assume you have $100 that you could put in a NOW account that earns 3% compounded annually, a savings account that earns 5% compounded annually, or a Certificate of Deposit (a CD) that earns 7% compounded annually. How would your results compare after 10 years?

An interest rate compounded annually of 5% means that at the end of each year, you earn 5% on the current value of your account. The interest is automatically deposited in your account, so from then on you earn 5% not only on your *principal* (the initial amount you invested), but also on the interest you have already earned. So the functions representing P_i, the current value of your account earning interest i, as a function of n, the number of years, would be:

$$P_{0.03} = 100 \cdot (1.03)^n$$
$$P_{0.05} = 100 \cdot (1.05)^n$$
$$P_{0.07} = 100 \cdot (1.07)^n$$

Remember that 3% expressed in decimal form is 0.03. To increase a number by 3% you multiply it by 1 + 0.03 or 1.03. Table 8.13 and Figure 8.13 compare the values of your account over 10 years. The $100 in the 3% account has risen to $134.39, while the $100 in the 7% account has almost doubled to $196.72.

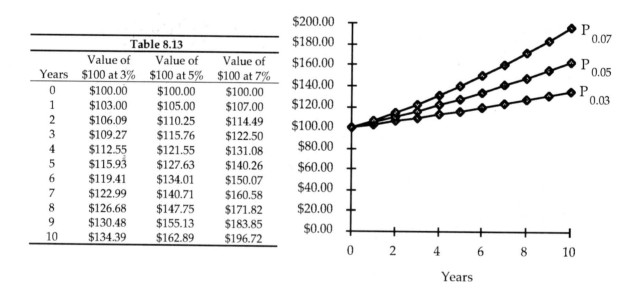

	Table 8.13		
Years	Value of $100 at 3%	Value of $100 at 5%	Value of $100 at 7%
0	$100.00	$100.00	$100.00
1	$103.00	$105.00	$107.00
2	$106.09	$110.25	$114.49
3	$109.27	$115.76	$122.50
4	$112.55	$121.55	$131.08
5	$115.93	$127.63	$140.26
6	$119.41	$134.01	$150.07
7	$122.99	$140.71	$160.58
8	$126.68	$147.75	$171.82
9	$130.48	$155.13	$183.85
10	$134.39	$162.89	$196.72

Fig. 8.13 $100 invested for 10 years at 3%, 4%, and 5%

Compound interest calculations are the same whether you are dealing with investments or inflation. For example, a 5% annual inflation rate would mean that what cost $1 today would cost $1.05 one year from today. If you think of the three percentages in the previous table as representing inflation rates, then how much would something that costs $100 today cost in 10 years? It would cost \$134.39 or \$162.89 or \$196.72 if the annual inflation rate were 3%, 5% or 7% respectively. So if you invest \$100 at 5% for 10 years it will be worth \$134.39. But inflation will drive up costs. So if during that 10 years inflation is also 5% a year, then what originally cost \$100 will now cost \$134.39. So in terms of purchasing power you will come out even, with no net gain or loss. This is why most economists usually use dollars adjusted for inflation, called *real* or *constant dollars*.

Inflation and the diminishing dollar

Inflation erodes the purchasing power of the dollar. If you have 1 dollar today and annual inflation is 5%, how much will it be worth in a year? In a year, what cost \$1.00 today will cost \$1.05 then. \$1.00 = 95.2% of \$1.05 (since $1.00 / 1.05 = 0.952$), so one dollar will only be worth 95.2 cents or 95.2 % of its original value. The decay factor is 0.952. The exponential function:

$$D = 1 \bullet (0.952)^n$$

gives D, the value (or purchasing power) of today's dollar, at year n in the future if there is a steady inflation rate of 5%. In economists' terms, it gives the real purchasing power in today's dollars. Table 8.14 and Figure 8.14 give an indication of how rapidly the value of the dollar declines.

Table 8.14	
Year	Value of dollar
0	\$1.00
1	\$0.95
2	\$0.91
3	\$0.86
4	\$0.82
5	\$0.78
6	\$0.74
7	\$0.71
8	\$0.67
9	\$0.64
10	\$0.61
11	\$0.58
12	\$0.55
13	\$0.53
14	\$0.50
15	\$0.48
16	\$0.46
17	\$0.43
18	\$0.41
19	\$0.39
20	\$0.37

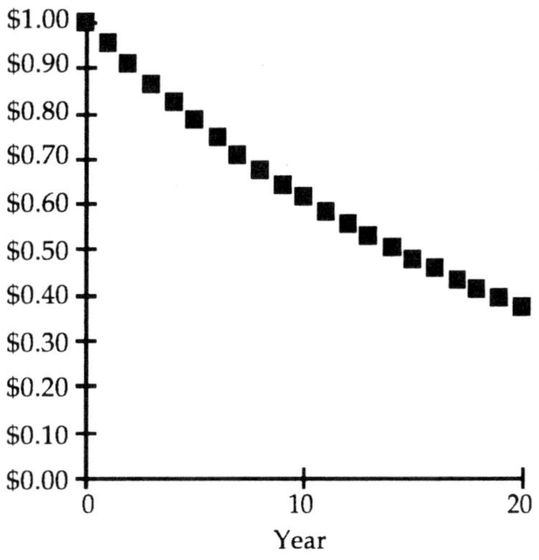

Fig. 8.14   Value of dollar with inflation at 5%

Every 14 years, the purchasing power of your money is cut in half. We can think of 14 years as the half-life of the dollar. So, in 14 years, your dollar is worth 50 cents. In 28 years, your dollar is worth only 25 cents. This is obviously a serious problem faced by those who retire on a fixed pension income.

Algebra Aerobics:[8]

1. Suppose you are planning to invest a sum of money. Determine the rate that you need so that your investment doubles in
    a) 5 years   b)   10 years   c)   7 years

2. Find the doubling time for an investment if the interest rate, compounded annually, is
    a) 3 %   b)   5 %   c)   7 %

Musical pitch

Exponential functions describe many different types of phenomena. In this example we use an exponential function to describe the relationship between musical octave and vibration frequency. The vibration frequency of Middle C is 263 hertz. The vibration frequency for each subsequent octave above Middle C is double that of the previous one.

Let N, the independent variable, be the number of octaves above Middle C and F the vibration frequency in hertz. Since the frequency doubles at each octave, the growth factor is 2. The initial frequency is 263 hertz. The function is then:

$$F = 263 \cdot 2^N$$

Table 8.15 and the graph in Fig. 8.15 show a few of the values for the vibration frequency.[9]

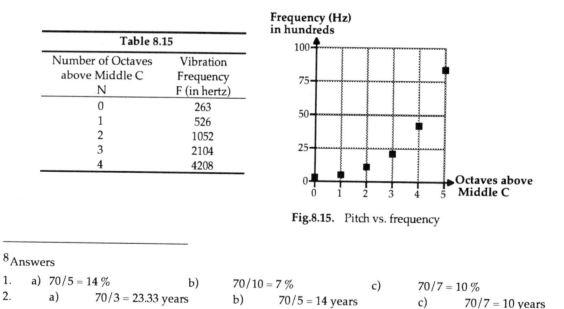

Table 8.15

Number of Octaves above Middle C N	Vibration Frequency F (in hertz)
0	263
1	526
2	1052
3	2104
4	4208

Fig.8.15.   Pitch vs. frequency

---

[8] Answers

1.   a) 70/5 = 14 %   b)   70/10 = 7 %   c)   70/7 = 10 %
2.   a)   70/3 = 23.33 years   b)   70/5 = 14 years   c)   70/7 = 10 years

[9] If you have access to a computer you might enjoy playing with a multimedia demonstration of this function in "E5: Keyboard" in *Exponential Graphs*

The Malthusian Dilemma

The most famous attempt to predict growth mathematically was made by a British economist and clergyman, Thomas Robert Malthus, in an essay published in 1798. He argued that the growth of the human population would overtake the growth of food supplies, because the population *multiplied* by a fixed amount each year, while food production only increased by *adding* a fixed amount each year. In other words, he assumed populations grow exponentially and food supplies grow linearly as a function of time. He concluded that humans are condemned always to breed to the point of misery and starvation, unless the population was reduced by other means, including war or disease.

We can frame his arguments algebraically by letting P_0 represent the original population size, a the fixed annual growth factor, and t the time in years. Then P, the population at time t is given by:

$$P = P_0 \bullet a^t$$

If F_0 represents the amount of food at time t=0, and q the constant quantity by which the food supply increases each year, then F, the total amount of food is:

$$F = F_0 + q \bullet t$$

Malthus believed the population of Britain, then about 7,000,000, was growing by 2.8% per year. ($P_0 = 7,000,000$ and a = 1.028). He counted food supply in units that he defined to be enough food for 1 person for a year, and since the food supply was at that time adequate, he assumed the Britons were at that time producing 7,000,000 food units. He thought they could increase food production by about 280,000 units a year. ($F_0 = 7,000,000$ and q=280,000.) So the two functions are:

$$P = 7,000,000 \bullet 1.028^t \qquad F = 7,000,000 + 280,000 \bullet t$$

Table 8.16 and Figure 8.16 of his predictions reveal that if the formulas were good models, then, after about 20 years, population would start to exceed food supply, and some people would starve.

Table 8.16		
Year	Population (millions)	Food units (millions)
0	7.00	7.00
5	8.04	8.40
10	9.23	9.80
15	10.59	11.20
20	12.16	12.60
25	13.96	14.00
30	16.03	15.40
35	18.40	16.80
40	21.13	18.20
45	24.25	19.60
50	27.85	21.00
100	110.77	35.00

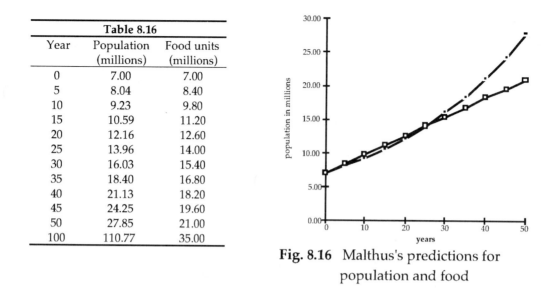

Fig. 8.16 Malthus's predictions for population and food

The two centuries since Malthus published his famous essay have not been kind to his theory. The population of the British Isles in 1995 is about 58 million and world population has grown from less than 1 billion in Malthus's time to about 5.5 billion today. Improved food production techniques and the opening of new lands to agriculture have kept food production in general growing faster than the population. Though famines still occur with unfortunate regularity in parts of the world, the mass starvation Malthus predicted has not come to pass.

8.5   Semi-log Plots of Exponential Functions

With exponential growth functions, we very often have the same problem that we did in Chapter 7 when we compared the size of atoms to the size of human beings and to the size of the solar system: the numbers go from very small to very large. For example, in our *E. coli* experiment, how can we determine whether the growth from 100 to 200 cells follows the same rule as the growth from 100 million to 200 million? It is impossible to display the entire data set on a standard graph.

One solution is to use a logarithmic (or order of magnitude) scale on the vertical axis. Recall that in Chapter 7 we used a logarithmic scale on a single horizontal axis. Moving a fixed distance up or down on a vertical logarithmic scale, corresponds to *multiplying* the variable by a constant factor, rather than to *adding* a constant as on a linear scale. Since in exponential growth, the dependent variable is multiplied by a constant factor when a constant is added to the independent variable (the *E. coli* population is multiplied by a factor of 2 every 20 minutes), exponential growth will always appear as a line of constant slope when a logarithmic scale is used on the vertical axis, and a standard "linear" scale on the horizontal axis. Since most graphing software will easily do such *log-linear* or *semi-log* plots, this is one of the easiest and most reliable ways to recognize exponential growth in a data set.

Table 8.17 repeats the *E. coli* data from Table 8.1. Figure 8.17 shows a graph of the data plotted on a log-linear graph. On the vertical axis, successive powers of 10 appear at equally spaced intervals: 100 million is just as "far" from 10 million as 100 is from 10. We can display the entire data set on a single graph, and its straight-line shape immediately tells us that it represents exponential growth.

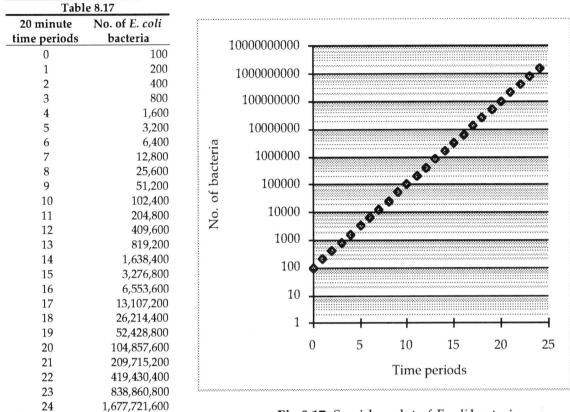

Table 8.17	
20 minute time periods	No. of *E. coli* bacteria
0	100
1	200
2	400
3	800
4	1,600
5	3,200
6	6,400
7	12,800
8	25,600
9	51,200
10	102,400
11	204,800
12	409,600
13	819,200
14	1,638,400
15	3,276,800
16	6,553,600
17	13,107,200
18	26,214,400
19	52,428,800
20	104,857,600
21	209,715,200
22	419,430,400
23	838,860,800
24	1,677,721,600

Fig.8.17 Semi-log plot of *E coli* bacteria

When an exponential functions is plotted using a standard linear scale on the horizontal axis and a logarithmic scale on the vertical axis, its graph is a straight line.
This is called a *log-linear* or *semi-log* plot.

Algebra Aerobics: From measuring on the graph, over what time interval does the population increase by a factor of 10? From the original expression for the population, $N = 100 \cdot 2^t$, over what time interval does the population increase by 8? By 16? Are these three answers consistent with each other?

Answers: Judging from the graph, the number of *E. coli* bacteria grows by a factor of 10 (for example, from 10 to 100, or 100,000 to 1,000,000) in a little over 3 time periods. From the equation $N = 100 \cdot 2^t$, we know that every 3 time periods, the quantity is multiplied by 2^3 or 8. Every 4 time periods, the quantity is multiplied by 2^4 or 16. So the answers are consistent, since somewhere between 3 and 4 times, the quantity should be multiplied by 10.

8.6 Modeling Epidemics: An Iterative Approach to Exponential Functions (optional)

The common feature of all growth models, whether of people, dollars, or diseases, is that growth comes from an initial source, and as it grows the source of growth is itself increasing. For instance, if 1 carrier of a disease infects 2 people in a week, then there are three people who can now infect others. If in the next week each of these three now infect 2 people each, then there are 6 new infected people plus the original 3, so there are now 9 carriers of the disease. To continue this sad story into the third week, 9 infectious people now infect 2 to make 18 new plus themselves, for a total of 27 infected, and infectious, people. A mathematical model of this process can be made using the variable t for time in weeks and P_t for infected population at time t. The constant P_0 represents the infected population at time t = 0. We start at $P_0 = 1$ at week t = 0.

	General Model	Specific Case
At week t = 1	$P_1 = P_0 + 2P_0 = P_0(1+2)$	
	$P_1 = P_0 \cdot 3$	$P_1 = 3$
Week t = 2	$P_2 = P_1 + 2P_1 = P_1(1+2) = P_1 \cdot 3$	
Substitute for P_1	$P_2 = (P_1 \cdot 3) \cdot 3$	$P_2 = 3^2$
	$P_2 = P_0 \cdot 3^2$	$P_2 = 9$
Week t=3	$P_3 = P_2 + 2P_2 = P_2(1+2) = P_2 \cdot 3$	
Substitute for P2	$P_3 = (P_0 \cdot 3^2) \cdot 3$	$P_3 = 3^3$
	$P_3 = P_0 \cdot 3^3$	$P_3 = 27$
At week t=4 verify that:	$P_4 = P_0 \cdot 3^4$	$P_4 = 81$

The pattern established for any week, t, is:

$$P_t = P_0 \cdot 3^t$$

where $P_t = P_0 =$ the initial population and $P_t =$ the infected population at time t.

Once again we have arrived at an exponential function through a slightly different route. The base is a constant 3, and each number in the time growth series 3, 9, 27, 81,... is 3 times the preceding number. A series of numbers of this type, where each number is multiplied by the same constant to get the next number, is called a *geometric series*. If the infection carried on at the same rate by the end of 13 weeks, a quarter of a year, over half a million people would be sick; and by the 19th week about 387,420,480 would be infectious. Since the US population is around 260,000,000, you can see that such a plague, unchecked, would spread throughout the world population fairly quickly. Of course this presumes that:

• no effective medicines are available to cure the sick and render them non-infectious

- the first infected people don't die quickly, and keep on infecting other people at the same rate
- there are no vaccines to render healthy people less vulnerable to infection
- there is no way to limit contact with the medium (insects, animals, blood, bacteria, semen) through which the disease is spread

Is this a realistic model? Can a disease spread this fast? The bubonic plague, sometimes called the black death, devastated Europe in the fourteenth century, and during a recurrence in the seventeenth century killed about 14% of the population of London. It was spread by fleas on hordes of rats, in unsanitary conditions that are fortunately now rare. The plague had a very fast rate of spread, but since many infected people died quickly, the plague finally decimated its source of infectors and the supply of healthy new people to infect, and eventually subsided. We still have concerns about epidemics today, like the AIDS and Ebola viruses.

Our model for disease growth describes a worst case scenario. By understanding how disease spreads we can find ways to fight it, such as:

- finding vaccines that confer immunity or medicines which cure or at least prevent the infected person from becoming a carrier.
- quarantining the infected or teaching them to practice effective prevention techniques.
- ensuring that healthy people understand the mechanism of infectious transmission so they can use prevention techniques.

Algebra Aerobics:[10]

Suppose that two cows on a ranch have Hoof-and-Mouth disease. Suppose each infected cow infects two other cows each day. If there are 10,000 cows on the ranch, how long will it take before they are all infected?

[10] Answers

day	number of infected cows
0	$2 = 2$
1	$2 + 2(2) = 6 = 2(3^1)$
2	$6 + 6(2) = 18 = 2(3^2)$
3	$18 + 18(2) = 54 = 2(3^3)$
4	$54 + 54(2) = 162 = 2(3^4)$
5	$162 + 162(2) = 486 = 2(3^5)$
6	$486 + 2(486) = 1458 = 2(3^6)$
7	$1458 + 1458(2) = 4374 = 2(3^7)$
8	$4374 + 4374(2) = 13,122 = 2(3^8)$

All of the cows will be infected on the eighth day.

Chapter 8 Summary

An *exponential function* has the form:

$$y = C \cdot a^x \quad (a > 0)$$

where a is the *base* and C is the *y-intercept* or *initial quantity* (when x = 0).
a is the *growth factor,* or the factor by which y is multiplied when x is increased by 1.
If $a > 1$, we have *exponential growth,* if $a < 1$, we hav e *exponential decay;.*

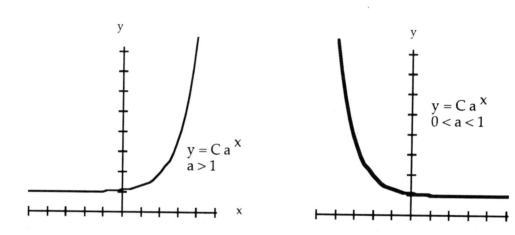

Fig. 8.18 Exponential growth: $a > 1$ Fig. 8.19 Exponential decay: $0 < a < 1$

Each exponential function can be described in terms of a percentage, r, called the *growth (or decay) rate*. If r is the *growth rate* (the percentage increase in decimal form), then a = 1 + r and

$$y = C \cdot a^x = C (1 + r)^x$$

If r is the *decay rate* (the percentage decrease in decimal form), then a = 1 – r and

$$y = C \cdot a^x = C (1 - r)^x$$

Exponential functions can be used to model any phenomenon that increases (or decreases) at a fixed percentage at regular intervals.

Exponential functions are commonly used to model the growth of populations. For example, the exponential function $N = 100 \cdot 2^t$ is an idealized model of the growth of *E. coli* bacteria. 100 is the number of initial bacteria, and 2, the base, is the growth factor. The bacteria are assumed to double every time period, t, of 20 minutes. Populations cannot sustain exponential growth for ever. Their long term growth often displays an S-shaped curve called a *sigmoid*, which is close to exponential in its early phases and then flattens out as it reaches a maximum value. The following sigmoid shows the results of monitoring the growth of E. coli in a lab for 8 hours.

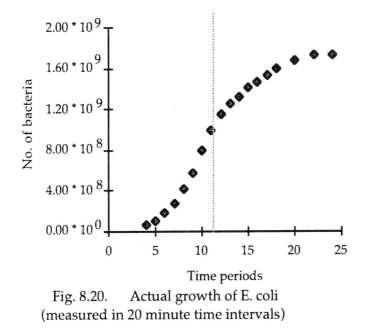

Fig. 8.20.   Actual growth of E. coli
(measured in 20 minute time intervals)

The *half-life* of an exponentially decaying function is the time for it to halve.

The *doubling time* of an exponentially increasing function is the time for it to double.

The "rule of 70" offers a simple way to estimate the doubling time. If some quantity is growing at R percent per year, then its doubling time is approximately $\dfrac{70}{R}$ years. You can substitute the appropriate time period for years. For example, if the quantity is growing at R percent per *month*, then $\dfrac{70}{R}$ gives its doubling time in *months*.

When an exponential functions is plotted using a standard scale on the horizontal axis and a logarithmic scale on the vertical axis, its graph is a straight line. This is called a *log-linear* or *semi-log;* plot.

EXERCISES

1. For each of the following exponential functions of the from $y = C \bullet a^x$:

   i) Identify C and a.
   ii) Specify whether the function represents growth or decay. In particular, for each unit increase in x what happens to y?
   iii) Generate a small table of values.
   iv) Make predictions about the graphs of these functions and then check you predictions by graphing the functions.
   v) Summarize what you learned from comparing these functions.

   a) $y = 2^x$
      $y = 5^x$
      $y = 10^x$

   b) $y = (0.5) \bullet 2^x$
      $y = 2 \bullet 2^x$
      $y = 5 \bullet 2^x$

   c) $y = 3^x$
      $y = (1/3)^x$
      $y = 3 \bullet (1/3)^x$

2. Given an initial quantity of 50 units, write an exponential function that represents quantity Q as a function of time t, if when t increases by 1:
   a) Q doubles
   b) Q increases by 5%
   c) Q is multiplied by 2.5

3. (Requires graphing calculator or computer.) Estimates for world population vary, but the following are reasonable estimates of the world population from 1800 to 1992.

   a) Enter the data table into the calculator or computer (you may wish to enter 1800 as 0, 1850 as 50, etc.)
   b) Generate a best fit exponential function. Record the equation and print out the graph if you can.
   c) Interpret each term in the function, and specify the domain and range of the function.
   d) What does your model estimate for the rate of growth?
   e) Using the graph of your function, estimate the following:

World Population	
Year	Millions
1800	910
1850	1130
1900	1600
1950	2510
1970	3578
1982	4600
1990	5300
1992	5480

   i) The world population in 1750, 1920, 2025 and 2050
   ii) The approximate year in which world population attains: 1 billion (i.e., 1,000 million), 3.2 billion, 4 billion, 8 billion
   iii) The length of time your model predicts it takes for the population to double in size

4. Determining how long ago an organism lived

Cosmic ray bombardment of the atmosphere produces neutrons, which in turn react with nitrogen to produce radioactive carbon-14. Radioactive carbon-14 enters all living tissues through carbon dioxide (via plants). As long as plant or animal is alive , carbon-14 is maintained in a living organism at a constant level. Once the organism dies however, carbon-14 decays exponentially into carbon-12. By comparing the amount of carbon-14 to the amount of carbon-12, one can determine approximately how long ago the organism died. (Willard Libby won a Nobel prize for developing this technique for use in dating archaeological specimens.).

The half-life of carbon-14 is about 5730 years. Assume that the initial quantity of carbon-14 is 500 milligrams.

   a)   Construct an exponential function that describes the relationship between C, the amount of carbon-14 and t, time. Be sure to specify the units in which you are measuring C and t.
   b)   Generate a table of values and plot the function. Choose a reasonable set of values for the domain. Remember that the objects we are dating may be up to 50,000 years old.
   c)   How many milligrams are left after 4000 years?   15,000 years? 45,000 years?

5.   Describe how a 6% inflation rate will erode the value of a dollar over time. Approximately when would a dollar be worth only 50 cents? This might be called the half-life of the dollar's buying power under 6% inflation.

   a)   Write an equation that describes the value of your account after n years if you invest $100 at 6% compounded annually. Plot the function
   b)   Write an equation that describes the value of your account after n years if you invest $200 at 3% compounded annually. Plot the function on the same graph.
   c)   Looking at your graph, will the amount in the first account ever exceed the amount in the second account? If so, approximately when?

6.   Tritium, the heaviest form of hydrogen, is a critical element in a hydrogen bomb. It decays exponentially with a half-life of about 12.3 years. Any nation wishing to maintain a viable hydrogen bomb has to replenish its tritium supply roughly every 3 years, so world Tritium supplies are closely watched. Construct an exponential function that shows the remaining amount of Tritium as a function of time as 100 grams of Tritium decays (about the amount needed for an average size bomb.)

7.   A bank compounds interest annually at 4%.
   a)   Write an equation for the value V of 100 dollars in t years.
   b)   Write an equation for the value V of 1000 dollars in t years.
   c)   After 20 years will the total interest earned on $1000 be ten times the total interest earned on $100?

8. Below are sketches of the graphs of four exponential functions. Match each sketch with the function which best describe the graphs.

$$P = 5 \bullet (0.7)^x \qquad R = 10 \bullet (1.8)^x$$
$$Q = 5 \bullet (0.4)^x \qquad S = 5 \bullet (3)^x$$

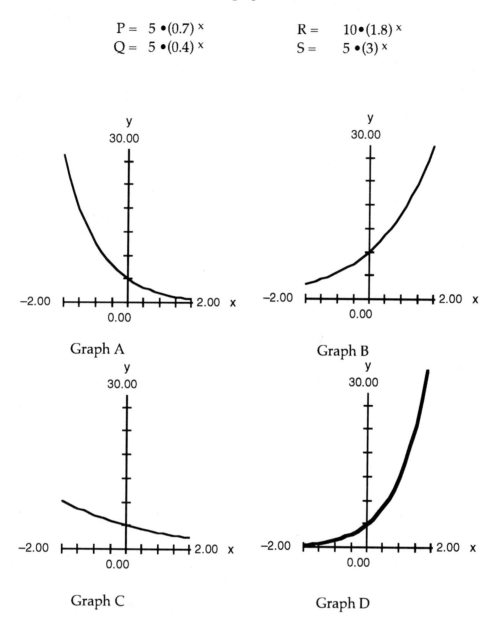

9. a) Will the graphs of $y = (1/10)^x$ and $y = 10^x$ intersect? If they do intersect, where will they intersect? Explain your reasoning. Check your predictions.

   b) Will the graphs of $y = 5 \bullet (1/10)^x$ and $y = 5 \bullet 10^x$ intersect? If they do intersect, where will they intersect? Explain your reasoning. Check your predictions.

10. (Requires a graphing calculator or function graphing program.)

Reliable data about Internet usage are hard to come by. While it is difficult to determine how many people use the Internet , it is easier to count the number of Internet hosts, since hosts must be registered. One host however could be a single user or an entire service such as American On Line to which several million people subscribe. Some of the best data, published by Network Wizards in Menlo Park, California, shows there were 1.2 million hosts in January 1993 , 4.9 million in January 1995, and 9.5 million in January 1996.

a) Give a rough estimate for the doubling time.
b) Enter the data on a graphing calculator and generate a best fit exponential function. (Set the year 1993 as 0.)
c) From your best fit function, what is the annual growth factor? The growth rate?
d) If this growth rate continues, approximately how long will it take for the number of Internet hosts to reach 20 million? 100 million? 1 billion?

11. (Some sections require a graphing calculator or computer)

Poliomyelitis, sometimes called infantile paralysis, was a major public health threat in the 1940's, but the development of the Salk vaccine brought the disease under control. The following data show the dramatic decline in polio cases after 1950 as the vaccine spread. In effect it shows a reverse epidemic.

year	1950	1960	1970	1980	1990
polio cases reported	33300	3190	33	9	0

Data from U.S. Dept. of Health and Human Services, Centers for Disease Control

a) Plot this data and rough in a curve through it.
b) Is this a growth or decay phenomenon? What generic formula might approximate this data?
c) Rewrite the data table so that the polio cases are listed in scientific notation, and years are listed in decades starting with 1950 as the zero year.

decades from 1950	0	1			
polio cases reported	$3.33 \bullet 10^4$				

d) Try to find a formula that approximates this data; if you have a computer or calculator see what it finds to fit the data.
e) Plot these formulae: $y = 3 \bullet 10^x$, $y = 3 \bullet 10^{-x}$, $y = 3000 \bullet 10^x$, $y = 3000 \bullet 10^{-x}$ for an x domain of 0 to 4.
f) If any of the previous graphs look like a fair match for the polio graph, what would you do to adjust the formula to use years rather than decades?

12. The body eliminates drugs by metabolism and excretion. To predict how frequently a patient should receive a drug dosage, the physician must determine how long the drug will remain in the body. This is usually done by measuring the half-life of the drug, the time required for the total amount of drug to diminish by one half.

   a) Most drugs are considered eliminated after five half-lives, because the amount remaining is probably too low to cause any beneficial or harmful effects. After 5 half lives, what percent of the original dosage is left in the body?

   b) The following graph shows a drug's concentration in the body over time starting with 100 milligrams.

       i)   Estimate the half-life of the drug.
       ii) Construct an equation that would approximate the curve.
       iii) How long would it take for five half-lives to occur and approximately how many milligrams of the original dosage would be left?
       iv) Write a paragraph describing your results to a prospective buyer of the drug.

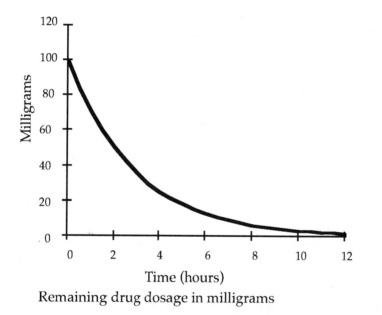

Remaining drug dosage in milligrams

13. If you have a heart attack and your heart stops beating, the amount of time it takes paramedics to restart your heart with a defibrillator is critical. According to a medical report on the evening news, each minute that passes decreases your chance of survival by 10%. From this wording it is not clear whether the decrease is linear or exponential. Assume that the survival rate is 100% if the defibrillator is used immediately.
a) Construct and graph a linear function that describes your chances of survival. After how many minutes would your chances of survival be 50% or less?
b) Construct and graph an exponential function that describes your chances of survival. Now after now many minutes would your chances of survival be 50% or less?

14. (Requires function graphing program or graphing calculator.)

In 1911 reindeer were introduced to St. Paul Island, one of the Pribilof Islands off the coast of Alaska in the Bering Sea. There was plenty of food and no hunting or reindeer predators. The size of the reindeer herd grew rapidly for a number of years as given below.

Year	Population Size	Year	Population Size
1911	17	1925	246
1912	20	1926	254
1913	42	1927	254
1914	76	1928	314
1915	93	1929	339
1916	110	1930	415
1917	136	1931	466
1918	153	1932	525
1919	170	1933	670
1920	203	1934	831
1921	280	1935	1186
1922	229	1936	1415
1923	161	1937	1737
1924	212	1938	2034

a) Plot the data

b) Find a best fit exponential function and use it to predict the size of the population in each year. (You may want to set 1911 = 0.)

c) How does the predicted population differ from the observed.

d) Does this give you any insights into why the model does not fit the observed data perfectly?

e) Estimate the doubling time of this population.

Source: Scheffer, V.B. 1951. "The rise and fall of a reindeer herd." *Sci. Monthly* 73:356-362.

15. (Optional use of graphing calculators)

According to Carlos Jarque, President of Mexico's National Institute of Geography, Information and Statistics, the population is growing at a rate of 6.46 percent per year in Mexico's Quintana Roo state, where the tourist industry in Cancun has created a boom economy. The population in Baja California, where many labor intensive border industries are located, is growing at a rate of 4.29 percent per year. But the nation's capital, Mexico City (which with more than 21 million inhabitants is already considered the world's largest metropolitan area) is growing at a rate of 0.5% per year.

a) If the growth rates continue, how long will it take for the population of Quintana Roo to double? How long for Baja California?

b) Assume the population of Mexico City continues to grow at the same rate.
   i) Write an equation that will predict its future population. (Hint: you may want to set your starting year as t=0.)
   ii) Using a calculator, generate a small table of values.
   iii) Graph the equation.
   iii) Approximately how long it will take for the population to increase by 1 million people? When would the population reach 25 million?

16. (Requires graphing calculator or computer) We have seen the following table and of the U.S. population a number of times before.

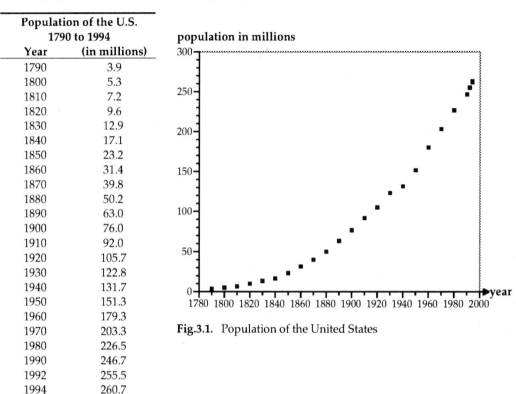

Year	(in millions)
1790	3.9
1800	5.3
1810	7.2
1820	9.6
1830	12.9
1840	17.1
1850	23.2
1860	31.4
1870	39.8
1880	50.2
1890	63.0
1900	76.0
1910	92.0
1920	105.7
1930	122.8
1940	131.7
1950	151.3
1960	179.3
1970	203.3
1980	226.5
1990	246.7
1992	255.5
1994	260.7

Population of the U.S. 1790 to 1994

population in millions

Fig.3.1. Population of the United States

a) Find a best fit exponential function and use it to estimate the size of the population in each year between 1790 and 1994.
b) Plot the estimated and actual U.S. population on the same graph. Describe how the estimated population size differs from the actual population size.
c) What would your model predict the population to be in the year 2000? In 2020?
d) In what ways is this exponential function a good model for the data? In what ways is it deficient?

17. Suppose you are offered two similar jobs. The Aerospace Engineering Group offers you a starting salary of $50,000 per year and raises of $1,000 every six months. The Bennington Corporation offers you an initial salary of $35,000 and a 10 % raise every year.
a) Make a table that shows your salary with each company during the first eight years.
b) On the same coordinate system, plot the graphs of the functions for years one to twenty.
c) For each corporation, write a function that gives your salary after t. years.
d) In what year did the salary at Bennington exceed the salary at Aerospace?
e) Using the models of part (c), determine your salary after ten years and after twenty years at each corporation.

18. The following data shows the total government debt for the U.S.

year	1970	1975	1980	1985	1990
national debt, billions $	370.1	533.2	907.7	1823.1	3233.3

Data from U.S. Dept. of the Treasury

a)   Plot this data and rough in a curve through it.

b)   Is this a growth or decay phenomenon? What generic formula might approximate this data?

c)   For 1975 to 1990 calculate the ratio of each debt to the one 5 years ago, such as 1975 debt/1970 debt. This is the growth factor for each particular 5 year period. Is the growth factor staying relatively constant? Average the ratios to get an average growth factor G.

d)   Plot $y = 370\, G^x$ for a domain of 0 to 4.

e)   What would G have to be for $3233 = 370\, G^4$? Use this new value of G to plot $y = 370\, G^x$ for a domain of 0 to 4.

f)   Which formula is the best fit to the data? Use the formula with the better fit to estimate national debts for 1995 and 2000.

g)   Predict the national debt using an exponential model. Compare the actual statistics available for years beyond 1990 with your predictions.

h)   Do you think the exponential model is an appropriate model for describing the growth of the national debt. Explain your reasoning.

EXPLORATION 8.1
Devising a Model for the Growth of Bacteria [11]

Objective

* to develop a model for bacterial growth by analyzing data on the growth of *Escherichia coli* cells. Either you will use previously generated data on bacterial growth, or you will generate original data in a laboratory experiment, then use these data to devise a model.

Materials/Equipment

* function graphing program with best fit function capabilities or graphing calculators with best fit function capabilities.

   This exercise can be carried out in two ways. First, you can analyze data provided. A disk and printout of data collected by a student at UMass/Boston is included for this purpose. Alternatively, you can measure bacterial growth in a laboratory experiment, then analyze the data generated from this experiment. For this experimental approach, equipment and materials are described in the Experimental Procedures Section at the end of this Exploration.

Introduction

In this laboratory, you will analyze the growth of a population of bacteria. You may actually collect data on the increase of a population over time, or such data will be provided. Then you will analyze the data and try to predict what kind of function defines the dynamics of population growth. The growth patterns you will observe apply to other populations besides those of bacteria. We are using population growth of bacteria, which is easy to measure, as a model to characterize population growth as a whole.

The particular bacterium we will use for this study is called *Escherichia coli*, or *E. coli* for short. *E. coli*, which is one of the most thoroughly studied of all organisms, is a rod-shaped bacterium that is approximately 1 micrometer (that's 10^{-6} meter) long. Normally, this organism inhabits the intestinal tract of humans. Only certain types of *E. coli* cause problems for the human host. Most types, in fact, aid in our digestion.

Cells of *E. coli* reproduce by simply splitting in half, so that one cell forms two "daughter cells." This form of cell division is called fission. It occurs at a rate that depends on the nutrients that are available. When conditions are ideal, *E. coli* can reproduce very quickly. On the other hand, in a restrictive environment, one in which the food supply is scarce or conditions are poor in other ways, cell division may slow down or stop altogether. To give you an idea of how fast *E. coli* can grow when conditions are unrestricted, consider a population of *E. coli* cells growing under ideal conditions with an unlimited supply of food. If we start with one cell, and cell division were to continue uninterrupted under these

---

[11] Developed by Rachel C. Skvirsky, Biology Department, University of Massachusetts/Boston

conditions, then after *only 36 hours*, there would be enough bacterial cells to cover the earth in a layer that's a foot deep (Campbell, 1993)!! Clearly growth conditions in nature are never so ideal!

In this laboratory, we investigate the effects of two environmental variables (nutrients and temperature) on the growth of a bacterial population. Because *E. coli* cells reproduce rapidly, we can plot at least a portion of the growth curve for *E. coli* under various environmental conditions. For example, we will compare growth of the bacteria using two different nutrient solutions -- one is called "LB," which is a standard nutrient broth, and the other is "Superbroth," which is an enriched nutrient broth. In addition, we will compare two temperature conditions. One is body temperature (37°C), and the other (30°C) is between body temperature and room temperature. Remember that *E. coli* normally grows inside the body of humans (or other mammals), hence at 37°C.

To monitor the growth of a population of bacteria, we must estimate the number of bacterial cells in the population at different times. Individual bacterial cells are too small to be seen with the naked eye. Therefore we can't simply count them, unless we examine a very small sample with a powerful microscope. Although counting is possible, we will use a much faster method. We take advantage of the fact that as bacteria multiply, they increasingly cloud the solution in which they are growing. The more bacteria, the more cloudiness. This cloudiness occurs because each tiny bacterium absorbs and/or scatters light to a tiny extent. We use this fact in our experiment: we measure the cloudiness of a sample of well-mixed solution by measuring the extent to which the solution absorbs light. Based on this, we can chart the increase in the concentration of bacteria in the solution.

For these measurements, we use a *spectrophotometer*. This is an instrument which measures the amount of light which is absorbed by a solution. Absorbance (or optical density) is related to the number of particles absorbing light. Specifically, optical density is equal to a constant times the concentration of particles. Therefore we use absorbance as a measure of the concentration of particles (bacteria) in the solution.

Biologists have found that an optical density of 0.01 is obtained with a concentration of 1 million bacteria per 5 milliliter (ml) sample. If the optical density were 0.02, then how many bacteria would there be per milliliter?

Procedure (based on provided data):

> The following are data from actual measurements of bacterial populations. To carry out this experiment, flasks containing 30 milliliters of either LB or Superbroth medium were inoculated with 0.3 milliliters of an overnight culture of *E. coli* cells (strain HB101). Measurements were taken with a "Klett-Summerson colorimeter," which is a simple type of spectrophotometer and measures the optical density or turbidity of the culture. Remember that optical density, represented in this case by a Klett value, is equal to a constant times the concentration of bacteria. However, a Klett colorimeter uses a different scale from other spectrophotometers. You can assume that a Klett value of 100 corresponds to 5×10^8 cells/milliliter.

	LB Cultures				Superbroth Cultures			
	Temp = 30° C		Temp = 37° C		Temp = 30° C		Temp = 37° C	
Elapsed time (min)	Klett Reading	# of cells/ml	Klett Reading	# of cells/ml	Klett Reading	# of cells/ml	Klett Reading	# of cells/ml
20	5	2.50×10^7	2	1.0×10^7	2	1.00×10^7	2	1.00×10^7
50	6	3.00×10^7	6	3.0×10^7	7	3.50×10^7	7	3.50×10^7
80	7	3.50×10^7	11	5.5×10^7	7	3.50×10^7	15	7.50×10^7
110	17	8.50×10^7	27	1.35×10^8	10	5.00×10^7	30	1.50×10^8
130	22	1.10×10^8	37	1.85×10^8	16	8.00×10^7	53	2.65×10^8
165	30	1.50×10^8	62	5.10×10^8	20	1.00×10^8	87	4.35×10^8
180	35	1.75×10^8	75	3.75×10^8	27	1.35×10^8	125	6.25×10^8
200	40	2.00×10^8	87	4.35×10^8	38	1.90×10^8	163	8.15×10^8
230	57	2.85×10^8	106	5.30×10^8	53	2.65×10^8	188	9.40×10^8

* Klett 100 = 5×10^8 cells/ml

Use these data as the basis for your analysis of bacterial growth. First, make a graph which plots either the Klett reading or the extrapolated cell density (this is the dependent variable and goes on the y-axis) versus time (this is the independent variable and goes on the x-axis). Note the maximum Klett value or cell density that was measured, as well as the maximum number of minutes elapsed, and use these numbers as guides in setting up the scales of the graph. Use different symbols and/or different colors for the four different cultures, plotting them all on the same graph.

Questions for thought

Re-read the introduction, and be sure you understand the relationship between optical density and concentration of bacteria. You will prepare a report on this lab, which should focus on two issues: 1.) The nature of population growth in your cultures of *E. coli*, and 2.) a comparison of population growth of *E. coli* in the 4 different environmental conditions. Be sure to address the following questions (a-g):

a) How does the concentration of bacteria in a 5 ml sample relate to the number of bacteria in the total population? Why is it valid to monitor population growth by measuring 5 ml samples?

b) Does population size of *E. coli* in the flasks change as time passes? How do you know?

c) Does the rate of *E. coli* population growth in the flasks change over time? How do you know?

d) The nutritional requirements of various types of bacteria are different; consequently, microbiologists use many different types of media to grow bacteria. "LB" medium supports the growth of a wide array of microorganisms. In this exercise, we use data from cells grown in LB medium, as well as nutrient broth that contains extra nutrients -- "Superbroth". In which of these growth media do *E. coli* cells grow better? Examine the recipes for these two growth media shown in the Experimental Procedures section. What are the differences between these two media?

e) Are the "test-tube" environments limited (so that the *E. coli* populations show restricted growth), or unlimited (so that *E. coli* shows unrestricted growth)?

f) How are the shapes of these growth curves likely to change if the experiment is continued for a substantially longer period of time? Why?

g) Does inspection of your graphs reveal differences in growth of *E. coli* populations under different environmental conditions? Do these differences make sense in terms of what you know of the natural habitat of *E. coli*?

Lab Report

Below is the format and some guidelines for preparing your lab report:

Objectives

Describe the purpose of the experiment. What were you trying to accomplish? How were you using data on flasks of growing bacteria to accomplish your objectives?

Materials and Methods

Describe the procedures that you followed, using the past tense to explain what was done. For example, what type of bacteria was used? Under what environmental conditions were the bacteria grown? At what intervals were measurements taken? If you followed the methods outlined in the Experimental Procedures section to take your own measurements of bacterial growth, then additional issues should be addressed. For example, how were the measurements taken? You may reference the lab handout itself, rather than repeating each and every detail. Perhaps most important, this section should describe any ways in which your procedure departed from the printed instructions, or any mistakes that were made. For example, did any flask of cells stay out of the water bath longer than planned? Do you have reason to believe that certain measurements were taken incorrectly? Often a great deal can be learned from the recording and analysis of mistakes or unexpected events.

Results

This section should include all of your data, presented in the form of a table. Enter your data in your function graphing program or graphing calculator and prepare a graph. Remember that optical density or cell density, which is the dependent variable, goes on the y-axis, and time (in minutes) which is the independent variable, goes on the x-axis. Also remember that optical density is an expression of cell density.

Analysis of Data

To analyze your data, use as a guide the questions asked here and the section Questions for Thought above.

- Can you predict a function that describes the growth of a population undergoing unrestricted growth? To develop your prediction, first do the following exercise: Pretend that you start with one bacterial cell. Remember that this cell reproduces by splitting in half. Now pick an arbitrary amount of time (such as one hour or whatever) that it takes this cell to divide. You can call this amount of time the "doubling time." If you start with one cell, how many cells would you have after 1 hour (1 doubling time)? How about after 2 or 3 or 4 hours (or doubling times)? Can you derive an expression that describes how many cells you will have after x hours or doubling times?

- Under each environmental condition, what's the average rate of change per hour of the population? Which environmental conditions best supported the growth of *E. coli*?

- Under the conditions that were used to generate the data, or the conditions that you actually used in the lab, was the rate of growth in the population constant? If not, how would you describe the change in rate? What do you think causes the change in growth rate? Can you predict what would happen to the population in these flasks if the experiment lasted another whole day? Justify your answer.

- Recall the basic difference between population growth in a restricted environment and growth in an unrestricted environment. In an unrestricted environment, growth is constant, whereas in an environment that is restricted by limited nutrients, growth declines as the population becomes more dense. How could we determine whether the specific environments in which you grew *E. coli* are restricted or not?

Experimental Procedure Section

(This section was modified from a biology laboratory manual from the University of Massachusetts Boston.)

Measuring the Growth of *E. coli* Cells in the Laboratory

Note to instructor: This laboratory experiment is best carried out in collaboration with a biology department. It requires bacterial cultures, controlled-temperature shakers, and a spectrophotometer or Klett-Summerson colorimeter. It also requires a three-hour laboratory period. Overnight cultures of *E. coli* should be used to inoculate fresh cultures (100 ml of growth media in 500 ml flasks) several hours before the laboratory, so that cells are actively growing during the measurement period. The materials and procedures are described in detail below.

Materials/Equipment

Spectronic 20 (or other spectrophotometer or Klett meter). If a Spectronic 20 or other
    spectrophotometer is used, the wavelength should be set at 686 nanometers
Tubes for Spectronic 20
500 ml Erlenmeyer flasks
2 controlled-temperature shakers or shaking water baths at 30°C and 37°C.
Broth cultures of *E. coli* cells in exponential phase
LB liquid medium ("nutrient broth")
Superbroth liquid medium ("enriched nutrient broth")
Test tubes
Pipets and pipet bulbs
Test tube racks
Parafilm
Markers for labeling tubes
Sheet of Spectronic 20 directions

Procedure

You must record all your data in your class notebook. Also remember to record the codes you use to label flasks and tubes, as well as anything unplanned or unexpected that happens in the course of the experiment.

Each group of students (three or four people to a group) should plan to work with four different "cultures." (A culture is a batch of cells grown in the lab, such as in a flask):
    a) LB, 30°C
    b) LB, 37°C
    c) Superbroth, 30°C
    d) Superbroth, 37°C

You will grow the four growth cultures (flasks of bacteria) described above, and you will use a Spectronic 20 to measure the population densities of *E. coli* about every 20 minutes. A Spectronic 20 is a specific kind of spectrophotometer. Be sure to read the instructions provided on how to use the Spectronic 20.

1. Each group should obtain six clean spectrophotometer (or "Spec") tubes, plus a test-tube rack to keep them in. These tubes are made of special optically clear glass and are expensive -- handle them with care. Label one of them "B-LB" (for blank - LB medium), another one "B-SB" (for blank - Superbroth medium) and the other four a, b, c, and d. Pour about 5 ml of the appropriate medium (without cells) in the proper blank tubes, and cover them with parafilm. You will need these blanks throughout the rest of the experiment. Do not discard! Keep the other four tubes ready for step 3 below.

2. Each group of four students will be given four cultures of *E. coli*, as described above, labeled with the appropriate letter (a, b, c, d). To begin, you should label these cultures with a code word that identifies your group so that you can easily retrieve them later from the appropriate water bath. You will maintain two of these cultures at 37°C, and the other two at 30°C.

3. Swirl each flask for a few seconds (the group should standardize their technique here -- the flasks should be swirled thoroughly so that the bacteria are evenly distributed) and pour approximately 5 ml from each flask into the appropriately labeled Spec tube (a,b,c, or d). This means that solution "a" goes in the tube labeled "a", etc. In order to estimate how much 5 ml will be, use the level to which the 5 ml came in either of the two blanks to get an approximate level (since all the tubes are the same diameter, 5 ml should come to the same height in each tube). The four tubes which you just filled with 5 ml samples from your flasks now contain your bacterial samples at time zero (that is, the beginning of the actual experiment). Record the exact time, using the lab clock or a watch.

4. Immediately place the four flasks in the appropriate water bath shakers (30°C or 37°C) and let them shake for approximately twenty minutes before you withdraw the next sample. (Why do the flasks need to be shaken?) One partner should keep an eye on the time and be alert for the next sampling time. From here on, keep your cultures in the water bath, with the shaker on, as much as possible. Why is this important?

5. Once your flasks are in the water bath, tell your instructor you are ready to measure the optical density (O.D.) of your time zero samples (prepared in step 3 above). You should have read the instructions for operating the Spectronic 20 instrument. The instructor will make sure you understand its operation; don't hesitate to ask for more instruction if you are not sure you are using the machine properly. Be sure to record in your note-book the absorbance (optical density) of each sample, not the "transmittance" which can also be read off the machine.

6. At approximately 20-minute intervals (samples can be taken at any time, provided that you record the exact time of sampling), remove each culture from the water bath, swirl

it, and take additional 5 ml samples. Then determine their optical density (O.D.). You may use the same Spec tubes as before (rinse them each time with distilled water), or you can use a new labeled set. Be sure to keep your blanks (B-LB and B-SB) throughout the experiment. Always use the correct blank to "zero" the machine for the culture you are going to measure. This means that you zero the Spec 20 machine with the B-LB blank when you read samples a and b, and zero with the B-SB blank when you read samples c and d. Why is this important?

7. While your cultures are incubating and sampling is progressing, each student should prepare a table with four columns that contain: (1) sample types (2) sampling times, (3) elapsed time (since experiment started), and (4) optical densities. Each student should also make a graph which plots the optical density (this is the dependent variable; it goes on the y-axis) versus time (this is the independent variable; it goes on the x-axis). The maximum optical density you will probably record is about 1.00, and the maximum time will probably be 3 hours or 180 minutes, so use these numbers as guides in setting up the scales of the graph. Start to plot your results as you receive them. Use different symbols and/or different colors for the four different cultures, plotting them all on the same graph.

8. From the Introduction section of this handout, you learned that optical density is equal to a constant times the concentration of bacteria. How could we verify that optical density and concentration of bacteria are truly related in this way? If you have time, try the following experiment:

    Obtain from your instructor a solution that has a high density of bacteria. Put 5 ml in a Spec tube and measure its optical density. Call the number of bacteria in there "x". Now prepare several different dilutions of the culture -- e.g. 1/2x, 1/4x, etc. Measure the optical density of each of the dilutions. Plot the following graph: concentration of bacteria (independent variable -- x-axis) vs. optical density (dependent variable -- y-axis). What should you see? What do you see?

9. Toward the end of the lab period, consult with other groups and with your lab instructor, comparing your graph with others. If no major problems of technique have been made, all results in the class should be similar, and you can concentrate on the results of your own group in doing an analysis of the results. Also, you should decide whether one member of your group can take another measurement later in the day or on the following day.

Nutrient Media Recipes:

LB: Per liter:
10 grams tryptone
5 grams yeast extract
5 grams NaCl

Superbroth: Per liter:
32 grams tryptone
20 grams yeast extract
5 grams NaCl
(Adjust pH to 7 with 1N NaOH)

EXPLORATION 8.2
Properties of Exponential Functions

Objective
- to explore the effects of a and C on the graph of the exponential function in the form

  $y = Ca^x$ where a>0 and a ≠ 1

Materials/Equipment
- computer and software on disk, "Exponential Functions" in *Exponential Graphs,* **or graphing** calculator
- graph paper

Procedure
We start by choosing values for **a** and **C** and graphing the resulting equations by hand. From these graphs we make predictions about the effects of **a** and **C** on the graphs of our equations. You can use graphing calculators or the "Exponential Functions" program in *Exponential Graphs* to test your predictions. Take notes on your predictions and observations, so you can share them with the class. Work in pairs and discuss your predictions with your partner.

Making Predictions

1. Start with the simplest case, where **C=1**. The equation will now have the form:

    $y = a^x$

    Make a data table and sketch on the same grid the graphs for $y = 2^x$ (note that a = 2 in this example) and $y = 3^x$ (note that a = 3 in this case). Use both positive and negative values for **x**. Make a prediction of where the graph of $y = 2.7^x$ would be located on your graph. Make a prediction of where the graph $y = 12^x$ would be located on your graph. Check your work and predictions with your partner.

x	$y = 2^x$	$y = 3^x$
-2		
-1		
0		
1		
2		
3		
4		

How would you describe your graphs? Do they have a maximum or a minimum value? What happens to **y** as **x** *increases*? What happens to **y** as **x** *decreases*? Which graph shows **y** changing the fastest compared to **x** ?

2. Now create three functions in the form $y = a^x$ where $0 < a < 1$. Create a data table and graph your functions on the same grid. Make predictions for other functions where C=1 and $0 < a < 1$.

x	y =	y =	y =
-2			
-1			
0			
1			
2			
3			
4			

3. Now consider the case where C has a value other than 1 for the general form of the exponential function:

$$y = Ca^x$$

Create a table of values and sketch the graphs of $y = (.3) 2^x$ (in this case C=.3 and a=2) and $y=(5)2^x$. (in this case C= 5 and a=2). What do all these graphs have in common? What do you think will happen when a=2 and C=10? What do you think will happen to the graph if a=2 and C =−5? Check your predictions with your partner.

x	$y=(.3)2^x$	$y = (5)2^x$	$y = (-5)2^x$
-2			
-1			
0			
1			
2			
3			
4			

How would you describe your graphs? Do they have a maximum or a minimum value? What happens to **y** as **x** *increases*? What happens to **y** as **x** *decreases*? Which graph shows **y** changing the fastest compared to **x** ? What is the y intercept?

Testing Your Predictions

We now want to test our predictions and explore the effects of changing the values for a and **C**. for equations in the form $y=Ca^x$. To do this, we will need a method for our investigation. For this part of the exploration, you can use a program called "Exponential Functions" in the *Exponential Graphing* software package or you can create graphs using a graphing calculator or function graphing program. The program "Exponential Functions" will allow you to choose values for a and C and explore the resulting graphs.

Summarizing your results

1) What effect does **a** have?

Make predictions for **a**>1 and 0< **a**< 1 based on the graphs you constructed by hand. Explore what happens when C = 1 and you choose different values for **a**. Check to see if your predictions about the effect of **a** hold true when C ≠ 1. You can check your prediction with the "Exponential Functions" program in *Exponential Graphing* or you can use a graphing calculator.

> If you are using a graphing calculator or function graphing program, check your predictions by graphing your equations on the same grid.

> If you are using the "Exponential Functions" program, drag the line in the **a** column. As you change the value of **a**, you are changing your equation and the corresponding changes in the graph of the equation will appear on your screen. This program will allow us to create more than one graph on our grid.

How does changing **a** change the graph? When does $y = a^x$ describe growth? When does it describe decay? When is it flat? Write a rule that describes what happens when you change the value for **a**. You only have to deal with cases when **a** > 0.

Rule:

2) What effect does **C** have?

Make a prediction based on the graphs you constructed by hand. Now choose a value for **a** and create a set of functions with different **C** values. Graph these functions on the same grid.

How does changing **C** change the graph? What does **C** tell you about the graph of functions in the form $y=Ca^x$? Describe your graphs when **C**>0 and when **C**<0.

Exploration Linked Homework

Reporting your results

> Verbally:
> Prepare a 60 second summary of your results to present to the class.
> In writing:
> Write a short paragraph summarizing your results.

EXPLORATION 8.3
Predictions Using Different Assumptions

Objectives
- to discover rules for finding growth and decay factors
- to compare linear and exponential growth

Materials/Equipment
- graph paper
- **graphing calculator or** spreadsheet and function graphing program **(optional)**

Procedure

Working in pairs
1. Using the exponential function, $f(x) = C \cdot a^x$ where $(a > 0$ and $a \neq 1)$,
   a) Write an expression for
      $f(n)$ the value of the function when $x = n$.
      $f(n+1)$ the value of the function when $x = n+1$.
      $f(n-1)$ the value of the function when $x = n-1$.
   b) The ratio from $x = n-1$ to $x = n$ is equal to $f(n)/f(n-1)$ and from $x = n$ to $x = n + 1$ it is equal to $f(n+1)/f(n)$. Calculate both of these ratios. For exponential functions, these ratios are called *growth or decay factors*.
   c) How do these ratios relate to the exponential function $f(x) = C \cdot a^x$. State your results in a general rule.

Using your results

1. Use the generalizations for exponential functions that you have found to analyze each of the following data tables.
   a) Find out if there is a growth or a decay factor.
   b) Explain in your owns words how to find $f(x)$ in terms of x.

x	f(x)
0	1
1	10
2	100
3	1,000
4	10,000

x	f(x)
0	3
1	6
2	12
3	24
4	48

x	f(x)
0	-3
1	-6
2	-12
3	-24
4	-48

x	f(x)
0	1.0
1	0.1
2	0.01
3	0.001
4	0.0001

x	f(x)
0	10.0
1	5.0
2	2.5
3	1.25
4	0.625

x	f(x)
0	25
1	100
2	400
3	1600
4	6400

c)) Use an equation to describe the pattern that you have found in the data tables. (Remember that $a^0 = 1$.) Does your equation predict the values in your data table? Explain in your own words how you found your equation.

d) Extend each of the data tables to include negative values for x using the rule you have found.

Making predictions with different assumptions

1. *The San Francisco Chronicle* (Thursday, Jan. 5, 1995, p.A6) reported that the number of poor children in California under 2 years of age grew from 184,753 in 1980 to 275,466 in 1990. Using time periods that represent ten year intervals and 1980 as the base year (or year zero), create models for each of the following assumptions:

   a) Assume that the number of poor children in California is increasing linearly with time. Find an equation relating number of poor children and time. Use your model to make predictions as to what will happen in the year 2000 and the year 2020.

   b) Assume that the number of poor children in California is increasing exponentially with time. Find an equation relating number of poor children and time. Use your model to make predictions as to what will happen in the year 2000 and the year 2020.

   c) Graph the two functions on the same grid. Use your graphs to estimate when each of these models will predict the number of poor children in California will be:

   i)   double the number of poor in 1980.

   ii) double the number of poor in 1990.

2. The federal government helps students to finance higher education through a grant program called Pell Grants (awards which students do not have to pay back) and through federal loans (which students do have to be pay back.) The National Education Association reported that the number of Pell Grant recipients fell from 4.2 million in fiscal year 1993 to 3.7 million in fiscal year 1994. (NEA Almanac of Higher Education, 1996. p.90). Using FY 1993 as the base year (or year zero) and FY 1994 as year one, create a model for each of the following assumptions:

   a) Assume that the number of recipients is declining linearly with time in years. Find an equation relating number of recipients and years.

   b) Assume that the number of recipients is declining exponentially with time in years. Find an equation relating number of recipients and years.

   c) Graph your functions. Use your graphs to estimate when each of these models will predict that funding will be half of what it was in FY 1993 . Check your estimates from your graphs by using these values in each of your functions.

   d) Use each of the functions to predict what would happen in FY 1995. Find out the number of recipients in FY 1995. Which prediction came the closest to the actual number?

Chapter 9

Functioning with Powers

Overview

In many physical systems, one variable varies as a constant power of another: the volume of an inflating balloon as a function of its radius, the gravitational force between two bodies as a function of their separation, and the volume of air in your lungs as a function of the depth to which you scuba dive. Combinations of power functions are useful in describing more complicated physical systems, and in analyzing experimental data.

This chapter presents *power functions*, both direct variation (positive exponent) and indirect variation (negative exponent). We compare them to exponential functions, and investigate their properties for odd and even exponents, and for varying values of the coefficient. We construct *polynomials* by adding together power functions with different exponents and study their properties, their zeros, and their use in fitting data.

In the first exploration, you will predict the properties of power and polynomial functions, and construct functions with specific properties. In the second, you will solve maximum area and constrained volume problems using polynomials. Finally, in the third you will learn to visualize power functions which represent inverse proportionality relations.

After reading this chapter you should be able to:

- predict the properties of a power function from its algebraic expressions
- construct power functions having the desired properties
- understand the differences between power and exponential functions
- be familiar with polynomial functions and their shapes
- find the intercepts of polynomials both graphically and algebraically

9.1   The Behavior of Gases

Question: If you fill a round balloon with helium, when will it begin to rise?

Discussion: The function

$$v = f(r) = \frac{4}{3}\pi r^3$$

describes the volume of a sphere as a function of its radius. The volume, v, equals a constant $\left(\frac{4}{3}\pi\right)$ times the cube of the radius, r. (The number π is a geometrical constant, the ratio of the diameter of a circle to its circumference, and it is approx. equal to 3.14.) If the radius increases, the volume will increase. If we inflate a spherical balloon and assume that the balloon does not compress the air, then Table 9.1 and Figure 9.1 show the relationship between the expanding radius, r, and the volume, v, of the balloon.

Table 9.1	
r	v
(cm.)	(cubic cm.)
0	0
2	33
4	267
6	902
8	2,138
10	4,176
12	7,216
14	11,459
16	17,106
18	24,356
20	33,410

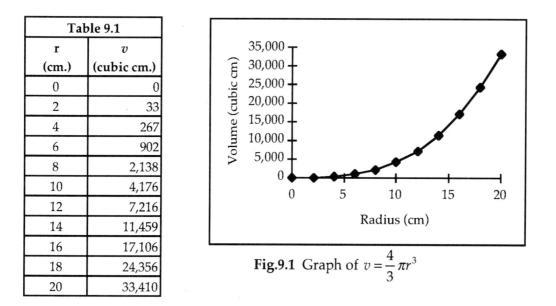

Fig.9.1 Graph of $v = \frac{4}{3}\pi r^3$

When the radius doubles from 10 to 20 cm, the volume increases from $(4/3)\pi(10)^3$ to $(4/3)\pi(20)^3 = (4/3)\pi(10)^3(2)^3$ cm. Hence the volume will become 2^3 or 8 times larger. In general if the radius doubles from R cm to 2R cm, then the volume increases from $(4/3)\pi(R)^3$ to $(4/3)\pi(2R)^3 = (4/3)\pi(R)^3(2)^3$, which is 2^3 or 8 times larger.

Similarly if the radius triples, the volume become 3^3 or 27 times larger.

How large will the radius have to be, before the balloon begins to rise? Archimedes' Law says that the balloon will float upwards when the total weight of the balloon, consisting of the weight of the rubber plus the weight of the helium inside, is less than the weight of an equal volume of air.

Suppose that the rubber of the balloon weighs 10 grams. A liter is 1000 cubic centimeters and the density of helium gas is 0.17 grams/liter (or 0.17 grams/1,000 cm^3). In order to simplify our computations, let's define the volume V to be the volume of the balloon in liters. So $V = v / 1000$. Hence the weight of the helium in the balloon with volume V (in liters) equals 0.17 V grams. So the total weight of the rubber plus the helium is

$$(10 + 0.17\, V)\ \text{grams}$$

where V is the volume, *in liters*, to which the balloon has been inflated.

Since the density of air is 1.2 grams/liter, then the weight of air with volume V equals 1.2 V grams. The balloon will begin to rise when it is light enough; that is, when

weight of rubber + weight of helium < weight of air
10   +   0.17 V <   1.2 V

rearranging terms $(1.2\ 0.17)\, V\ >\ 10$
gives us $V\ >\ 10 / (1.2 - 0.17) = 9.7$ liters or 9,700 cm^3

Estimating from Table 9.1, when the radius is somewhere between 12 and 14 cm, the balloon's volume will become > 9,700 cm^3 (or 9.7 liters). At that point the total weight of the balloon is less than the comparable volume of air, and the balloon will ascend.

Algebra Aerobics[1]

1. Use the equation $v = \dfrac{4}{3}\pi r^3$ to calculate a more precise value for the radius at which the balloon will begin to rise.
2. Show that if the radius triples, the volume of a sphere becomes 27 times larger.
3. Assume that $v = (4/3)\,\pi\, r^3$ represents a relationship between two abstract variables r and v, so that r could be any real number. When r < 0, will v be positive or negative? Draw a rough sketch of the graph.

Something to think about

The radius of the core of the earth is over half the radius of the earth as a whole, yet it is only about 16% of the total volume of the earth. How is this possible?

[1] Answer:
1. Solving $9700 = (4/3)\,\pi r^3$ for r , gives $r^3 = 9700/((4/3)\,\pi\,) \approx 2{,}316$, so $r \approx 13.2$ cm.
2. If the radius triples from R to 3R, then the volume increases from $(4/3)\,\pi\, R^3$ to
   $(4/3)\,\pi\,(3\,R)^3 = (4/3)\,\pi\,(R)^3\,(3)^3$, which is 3^3 or 27 times larger.
3. If r < 0, then $v < 0$. The graph will look roughly like:

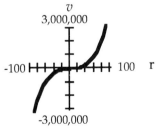

Be sure to notice the difference in scales on the two axes.

Question: The most important rule in scuba diving is "Never, ever hold your breath." Why?

Discussion:

According to Boyle's Law, if the temperature is held fixed, then the volume, **V**, of a gas equals a constant divided by the pressure, **P**. That is, the volume is a function, $g(P)$, of the pressure:

$$V = g(P) = \frac{k}{P}$$

for some constant k. As the pressure increases, the volume of gas decreases. Pressure is measured in *atmospheres*, where 1 atmosphere (abbreviated 1 ATM) = 15 lbs/in^2, which is the pressure applied by the column of air above 1 square inch of the earth.

For the particular case where the volume, **V**, is 1 cubic foot and the pressure, **P**, is 1 ATM then, k = 1 and we have

$$V = \frac{1}{P}$$

The units of the constant in the numerator depend on the units being used to measure the volume. In this case they are: (ft^3) (lb)/(in^2) or (ft^3) (ATM). These are the units needed to make the units of both sides of the equation the same.

$$1 \text{ ft}^3 = \frac{1 \text{ (ft}^3\text{)(ATM)}}{1 \text{ (ATM)}} = 1 \text{ ft}^3$$

If we double the pressure to 2 ATMs, the volume of air will drop in half to 0.5 cubic feet. At 4 ATMs, the volume will drop to 1/4th of the original size (or to 0.25 cubic feet). A small table of values and graph would look like:

Table 9.2	
P	**V**
0.25	4
0.5	2
1	1
2	0.5
4	0.25

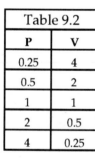

Why does this matter to divers?

Fig. 9.2 Graph of V = 1/P

Suppose you are swimming in a pool, take a lung full of air at the surface and then dive down to the bottom. As you descend, according to Boyle's Law, the build-up of pressure will decrease the volume of air in your lungs. When you ascend back to the surface, the volume of air in your lungs will expand back to its original size and everything is fine.

But when you are scuba diving, you are constantly breathing air that has been pressurized to equal the surrounding water pressure. Starting with 1 ATM at the surface of the water, each additional 33 ft of water adds 1 ATM to the pressure. So if you are scuba diving 33 feet below the surface of the water, the surrounding water pressure is at 2 ATMs, twice that at the surface. What will happen then if you fill your lungs from your tank, hold your breath, and ascend to the surface? When you reach the surface, the pressure will drop in half, from 2 ATMs down to 1 ATM, so the volume of air in your lungs will double, rupturing your lungs! Hence the first rule of scuba diving: "Never, ever hold your breath."

Something to think about

Lest we scare you off from taking up scuba diving, what simple strategy can you adopt to avoid the nasty consequences of Boyle's Law when diving?

Algebra Aerobics[2]

1. If you are scuba diving at 99 feet below the surface of the water, and you use your tank to inflate a balloon with 1 cubic foot of compressed air, by how much will the volume have increased by the time it reaches the surface (assuming it doesn't burst).

2. Think of the function $V = 1/P$ as relating two abstract variables P and V. What would be the constraints on the domain? What would a graph of the function look like?

---

[2] Answers:

1. At 99 feet below the surface, the pressure is 4 ATMS. So moving from 99 feet to 0 feet, the pressure would decrease from 4 ATMs to 1 ATM, so the volume will become 4 times larger, or expand to 4 cubic feet.

2. P can not equal 0. The graph would look roughly like:

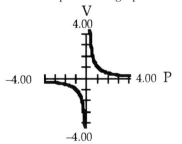

The graphs of the functions $v = \frac{4}{3}\pi r^3$ and $V = \frac{1}{P}$ look quite different. However, both functions share a common general form:

dependent variable = (constant) (independent variable) ^{power}

$$y = k x^{p} \qquad \text{where k and p are constants.}$$

In $\quad v = \frac{4}{3}\pi r^3 \qquad$ *r* and *v* are the variables, $\quad k = (4/3)\pi, \quad$ and $\quad p = 3$

In $\quad V = \frac{1}{P} = P^{-1} \qquad$ **P** and **V** are the variables, $\quad k = 1, \qquad$ and $\quad p = -1$

These are both examples of *power functions*.

9.2 Power Functions

A *power function* has the form
$$y = kx^{p}$$
where k and p are any constants.

A power function is one in which the dependent variable is a constant times a power of the independent variable. Note that the exponent, p, could be any real number, positive, negative or zero, as the following examples illustrate. In most scientific applications, k is positive.

$A = \dfrac{t^2}{40}$ $\qquad$ or equivalently $A = \left(\dfrac{1}{40}\right)t^2$, is the equation derived in Section 2.4 that describes the area of a *sordaria finicola* colony as a constant times t^2. So A is a power function, where k = 1/40 and p = 2.

$S = \sqrt{30d} \approx 5.48 d^{\frac{1}{2}}$ $\qquad$ is a function that describes the relationship between S, the speed of a car, as a constant times the square root of d, the distance a car skids after applying the brakes on a dry tar road. So S is a power function, with k ≈ 5.48 and p = 1/2.

$F = \dfrac{k}{d^2} = kd^{-2}$ $\qquad$ describes F, the gravitational force between you and the earth, as a constant k times d^{-2}. F is a power function, where k is the constant and p = -2.

A note about the domains of power functions

For certain values of the power p, the domain may need to be constrained. For example, if $p = 1/2$, as in the equation $S = 5.48\,d^{1/2}$, then the implied domain is values of $d \geq 0$. In most power functions representing physical situations, the implied domain is non-negative numbers and the constant k is positive.

Direct and inverse variation

In Chapter 4, we said that y was *directly proportional to* or *varied directly with* x , if y equaled a constant times x. For example, if $y = 4x$, then y is directly proportional to x. We can extend the same concept to power functions with positive exponents. For instance, if $y = 4x^3$, we say that y *is directly proportional to* or *varies directly with* x^3. In particular, for positive values of x, when x increases, y increases. In the previous examples, the equation $A = \left(\dfrac{1}{40}\right)t^2$ indicates that A is directly proportional to t^2 and the equation $S = 5.48\,d^{\frac{1}{2}}$ says that S is directly proportional to $d^{1/2}$. As time and distance skidded increase, so do Area and Speed, respectively.

What if the power is negative? If say, $y = 4x^{-3} = \dfrac{4}{x^3}$? We say that y *is inversely proportional to* or *varies inversely with* x^3. For positive values of x, as x increases, y decreases. For example, the force, F, given by the equation, $F = \dfrac{k}{d^2} = kd^{-2}$ is *inversely proportional to* d^2. When distance increases, force decreases.

Let p be a positive number.

If $\qquad y = kx^p$

we say that y *is directly proportional to* or *varies directly with* x^p

If $\qquad y = \dfrac{k}{x^p} = kx^{-p}$

we say that y is *inversely proportional to* or *varies indirectly with* x^p

---

Something to think about

In our discussions of direct and inverse proportionality we have only mentioned the relationship between two variables at a time. But the cases may be extended to include more than two variables. For example, the gravitational force, F, between two objects varies directly with the masses of the two objects, m_1 and m_2, and inversely with the square of the distance, d, between the two objects. Thus:

$$F = \frac{km_1 m_2}{d^2}$$

for some constant k. Increasing the masses increases the force. Increasing the distance decreases the force.

Algebra Aerobics[3]

1.   In each case, indicate whether or not the function is a power function. If it is, identify the values for k and p, the implied domain, and whether the function represents direct or inverse variation.

a) The area of a circle is given by the formula $A = \pi r^2$
b) $y = x^5$ c) $y = x^5 + 10$ d) $y = 5^x$

2.   Note: x varies jointly as y and z means that x = kyz for some constant, k.
x varies directly as y and inversely as z means x = ky/z for some constant, k.

Give an equation of variation for each of the following:

a) The pressure of a gas enclosed in a container varies directly as the temperature, T, and inversely as the volume of the container, V.

b) The force between two planets of mass m_1 and m_2 varies jointly as their masses and inversely as the square of the distance between them.

---

[3]   Answers
1 a) Yes, area is a power function, where $k = \pi$ and $p = 2$. The domain is values of $r > 0$. A is directly proportional to r^2.
  b) Yes, y is a power function, where $k = 1$ and $p = 5$. Domain is all real numbers.; y is directly proportional to x^5.
  c) No, y is not a power function, because of the 10 added on.
  d) No, y is not a power function. The independent variable x is in the power, not the base.
2. a) $P = kT/V$ b) $F = km_1m_2/d^2$

9.3   Comparing Power and Exponential Functions

Question:
Which eventually grows faster, a power function or an exponential function?

Discussion:

Although power and exponential functions may appear to be similar in construction, in each function the independent variable assumes a very different role. For power functions the independent variable, x, is the *base* which is raised to a fixed power. Power functions have the form:

$$y = kx^{power}$$

$$y = kx^p \qquad \text{where k and p are any constants}$$

For exponential functions the independent variable, x, is the *exponent* which is applied to a fixed base. Exponential functions have the form:

$$y = C \bullet (base)^x$$

$$y = C \bullet a^x \qquad \text{where C and a are constants with } a > 0 \text{ and } a \neq 1$$

Table 9.3 compares the role of the independent variable, x, in a linear, power and exponential function.

Table 9.3

x	linear function $y = x + 3$	power function $y = x^3$	exponential function $y = 3^x$
	x is added to 3	x is the *base* raised to third power	x is the *exponent* for base 3
0	0 + 3-= 3	0•0•0 = 0	$3^0 = 1$
1	1 + 3 = 4	1•1•1 = 1	$3^1 = 3$
2	2 + 3 = 5	2•2•2 = 8	$3^2 = 9$
3	3 + 3 = 6	3•3•3 = 27	$3^3 = 27$
4	4 + 3 = 7	4•4•4 = 64	$3^4 = 81$
5	5 + 3 = 8	5•5•5 = 125	$3^5 = 243$

Visualizing the difference

Table 9.3 and Figure 9.3 show that the power function $y = x^3$ and exponential function $y = 3^x$, grow very quickly relative to $y = x + 3$, a linear function.

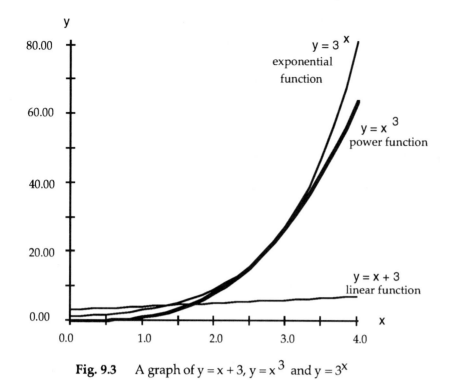

Fig. 9.3 A graph of $y = x + 3$, $y = x^3$ and $y = 3^x$

Yet, there is a vast difference between the growth of an exponential function and a power function. In Figure 9.4 we zoom out on our graph.

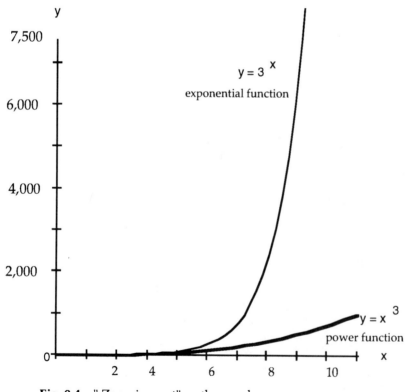

Fig. 9.4 "Zooming out" on the graph

Notice that now the scale on the x-axis goes from 0 to 10 (rather than just 0 to 4 as in Figure 9.3) and the y-axis extends to 7,500, which is still not large enough to show the value of 3^x when $x > 8$.

Now, the exponential function $y = 3^x$ clearly dominates the power function x^3. The exponential function continues to grow so rapidly that its graph appears almost vertical relative to the graph of the power function.

What if we had picked a larger exponent for the power function? Would the exponential function still overtake the power function? The answer is yes. Let's compare for instance the graphs of $y = 3^x$ and $y = x^{10}$. If $x = 2$, then $3^2 < 2^{10}$. So at least for a while the graph of $y = 3^x$ lies below the graph of $y = x^{10}$. Are there eventually values for x such that $3^x > x^{10}$? Again, the graph needs to have a large enough scale on the axes in order to see what eventually happens. But in Figure 9.5 we can see that somewhere after $x = 30$, the values for 3^x start to become substantially larger than the values for x^{10}.

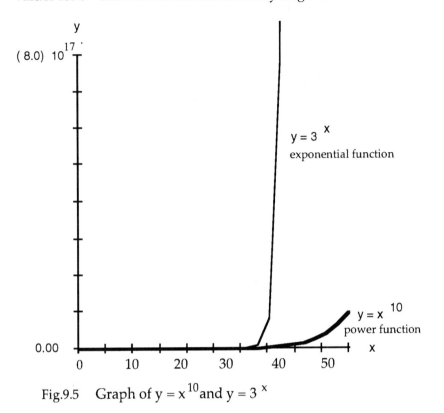

Fig.9.5   Graph of $y = x^{10}$ and $y = 3^x$

In summary,

Any increasing exponential function will eventually dominate any power function.

Algebra Aerobics

1. Construct tables and graphs using non-negative values of x for each of the following functions. Plot the graphs on the same grid.

  a) $y = 2^x$ b) $y = x^2$

2. Using the table and graph, where will the graphs for $y = 2^x$ and $y = x^2$ intersect for positive values of x? Does one function eventually dominate? If so, after what value of x?

3. Which function eventually dominates?

  a) $y = x^4$ or $y = 2^x$?   b) $y = (1.000005)^x$ or $y = x^{1,000,000}$?

Answers:

1.

x	$y = x^2$	$y = 2^x$
0	0	1
1	1	2
2	4	4
3	9	8
4	16	16
5	25	32
6	36	64

2. For positive values of x, the graphs for $y = 2^x$ and $y = x^2$ will intersect at the points (2, 4) and (4, 16).

When $x > 4$, then $2^x > x^2$, so the function $y = 2^x$ will dominate the function $y = x^2$ after $x = 4$.

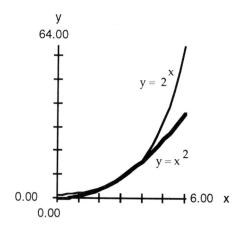

Comparison of $y = x^2$ and $y = 2^x$

3. An increasing exponential function will *always* eventually dominate a power function. So

  a) $y = 2^x$ will eventually dominate $y = x^4$

  b) $y = (1.000005)^x$ will eventually dominate $y = x^{1,000,0000}$

9.4   Visualizing Positive Integral Powers

Visualizing odd vs. even power functions

If we graph the simplest power functions: $y = x$, $y = x^2$, $y = x^3$, $y = x^4$, $y = x^5$, $y = x^6$, and so on, we can quickly see that the graphs of these power functions seem to fall into two groups: the odd powers and the even powers.

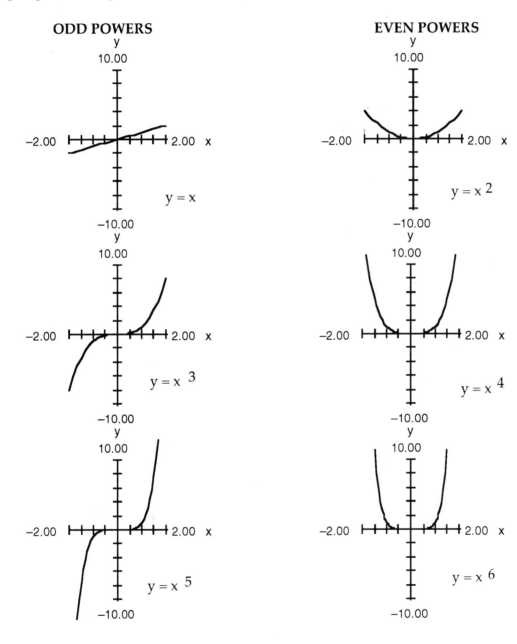

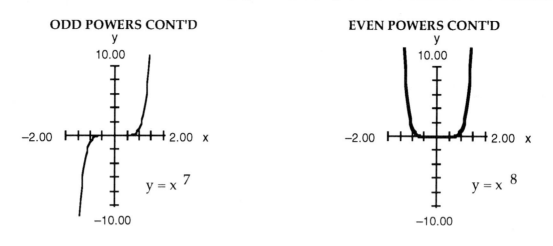

Fig. 9. 6 Graphs of odd and even power functions

Something to think about

Take a moment and try to describe in your own words the differences between the graphs of odd and even power functions.

For all the odd positive powers, x^1 (usually written as just x), x^3, x^5, x^7, .as x increases, y increases. The graphs are rotationally symmetric about the origin; that is, if you hold the graphs fixed at the origin, the point (0,0), and then rotate any of the function graphs 180 °, you'd end up with the same graph. All the graphs of odd powers above n = 1 appear to "bend" near the origin.

The graphs of even powers, x^2, x^4, x^6, x^8, ... are U shaped. For all the positive power functions of even degrees, as x increases, first y decreases and then increases. The graphs are symmetric about the y-axis. If you think of the y-axis as a dividing line, the "left" side of the graph is a mirror image of the "right" side.

Algebra Aerobics[4]

1. Do a quick sketch by hand of the power functions: $y = x^9$ and $y = x^{10}$.
2. If possible, check your work using a graphing calculator or computer. Try switching among various window sizes to compare different sections of the graph.

---

4. Answers

1.

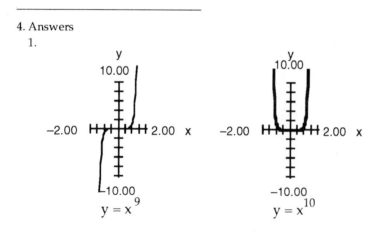

The effect of adding a coefficient to a power function

We can introduce coefficients to the powers of x and ask what effect this has on the function and its graph. In other words, what do power functions of the form:
$$y = a x^n$$
look like? (where a is a real number, and x a positive integer)

Case 1: $a > 0$ How does the power function $y = a x^n$ (where n is a positive integer) compare to $y = x^n$ if a is positive?

We know from our work with power functions of degree 1 (that is, linear functions of the form $y = a x^1$, or more simply $y = ax$), that when a is positive, multiplying x by a, changes the steepness of the line, but not its basic form. For values of $a > 1$, multiplying x by a increases the steepness of the graph, forcing the function to "hug" the y-axis more closely. The larger the value for a, the more closely the graph "hugs" the y-axis. The same is true for general power functions of the form $y = a x^n$ (where n is a positive integer) .

Similarly when $0 < a < 1$, multiplying x^n by a, decreases the steepness, moving the graph closer to the x-axis.

In general, for any positive a, the larger the value for a, the steeper the graph of the power function $y = a x^n$. Tables 9.4 and 9.5, and the accompanying Figures 9.7 and 9.8 illustrate this effect for power functions of degrees 3 and 4.

Table 9.4					
x	$y = x^3$	$y = 5 x^3$	$y = 10 x^3$	$y = 0.5 x^3$	$y = 0.25 x^3$
-2.0	-8.00	-40.00	-80.00	-4.00	-2.00
-1.5	-3.38	-16.88	-33.75	-1.69	-0.84
-1.0	-1.00	-5.00	-10.00	-0.50	-0.25
-0.5	-0.13	-0.63	-1.25	-0.06	-0.03
0.0	0.00	0.00	0.00	0.00	0.00
0.5	0.13	0.63	1.25	0.06	0.03
1.0	1.00	5.00	10.00	0.50	0.25
1.5	3.38	16.88	33.75	1.69	0.84
2.0	8.00	40.00	80.00	4.00	2.00

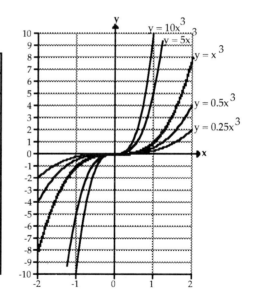

Fig. 9.7 When $a > 0$, the larger the value for a, the more closely the odd power function $y = a x^3$ "hugs" the y-axis.

Table 9.5					
x	$y = x^4$	$y = 5x^4$	$y = 10x^4$	$y = 0.5x^4$	$y = 0.25x^4$
-2.0	16.00	80.00	160.00	8.00	4.00
-1.5	5.06	25.31	50.63	2.53	1.27
-1.0	1.00	5.00	10.00	0.50	0.25
-0.5	0.06	0.31	0.63	0.03	0.02
0.0	0.00	0.00	0.00	0.00	0.00
0.5	0.06	0.31	0.63	0.03	0.02
1.0	1.00	5.00	10.00	0.50	0.25
1.5	5.06	25.31	50.63	2.53	1.27
2.0	16.00	80.00	160.00	8.00	4.00

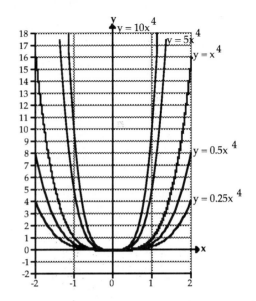

Fig. 9.8 Again, when $a > 0$, the larger the value for a, the more closely the even power function $y = ax^4$ "hugs" the y-axis.

Case 2: $a < 0$ How does the power function $y = ax^n$ (where n is a positive integer) compare to $y = x^n$ if a is negative?

We know from our work with linear power functions of the form $y = ax$, that when a is negative, multiplying x by a, not only changes the steepness of the line, but also "flips" the line across the x axis. The graphs of $y = ax$ and $y = -ax$ are mirror images across the x-axis. Similarly, the graphs of $y = ax^n$ and $y = -ax^n$ (where n is a positive integer) are mirror images of each other across the x-axis. Hence $y = -2x$ is the mirror image of $y = 2x$, and $y = -4x^5$ is the mirror image of $y = 4x^5$. The following four graphs show various pairs of power functions of the type $y = ax^n$ and $y = -ax^n$.

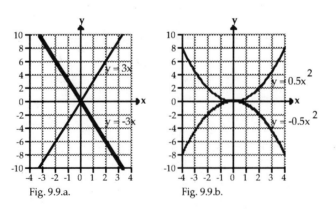

Fig. 9.9.a.   Fig. 9.9.b.

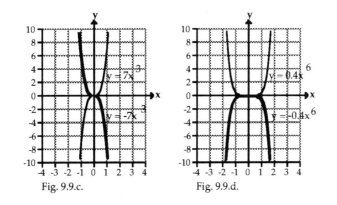

Fig. 9.9.c.   Fig. 9.9.d.

Fig. 9.9 In each case $y = a\,x^n$ and $y = -a\,x^n$ are mirror images of each other across the x-axis.

Algebra Aerobics [5]

a) Do a rough hand sketch of:
    i) $y = x$, $y = 4\,x$ and $y = -4\,x$ (all on the same graph)
    ii) $y = x^4$, $y = 0.5\,x^4$, and $y = -0.5\,x^4$ (all on the same graph)
b) Check by graphing the same sets of functions using a function graphing program.

[5] Answers:
  a) i)   ii)

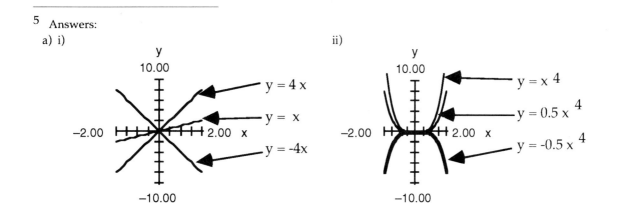

9.5   Polynomials: Adding Power Functions Together

Suppose we lay a spring flat on a table, attach a weight to one end of the spring and attach the other end to the table. Now imagine stretching out the spring, and then releasing it. Disregarding the effect of friction, the weight will slide back and forth on the table top, alternately compressing and stretching the spring. To push the weight away from its resting position in either direction, work must be performed on it. Using the terminology of physics, the potential energy is given by a power function:

$$E_{spring} = k_1 x^2$$

where x measures the distance that the spring is stretched or compressed away from its neutral starting position (so x could be positive or negative), and k_1 is the "spring constant." So E_{spring} is directly proportional to the square of the position. This function has a minimum value when x = 0, the most stable position of the weight.

Now suppose we tilt the table. This adds an additional force, due to gravity, which makes the weight want to slide down the table. To hold the weight in position, we must supply a constant force; this means that the energy has an additional term that is directly proportional to the position of the weight.

$$E_{gravity} = -k_2 x$$

where k_2 is a positive constant that depends on gravity.

The total energy is the sum of the two energies:

$$E_{total} = k_1 x^2 - k_2 x$$

E_{total} is the sum of the two power functions E_{spring} and $E_{gravity}$, and is called a *polynomial function* in x, the position of the weight.

Something to think about

If you graph E_{total} using different values of k_1 and k_2, you can verify that the function has a minimum at $x_{min} = k_2 / (2 k_1)$. That is, the weight now wants to come to rest at a different position, in which the spring is stretched by gravity.

In general, we can add together one or more power functions to create a polynomial function. For example, if we add together the three power functions $y_1 = 16 x^3$ and $y_2 = -5 x$ and $y_3 = 10$, we create the polynomial function $y = y_1 + y_2 + y_3$ where $y = 16 x^3 - 5 x + 10$. We say that y is a polynomial of degree 3, since the highest power of the independent variable x is 3.

> A *polynomial function* has the form
> $$y = a_n x^n + a_{n-1} x^{n-1} + \ldots + a_1 x^1 + a_0$$
> where each a_i is a constant and
> n is a positive integer, called the *degree* of the polynomial (provided $a_n \neq 0$).

Polynomials of certain degrees are given special names:

Polynomials of degree	are called
2	*quadratics*
3	*cubics*
4	*quartics*
5	*quintics*

Algebra aerobics [6]

1. What are the degrees of the following polynomials?

  a) $y = 11 x^5 + 4 x^3 - 11$ b) $f(x) = 4 x^3 + 11 x^5 - 11$ c) $g(x) = (x^2 + 3)(x^2 - 1)$

2. Let $f(x) = x^4 - 3 x^2$ and $g(x) = 2 x^2 + 4$,.

  a) Write out the function $h(x)$ if $h(x) = f(x) + g(x)$.

  b) What is $h(0)$? $h(2)$? $h(-2)$?

---

[6]   a) 5   b) 5 (it's the same polynomial as in part a, the terms have just been rearranged)

  c) 4 (multiply the terms out first in order to determine the degree, i.e., $y = x^4 + 2 x^2 - 3$)

2. a) Combining like terms, $h(x) = x^4 - x^2 + 4$. So $h(0) = 4$; $h(2) = 2^4 - 2^2 + 4 = 16$;

  $h(-2) = (-2)^4 - (-2)^2 + 4 = 16$

What we can predict about the graph of a polynomial function given its equation?

Examine the graphs of polynomials of different degrees in Figures 9.10 - 9.13..

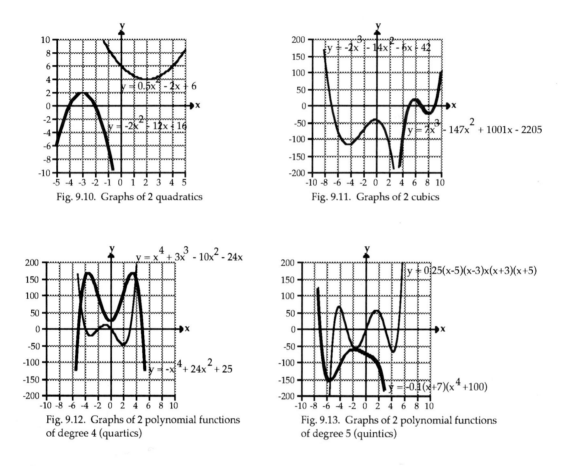

Fig. 9.10. Graphs of 2 quadratics

Fig. 9.11. Graphs of 2 cubics

Fig. 9.12. Graphs of 2 polynomial functions of degree 4 (quartics)

Fig. 9.13. Graphs of 2 polynomial functions of degree 5 (quintics)

What can we observe from each of these pairs of polynomial functions?

1.   The first thing you might notice is the number of times each graph "wiggles" or bends into a new direction. The quadratics bend once, the cubics seem to bend twice, the quartics three times, and one quintic seems to bend four times, and the other appears to bend twice. In general, *a polynomial function of degree n will bend at most n-1 times.*

2.   Secondly, you might notice the number of times the graph crosses the x-axis. Each quadratics crosses 2 or 0 times; the cubics each cross 1 or 3 times; the quartics cross 2 or 4 times; and the quintics cross 1 or 5 times. In general *a polynomial function of degree n will cross the x-axis at most n times.*

3. Finally imagine zooming way out on the graph, to look at it on a global scale. After all, the sections of the x-axis displayed in the previous four Figure are really quite small, the largest containing values of x only between –10 and +10. Suppose we consider values of x between –1000 and + 1000 or between –1,000,000 and + 1,000,000. What will the graphs look like?

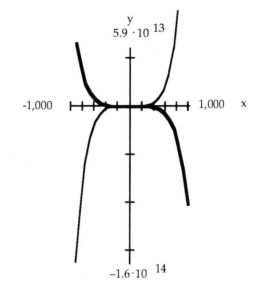

Fig.9.14 Zooming out on the graphs of the two quintics

Figure 9.14 displays a graph of the two quintics in Figure 9.13. Here the values of x extend between -1000 and +1000. The wiggles are no longer noticeable. The dominant feature to notice now is whether the two "arms" of the function stretch indefinitely up or indefinitely down. For even power functions, both "arms" will stretch in the same direction, either both up or both down. For odd power functions like the quintics in Figure 9.14, the two "arms" stretch in opposite directions, one up and one down. This is a result of the fact that given a polynomial

$$y = a_n x^n + a_{n-1} x^{n-1} + \ldots + a_1 x^1 + a_0$$

of degree n, for large values of x, the values of the leading term $a_n x^n$ will dominate the values of the other terms of smaller degree. In other words, for large values of x the function behaves a lot like the simple power function $y = a_n x^n$. *The degree of a polynomial determines its global shape.*

In summary

> The degree n of a polynomial function determines
> - that the graph will bend in the middle at most n-1 times
> - that the graph will cross the x-axis at most n times
> - the global shape of the polynomial

9.6   Finding the Intercepts of Polynomial Functions

Finding the y (or vertical) intercept

The y-intercept is the point at which the graph crosses the y-axis. Given any general polynomial function of the form $y = f(x) = a_n x^n + a_{n-1} x^{n-1} + \ldots + a_1 x^1 + a_0$, the y-intercept will occur when $x = 0$. So the graph will cross the y-axis at $f(0) = a_n(0)^n + a_{n-1}(0)^{n-1} + \ldots + a_1(0)^1 + a_0 = a_0$. Hence the coordinates of the y-intercept are $(0, a_0)$. We often shorten this to say that the y-intercept is a_0

Example:

The function $y = x^4 - 2x^2 - 3$ has a y-intercept at $(0, -3)$. Or we say simply that the y-intercept is -3.

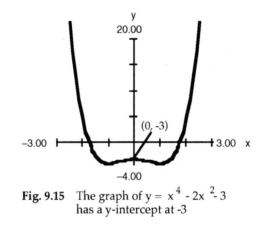

Fig. 9.15   The graph of $y = x^4 - 2x^2 - 3$
has a y-intercept at -3

The y-intercept is the point $(0, a_0)$ for any polynomial function of the form

$$y = f(x) = a_n x^n + a_{n-1} x^{n-1} + \ldots + a_1 x^1 + a_0$$
(where n is a positive integer and $a_n \neq 0$)

We often abbreviate this to say that the y-intercept is a_0.

Finding the x-intercepts

The x-intercepts are the points at which the curve crosses the x-axis. Since any point on the x-axis will have a y-coordinate of 0, we often refer to the x-intercept only by its x coordinate which we call a *zero* of the function.

Given a polynomial function

$$y = f(x) = a_n x^n + a_{n-1} x^{n-1} + \ldots + a_1 x^1 + a_0$$
$$(\text{where } n \text{ is a positive integer and } a_n \neq 0)$$

the *zeros* of the function are the values of x that make y equal to 0.

For example, if $f(x) = x^2 - 9$, then both 3 and –3 are zeros of the function, since $f(3) = 3^2 - 9 = 0$ and $f(-3) = (-3)^2 - 9 = 0$. So the graph crosses the x-axis at (3, 0) and (-3, 0).

The zeros can also be thought of as the solutions (called *roots*) of the equation

$$0 = a_n x^n + a_{n-1} x^{n-1} + \ldots + a_1 x^1 + a_0$$

For example, we say that 3 and –3 are the *roots* of the associated equation $0 = x^2 - 9$.

Given a polynomial equation

$$0 = a_n x^n + a_{n-1} x^{n-1} + \ldots + a_1 x^1 + a_0$$
$$(\text{where } n \text{ is a positive integer and } a_n \neq 0)$$

the solutions of the equation are called *roots*.

Graphical strategies for estimating x-intercepts: zooming in

A graphing calculator or computer (preferably with a zoom function) is very useful in helping to identify the number of x-intercepts and their <u>approximate</u> values. Remember that only algebraic strategies can give us exact solutions.

For example, if we graph the function $y = x^4 - 2x^2 - 3$, it appears to cross the x-axis twice, suggesting the existence of 2 x-intercepts.

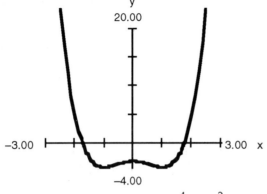

Fig. 9.16   The graph of $y = x^4 - 2x^2 - 3$
suggests 2 x-intercepts

If we zoom in on the x-intercept to the right of the y-axis, we can obtain increasingly accurate, though still approximate, values for that intercept.

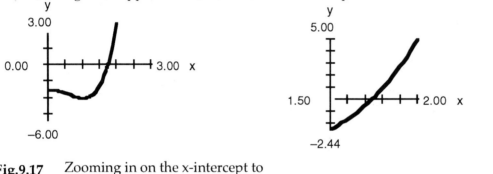

Fig.9.17   Zooming in on the x-intercept to the right, tells us that the value is somewhere between 1.5 and 2.0

Fig.9.18   Zooming in again, indicates that the x-intercept occurs between 1.7 and 1.8.

Algebra aerobics: [7]

Use a function graphing program (and its zoom feature) to estimate the number of x-intercepts and their approximate values for: a) $y = 3x^3 - 2x^2 - 3$ b) $f(x) = x^2 + x + 3$

----

[7] Answers

a) There is one x-intercept or zero at approx. 1.28

b) The graph does not cross the x-axis, so there are <u>no</u> x intercepts.

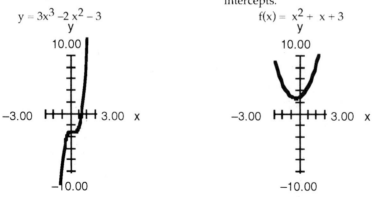

$y = 3x^3 - 2x^2 - 3$

$f(x) = x^2 + x + 3$

Algebraic strategies for finding x-intercepts: factoring

For polynomials of degree 4 or less, there are formulas that give us the exact x-intercepts or zeros. We calculated x-intercepts for lines in Part I, and in the chapter "Parabolic Reflections" we find the x-intercepts for polynomials of degree 2. Life isn't so easy when dealing with polynomial functions of higher degrees. The formulas for the zeros of third and fourth degree polynomials are extremely complicated. It has been proved that there are <u>no</u> general algebraic formulas for the zeros of polynomials of degree 5 or higher. There are however algebraic approximation methods (rather like the zoom feature on the calculator), that allow us to calculate values for the x-intercepts of functions accurate to as many decimal places as we wish.

The basic strategy for finding the zeros of a function is always the same: set $y = 0$, and try to solve the resulting equation for x.

Example 1: The x-intercepts of a cubic (finding zeros by factoring)

If a function y can be written as product of linear factors, then each factor corresponds to a real zero of the function. For example, the cubic function $y = x^3 + x^2 - 2x$ can be factored into 3 linear terms.

given	$y = x^3 + x^2 - 2x$
if we factor out x. we get	$= x(x^2 + x - 2)$
then factor the quadratic	$= x(x+2)(x-1)$
if we set y = 0 we get	$0 = x(x+2)(x-1)$

The *Zero Product Rule* tells us that whenever the product of terms equals 0, one or more of the terms must equal 0.

Zero Product Rule

For any two real numbers *r* and *s*, if the product $rs = 0$, then *r* or *s* or both must equal 0.

---

if $\qquad 0 = x(x+2)(x-1)$

then $\qquad x = 0 \qquad$ or $\qquad x + 2 = 0 \qquad$ or $\qquad x - 1 = 0$

Solve each equation $\quad x = 0 \qquad$ or $\qquad x = -2 \qquad$ or $\qquad x = 1$

The three linear factors x, x + 2 and x − 1 give us the corresponding zeros: 0, −2 and 1, which we can see in Figure 9.19.

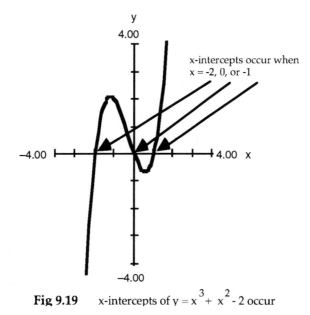

Fig 9.19   x-intercepts of $y = x^3 + x^2 - 2$ occur

Example 2: Find the zeros of the fifth degree polynomial function:

$$y = 0.25 (x - 5)(x - 3) x (x + 3)(x + 5)$$

given	$y = 0.25 (x - 5)(x - 3) x (x + 3)(x + 5)$
set $y = 0$	$0 = 0.25 (x - 5)(x - 3) x (x + 3)(x + 5)$

use the Zero Product

which tells us that $0 = x - 5$ or $0 = x - 3$ or $0 = x$ or $0 = x + 3$ or $0 = x + 5$

solve each

equation for x; $x = 5$ or $x = 3$ or $x = 0$ or $x = -3$ or $x = -5$

So the polynomial has zeros or x-intercepts when x is 5, 3, 0, –3, or –5

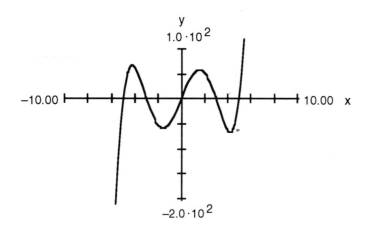

Fig. 9.20 Graph of $y = 0.25(x-5)(x-3)x(x+3)(x+5)$
with zeros or intercepts when x = -5, -3, 0, 3, 5

Algebra Aerobics

1. Use algebraic strategies to find the x-intercepts of the following functions, then graph the functions to check your work.

   a) $y = 3x + 6$
   b) $y = x^2 + 3x - 4$
   c) $y = (x + 5)(x - 3)(2x + 5)$

2. Match each of the following functions with its graph.

   a) $f(x) = x^2 + 3x + 1$ b) $f(x) = 2 - x$ c) $f(x) = -\frac{1}{2}x^3 + x - 3$

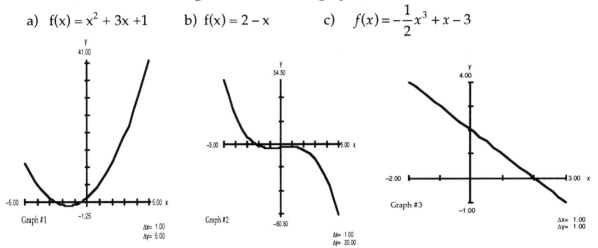

Answers:

1.

a) $y = 3x + 6$
   $y = 3(x + 2)$
   $x + 2 = 0$
   $x = -2$

b) $y = x^2 + 3x - 4$
   $y = (x+4)(x-1)$
   $x+4 = 0$ $x = -4$
   or $x - 1 = 0$ $x = 1$

c) $y = (x + 5)(x - 3)(2x + 5)$
   $x + 5 = 0$ $x = -5$
   or $x - 3 = 0$ $x = 3$
   or $2x + 5 = 0$ $x = -2.5$

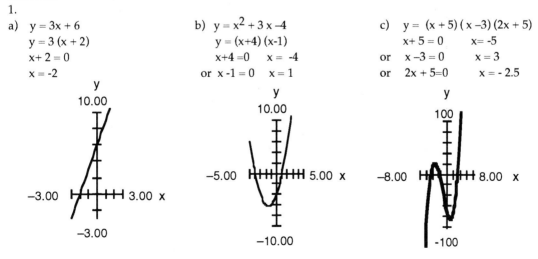

2. a) graph #1   b) graph #3   c) graph #2

Chapter 9 Summary

A *power function* has the form $y = kx^p$ where k and p are any constants.

Let p be a positive number. If $y = a\,x^p$ we say that y *is directly proportional to* x^p.

If $y = \dfrac{a}{x^p} = ax^{-p}$ we say that y is *inversely proportional to* x^p.

For graphs of power functions of the form $y = x^n$ where n is an odd positive number: as x increases, y increases; graphs are symmetric about the origin; and "bend" near the origin when $n > 1$.

The graph of power functions of the form $y = x^n$ where n is an even positive number: as x increases, y decreases and then increases; graphs are symmetric about the y-axis; and are U shaped.

If $a > 0$, the larger the value for a, the steeper the graph of $y = ax^n$, hence the more closely the graph "hugs" the y-axis.

The graph of $y = a\,x^n$ and $y = -a\,x^n$ are mirror images of each other across the x-axis.

A *polynomial function* has the form

$$y = \ (f(x) = a_n x^n + a_{n-1} x^{n-1} + \ldots + a_1 x^1 + a_0$$

where each a_i is a constant and n is a positive integer, called the *degree* of the polynomial (provided $a_n \neq 0$.)

Polynomials of degree	are called
2	*quadratics*
3	*cubics*
4	*quartics*
5	*quintics*

The degree n of a polynomial function determines: the global shape of the polynomial; that the graph will cross the x-axis at most n times; and that the graph will bend in the middle at most n-1 times.

Given a polynomial *function* $\quad y = a_n x^n + a_{n-1} x^{n-1} + \ldots + a_1 x^1 + a_0$
(where n is a positive integer and $a_n \neq 0$) the *x-intercepts* or *zeros* of the function are the values of x that make y equal to 0.

Given a polynomial *equation* $\quad 0 = a_n x^n + a_{n-1} x^{n-1} + \ldots + a_1 x^1 + a_0$
(where n is a positive integer and $a_n \neq 0$) the solutions of the equation are called *roots*.

We can estimate the zeros of a function by using a graphing calculator to zoom in on the x-intercepts. We can use factoring to help us find the exact roots algebraically.

EXERCISES

1. a) If the radius of a sphere becomes four times larger, how many times larger is the volume?

    b) If the radius is n times larger, how many times larger is the volume?

2. In the *Chicago Reader*, April 5, 1996, "Love That Dirty Water," Scott Berinato interviewed Ernie Vanier, Captain of the towboat "Debris Control." The Captain said "We've found a lot of bowling balls. You wouldn't think they'd float, but they do?"

    When will a bowling ball float in water? The bowling rule book specifies that a regulation ball must have a circumference of exactly 27 inches. Recall that the circumference, C, of a circle is directly proportional to its radius, r, by the formula $C = 2 \pi r$.

    a) What is a bowling ball's radius in inches? In centimeters? (Note: 1 inch = 2.54 cm)

    b) What is the *volume* of a bowling ball in inches? In cubic centimeters? (Neglect the finger holes.)

    c) What is the weight in grams of a volume of water equivalent to a bowling ball? (water weighs $1 \, gram/cm^3$.)

    d) Bowling balls are made in weights between 6 and 16 pounds. What is the range of weights in kilograms? (1 kilogram = 2.205 pounds)

    e) What is the heaviest weight of a bowling ball that will float in water?

    f) Typical men's bowling balls are 15 or 16 pounds. Women commonly use 12 pound balls. Will these bowling balls sink or float?

3. Boyle's Law says that if the temperature is held constant, then the volume, V, of a gas is inversely proportional to the pressure, P. That is,
    $$V = k/P \qquad \text{for some constant k.}$$

    a) If the pressure doubles, what happens to the volume?

    b) If the pressure triples, what happens to the volume?

    c) If the pressure is multiplied by a constant K, what happens to the volume?

4   The pressure of the atmosphere around us is relatively constant at 15 pounds per square inch at sea level, or 1 atmosphere of pressure (1 ATM). In other words, the column of air standing above one square inch of the earth's surface is exerting 15 pounds of force on that sq. in. of earth. Water is considerably more dense. Pressure increases at a rate of 1 ATM for each additional 33 feet of water. The following table shows a few corresponding values for water depth and pressure.

Water Depth (feet)	Pressure (in ATMs)
0	1
33	2
66	3
99	4

a) What type of relationship does the table describe?

b) Construct an equation that describes pressure, P as a function of depth, D.

c)   In Section 9.2 we looked at a case of Boyle's Law for the behavior of gases, P = 1/V, (where when V = 1 cubic foot, P = 1 ATM.)

   i)   Use this law to construct an equation for volume, V, as a function of depth, D. (Note: D is now the *independent* variable.)

   ii) When D = 0 feet, what is V?

   iii) When D = 66 feet, what is V?

   iv) If a snorkeler takes a lung full of air at the surface, dives down to 10 feet, and returns to the surface, describe what will happen to the volume of air.

   v)   If a large balloon is filled with a cubic foot of compressed air from a scuba tank at 99 feet, sealed tight, and allowed to ascend to the surface, describe the change in volume of air.

5.   Suppose you are traveling in your car at speed S, and you suddenly brake hard, leaving skid marks on the road. Then a "rule of thumb" for the distance, D, that the car would skid is given by:

$$D = \frac{S^2}{30f}$$

where D = the distance the car skidded in feet; S = speed of the car in miles per hour; and f is a number called the coefficient of friction that depends upon the road surface and condition. For a dry tar road, f ≈ 1.0. For a wet tar road, f ≈ 0.5.

a)   What is the function relating distance skidded and speed for a dry tar road? For a wet tar road?

b)   Generate a small table of values for both functions, including speeds between 0 and 100 miles per hour.

c)   Plot both functions on the same graph.

d) Why is the coefficient of friction less for a wet road than a dry road? What effect does this have on the graph?

e) In the following table, estimate the speed given the following distances skidded on dry and on wet tar roads. Describe the method you used to find these numbers.

f) If one car is going twice as fast as another when they both jam on the brakes, how much farther will the faster car skid? Explain. Does your answer depend on whether the road is dry or wet?

Distance skidded (feet)	Estimated speed (in mph)	
	dry tar	wet tar
25		
50		
100		
200		
300		

6. A scaling factor is used for enlarging and shrinking pictures and three dimensional objects. The scaling factor is the number by which each linear dimension of an object is multiplied. When pictures or objects are enlarged or shrunk they are changed by the scaling factor in all directions.

a) For the pictures below, the scaling factor, F, is equal to 3. Show that the following relationship is true

$$\text{Area of scaled-up object} = \text{Area of original object} * F^2$$

$A_{original}$

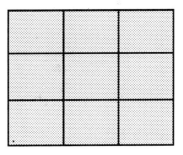

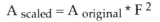

$$A_{scaled} = A_{original} * F^2$$

b) Show that the area of any scaled-up rectangular figure is directly proportional to the square of the scaling factor.

c) Show that the surface area of a scaled-up cube is directly proportional to the square of any scaling factor.

d) If the scaling factor is F, what will happen to the volume of a scaled-up cube? What will happen to the volume of a scaled-up rectangular solid? Write equations that describe your findings.

7.   Using a graphing calculator or a function graphing program, graph $y = x^4$ and $y = 4^x$ on the same grid.

a) Describe what happens to the graphs if x is greater than zero but relatively small. What happens as x increases? You may need to change the scales on the axes in order to see what is happening to the graphs.
b) Where do your graphs intersect? Do they intersect more than once? For what values of x are the two functions equal?

8.   Match each of the following with its graph:

a) $y = 2x - 3$ b) $y = 2 - x$
c) $y = 3(2^x)$ d) $y = (x^2 + 1)(x^2 - 4)$

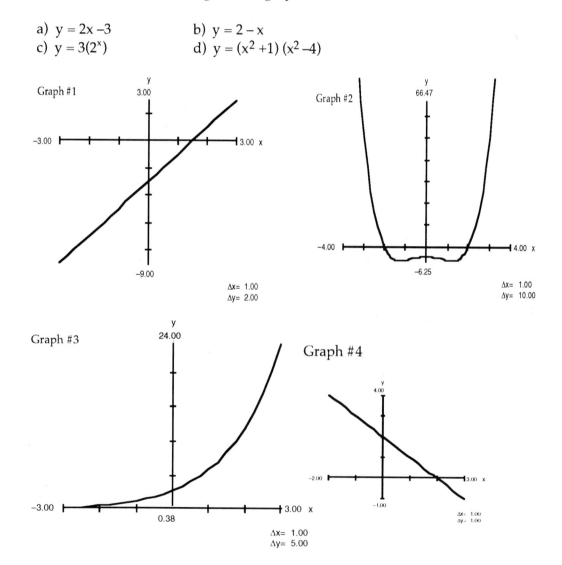

EXPLORATION 9.1
Predicting Properties of Power and Polynomial Functions

Objectives
- construct functions with certain characteristics
- find patterns in the graphs of power and polynomial functions
- compare graphs of power functions and exponential functions

Materials
- graphing calculator or function graphing program
- graph paper

Procedure
Working in pairs
Constructing power functions in the form $y = kx^p$ where k and p are constants

1. For each of the following construct at least two power functions. Construct your functions such that the graphs of the functions have the following characteristics:
   a) both functions are symmetrical around the y axis
   b) both functions are symmetrical around the origin
   c) the two functions are mirror images of each other around the x axis
   d) the two functions are mirror images of each other around the y axis

   Graph your functions using a graphing calculator or function graphing program to check if your functions have these characteristics.

2. Fix **p** and construct two different power functions with the following characteristics:
   a) both functions are even powers and the graph of one function "hugs" the y axis more closely than the other function
   b) both functions are odd powers and the graph of one function "hugs" the y axis more closely than the other function

   Now fix **k** and construct two different power functions with the following characteristics:
   c) both functions are even powers and the graph of one function "hugs" the y axis more closely than the other function
   d) both odd powers and one function "hugs" the y axis more closely than the other function

Constructing Polynomials
1. Construct polynomial functions with the following characteristics:
   a) crosses the x axis at least three times
   b) crosses the x axis at –1, 3 and 10
   c) has a y intercept of 4 and degree of 3
   d) has a y intercept of – 4 and degree of 5

Class Discussion
Compare your findings. As a class come up with a 60 Second Summary" for each type of function that you have examined.

EXPLORATION 9.2
Constructing under Constraints

Objectives
- construct polynomial functions that represent physical phenomena
- explore the effects of various constraints on constructing geometric figures

Materials
- graphing calculator or function graphing program (optional)
- graph paper

Procedure
Class demonstration

Finding maximum area holding perimeter constant
Assume you wish to enclose a rectangular region and are constrained by a fixed perimeter of 24 meters. (The classic problem is a farmer who has a fixed amount of fencing material with which to enclose a rectangular grazing area) Find the dimensions of the rectangle that will contain the greatest area.

1. *Solving by collecting and graphing data*
    Fill in the data table below. Assume that the perimeter of the rectangle must always equal 24 meters. You may wish to use graph paper to sketch some of the rectangles.

Dimension of one side of rectangle (in meters) x	Dimension of other side of rectangle (in meters)	Area of Rectangle (in square meters) A
0	12	0
1	11	11
2		
3		
4		
5		
6		
7		
8		
9		
10		
11		

a) From the table, what appears to be the maximum value for the area?

b) Plot your data by hand, using the x as the independent and A as the dependent variable. From the shape of the graph, what kind of function might A be?

c) If you have a graphing calculator or function graphing program enter the values for x and A, plot the data, and generate a best fit function.

2. *Solving by constructing equations*

Let

      $x =$ dimension of one side of the rectangle

Then

      ? =   dimension of other side of the rectangle(expressed in terms of x)
          Hint: remember that the perimeter is fixed at 24 meters.

Now write the area, A, as a function of x:

    $A = \,?$

Graph this function (using a function graphing program if you have one). Estimate the maximum value for A. How does this function compare to the best fit function you generated in part 1?

Analysis of Results:

- Of all the possible rectangles with a perimeter of 24 meters, what are the dimensions of the rectangle with the greatest area?
- If the perimeter of the rectangle were held constant at 200 meters, what would your equation be? How would you find the dimensions of the rectangle with the greatest area?
- If the perimeter of the rectangle were held constant at P meters, what would your equation be?

On your own or in a small group:

Choose one of the following problems to solve. Use the strategy of constructing equations and then graphing the equations to solve each problem.

1. A box has a base of 10 feet by 20 feet, and a height of h feet. If we decrease each side of the base by h feet, for what value of h will the volume be maximized ?

2. Find the dimensions of a cylinder with the maximum volume if the surface area (including the two bases) is held constant at 80 sq. meters.

Formulas:
    Volume of rectangular solid = Area of Base • Height
    Volume of cylinder =   Area of Base • Height
             = πr^2 • Height
    Surface Area of Cylinder = (Circumference of Cylinder • Height) + (2 • Area of Base)

EXPLORATION 9.3
Visualizing Negative Integral Power Functions

Objectives
- construct negative integral power functions
- find patterns in the graphs of negative integral power functions

Materials
- graphing calculator or function graphing program
- graph paper

Procedure

Class demonstration

Constructing negative integral power functions
1. A power function has the form $y = kx^p$ where k and p are constants. Consider the two power functions with negative integral exponents:
$$y = x^{-1} \qquad \text{and} \qquad y = x^{-2}.$$
    We can also write these as
$$y = \frac{1}{x} \qquad \text{and} \qquad y = \frac{1}{x^2}$$
    What is the value for k and p in each case? What are the constraints on the domain for each of these functions? What are the constraints on the range for each?
2. Construct a table of values (using positive and negative values) and hand sketch a graph for each function. (Note: you have seen the graph of y = 1/x before.)
3. Use a function graphing program or graphing calculator to check your graph.
4. Describe the overall behavior of these graphs. In each case when is y increasing? Decreasing? How do the graphs behave for values of x near 0?

Working in pairs
Try each of the following explorations and compare your findings with each other. After you compare your findings write down your observations in your notebook.

1. What is the effect of "p" on the graph?

    a)   Let k = 1. Construct equations of the form $y = x^{-p}$ or equivalently $y = \frac{1}{x^p}$

        where p = 3, 4, 5, 6, etc. What patterns start to emerge? Make a prediction as to what will happen to the graphs when p is odd and when p is an even.

    b)   Use your predictions to hand draw a sketch for the graphs of $y = \frac{1}{x^{11}}$ and

        $y = \frac{1}{x^{12}}$

Double check your graph with your graphing calculator or function graphing program.

c)   Describe the behavior of the graphs when p is odd. When is the function increasing? Decreasing? How do the graphs behave for values of x near 0?

d)   Describe the behavior of the graphs when p is even. When is the function increasing? Decreasing? How do the graphs behave for values of x near 0?

In you own words describe the effect of "**p**" on the graph of functions of the form $y = \dfrac{1}{x^p}$ where p is a positive integer.

2.   What is the effect of "**k**" on the graph?

a)   Fix a value for **p** in the function $y = \dfrac{k}{x^p}$ where p is a positive integer.

Now write down several functions where **p** has the same value but **k** assumes different *positive* values. Graph the functions on your graphing calculator. As you choose larger and larger values for k, what happens to the graphs? Try choosing values for k between 0 and 1. What happens to the graphs?

b)   Now keep the same value for p as in part 2a, but write down several functions with *negative* values for k. Graph the functions on your graphing calculator. Describe the effect on your graphs. Do you think your observations about k will hold for any value of p?

c)   Choose a new value for **p** and repeat your experiment. Are your observations still valid? Compare your observations with your partner. Have you examined both odd and even negative integral powers?

In your own words describe the effect of "k" on the graphs of functions of the form $y = \dfrac{k}{x^p}$ where p is a positive integer.

Class discussion:

Compare the findings of each of the small groups. As a class, develop 60 second summaries for the effect of p and for the effect of "k" on power functions of the form $y = \dfrac{k}{x^p}$ where p is a positive integer.

Chapter 10

The Mathematics of Motion

Overview

Scientists probe the functioning of our universe with carefully designed and conducted experiments. Observation, experimentation, and the discovery of theories are at the core of the scientific method. Mathematics provides the language needed to express these theories. Though we now take this process for granted, it is a relatively recent development. Four hundred years ago, in Galileo's time, it was revolutionary to study physical phenomena by attempting to measure observed behavior and to describe it mathematically.

In this chapter we use the laboratory methods of the modern physicist to collect and analyze data for freely falling bodies and then examine the questions that Galileo asked about bodies in motion. In generating questions that could be answered by experimentation, Galileo discovered the rules of motion for freely falling bodies. The free fall experiment is a classic, because of its historical significance in the development of science, and because it investigates gravity, the prime force of celestial mechanics. The mathematical focus of the chapter is *quadratic functions* (polynomials of degree 2), which provide useful models for describing the motion of moving bodies.

In the explorations you will collect and analyze your own free fall data.

After reading this chapter you should be able to:

- understand the importance of "The Scientific Method"
- construct a quadratic model of the relationship between distance and time for freely falling bodies
- derive equations describing the velocity and acceleration of a freely falling body
- apply motion equations under differing conditions

10.1 The Scientific Method

In the twentieth century it is taken for granted that scientists study physical phenomena in laboratories, using sophisticated equipment. But in the early 1600's, when Galileo did his experiments on motion, the concept of accurately measured laboratory experiments was unknown. In his attempts to understand nature, Galileo asked questions that could be tested directly by experiment. His use of observation and direct experimentation and his discovery that aspects of nature were subject to quantitative laws was of decisive importance not only for science but in the broad history of human ideas.

The Greeks and medieval thinkers believed that basic truths exist within the human mind and that these truths could be uncovered through reasoning, not empirical experimentation. Their scientific method has been described as a "qualitative study of nature." The Greek and medieval scientists were interested in "why" objects fall. They believed that a heavier object falls faster than a lighter one because "it has weight and it falls to the Earth because it, like every object, seeks its natural place, and the natural place of heavy bodies is the center of the Earth. The natural place of a light body, such as fire, is in the heavens, hence fire rises."[1].

Galileo changed the question from "why" things fall to "how" things fall. This question suggested other questions that could be tested directly by experiment. "By alternating questions and experiments, Galileo was able to identify details in motion no one had previously noticed or tried to observe" [2] His quantitative descriptions of objects in motion led not only to new ways of thinking about motion, but new ways of thinking about science. His process of careful observation and testing began the critical transformation of science from a qualitative to a quantitative study of nature.[3] Galileo's search for quantitative descriptions "was the most profound and the most fruitful thought that anyone has had about scientific methodology."[4] This approach became known as "The Scientific Method."

---

[1] Morris Kline, *Mathematics for the Nonmathematician*" Dover Publications, Inc., New York, 1967, p. 287.
[2] Elizabeth Cavicchi', "Watching Galileo's Learning", in the Anthology of Readings.
[3] Galileo's scientific work was revolutionary, not only in science, but in the politics of the time; his work was condemned by the ruling authorities and he was arrested.
[4] Morris Kline, Ibid. p.288.

10.2 Interpreting Data from a Free Fall Experiment;

To gain some understanding of how modern science approaches the search for quantitative natural laws, we examine a modern version of Galileo's free fall experiment. This classic experiment can be performed either with a graphing calculator connected to a motion sensor, or in a physics lab with an apparatus that drops a heavy weight and records its position on a tape each 1/60 of a second. The adjacent sketch of a tape gives data collected by a group of students from a falling object experiment. Each dot represents how far the object fell in each succeeding 1/60 of a second.

Since the first few dots are too close together to get accurate measurements, measurements are started instead at the sixth dot, which we call dot_o. At this point, the bob is already in motion. This dot is considered to be the starting point, and the time, t, at dot_o, is set at 0 seconds. The next dot represents the position of the object 1/60 of a second later. Time increases by 1/60 of a second for each successive dot. In addition to assigning a time to each point, we also measure the total distance fallen, d, from the point designated dot_o. For every dot we have two values- -the time, t, and the distance fallen, d. At dot_o , t = 0 and d = 0.

The time and distance measurements from the tape are recorded in Table 10.1 and plotted on the graph in Fig. 10.1. Time, t, is the independent variable, and distance, d, is the dependent variable. The graph gives a representation of the data collected on distance fallen over time, not a picture of the physical motion of the object.

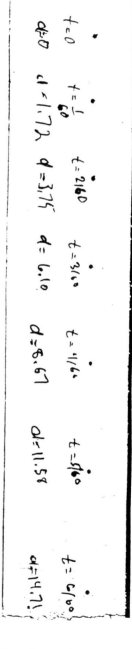

Fig. 10.2 Tape from a free fall experiment

Table 10.1	
Time Seconds	Total Distance Fallen (cm)
0.0000	0.00
0.0167	1.72
0.0333	3.75
0.0500	6.10
0.0667	8.67
0.0833	11.58
0.1000	14.71
0.1167	18.10
0.1333	21.77
0.1500	25.71
0.1667	29.90
0.1833	34.45
0.2000	39.22
0.2167	44.22
0.2333	49.58
0.2500	55.15
0.2667	60.99
0.2833	67.11
0.3000	73.48
0.3167	80.10
0.3333	87.05

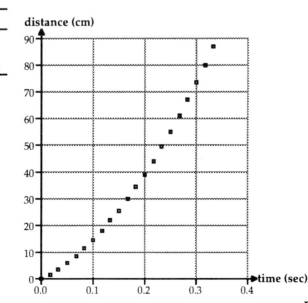

Fig.10.1. Free Fall: distance vs. time

The graph of the data looks more like a curve than a straight line, so we expect the average rates of change found between different pairs of points to be different. We know how to calculate the average rate of change between two points, and that it represents the slope of a line segment connecting the two points.

average rate of change = $\dfrac{\text{change in distance}}{\text{change in time}}$ = slope of line segment

Table 10.2 and Figure 10.3 show the increase in the average rate of change over time for three different pairs of points. The time interval nearest the start of the fall shows a relatively small change in the distance for each time step and therefore a fairly gentle slope of 188 cm/sec.. The time interval farthest from the start of the fall shows a greater change of distance per time step and a much steeper slope of 367 cm/sec..

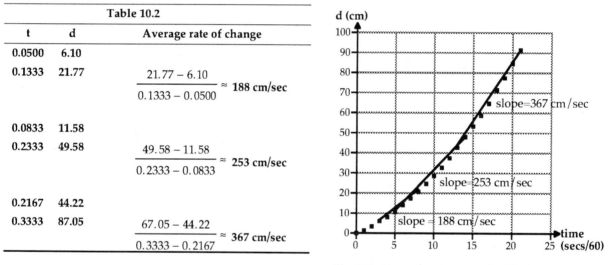

Table 10.2

t	d	Average rate of change
0.0500	6.10	
0.1333	21.77	$\dfrac{21.77 - 6.10}{0.1333 - 0.0500} \approx 188$ cm/sec
0.0833	11.58	
0.2333	49.58	$\dfrac{49.58 - 11.58}{0.2333 - 0.0833} \approx 253$ cm/sec
0.2167	44.22	
0.3333	87.05	$\dfrac{67.05 - 44.22}{0.3333 - 0.2167} \approx 367$ cm/sec

Fig.10.3. Slopes (or average velocities) between 3 pairs of end points

In this experiment the average rate of change has an additional important meaning. For objects in motion, the change in distance divided by the change in time is also called the *average velocity* over that time period. So in the first set of calculations in Table 10.2, the average rate of change of 188 cm/sec represents the average velocity of the falling bob between 0.0500 and 0.1333 seconds.

average velocity = $\dfrac{\text{change in distance}}{\text{change in time}}$

Important Questions

Do objects fall at a constant speed? The rate of change calculations and their graphical representations indicate that the average rate of change of position with respect to time (i.e., velocity) of the falling object is not constant. Moreover the average velocity appears to be increasing over time. In other words, as the object falls, it is moving faster and faster. Our calculations agree with Galileo's observations. He was the first person to show that the velocity of freely falling objects is not constant.

This finding prompted Galileo to ask more questions. One of these questions was: If the velocity of freely falling bodies is **not constant**, is it increasing at a **constant rate**? Galileo discovered that the velocity of freely falling objects does increase at a constant rate. If the rate of change of velocity with respect to time is constant, then the graph of velocity vs. time is a straight line. Equivalently, the rate of change of the average rate of change is constant. A theory of gravity has been built around Galileo's discovery of a constant rate of change for the velocity of freely falling bodies. This constant of nature, the gravitational constant of the Earth, is denoted by g.

In Sec. 10.4 we will show that for the free fall data in Table 10.1, the graph of velocity vs. time is indeed a straight line and the slope of that line is approximately 980 cm/sec^2, which is the conventional value for g. It tells us that the velocity of the freely falling object increases 980 cm/sec during each second of fall.

Algebra Aerobics:[5]

1. Complete the table below. What happens to the average velocity of the object as it falls?

Time (sec)	Distance fallen (cm)	Average rate of change (average velocity)
0.0000	0.00	N/A
0.0333	3.75	(3.75-0.00)/(0.0333-0.0000)=112.61 cm/sec
0.0667	8.67	(8.67-3.75)/(0.0667-0.0333)=147.31 cm/sec
0.100	14.71	
0.1333	21.77	
0.1667	29.90	

---

[5]Answers   As the object falls, its average velocity increases.

Time(sec)	Distance fallen(cm)	Average rate of change (average velocity)
0.0000	0.00	N/A
0.0333	3.75	(3.75-0.00)/(0.0333-0.0000)=112.61 cm/sec
0.0667	8.67	(8.67-3.75)/(0.0667-0.0333)=147.31 cm/sec
0.1000	14.71	(14.71-8.67)/(0.1000-0.0667)=181.38 cm/sec
0.1333	21.77	(21.77-14.71)/(0.1333-0.1000)=212.01 cm/sec
0.1667	29.90	(29.90-21.77)/(0.1667-0.1333)=243.41 cm/sec

10.3 Deriving an Equation Relating Distance and Time

Galileo wanted to be able to describe the distance an object falls mathematically. Using mathematical and technological tools not available in Galileo's time, we can describe the distance fallen in the free fall experiment using a "best fit" function for our data. Galileo had to describe his finding in words. Galileo described the free fall motion, first by direct measurement and then abstractly with a time-squared rule. ("Watching Galileo's Learning" in the Anthology of Readings describes how Galileo arrived at this rule, and Exercise 16 gives you the opportunity to follow the path he took.) "This discovery was revolutionary, as the first evidence that motion on Earth was subject to mathematical laws."[6]

Using Galileo's finding that distance is related to time in terms of a time-squared rule, we find a best fit polynomial function of degree 2 (a quadratic) for the free fall data. A function graphing program gives the following best fit quadratic function for the free fall data in Table 10.1 as:

$$d = 487.92\ t^2 + 98.6\ t + 0.0027$$

Figure 10.4 shows a plot of the data and the function. If your curve fitting software program or graphing calculator does not provide a measure of closeness of fit, like the correlation coefficient for regression lines, you may have to rely on a visual judgment. Rounding the coefficients to the nearest unit, we obtain the following equation:

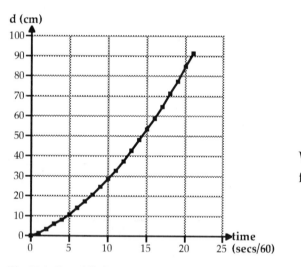

$$
\begin{aligned}
d &= 488\ t^2 + 99\ t + 0 \\
  &= 488\ t^2 + 99\ t \qquad (1)
\end{aligned}
$$

We now have a mathematical model for our free fall data.

Fig.10.4. Free fall distance vs. time

___
Something to think about
The graph of distance fallen vs. time looks as if it could possibly be an exponential function. Why is the data set not exponential? (Hint: What does the graph of the average rate of change of an exponential function look like?)

___
[6]Elizabeth Cavicchi, "Watching Galileo's Learning"

What if we ran the experiment again? How would the results compare? In one class, four small groups did the free fall experiment, plotted the data, and found a corresponding best fit second-degree polynomial. The functions are listed below, along with the equation we found in (1).

$$d = 488\,t^2 + 99\,t + 0.00 \quad (1)$$
$$d = 486\,t^2 + 72\,t + 0.02 \quad (2)$$
$$d = 484\,t^2 + 173\,t + 0.01 \quad (3)$$
$$d = 486\,t^2 + 73\,t + 0.00 \quad (4)$$
$$d = 495\,t^2 + 97\,t + 0.00 \quad (5)$$

All the functions have similar coefficients for the t^2, different coefficients for the t term, and approximately 0 for the constant term (you could think of this as the coefficient for the t^0 term). Why is this the case? Using concepts from physics, we can describe what each of the coefficients represent.

The coefficients of the t^2 term found in equations (1) - (5) above are all close to one half of 980 cm/sec^2 or half of g, the Earth's gravitational constant. We leave it up to you to look into why this is so in Exploration 10.2. It is interesting that data from this simple experiment gives very good estimates for 1/2 of g.

The coefficient of the t term represents the initial velocity, v_o, of the bob when t = 0. In equation (1), v_o = 99 cm/sec. Recall that we didn't start to take measurements until the sixth dot, the dot we called dot_o. At dot_o, where we set t = 0, the object was already in motion with a velocity of approximately 99 cm/sec. The initial velocities or v_o values in equations (2)–(5) range from 72 to 173 cm/sec. Each v_o represents approximately how fast the bob was moving when t = 0, the point chosen to begin recording data in each of the various experiments.

The constant term is close to or equal to zero in each of the equations (1)–(5). In the original data, when time, t = 0, distance, d = 0. So we expect that if we substitute in the equations, t = 0, then the value for d (which in this case is equal to the value of the constant term) will be 0. We can expect some variation in the constant term, such as in equation (3) above, due to error in measuring.

Galileo's discoveries are the basis for the following equations relating distance and time.

The general equation of motion of freely falling bodies that relates distance fallen, d, to time of falling, t, is

$$d = (^1/_2)\,g\,t^2 + v_o\,t \qquad \text{where}$$

v_o = initial velocity

g = acceleration due to gravity on Earth (approximately 980 cm/sec^2)

In our particular model the equation is $d = 488\, t^2 + 99\, t$ where 488 approximates half of g in cm/sec^2, and 99 approximates the initial velocity in cm/sec when $t = 0$.

Algebra Aerobics:[7]

1. In the equation of motion, $d = (1/2)gt^2 + v_o t$, distance is measured in centimeters, velocity in centimeters per second, and time in seconds. Rewrite this as an equation that shows only units of measure. If you perform calculations on units in the distance equation, do you get cm= cm?

2. Suppose that the above equation could also be written using meters rather than centimeters. Rewrite the equation as an equation that shows only units of measure and show that you get meters = meters.

3. Now suppose that the above equation could also be written using feet rather than centimeters. Rewrite the equation as an equation that shows only units of measure and show that you get feet = feet.

---

[7]Answers

1. centimeters $= (cm/sec^2)(sec^2) + (cm/sec)(sec)$ $\Rightarrow$ centimeters = centimeters
2. meters $= (meters/sec^2)(sec^2) + (meters/sec)(sec)$ $\Rightarrow$ meters = meters
3. feet $= (feet/sec^2)(sec^2) + (feet/sec)(sec)$ $\Rightarrow$ feet = feet

10.4 Deriving Equations for Velocity and Acceleration

Galileo's Question

If the velocity for freely falling bodies is not constant, is it increasing at a **constant rate**? Galileo discovered that the rate of change of the velocity of a freely falling object is constant. In this section we confirm his finding with data from the free fall experiment.

Velocity: change in distance over time

If the rate of change of the average velocity is constant, then the graph of velocity vs time should be a straight line. Previously we calculated the average rates of change of distance with respect to time (or average velocities) for three arbitrarily chosen pairs of points. Now, in Table 10.3 we calculate the average rates of change for all the pairs of adjacent points in our free fall data. The results are in column four. For increased precision since each computed velocity is the average over an interval, we associate each velocity with the midpoint time of the interval. In Figure 10.5, we plot velocity from the fourth column against the midpoint times from the third column. The graph is strikingly linear.

Table 10.3			
t = time (sec)	d = distance fallen (cm)	t=midpoint time (sec)	v=velocity (cm/sec)
0.0000	0.00		
		0.0084	103.2
0.0167	1.72		
		0.0250	121.8
0.0333	3.75		
		0.0417	141.0
0.0500	6.10		
		0.0584	154.2
0.0667	8.67		
		0.0750	174.6
0.0833	11.58		
		0.0917	187.8
0.1000	14.71		
		0.1084	203.4
0.1167	18.10		
		0.1250	220.2
0.1333	21.77		
		0.1417	236.4
0.1500	25.71		
		0.1584	251.4
0.1667	29.90		
		0.1750	273.0
0.1833	34.45		
		0.1917	286.2
0.2000	39.22		
		0.2084	300.0
0.2167	44.22		
		0.2250	321.6
0.2333	49.58		
		0.2417	334.2
0.2500	55.15		
		0.2584	350.4
0.2667	60.99		
		0.2750	367.2
0.2833	67.11		
		0.2917	382.2
0.3000	73.48		
		0.3084	397.2
0.3167	80.10		
		0.3250	417.0
0.3333	87.05		
		0.3417	430.8
0.3500	94.23		

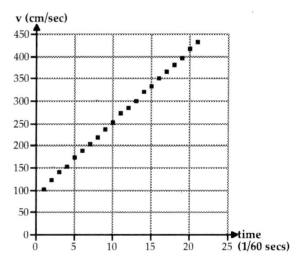

Fig.10.5. Free fall: average velocity vs. time

Generating a best fit linear function, we obtain the following equation:

Average velocity = 977 t + 98.2

where t is time in seconds, and average velocity is in cm/sec. at time, t. The graph of this function appears in Figure 10.6.

The coefficient of t represents the acceleration of gravity g, conventionally given as 980 cm/sec^2. Using 980 for g and rounding 90.1 to 90 we can simplify the velocity equation to:

average velocity = 980 t + 98

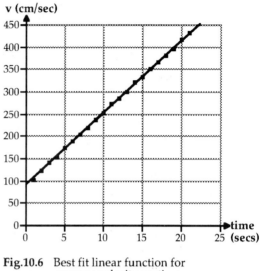

Fig.10.6   Best fit linear function for average velocity vs. time

With this equation we can estimate the velocity at any given time t. In particular, when t = 0, then v = 98 cm/sec This means that the object was already moving at about 98 cm/sec when we set t = 0. In our experiment, the velocity when t = 0 depends on where we chose to start measuring our dots. If we chose a dot closer to the start, we would have an initial velocity smaller then 98 cm/sec. If we chose a dot further away from the start, we would have an initial velocity larger than 98 cm/sec. Note that 98 cm/sec closely matches the value of 99 cm/sec that we found the initial velocity in Section 10.3.

The general equation that relates v, the velocity of freely falling objects, to t, time, is

$v = g\,t + v_0$ where v_0 = initial downward velocity   (velocity at time t = 0)
and g is the acceleration due to gravity
(approximately 980 cm/sec^2)

Acceleration: change in velocity over time

Acceleration means a change in velocity or speed. If you push the accelerator pedal in a car down just a bit, the speed of the car increases slowly, and if you floor the pedal, the speed increases rapidly. The rate of change of velocity with respect to time is called *acceleration*. Calculating the average rate of change of velocity with respect to time gives an estimate of acceleration. For example, if a car is traveling 20 miles/hr and one hour later the car has accelerated to 60 miles/hr, then

$$\underline{\text{change in velocity}} = \underline{60 - 20 \text{ miles/hr}} = (40 \text{ miles/hr})/\text{hr} = 40 \text{ miles/hr}^2$$
$$\phantom{\underline{\text{change in time}}}\quad\quad\quad 1 \text{ hr}$$
change in time

In 1 hour, the velocity of the car changed from 20 to 60 miles per hour, so its average acceleration was 40 miles per hour, per hour, or as usually written, 40 miles/hr/hr, or 40 miles/hour^2.

$$average\ acceleration\quad = \quad \underline{\text{change in velocity}}$$
$$\text{change in time}$$

Table 10.4 uses the average velocity data from Table 10.3 to calculate average accelerations. Figure 10.7 shows the plot of time in seconds (the first column) vs. average acceleration in cm/sec^2 (the third column).

t time (sec)	v aver. velocity *cm/sec	a aver. accel. (cm/sec/sec)
0.0084	103.2	n.a.
0.0250	121.8	1120.5
0.0417	141.0	1149.7
0.0584	154.2	790.4
0.0750	174.6	1228.9
0.0917	187.8	790.4
0.1084	203.4	934.1
0.1250	220.2	1012.0
0.1417	236.4	970.1
0.1584	251.4	898.2
0.1750	273.0	1301.2
0.1917	286.2	790.4
0.2084	300.0	826.3
0.2250	321.6	1301.2
0.2417	334.2	754.5
0.2584	350.4	970.1
0.2750	367.2	1012.0
0.2917	382.2	898.2
0.3084	397.2	898.2
0.3250	417.0	1192.8
0.3417	430.8	826.3

Table 10.4

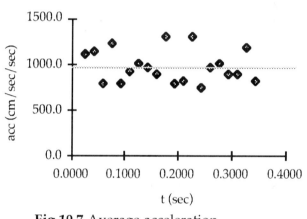

Fig.10.7 Average acceleration

The data lie along a roughly horizontal line. The calculated average acceleration values vary up and down between 790.4 and 1301.2 cm/sec/sec, with a mean of 983.3. Rounding off we find:

$$acceleration = 980 \text{ cm/sec}^2$$

The equation means that for each additional 1 second of fall, the velocity of the falling object increases by approximately 980 cm/sec. The longer it falls the faster it goes. We have verified a characteristic feature of gravity near the surface of the Earth: it causes objects to fall at a velocity that increases every second by about 980 cm/sec. We say that the acceleration due to gravity near the Earth's surface is 980 cm/sec^2.

There are 2.54 centimeters per inch and 12 inches per foot, so the conversion for centimeters to feet is $\left(\dfrac{2.54\text{cm}}{\text{in}} \right)\left(\dfrac{12\text{in}}{\text{ft}} \right) = 30.48\text{cm}/\text{ft}$. If we divide 980 cm by 30.48 cm/ft to convert centimeters to feet, we get an equivalent value for g of approximately 32 feet/sec^2.

$$\frac{980\ \text{cm}}{30.48\ \text{cm}/\text{ft}} \approx 32.15\ \text{ft} \approx 32\ \text{ft}$$

The conventional values for g, the acceleration due to gravity near the surface of the Earth, are:

$$g = 32\ \text{ft}/\text{sec}^2$$

or equivalently $\qquad g = 980\ \text{cm}/\text{sec}^2.$

The numerical value used for the constant g depends on the units being used for the distance, d, and the time, t. The exact value of g also depends on where it is measured.[8]

The general equations for the motion of freely falling bodies
that relate distance, d, and velocity, v, to time, t, are

$$d = (^1/_2)gt^2 + v_o t$$

$$v = gt + v_o$$

where $v_o =$ initial velocity $\qquad$ and $\qquad g =$ acceleration due to gravity

---

[8] Because the earth is rotating, is not a perfect sphere and is not uniformly dense, there are variations in g according to latitude and elevation. Here are a few examples of local values for g:

Location	North latitude°	Elevation (m)	g (cm/sec^2)
Panama Canal	9	0	978.243
Jamaica	18	0	978.591
Denver, Co.	40	1638	979.609
Pittsbsurgh, Pa.	40.5	235	980.118
Cambridge, Mass	42	0	980.398
Greenland	70	0	982.534

Source: Hugh D. Young, *University Physics*, Vol I, eighth edition, Addison Wesley, 1992. p. 336.

Algebra Aerobics: (answers on the next page)

1. Suppose an object is moving with constant acceleration, a, and its motion was initially observed at a moment when its velocity was v_o. Then its velocity t seconds after the initial observation is $v(t) = at + v_o$. (Note that the product of acceleration and time is velocity.) Now suppose we want to find its average velocity for the t seconds. The average velocity can be measured in two ways. First, we can find the average of the initial and final velocity by calculating a numerical average: add and divide by 2. Thus,

    average velocity after t seconds is $(v_o + v(t))/2$

    We can also find average velocity by determining the distance traveled by the object and the number of seconds it traveled, then dividing. Thus,

    average velocity $=$ distance/time $= d/t$

    For a quadratic equation both of these expressions for the average velocity should give the same value. Setting them equal we get:

    $$d/t = (v_o + v(t))/2$$

    But we also know that $v(t) = at + v_o$.   Substituting for $v(t)$, we have

    $$d/t = [v_o + (at + v_o)]/2 \qquad \text{or} \qquad d/t = [at + 2v_o]/2 = (1/2)at + v_o$$

    Solve the equation $d/t = (1/2)at + v_o$ for d.

2. An object that is moving on the ground is observed to have (initial) velocity of 60 cm/sec and is accelerating at 10 cm/sec².
    a) Determine its velocity after 5 seconds; one minute; t seconds.
    b) Find the average velocity for the object after 5 seconds.

3. Find the distance traveled by the object described in #2 above after 5 seconds by:
    a) multiplying the average velocity by 5 sec
    b) writing an equation of motion $d = (1/2)at^2 + v_o t$ using a = 10 cm/sec² and v_o =60 cm/sec with t = 5

4. An object is observed to have (initial) velocity of 200 m/sec and accelerates at 60 m/sec².
    a) Write an equation for its velocity after t seconds.
    b) Write an equation for the distance traveled after t seconds.

Answers

1. $d = (1/2)at^2 + v_o t$ This is one way of deriving the distance equation given constant acceleration.

2a) $v_o = 60$ cm/sec $a = 10$ cm/sec²
   if t = 5, then $v = (10)(5) + 60 = 110$ cm/sec
   if t = 60, then $v = (10)(60) + 60 = 660$ cm/sec
   $v = 10\,t + 60$
  b) if t = 0, then v = 60; if t = 5, then v = 110
   average velocity from 0 to 5 sec = (60 + 110)/2 = 85 cm/sec

3. a) d = (85 cm/sec)(5 sec) = 425 cm
   b) d = (1/2)(10 cm/sec²)(25 sec²) + (60 cm/sec)(5 sec) = 125 + 300 = 425 cm

4. $v_o = 200$ m/sec $a = 60$ m/sec²
   a) $v = at + v_o$ $v = 60t + 200$
   b) $d = (1/2)\,at^2 + v_o\,t$ $d = 30\,t^2 + 200t$

10.5 Using the Free Fall Equations

Applying the model

Example 1: Interpreting a free fall equation

The data from a class free fall tape generates this equation relating distance fallen in cm. and time in seconds: $d = 485.7\,t^2 + 72.6\,t$

  a) Give a physical interpretation of each of the coefficients along with their appropriate units of measurement.

   485.7 is in cm/sec² and approximates half the acceleration due to gravity. 72.6 is in cm/sec. and represents the initial velocity of the object.

  b) How far has the bob fallen after 0.05 seconds? 0.1 seconds? 0.3 seconds?

   We need to evaluate d for each value of t:

Table 10.5		
t (sec)		d (cm)
0.05	485.7•(0.05)^2 + 72 •(0.05) =	4.81
0.10	485.7•(0.1)^2 + 72 • (0.1) =	12.1
0.30	485.7• (0.3)^2 + 72 • (0.3) =	65.3

   After 0.1 second, for instance, the object would have fallen 12.1 cm.

Example 2: Translating units

What would the free fall equation $d = 490\, t^2 + 90\, t$ become if d were measured in feet instead of centimeters?

In the equation, $d = 490t^2 + 90\, t$, the number 490 represents $(1/2)g$ and is in cm/sec^2.

Since
$$1\text{ centimeter} = 0.0328\text{ feet}$$
$$490\text{ cm/sec}^2 = (490)(0.0328) \approx 16.1\text{ feet/sec}^2$$

As a check, recall that since g is approximately 32 feet/sec², then $(1/2)g$ is approximately 16 feet/sec².

Similarly
$$90\text{ cm/sec} = (90)(0.0328) \approx 3.0\text{ feet/sec}$$

So we can rewrite the original equation as: $d = 16.1\, t^2 + 3.0\, t$, where t is still in seconds but d is now in feet.

Example 3: Translating from distance to height measurements

Assume you have the following motion equation relating distance fallen in centimeters and time in seconds:

$$d = 490\, t^2 + 45\, t$$

and assume that when t = 0 the height, h, of the object was 110 cm above the ground. Until now, we have considered the distance from the point the bob was dropped, a value that *increases* as the bob falls down. How can we describe a different distance, the *height above ground* of an object, as a function of time, a value that *decreases* as the object falls?

At time zero, the distance fallen is zero and the height is 110 centimeters. After 0.02 seconds, the object has fallen 0.89 cm, so its height would be 110–0.89 = 109.11 cm. In general h = 110 - d. The following table gives associated values for time, t, distance, d, and height, h. The graphs in Fig. 10.8 show distance versus time and height versus time.

	Table 10.6	
time (sec)	distance fallen $=490t^2 + 45t$	height above landing pad
0.00	0.00	110.00
0.02	1.10	108.90
0.06	4.46	105.54
0.10	9.40	100.60
0.14	15.90	94.10
0.18	23.98	86.02
0.22	33.62	76.38
0.26	44.82	65.18
0.30	57.60	52.40
0.34	71.94	38.06
0.38	87.86	22.14
0.42	105.34	4.66

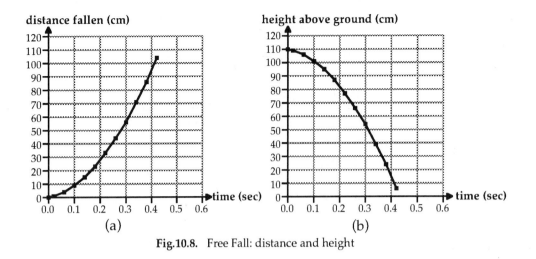

Fig.10.8. Free Fall: distance and height

The equation $d = 490\,t^2 + 45\,t$ relates distance fallen and time. What is the comparable algebraic expression relating height and time?

We know that the relationship between height and distance is $h = 110 - d$. We can substitute the expression for d into the height equation.

Substitute $d = 490\,t^2 + 45\,t$ into $h = 110 - d$.

$$\begin{aligned}
h &= 110 - (490\,t^2 + 45\,t) \\
&= 110 - 490\,t^2 - 45\,t \\
&= -490\,t^2 - 45\,t + 110 \qquad\qquad (1)
\end{aligned}$$

Since we can switch the order in which we add terms, we can write this equation in many equivalent forms. The constant term, here 110 cm, represents the initial height, when $t = 0$. Recall that often linear equations are written in the form $y = b + mx$ to emphasize the constant b as the base or starting value. Similarly height equations often appear in the form $h = c + b\,t + a\,t^2$ to emphasize the constant c as the starting height. Hence we could equally well write equation (1) as:

$$h = 110 - 45\,t - 490\,t^2$$

Note that in our equation for height, the coefficients of both t and t^2 are negative. If we consider what happens to the height of the object, this makes sense. As time increases, the height decreases. (See Table 10.6 and Figure 10.8..) When we were measuring the increasing distance an object fell, we did not take into account the direction in which it was going (up or down). We cared only about the magnitudes (the absolute values) of distance and velocity, which we are positive. But when we are measuring a decreasing height, we have to worry about directions. In these cases we define downwards motion to be negative and upwards motion to be positive. In the height equation, $h = 110 - 45t - 490\,t^2$, the constant term, initial height, is 110 cm; the change in height due to initial velocity, $-45\,t$, is negative because the object was moving down when we started to measure it; and the change in height due to acceleration, –490 t^2, is negative because the effect of gravity is downward motion-- it reduces the height of a falling object as time increases.

Once we have introduced the notion that downward motion is negative and upward motion is positive, we can also deal with situations in which the initial velocity is upwards and the acceleration is downwards.

Example 4: Working with an upwards initial velocity

Write an equation relating height and time if an object at an initial height of 87 cm is thrown upwards with an initial velocity of 97 cm/sec.

The initial height is 87 cm when t = 0, so the constant term is 87. The coefficient of the initial velocity term is positive 97 since the initial motion is upward. The coefficient of the gravity term is negative 490.

Substituting these values into the equation for height, we get

$$h = 87 + 97\,t - 490\,t^2$$

Table 10.7 gives a series of values for heights corresponding to various times.

Table 10.7	
t (sec)	h (cm)
0.00	87.00
0.05	90.63
0.10	91.80
0.15	90.53
0.20	86.80
0.25	80.63
0.30	72.00
0.35	60.93
0.40	47.40
0.45	31.43
0.50	13.00

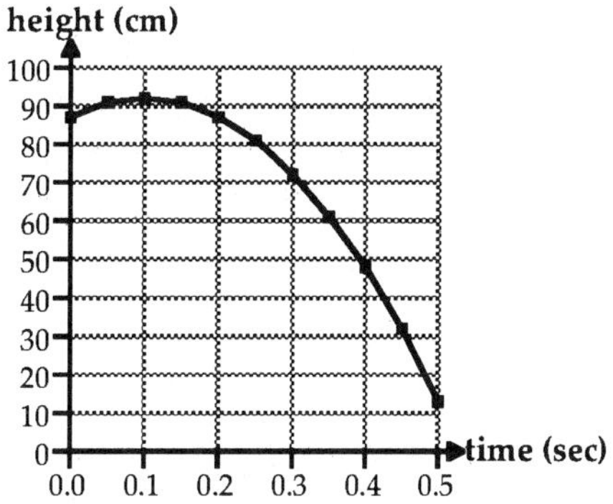

Fig.10.9.   Height of a thrown ball

The graph of the heights at each time(Fig. 10.9) should not be confused with the trajectory of a thrown object. The actual motion we are talking about is a purely vertical motion- -straight up and straight down. The graph shows that the object travels up for a while before it starts to fall. This corresponds with what we all know from practical experience throwing balls. The upward (positive) velocity is decreased by the pull of gravity until the ball stops moving up and begins to fall. The downward velocity (negative) is increased in magnitude by the pull of gravity until the object strikes the ground.

The velocity equation is $v = -gt + v_0$
where v_0 could be either positive or negative, depending on whether the object is thrown upward or downward, and the sign for the g term is negative because gravity accelerates

downward in the negative direction. So, in describing the *height* of a freely falling object rather than the distance fallen, we have extended the set of motion equations.

> The general equations of motion of freely falling bodies that relate height, h, and velocity, v, to time, t, is
>
> $$h = h_o + v_o t - (^1/_2)gt^2.$$
> $$v = -gt + v_o$$
>
> where h_o = initial height,
> v_o = initial velocity,
> g = acceleration due to gravity

Algebra Aerobics:

1. In the equation $d = 4.9\,t^2 + 500\,t$, time is measured in seconds and distance in meters. What does the 500 represent?
2. In the height equation $h = 300 + 50\,t - 4.9\,t^2$, time is measured in seconds and height in meters.
   a) What does the 300 represent?
   b) What does the 50 represent? In what direction?
3. The height of an object that was projected vertically from the ground with initial velocity of 200 meters per second is $h = 200\,t - 4.9\,t^2$.
   a) Find the height of the object after 0.1, 2, and 10 seconds.
   b) Sketch a graph of height versus time.
   c) Use the graph to determine the approximate number of seconds that the object travels before hitting the ground and maximum height of the projectile.
4. The height of an object that was shot downwards from a 200 meter platform with an initial velocity of 50 m/sec is $h = -4.9\,t^2 - 50\,t + 200$. Sketch the graph of height versus time. Use the graph to determine approximately the number of seconds that it travels before hitting the ground.

Answers

1. Initial velocity of 500 meters per second in the same direction as gravity (downwards)
2. a) Initial height of 300 meters b) Initial velocity of 50 m/sec in the opposite direction of gravity (i.e., upwards)
3. a) $t = 0.1$ $h = 200(0.1) - 4.9(0.01) = 19.951$ m
   $t = 2$ $h = 200(2) - 4.9(4) = 380.4$ m
   $t = 10$ $h = 200(10) - 4.9(100) = 1510$ m

4.

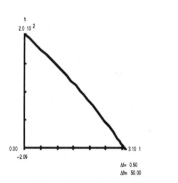

approx. 3.1 sec before hitting the ground

   c) approx. 41 seconds; max. height is approx. 2,000 meters

Chapter 10 Summary

Galileo discovered that while the velocity of freely falling bodies is *not constant*, it is increasing at a *constant rate*. His process of careful observation and testing began the critical transformation of science from a qualitative to a quantitative study of nature. This approach became known as "The Scientific Method."

We can replicate his experiments by taking distance vs. time measurements for a freely falling body. We can look at the

$$average\ velocity\ =\ \frac{change\ in\ distance}{change\ in\ time}$$

and

$$average\ acceleration\ =\ \frac{change\ in\ velocity}{change\ in\ time}$$

The general equations of motion that relate distance, velocity and time for freely falling bodies are:

$$d = (1/2)\,g\,t^2 + v_o\,t \qquad \text{and} \qquad v = g\,t + v_o$$

where d = distance, t = time, v_o = initial velocity and g = acceleration; near the surface of the earth due to gravity.

The conventional values for g are $32\ ft/sec^2$ or equivalently $980\ cm/sec^2$.

The general equations of motion of freely falling bodies that relate height, velocity and time are:

$$h = h_o + v_o t - (1/2)g t^2. \quad \text{and} \qquad v = -gt + v_o$$

where h_o = initial height, v_o = initial velocity and g = acceleration due to gravity. In these equations downward motion is considered negative, upwards motion is positive.

EXERCISES

1. "Watching Galileo's Learning" in the Anthology of Readings examines the learning process that Galileo went through to come to some of the most remarkable conclusions in the history of science. Write a summary about one of Galileo's conclusions about motion. Include in your summary the process by which Galileo made this discovery and some aspect of your own learning or understanding of Galileo's discovery.

2. The equation $d = 490 t^2 + 50 t$ describes the relationship between distance fallen, d, in centimeters, and time, t, in seconds, for a particular freely falling object.

   a) Interpret each of the coefficients and specify their units of measurement.
   b) Generate a table for a few values of t between 0 and 0.3 seconds.
   c) Graph distance versus time by hand and verify with a computer or calculator if available.

3. The equation $d = 4.9 t^2 + 1.7 t$ describes the relationship between distance fallen, d, in meters, and time, t, in seconds, for a particular freely falling object.

   a) Interpret each of the coefficients and specify their units of measurement.
   b) Generate a table for a few values of t between 0 and 0.3 seconds.
   c) Graph distance versus time by hand and verify with a computer or calculator if available.
   d) Relate your results to earlier results in this chapter.

4. Let $d = 16 t^2 + 12 t$ be a motion equation describing distance fallen in feet versus time in seconds.

   a) Interpret each of the coefficients and specify their units of measurement.
   b) Generate by hand a table for a few values of t between 0 and 5 seconds.
   c) Graph distance versus time by hand and verify with a computer or calculator if available.

5. Let $h = 85 - 490 t^2$ be a motion equation describing height, h, in centimeters versus time, t, in seconds.

   a) Interpret each of the coefficients and specify their units of measurement.
   b) Generate a table for a few values of t between 0 and 0.3 seconds.
   c) Graph height versus time by hand and verify with a computer or calculator if available.
   d) What is the initial velocity?

6.   Let h = 85 + 20 t – 490t² be a motion equation describing height, h, in centimeters versus time, t, in seconds.

    a)   Interpret each of the coefficients and specify their units of measurement.
    b)   Generate a table for a few values of t between 0 and 0.3 seconds.
    c)   Graph height versus time by hand and verify with a computer or calculator if available.

7.   At t = 0, a ball is thrown upwards at a velocity of 10 ft/sec, from the top of a building 50 feet high. Its height is measured in feet above the ground. Recall that the magnitude of the acceleration due to gravity is 32 ft/sec².

    a)   Is the initial velocity positive or negative? Why?
    b)   Write the motion equation that describes height, h, at time, t.

8.   The relationship between the velocity of a freely falling object and time is given by

$$v = \ -gt - 66$$

where g is the acceleration due to gravity and the units for velocity are cm/sec.

    a)   Generate a table of values for t and v, letting t range from 0 to 4.
    b)   Graph velocity versus time and interpret your graph.
    c)   What was the initial condition? Was the object dropped or thrown upward. Give your reasoning.

9.   The concepts of velocity and acceleration are useful in the study of human childhood development. Figure (a) shows a standard growth curve of weight over time. Figure (b) shows the rate of change of weight over time (the *growth rate* or *velocity*.) Figure (c) shows the rate of change of the growth rate over time, or *acceleration*.. Describe in your own words what each of the graphs shows about a child's growth.

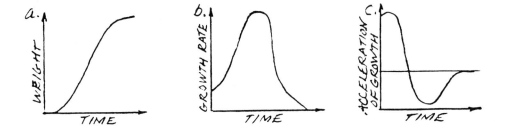

Graphs adapted from "The Evolution of Human Childhood" by Barry Bogin, *BioScience*, vol. 40, p.16.

10. (This exercise requires a free fall data tape collected using a spark timer.)
   a) Making a graph from your tape: Cut the tape with scissors crosswise at each spark dot, so you have a set of strips of paper which are the actual lengths of the distances fallen by the object during each time interval. Arrange them in increasing order, with one end on a line, like steps. You could paste or tape them down on a big piece of paper or newspaper.
   b) Lay down a straight edge so it passes through the center of the top of the strips. Can you draw a straight line through all the strips? The separate strips represent the object's speed in each interval. Interpret the graph of the line you have constructed in terms of the free fall experiment.

11. A certain baseball is at height, $h = 4 + 64\,t - 16\,t^2$ feet, at time, t, in seconds. Compute the average velocity for each of the following time intervals and indicate for which intervals the baseball is rising and for which it is falling. In which interval was the average velocity the greatest?

   a) $t = 0$ to $t = 0.5$
   b) $t = 0$ to $t = 0.1$
   c) $t = 0$ to $t = 1.0$
   d) $t = 1$ to $t = 2.0$
   e) $t = 2$ to $t = 3.0$
   f) $t = 1$ to $t = 3.0$
   g) $t = 4$ to $t = 4.01$

12. At $t = 0$, an object is in free fall 150 cm above the ground, falling at a rate of 25 cm/sec. Its height, h, is measured in centimeters above the ground.

   a) Is its velocity positive or negative? Why?
   b) Write an equation that describes its height h at time t (until it hits the ground).
   c) What is the average velocity from $t = 0$ to $t = {}^1/_2$? How does it compare to the initial velocity?

13. An object is released at time $t = 0$.

   a) Solve the distance equation, $d = 16\,t^2$, for t.
   b) What values of d make physical sense?
   c) Write t as a function of d. Remember that for t to be a function, there must be a *unique* value of t for each value of d.
   d) Generate a table of values for this function.

   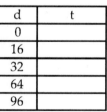

d	t
0	
16	
32	
64	
96	

   e) Sketch by hand the graph of the function you found. Note that the horizontal axis (the independent variable) is now d and the vertical axis (the dependent variable) is t.

14.. The force of acceleration on other planets.

We have seen that the function: $d = (1/2) g \bullet t^2 + v_0 \bullet t$ (where g is the force of acceleration due to earth's gravity and v_0 is the object's initial velocity) is a mathematical model for the relationship between time and distance fallen by freely falling bodies near the earth's surface. This relationship works for freely falling bodies near the surface of other planets. We need to replace g, the force of acceleration of earth's gravitational field, with the force of acceleration for the planet under consideration. The following table gives the force of acceleration due to gravity for the planets in our solar system.

| | Acceleration due to gravity | |
	meters/sec^2	feet/sec^2
Mercury	3.7	12.1
Venus	8.9	29.1
Earth	9.8	32.1
Mars	3.7	12.2
Jupiter	24.8	81.3
Saturn	10.4	34.1
Uranus	8.5	27.9
Neptune	11.6	38.1
Pluto	0.6	2.0

Source of data: The Astronomical Almanac.
U.S. Naval Observatory, 1981.

a) Choose units of measurement (meters or feet) and three of the planets (other than Earth). Find the equation for the relationship between distance an object falls and time on this planet. Assume for the moment that the initial velocity of your freely falling object is 0.

Name of Planet	Function relating distance and time	Units for distance

b) Using a function graphing program or a graphing calculator, plot the three functions, with time on the horizontal axis and distance on the vertical axis. Choose a domain that includes negative and positive values of t.

c) On which of your planets will an object fall the farthest in a given time? On which will it fall the least distance in a given time? What part of the domain makes physical sense for your model. Why?

d) Examine the graphs and think about the *similarities* that they share. Describe their general shape. What happens to d as the value for t increases? as the value for t decreases?

e) Think about the *differences* among the three curves. What effect does the coefficient of the t^2 term have on the shape of the graph? When the coefficient gets larger (or smaller) how is the shape of the curve affected? Which graph shows d increasing the fastest compared to t? What happens to d as the value for t decreases?

15. Use the table for the acceleration of gravity on other planets in Exercise 14 to answer the following questions:

    a)   In 1974 in Anaheim, California, Nolan Ryan threw a baseball at just over 100 mph. If he had thrown the ball straight upward at this speed, it would have risen to a height of over 336 feet and taken just over 9 seconds to fall back to Earth. Choose another planet and see what would have happened if he had been able to perform this experiment on this planet.

    b)   At the 1968 Olympics in Mexico City, Bob Beamon made an extraordinary jump of 29 feet 2 1/2 inches, bettering existing marks by almost a yard. If his forward velocity was that of a 100 meter dash record holder (10 m/sec) his 8.9 meter jump required an upward velocity of 4.36 m/sec on take off. How far might Bob Beamon have been able to jump on Mars, assuming his performance (in term of take off velocities) is the same as on Earth?

16. In the Anthology reading "Watching Galileo's Learning", Cavicchi notes that Galileo generated a sequence of odd integers from his study of falling bodies. Show that in general the odd integers can be constructed from the difference of the squares of successive integers; that is, the terms $(n+1)^2 - n^2$ (where n = 1, 2, 3, ...) generate a sequence of odd integers.

EXPLORATION 10.1.
Free Fall Experiment

Objective
- to determine the relationship between time and distance fallen by a freely falling object.

Equipment/materials
- graphing calculators with best fit function capabilities or
  spreadsheet and function graphing program
- Notebook for recording measurements and results

  Equipment needed for collecting data in physics laboratory
  Free fall apparatus
  Meter sticks two meters long
  Masking tape

  Equipment needed for collecting data with CBL[T] (Calculator Based Laboratory System[T])
  CBL [T]unit with AC-9201 power adapter
  Graphing calculator
  Vernier CBL[T] ultrasonic motion detector
  Extension cord and some object to drop such as a pillow or rubber ball

  Data from a sample free fall experiment is also on disk included with text.

Related Readings
Elizabeth Cavicchi's "Watching Galileo's Learning"

Related Software
"Q10: Falling Objects" and "Q11: Average Rate of Change" in *Quadratic Graphs*
"L8: Average Velocity and Distance" in *Linear Graphs;*

Preparation
If collecting data in Physics Lab, schedule a time for doing the experiment and have the lab assistant available to set up the equipment and assist with the experiment. If collecting data with a CBL[T] unit with graphing calculators, instructions for using a CBL[T] unit are in the Teacher's Manual .

Procedure
Students work in pairs to collect and record data.
The following procedures can be used for collecting data in a Physics Lab.[9] If you are collecting data with a CBL[T], collect the data and go to the **Results** section.

---

[9] These procedures are adapted from Laboratory Notes for Experiment 2: The Kinematics of Free Fall. UMass/Boston, Elementary Physics 181

Collecting the data:

Since the falling times are too short for you to record with a stopwatch, we use a free fall apparatus. Every 1/60 second (this is the 60 Hertz of the alternating electric current) a spark jumps through a small gap of air inside the timer and records the position of the bob by a spark dot on the tape.

1.   Position the bob at the top of the column in its holder.
2.   Pull the tape down the column so that a fresh tape is ready to receive spark dots.
3.   Be sure that the bob is motionless before turning on the apparatus.
4.   Turn on the spark switch and bob release switch as demonstrated by the technician.
5.   Tear off the length of tape recording the fall of the bob.

Obtaining and recording measurements from the tapes:

The tape is a record of the distance fallen by the bob between each 1/60 th of a second spark dot. Each pair of students should measure and record the distance between the dots on the tape. Let **d** = the distance fallen in cm. and **t** = time in seconds.

1.   Fasten the tape to the table using masking tape.
2.   Inspect the tape for missing dots. *Caution:* the sparking apparatus sometimes misses a spark. If this happens, take proper account of it in numbering the dots.
3.   Position the two-meter stick *edgewise over the tape.* Use masking tape to fasten the meter stick to the table, making sure that the spots line up in front of the bottom edge of the meter stick so you can read their positions off the stick.
4.   Beginning with the sixth visible dot, mark the time for each spot on the tape, i.e. write t = 0/60 sec. under the sixth dot, t = 1/60 under the next dot, t = 2/60 under the next dot and so on until you reach the end of the tape.

Note: The first five dots are ignored in order to increase accuracy of measurements (one cannot be sure that the object is released exactly at the time of the spark instead of between sparks, and the first few dots are too close together to get accurate measurements). When the body passes the sixth dot it already has some velocity, which we call v_0, and this point is arbitrarily taken as the initial time, t = 0.

Sample Sketch of Tape

5.   Measure the distance from the sixth dot to each of the other dots within a fraction of a millimeter and record this number under the spot on the tape.
6.   Recheck your measurements.
7.   Clean your work area.

Results

Use your notebook to keep a record of your data, observations, graphs and analysis of the data.

1. Record the data obtained from your measurements on the tape or from using a CBL^T unit. If you are entering your data into a function graphing program or a spreadsheet, you can use a printout of the data and staple it into your lab notebook.

    a) Record the data obtained for each equal time interval. (We assume here an interval size of 1/60 second. Check your equipment to see if it uses a different interval size.) Your data from the free fall experiment should be recorded in a data table with **t**= time and **d** = distance fallen.

t (sec)	d (cm)
0/60	0
1/60	_____
2/60	_____
....	
to last record	_____

    b) Note at which dot on the tape you started to make your measurements

Analysis of Results

Analysis of data
1. Generate by hand

    a) A graph of your data, using the vertical axis for distance fallen, d, in cm and the horizontal axis for time, t, in seconds.
    b) Average rate of change calculations for distance (d) with respect to time (t) for three pairs of points from your data table. Show your work.

    $$\text{average rate of change} = \frac{\text{change in distance}}{\text{change in time}} = \frac{\Delta d}{\Delta t} =$$

    This average rate of change is called the *average velocity* of the falling bob between these two points.
    c) Jot down your observations from your graph and calculations in your notebook. Staple your graph into your notebook. From your graph, how do you think the change in **d** compares to the change in **t** ? What do your hand calculations show?

2.   Generate with graphing calculators or computers

   a) A graph of your data for the free fall experiment. Plot time, t, on the horizontal axis
       and distance fallen, d, on the vertical axis.
   b) Find a best fit function for distance fallen.
   c) Using a spreadsheet or graphing calculator calculate the average rate of change in
       distance for each 1/60 th of a second time interval. This average rate of change is
       called the average velocity over these time intervals.
   d) Plot average velocity vs. time. Plot time on the horizontal axis and average velocity on
       the vertical axis.
   e) Jot down your observations from your graphs and calculations in your notebook. Be
       sure to specify the units for any numbers you have recorded in your notebook.

Conclusions

Summarize your conclusions from the experiment.

   • Describe what you found out from your graph of distance vs. time and your
      calculations for the average rate of change of distance with respect to time. Do your
      results support your prediction? Is the average rate of change of distance the same for
      each 1/60 of a second?
   • What does your graph of the average velocity vs. time tell you about the average
      velocity of the freely falling body? Is the average change in velocity from one interval
      to the next roughly constant?
   • Interpret your graphs for distance and average velocity and interpret the coefficients in
      the equation you found for distance in light of the readings and class discussion.

EXPLORATION 10.2.
Constructing Equations of Motion

Objectives
- to find an equation for height of object in free fall
- to interpret the height and velocity equations for the free fall experiment

Materials/Equipment
- graph paper
- graphing calculators or spreadsheet and function graphing program (optional)

Procedure
For this exploration, we find an equation for the height or position of the object which we call y. You can use data from Table 10.1 or data that you have collected from a free fall experiment. To find the height or position of the object, subtract the distance fallen from the initial height at which the object was dropped. For the data set in Table 10.1 the initial height was approximately 90.5 cm.

1. Enter your data either into a graphing calculator or a spreadsheet, in four columns, as follows:

Time (t) sec	Height (y) cm	Average velocity from t_{n-1} to t_n
$t_0 = 0.0000$	$y_0 = (90.5 - 0.00)$	
$t_1 = 0.0167$	$y_1 = (90.5 - 1.72)$	

For the heights, y_0, y_1, y_2, etc., enter the appropriate measurements using Table 10.1 or your own data.

> Note that the average velocities you compute will be negative. This is because the object is falling down, so its position or height is decreasing. The direction that the object is falling is now being considered. (See Example 3 in sec. 9.5)

2. Construct graphs for :
    a) Height versus time
    b) Average velocity from t_{n-1} to t_n versus time

3. Write the equation of a line approximating the nearly linear graph. Do it either by hand, or by determining the regression line through these points. Call this equation **(1).**

4. The equation of the motion is an equation giving height, y , at time, t. It can be derived from the average velocity equation from the starting time t_0 to time t:

Average velocity between time 0 and time t =   change in height
                                                change in time

$$= (y - y_0)/ (t - 0)$$

$$= (y - y_0) / t$$

5. Solve this equation for y and you have an equation in terms of the height y and call this equation (2)

$$y = y_0 + t \text{ (average velocity between 0 and t)} \qquad (2)$$

6. We need the average velocity from time 0 to time t. The graph of velocity is a straight line. So it's plausible and justifiable by more advanced means (calculus) to take the average velocity as just the velocity halfway between the extremes, at time t/2. (see the figure below)

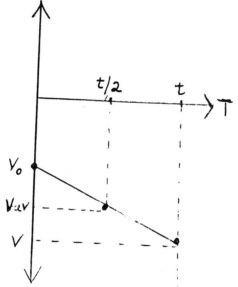

Find the average velocity from time 0 to time t by using t/2 for time in equation (1). Call this equation (3)

7. Substitute equation (3) for average velocity (from time 0 to time t) into equation (2). Express in your own words what this equation tells you about the height (y).

Analysis of Results

- In you own words explain what you did to find the average velocity from time 0 to time t.
- How close did you come to the conventional value for g ?
- Use your equation for height of object in free fall to compute the height of object for time 1/60, 2/60.. and compare to those from your data table.
- Interpret the coefficients for each of the terms of the equation you found for height.

Chapter 11

Parabolic Reflections

Overview

The free fall experiment in Chapter 10 introduced quadratic relationships between distance and time. The graph of a quadratic function is called a *parabola*. A typical parabolic curve is easily seen by anyone who watches a thrown ball in slow motion. *Balli*stics, literally the study of dropped, thrown, or projected balls, gives many variations of quadratic data, all of which appear as parabolic curves when vertical position is plotted vs. time. Mirrors in the shape of parabolas have the property of focusing or reflecting parallel rays, so that physical uses of parabolas are found as reflectors or concentrators of rays in devices such as light fixtures, heaters, or satellite dishes.

In this chapter we study the general properties of *quadratic functions*. We develop direct ways to identify critical properties of quadratics and sketch their graphs.

The Explorations build an intuitive understanding of quadratic functions and show some of their practical applications.

After reading this chapter you should be able to:

- recognize a quadratic function
- determine the vertex, axis of symmetry, and the intercepts for a quadratic functions
- construct equations and graphs for quadratic functions

11.1. Visualizing Quadratic Functions

The definition of a quadratic function

In Chapter 10 we needed a time squared term, t^2, giving us a quadratic of the form $d = a t^2 + b t + c$, to model distance fallen versus time in the free fall experiment. As we saw in Chapter 9, quadratics form a subset of the family of polynomial functions.

A quadratic function or *polynomial function of degree two;*, has the form

$$y = ax^2 + bx + c$$

where a, b and c are constants and $a \neq 0$.
The graph of a quadratic function is a *parabola;*.

Since the order in which you add terms does not matter, the general quadratic function could equally well be described as having the form $y = c + bx + ax^2$.

Graphs of quadratic functions

We have seen that the quadratic function $d = 488 t^2 + 99 t$ models the relationship between distance fallen (d in centimeters) and time (t in seconds) of a freely falling body. Now, divorce this function from its physical context and think of it as describing a relationship between two abstract variables d and t. The domain is then all the real numbers. Since the function is a polynomial of degree 2 we would expect the graph to be U- shaped, bending only once. Table 11.1 and Figure 11.1 show some points that satisfy the abstract equation.

Table 11.1	
t	$d = 488 t^2 + 99t$
-0.4	38.5
-0.3	14.2
-0.2	-0.3
-0.1	-5.0
0.0	0.0
0.1	14.8
0.2	39.3
0.3	73.6
0.4	117.7

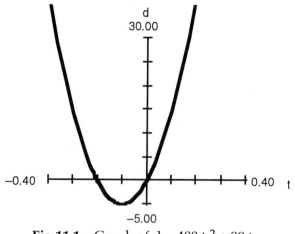

Fig.11.1   Graph of $d = 488 t^2 + 99 t$

When we graph points corresponding to both positive and negative values for t, we can see the basic form shared by all these parabolas. Figure 11.1 contains only a small portion of the entire graph; the whole parabola extends indefinitely upwards and outwards.

If you have access to a graphing calculator or the course software, try graphing a number of different quadratics, like those illustrated below.

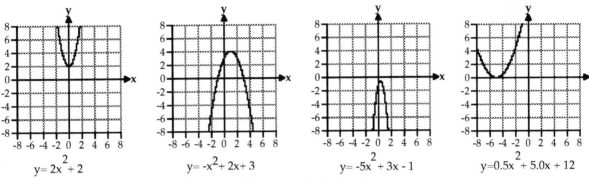

$y = 2x^2 + 2$ $y = -x^2 + 2x + 3$ $y = -5x^2 + 3x - 1$ $y = 0.5x^2 + 5.0x + 12$

Fig.11.2.. Graphs of four different quadratic functions

All parabolas share a distinctive bowl–like shape, some wide, some narrow, some rightside up, some upside down.

Symmetry

Each of the graphs in Figure 11.2 seems to have a "middle point," which might be either the lowest or the highest point on the curve, depending on whether the parabola opens up or down. Imagine drawing a vertical line through this middle point and folding the parabola along that line. The right half of the curve would fall exactly on the left half: the right half of the curve is a mirror image of the left half. Such a graph is said to be *symmetrical*. The vertical line of the fold is called the *axis of symmetry*, and the middle point is called the *vertex*.

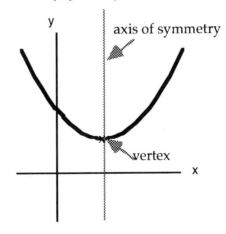

Fig. 11.3   Each parabola has a vertex that lies on an axis of symmetry

Algebra Aerobics: [1]

Using a graphing tool, plot each of the following. Then, give the equation of the axis of symmetry and the coordinates of the vertex for each parabola.

1. $y = x^2 - 2$

2. $y = (x+1)^2$

3. $y = 0.5(x-2)^2 + 1$

Predicting what the graph of a quadratic function will look like [2]

Given any quadratic function, $y = ax^2 + bx + c$, it is fairly easy to predict the general appearance of its parabolic graph. By examining the coefficients a, b, and c, it is possible to tell if the parabola opens up or down; where it crosses the y-axis, and whether the curve is wide or narrow.

The effect of the coefficient a

Case 1: $a > 0$. What can we predict about the shape of the parabola if a is positive? If we increase the size of a?

We start by examining the simplest quadratic, $y = x^2$, where $a = 1$, and b and c are 0. Figure 11.4 shows a parabola that opens upwards. Now consider $y = a x^2$ when $a = 2$ and then 4. Examine Table 11.2 for the effects of a larger value of a. Except when $x = 0$, the larger the value of a, the larger the value for y.

Table 11.2

x	$y = x^2$	$y = 2x^2$	$y = 4x^2$
−2	4	8	16
−1	1	2	4
0	0	0	0
1	1	2	4
2	4	8	16

[1] Answers

1.	vertex:	(0,-2)	axis:	$x = 0$
2.	vertex:	(-1,0)	axis:	$x = -1$
3.	vertex:	(2,1)	axis:	$x = 2$

[2] We strongly suggest that you take a look at the "Quadratic graphs with sliders" program in the *Quadratic Graphs* or use your graphing calculator to try out for yourself the steps in the text.

Compare the graphs of $y = x^2$, $y = 2 x^2$ and $y = 4 x^2$ in Figure 11.4. Each parabola opens upward, but the larger the value of a, the steeper the graph. Thus the graph of $y = 2x^2$ hugs the y-axis more closely than the graph of $y = x^2$ The graph of $y = 4x^2$ is even closer. Imagine the graph of $y = 3x^2$. It should lie between the graphs of $y = 2x^2$ and $y = 4x^2$.

Try setting b and c to values other than 0, while keeping $a \geq 1$. Though the vertex may change, varying a still has the same effect. The parabola still opens up and gets steeper as the value for a increases.

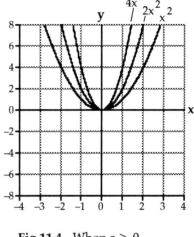

Fig.11.4 When a ≥ 0

When we look at the quadratic function $y = 7x^2 + 5x$ we know, before calculating any values or plotting any points, that since a = 7, the graph must open up and would be steeper, for example, than the graph of $y = 3 x^2 + 5x$, since 7 is greater than 3.

What happens to the parabola if a is between 0 and 1 ?

	Table 11.3		
x	$y = x^2$	$y = 0.5x^2$	$y = 0.25x^2$
-2	4	2.0	1.00
-1	1	0.5	0.25
0	0	0.0	0.00
1	1	0.5	0.25
2	4	2.0	1.00

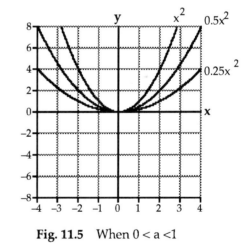

Fig. 11.5 When 0 < a < 1

Again we can use the function $y = x^2$ as a basis of comparison. In Table 11.3 compare the values for $y = x^2$ to those for $y = 0.5x^2$ and $y = 0.25x^2$. (Remember that when $0 < a < 1$, then ax^2 is less than x^2.) Except when x = 0, the smaller the value of a, the smaller the value for y. From Figure 11.5, we see that all the parabolas still open up, but the smaller the value of a, the more the parabola flattens out. The parabola for $y = 0.5x^2$ is broader than $y = x^2$ and the graph of $y = 0.25x^2$ lies even closer to the x-axis. Imagine the graph of $y = 0.4x^2$. It will lie between the graphs of $y = 0.25x^2$ and $y = 0.5x^2$.

When we look at the function $A = (1/40)t^2$, a description of *sordaria finicola* fungus growth (from Section 2.4), we know that, since a = 1/40, the graph opens upward and is flatter than the graph of the function $A = t^2$, where a = 1.

Try setting b and c to values other than 0. Varying the values of a between 0 and 1 still has the same effect.

Whether $a > 1$ or $0 < a < 1$, the larger the value for a, the narrower the parabola.

In general, when a is positive the graph of $y = ax^2 + bx + c$ opens upward; and when a increases, y increases, narrowing the curve.

Algebra Aerobics: [3]

Without drawing the graph, list these parabolas in order, from the narrowest to the broadest.

a) $y = x^2 + 20$

b) $y = 0.5x^2 - 1$

c) $y = \dfrac{1}{3}x^2 + x + 1$

d) $y = 4x^2$

e) $y = 0.1x^2 + 2$

Case 2: $a < 0$. What happens to the parabola if a is negative?

Compare the values in Table 11.4 for $y = x^2$ (where $a = 1$) and for $y = -x^2$ (where $a = -1$).

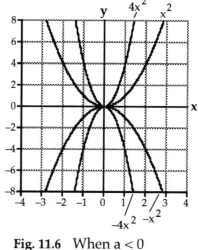

Table 11.4			
x	$y = x^2$	$y = -x^2$	$y = -4x^2$
-2	4	-4	-16
-1	1	-1	-4
0	0	4	0
1	1	-1	-4
2	4	-4	-16

Fig. 11.6 When a < 0

Since $x^2 \geq 0$ for any x, then $-x^2 \leq 0$ for any x. In Figure 11.6 we can see that while the graph of $y = x^2$ opens upward, the graph of $y = -x^2$ opens down. The two parabolas are mirror images across the x–axis. Similarly while the graph of $y = 4x^2$ opens upward, its mirror image, the graph of $y = -4x^2$, opens down.

Try graphing some parabolas where b and c are non–zero. If a is positive the parabola opens upward, if a is negative the parabola opens down. For example, in the quadratic function $h = -490\,t^2 - 45t + 110$ the value for a is negative ($a = -490$), so we know that its graph opens down.

In general, when a is negative, the graph of $y = ax^2 + bx + c$ opens down, and when $|a|$ (the absolute value or magnitude of a) increases, the curve narrows.

---

[3] Answer: d,a,b,c,e

Algebra Aerobics: [4]

Without drawing the graph, describe each parabola as

a)   opening up or down

b)   narrower or broader than $y = x^2$

   1. $y = 2x^2 - 5$

   2. $y = 0.5x^2 + 2x - 10$

   3. $y = 3 + x - 4x^2$

   4. $y = -0.2x^2 + 11x + 8$

Something to think about

So far we have ignored the case when $a = 0$. If a is 0, what would the equation $y = ax^2 + bx + c$ become and what sort of graph would it have?

The effect of the constant c

What happens to the graph of $y = ax^2 + bx + c$ when c varies?

Table 11.5			
x	$y = x^2$	$y = x^2 + 4$	$y = x^2 - 4$
−2	4	8	0
−1	1	5	−3
0	0	4	−4
1	1	5	−3
2	4	8	0

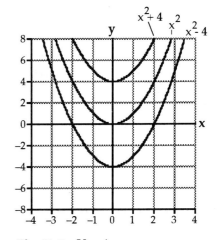

Fig. 11.7   Varying c

Table 11.5 shows that when $y = x^2$ ($a = 1$, b and $c = 0$), changing the equation to $y = x^2 + 4$ means increasing the value of y by 4. On the graph this raises the parabola $y = x^2$ up 4 units to become the graph of $y = x^2 + 4$. Similarly changing $y = x^2$ to $y = x^2 - 4$ decreases each value of y by 4. The graph of $y = x^2 - 4$ can be generated by lowering the graph of $y = x^2$ down 4 units.

Changing c changes the vertical position of the graph, not its shape.

---

[4]   Answers

1.	a)	up	b)	narrower
2.	a)	up	b)	broader
3.	a)	down	b)	narrower
4,	a)	down	b)	broader

c has the same effect when $b \neq 0$. For instance, the graphs of two height functions $h_1 = -16t^2 - 8t + 25$ and $h_2 = -16t^2 - 8t + 30$ have the identical shape. But the graph of h_2 is 5 units above the graph of h_1 since 30 is 5 more than 25. In height equations for freely falling bodies the constant term c represents the value for the *initial height* when $t = 0$. So the initial height for h_1 is 25 units and for h_2 is 30 units.

Think of c as an "elevator" term. Without changing the shape of the curve, c raises the parabola if c is positive and lowers the parabola if c is negative.

The effect of the coefficient b

The most difficult case to examine is the effect of changing the b term. We will do this by looking at some graphs, since it's harder to discern what is happening through data tables or equations. Figure 11.8 shows the graph of the parabolas.

$$y = x^2$$
$$y = x^2 + 4x$$
$$y = x^2 - 4x$$

In each case the constants a and c are fixed ($a = 1$ and $c = 0$). The equations differ only in the b value. Changing b repositions the original parabola, but does not change its shape.

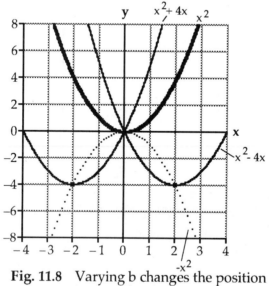

Fig. 11.8   Varying b changes the position
but not the shape of a parabola

Something to think about

By looking at Figure 11.8 or by playing with "Quadratic graphs with sliders" in *Quadratic Graphs* in the course software, can you describe the movement of the vertex of the original parabola as you change the value for b?

In summary

Graphical Characteristics of Quadratic Functions

Given a quadratic function $\quad y = ax^2 + bx + c$
where a, b and c are real numbers and $a \neq 0$
its graph, called a parabola, will have the following properties:

- It is in the shape of a symmetric "bowl."

- If $a > 0$ the parabola opens upwards (holds water).
- If $a < 0$ the parabola opens downwards (sheds water).
- As $|a|$ (the absolute value or magnitude of a) increases, the parabola becomes narrower.

- Changing c does not affect the shape of the parabola.
- As c increases the parabola shifts up.
- As c decreases the parabola shifts down.

- Changing b does not affect the shape of the parabola.
- As b varies, the parabola moves both vertically and horizontally.

Algebra Aerobics: [5]
Without drawing the graphs, compare the graph of (b) to the graph of (a)

1.	a)	$y = x^2 + 2$	b)	$y = 2x^2 + 2$
2.	a)	$y = x^2 + 3x + 2$	b)	$y = x^2 + 3x + 8$
3.	a)	$y = x^2 + 5$	b)	$y = -x^2 + 5$
4.	a)	$y = 5x^2$	b)	$y = 0.5x^2$

---

[5] Answers

1. b is narrower than a; both open up
2. b is 6 units higher than a; both have the same shape and both open up
3. have same shape; a opens up; b opens down
4. b is much broader than a; both open up

11.2   Finding the Intercepts of a Quadratic Function

Given the equation of a quadratic function, how can we identify key points that anchor the graph? The answer will lie in algebraic strategies for finding the x and y-intercepts and the vertex discussed in this and the next section.

Finding the y (or vertical) intercept

The y-intercept is the point at which the graph crosses the y-axis. Algebraically, it is the value for y in a quadratic function when x = 0.

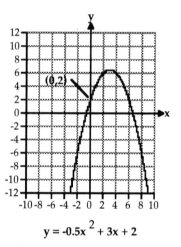

Example 1:   To find the y-intercept of the quadratic function $y = 0.5x^2 + 3x + 2$, we would evaluate y when x = 0, getting $y = 0.5(0)^2 + 3(0) + 2 = 2$. So the y-intercept, the point where the parabola crosses the y–axis, is (0,2). We also abbreviate this and say that the y–intercept is 2. The graph is shown in Figure 11.9.

$$y = -0.5x^2 + 3x + 2$$

To find a general expression for the y-intercept for the function $y = ax^2 + bx + c$, we substitute x = 0 into $y = ax^2 + bx + c$.

Fig.11.9. The y-intercept indicates where the function crosses the y axis.

For x = 0 $y = a(0)^2 + b(0) + c$

$y = c$

The y–intercept is the point (0,c) for any quadratic function of the form

$y = ax^2 + bx + c$.

Simply stated, the y-intercept is c.

Example 2: In many mathematical models, especially when time is the independent variable, the vertical intercept represents an initial or starting condition. In the motion equation $h = 23 - 15t - 16t^2$ (h = height in feet, t = time in seconds), the vertical or h intercept is 23 feet. Here 23 feet represents the initial height when t = 0.

Algebra Aerobics: [6]

1. Find the y-intercept of each parabola.

   a) $y = 2x - 3x^2$

   b) $y = 5 - x - 4x^2$

   c) $y = \frac{2}{3}x^2 + 6x - \frac{11}{3}$

2. Find and interpret the vertical intercept for the given height equations.

   a) $h = -4.9t^2 + 50t + 80$

   b) $h = 150 - 80t - 490t^2$

Finding the x (or horizontal) intercepts

The x-intercepts are the points where a curve crosses the horizontal axis. Algebraically, they correspond to values of x that make y equal to zero. These values are called the *zeros* of the function. A quadratic function may have zero, one, or two x-intercepts. In Figure 11.10 we can see that there are no x-intercepts if either (1) the vertex is above the x-axis (k>0) and the parabola opens upward (a>0), or (2) the vertex is below the x-axis (k<0) and the parabola opens downward (a<0).

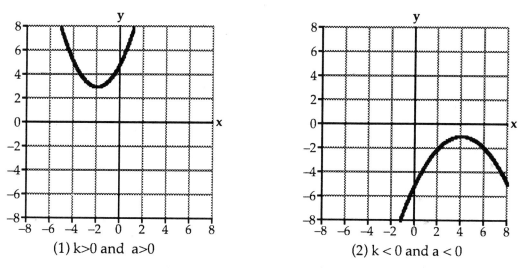

(1) k>0 and a>0   (2) k < 0 and a < 0

Fig. 11.10 Graphs of quadratics with no x-intercepts

There is one x-intercept if the vertex is on the x-axis (k=0), regardless of which direction the parabola opens; then the x-intercept is the same as the vertex as shown in Figure 11.11.

---

[6] Answers

1. a) (0,0)   b) (0,5)   c) (0,-11/3)

2. a) (0,80)   initial position: 80 meters

   b) (0,150)   initial position: 150 centimeters

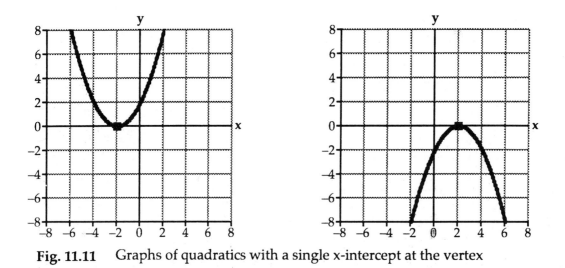

Fig. 11.11   Graphs of quadratics with a single x-intercept at the vertex

There are two x-intercepts if either (1) the vertex is below the x-axis (k<0) and the parabola opens upward (a>0), or (2) the vertex is above the x-axis (k>0) and the parabola opens downward (a<0).

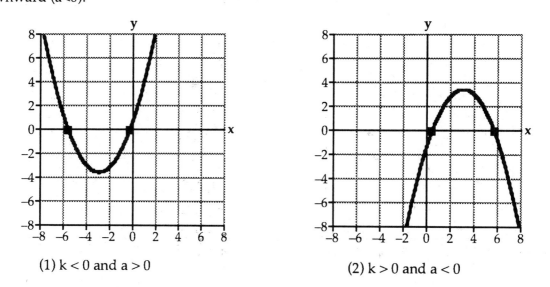

(1) k < 0 and a > 0   (2) k > 0 and a < 0

Fig.11.12 Graphs of quadratics with two x-intercepts, the maximum number possible

Estimating zeros with a function graphing program

If you have access to a function graphing program on a computer or calculator, you can use it to approximate the x-intercepts (or zeros) of a quadratic.

For example, if we plot the function $y = -1.7x^2 + 2.3x + 0.9$ it is easy to see that the resulting graph has two x-intercepts, one between 0 and -1, and one between 1 and 2.

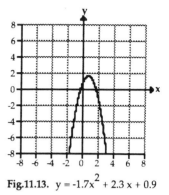

Using either a *trace* or *root finder* function the calculator can estimate the two x coordinates where the curve crosses the x-axis. You should get something like:
$$x = -0.32 \text{ and } x = 1.67$$

Fig.11.13. $y = -1.7x^2 + 2.3x + 0.9$

The technological approach gives good estimates of intercepts but they may not be exact. With algebra it is possible to find exact values.

Algebra Aerobics: [7]
Using a function graphing program plot the functions and estimate values for their zeros.
$$1) \quad y = x^2 + 4x - 7 \qquad\qquad 2) \quad y = 4 - x - 2x^2$$

Finding zeros algebraically some of the time: factoring

The basic question is: "What value of x makes y zero?" The algebraic strategy for finding the zeros of a quadratic function $y = ax^2 + bx + c$ is to set $y = 0$ and solve (find the roots of) the related quadratic equation $0 = ax^2 + bx + c$.

In real physical examples, few quadratic equations have integer roots, so they do not factor neatly. But when factoring is easy, it offers a good shortcut.

Example 1: Find the zeros of the function $y = 5700x^2 + 3705x$.

Solution: To finding the zeros or x-intercepts, set y equal to 0 and find the roots of the quadratic equation:
$$0 = 5700x^2 + 3705x$$

---

[7] Answers:

1) Estimated zeros at x= -5.32 and x = 1.32   2) Estimated zeros at x = -1.69 and x = 1.19

Both terms have x in common so you can factor x out.

$$0 = x(5700x + 3705)$$

By the Zero Product Rule:

$$x = 0 \quad \text{or} \quad 5700x + 3705 = 0$$

So one root occurs when x = 0 and the other root occurs when

$$5700x = -3705$$
$$x = -3705/5700$$
$$= -0.65$$

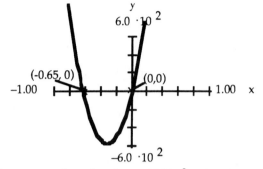

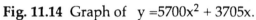

Fig. 11.14 Graph of $y = 5700x^2 + 3705x$.

The function crosses the x–axis when x = - 0.65, and when x = 0 .

Example 2: Find the zeros of $y = x^2 + 2x - 15$.

Solution: To find the zeros, set y = 0 and solve the equation.

$$0 = x^2 + 2x - 15$$

With a few tries, you would probably discover you could factor the polynomial on the right hand side:

$$0 = (x + 5)(x - 3)$$

Since both (x + 5) and (x – 3) represent real numbers, we can apply the Zero Product Rule.

So

$$x + 5 = 0 \quad \text{or} \quad x - 3 = 0.$$

Then $\quad x = -5 \quad$ or $\quad x = 3.$

Thus x = –5 or x = 3 are the roots of the equation $0 = x^2 + 2x - 15$. So the function $y = x^2 + 2x - 15$ crosses the x-axis at (–5,0) and (3,0)

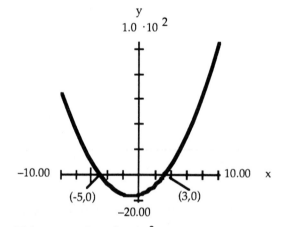

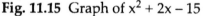

Fig. 11.15 Graph of $x^2 + 2x - 15$

Algebra Aerobics: [8]

Find the zeros of each function.

a) $\quad y = 2x^2 + x - 5$ b) $\quad y = -16t^2 + 50t$

[8] Answers:.   a)   approx. -1.85 and 1.35   b)   0 and 25/8

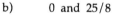

Finding zeros algebraically all of the time: the Quadratic Formula

In the next few pages we'll show that we can use the quadratic formula to find the zeros for any quadratic function.

The Quadratic Formula;

For any quadratic function of the form $y = ax^2 + bx + c$ the *x–intercepts*, or *zeros* of the function are given by:

$$x = \frac{-b \pm \sqrt{b^2 - 4ac}}{2a}$$

Example: What are the zeros (if any) for the quadratic height function, $h = 34 + 32t - 16t^2$ (where t is in seconds and height in feet)? What significance do they have?

Here a = -16, b = 32 and c = 34. Since a is negative, we know that the graph opens down. We previously found the vertex to be at (1, 50) so the function must cross the t-axis twice, and hence have two zeros.

We can use the quadratic formula to find the zeros.

$$\frac{-b \pm \sqrt{b^2 - 4ac}}{2a} = \frac{-32 \pm \sqrt{32^2 - 4(-16)(34)}}{(2)(-16)}$$

$$= \frac{-32 \pm \sqrt{1024 + 2176}}{-32}$$

$$= \frac{-32 \pm \sqrt{3200}}{-32}$$

$$\approx \frac{-32 \pm 56.6}{-32} \qquad \text{(using a graphing calculator)}$$

So there are two zeros, one at $\dfrac{-32 + 56.6}{-32} = \dfrac{24.6}{-32} \approx -0.77$ and the other at

$\dfrac{-32 - 56.6}{-32} = \dfrac{-88.6}{-32} = 2.77$ So the parabola crosses the horizontal axis at (-0.77, 0) and

(2.77, 0) as shown in Figure 11.16.

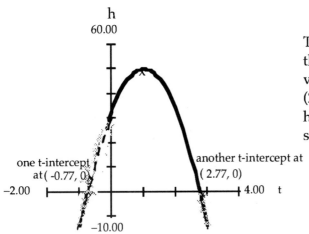

The t-intercept at (-0.77, 0) lies outside the model, since it represents a negative value for time, t. The second t-intercept at (2.77, 0)says that when t = 2.77 seconds, h = 0 feet. In other words after 2.77 seconds, the object has hit the ground!

Fig. 11.16 Graph of the equation h = $34 + 32t - 16 t^2$
(with two t-intercepts)

Algebra Aerobics:

Find the zeros of the following function. Write the coordinates of the x-intercept(s), if any.

1. $y = 4 - x - 5x^2$

2. $y = 4x^2 - 28x + 49$

3. $y = 2x^2 + 5x + 4$

Proving the quadratic formula (optional)

In Section 11.3 we will see that any quadratic function can be written in what is called the "a-h-k form"; that is, we can transform a quadratic of the form $y = ax^2 + bx + c$ into the form $y = a (x - h)^2 + k$ where (h, k) are the coordinates of the vertex. To find the zeros of a quadratic in the a-h-k form, we look for values of x which satisfy the equation:

$$a (x - h)^2 + k = 0.$$

First, we subtract k from both sides of the of the equation, and divide by a to get:

$$(x - h)^2 = -\frac{k}{a}$$

Answers:

1. zeros: -1 and 4/5   x-intercepts: (-1,0) and (4/5, 0)

2. zero: 7/2   x-intercept: (7/2, 0)

3. zeros: $\dfrac{-5 \pm \sqrt{25 - 4(2)(4)}}{2(2)} = \dfrac{-5 \pm \sqrt{25 - 32}}{4} = \dfrac{-5 \pm \sqrt{-7}}{4}$ undefined in the reals, so no x-intercepts

Now, since the square of either a positive or a negative number is positive, this equation can only be satisfied if $-k/a \geq 0$, that is, if $k/a \leq 0$ (remember that multiplying through by a negative number, reverses the inequality.) For there to be any values of x which make this equation true, we must have k and a of opposite signs: either $k \geq 0$ and $a<0$, or $k \leq 0$ and $a>0$, just as we had guessed above by looking at the graphs in Figures 11.17 and 11.18. The case $k/a=0$ is somewhat special, and for now we shall suppose that $k/a < 0$.

There are then *two* values for x - h which make this equation true, a positive square root and a negative square root. We write

$$x - h = \pm\sqrt{-\frac{k}{a}}$$

where the symbol $\pm$ lets us write both roots with one formula: replacing it by + gives us one root and replacing it by - gives us the other root. Finally we may write the two solutions as

$$x = h \pm\sqrt{-\frac{k}{a}}$$

Notice that the two solutions are symmetric about the vertex location h: one of them is h plus $\sqrt{-\frac{k}{a}}$, and the other is h minus $\sqrt{-\frac{k}{a}}$. Since the whole graph is symmetric, clearly the positions of the zeros must be as well.

If the original quadratic function is given in the form $y = a x^2 + b x + c$, then we can use the same formula and substitute the expressions for h and k in terms of a, b and c; that is, $h = -\frac{b}{2a}$ and $k = -\frac{b^2}{4a} + c$. The two zeros are then

$$x = -\frac{b}{2a} \pm \sqrt{-\frac{-\frac{b^2}{4a} + c}{a}}$$

combining terms in the numerator

$$= -\frac{b}{2a} \pm \sqrt{-\frac{\frac{-b^2 + 4ac}{4a}}{a}}$$

simplifying

$$= -\frac{b}{2a} \pm \sqrt{-\left(\frac{-b^2 + 4ac}{4a^2}\right)}$$

$$= -\frac{b}{2a} \pm \sqrt{\frac{b^2 - 4ac}{4a^2}}$$

$$= -\frac{b}{2a} \pm \frac{\sqrt{b^2 - 4ac}}{2a}$$

taking the square root of $4a^2$

$$= \frac{-b \pm \sqrt{b^2 - 4ac}}{2a}$$

This final expression is the *quadratic formula*.

11.3   Finding the Vertex of a Quadratic Function

Why the vertex is important

Since the vertex is the point at which a parabola assumes a maximum or minimum value (if the parabola opens down or up respectively), it often assumes particular significance in a model.

Example 1: Biologists have discovered a quadratic relationship between species diversity and ocean depth.

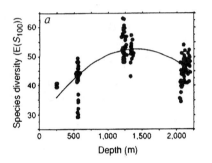

Source: "Patterns of species diversity in the deep sea as a function of sediment particle size diversity", by Ron J. Etter & Frederick Grassle, *Nature*, Vol. 360, Dec. 10, 1992. Reprinted with permission from Macmillan Magazines Limited.

Fig. 11.17 A best fit quadratic function for species diversity as a function of depth in Northern regions

The graph shows the number of different species as a function of the depth of the water in cold Northern oceans. As the depth increases, the number of species initially increases, reaches a maximum at the vertex (at a depth of approximately 1,250 meters) and then decreases. [9]

Example 2:   Urban planners and highway designers are very interested in knowing how many cars can pass along a section of roadway in a certain amount of time, and how the number of cars moving on the road might be increased by controlling traffic flow. Observations indicate that the primary controlling variable (hence a choice for the independent variable) is the density of cars on the roadway; the closer each driver sees the car ahead, the slower he or she drives. The graph in Fig. 11.18 shows a quadratic model of traffic flow rate (the number of cars per hour) as a function of the density (the number of cars per mile). It was derived from observing traffic patterns in the Lincoln Tunnel.

---

[9]   According to the authors of the article, each point represents data from a single box core with a cross section of 0.25m². Each box core costs more than $100,000 to collect and analyze!

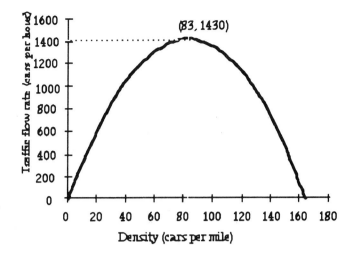

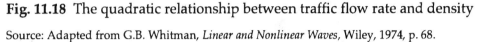

Fig. 11.18 The quadratic relationship between traffic flow rate and density

Source: Adapted from G.B. Whitman, *Linear and Nonlinear Waves*, Wiley, 1974, p. 68.

We can estimate the domain (possible values for the independent variable density) to lie between 0 and 165. The range (possible values for the dependent variable traffic flow) appears to lie between 0 and 1,430. The function increases until Density = 83 cars per mile, when the traffic flow appears to reach a maximum rate of 1430 cars per hour. This point, (83, 1430), corresponds to the vertex of the function. Note that at 83 cars per mile or equivalently 83 cars per 5280 feet, cars are spaced about 64 feet apart (since 64 = 5280/83). This spacing apparently maximizes the traffic flow, the number of cars per hour. When the density is above 83 cars per mile, the traffic flow decreases.

Finding the vertex of a parabola

The next few pages show that:

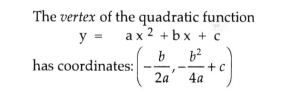

> The *vertex* of the quadratic function
> $$y = a x^2 + b x + c$$
> has coordinates: $\left(-\dfrac{b}{2a}, -\dfrac{b^2}{4a} + c \right)$

In order to do this we transform a quadratic function in the form $y = ax^2 + bx + c$ (which we'll call the "*a-b-c*" *form*) into what we call the "*a-h-k*" *form,* from which we can easily recognize the coordinates of the vertex.

Getting to the a-h-k form: shifting a parabola horizontally h units

The simplest quadratic function is the power law

$$y = a x^2. \qquad (1)$$

We have seen that the graph of this function is just a parabola whose vertex is at the origin (0,0); it opens upward if $a>0$, downward if $a<0$, and it is narrow and steeply curved if the magnitude (the absolute value) of a is large, and is wide and flat if a is close to zero. The graph is symmetric about the vertical y-axis, the line x = 0. All quadratic functions are just as simple as this one, if we write them in the right way.

Suppose we modify the function $y = a x^2$ by changing x to x - h; that is, we consider functions of the form

$$y = a (x - h)^2. \qquad (2)$$

For example, we could start with the function $y = 2 x^2$ and look at the related functions $y = 2 (x - 3)^2$ and $y = 2 (x+5)^2$.

The functional expression in (2) tells us to subtract h from x, square the result, and multiply by a. The graph of this function is the same as the graph of the original function $y = a x^2$, except that it is shifted to the right by a distance h if h is positive, or to the left if h is negative.

Compare the graphs in Fig. 11.19

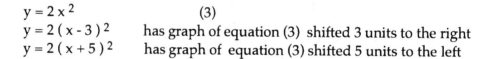

$$y = 2 x^2 \qquad\qquad (3)$$
$$y = 2 (x - 3)^2 \qquad \text{has graph of equation (3) shifted 3 units to the right}$$
$$y = 2 (x + 5)^2 \qquad \text{has graph of equation (3) shifted 5 units to the left}$$

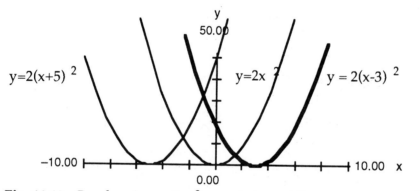

Fig. 11.19 Graphs of $y = 2 x^2$, $y = 2 (x - 3)^2$ and $y = 2 (x+5)^2$

Setting $x = h$ in the new function $y = a(x-h)^2$ gives the same value for y as setting $x = 0$ in the original function, $y = ax^2$. In general, setting $x = h + r$ in the new function gives the value for y, namely $y = a((h+r)-h)^2 = ar^2$, as setting $x = r$ in the original function. [10]

Therefore, by simply shifting our knowledge of the original function, we may say that the graph of $y = a(x-h)^2$ is a parabola whose vertex is on the x-axis at (h,0), opening up or down depending on the value of a. It is symmetric about the vertical line x = h passing through the vertex (h,0).

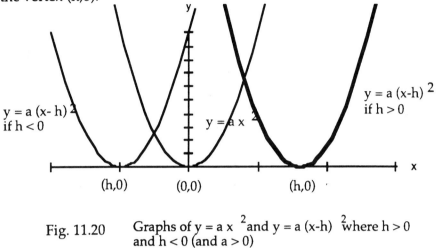

$y = a(x-h)^2$
if $h < 0$

$y = ax^2$

$y = a(x-h)^2$
if $h > 0$

(h,0)   (0,0)   (h,0)

Fig. 11.20   Graphs of $y = ax^2$ and $y = a(x-h)^2$ where h > 0 and h < 0 (and a > 0)

Algebra Aerobics:

Plot a, b, and c on the same graph. Compare the position of the vertices of (b) and (c) to (a).

1.   a) $y = x^2$ b) $y = (x+3)^2$ c) $y = (x-2)^2$

2.   a) $y = 0.5x^2$ b) $y = 0.5(x-1)^2$ c) $y = 0.5(x+4)^2$

Answers (continued on bottom of next page)

1.   b: 3 units to the left   c: 2 units to the right   2.   b: 1 unit to the right   c: 4 units to the left

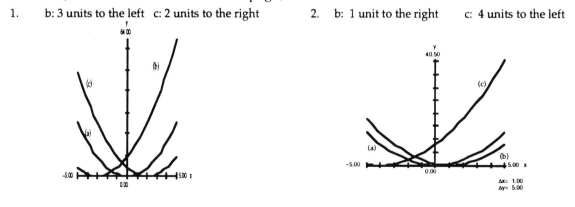

---

[10] We have used this "shifting" operation before in an example in Section 4.3, where we considered a linear model for recommended weight as a function of height. There, we let the independent variable x represent inches of height in excess of 5 feet, rather than height itself.

Something to think about

How do you think the graph of $y = 2(x-4)^3 + 7$ would compare to the graph of $y = 2x^3$? And the graph of $y = 2(x-4)^3 - 5$? Check your guesses by plotting the functions on a graphing calculator.

We now consider a function of the form:

$$y = a(x-h)^2 + k$$

containing three coefficients: a, h, and k.

The graph of this function is a parabola whose vertex is at (h,k), and whose narrowness and opening direction are given by a. It has exactly the same graph as the function $y = x^2$, except that it has been shifted by h units in the x-direction and k units in the y-direction. As shown in Fig.11.21, it is symmetric about the vertical line x = h that passes through the vertex (h,k).

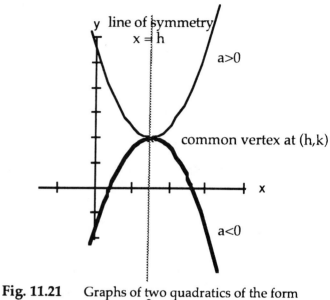

Fig. 11.21   Graphs of two quadratics of the form $y = a(x-h)^2 + k$, one where a > 0, the other where a < 0

Answers (continued from Algebra Aerobics from previous page)

1. b: 2 to the right of (a)
    c: 2 to the right and 4 up from (a)
    d: 2 to the right and 3 down from (a)

2. b: 3 to the left of (a)
    c: 3 to the left and 1 down from (a)
    d: 3 to the left and 4 up from (a)

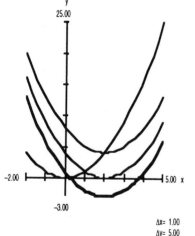

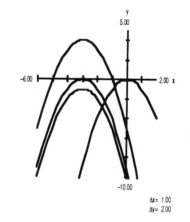

Hence

> A quadratic function in the "a-h-k" form
>
> $$y = a(x-h)^2 + k$$
>
> has a vertex at (h, k)

Algebra Aerobics:

Sketch the graph and identify the vertex of the following functions..

1. $y = \dfrac{1}{2}(x-2)^2 + 3$ 2. $y = -2(x+1)^2 + 5$ 3. $y = -0.4(x-3)^2 - 1$

Answers

1. vertex: (2,3)

2. vertex: (-1,5)

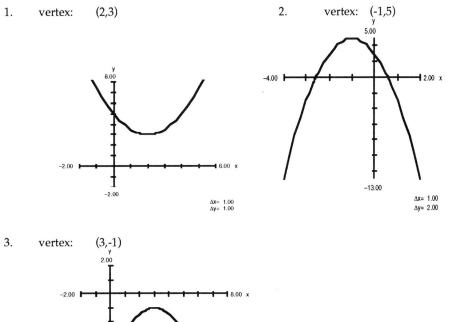

3. vertex: (3,-1)

Moving back and forth between the two forms

Getting from the a-h-k form to the a-b-c form of a quadratic

Are these functions, written in terms of h and k, the same as the quadratic functions we have been considering, with coefficients a, b, and c? It is easy to see that every function written in a-h-k form is the same as a quadratic function in a-b-c form, if we multiply it out and group terms with the same power of x:

write out the factors	$a(x-h)^2+k =$	$a(x-h)(x-h)+k$
multiply the factors	$=$	$a(x^2-2hx+h^2)+k$
multiply through by a and group the constant terms	$=$	$ax^2-2ahx+(ah^2+k).$

So if we set

$$b = \quad -2ah \qquad \text{and} \quad c = \quad ah^2+k,$$

then this expression is in the form ax^2+bx+c. That is, the "shifted parabola" expression is just a different way of writing a quadratic function. Notice that, while we are using h and k in one version and b and c in the other, the coefficient a is the same in both forms.

Getting from the a-b-c form to the a-h-k form of a quadratic: ("completing the square")

Now suppose we want to go the other way. Suppose we are given a quadratic function in the form $y = ax^2+bx+c$. Can we find values of h and k so that $y = a(x-h)^2+k$ is the same quadratic function? If we can do this, then we will have a simple way to understand any quadratic function, since parabolas in the a-h-k form are easy to understand.

Sometimes this is easy to do. Suppose the given values of a, b, and c were such that

$$c = \frac{b^2}{4a}$$

There is no reason that they should be like this, but for just a moment let us suppose that they are. Then we could simply calculate that

$$y = ax^2 + bx + \frac{b^2}{4a}$$

factoring out a
$$= a\left(x^2 + \frac{bx}{a} + \frac{b^2}{4a^2} \right)$$

factoring again
$$= a\left(x + \frac{b}{2a} \right)^2$$

This expression is in a-h-k form, with $h = -\dfrac{b}{2a}$ and $k = 0$. So if the constant c happens to have this special value $\left(\dfrac{b^2}{4a} \right)$, then we can write the quadratic equation in a-h-k form.

But if we start with any values of a, b, and c, we can use the above calculation simply by adding and subtracting the special value $\dfrac{b^2}{4a}$. Starting with any quadratic function with coefficients a, b, and c, we can write

$$y = ax^2 + bx + c.$$

adding 0 in the form of $\dfrac{b^2}{4a} - \dfrac{b^2}{4a}$
$$= ax^2 + bx + \left(\frac{b^2}{4a} - \frac{b^2}{4a} \right) + c$$

regrouping terms
$$= \left(ax^2 + bx + \frac{b^2}{4a} \right) - \frac{b^2}{4a} + c$$

$$= a\left(x + \frac{b}{2a} \right)^2 - \frac{b^2}{4a} + c$$

factoring as above gives us

This powerful strategy is called "completing the square." The final expression is in a-h-k form, with

$$h = -\frac{b}{2a} \qquad \text{and} \qquad k = -\frac{b^2}{4a} + c$$

That is, any quadratic function in a-h-k form may be converted to a quadratic function in a-b-c form, and vice versa. By doing this simple computation, we now know the location of the vertex of the graph of any quadratic function, and that the graph is symmetric about the vertical line through the vertex.

The relationship between the a-h-k and the a-b-c forms and the vertex

---

The quadratic function given a- b-c form
$$y = a x^2 + b x + c$$
may be written as a shifted parabola in a-h-k form
$$y = a (x - h)^2 + k$$
by setting $\qquad h = -\dfrac{b}{2a} \qquad$ and $\qquad k = -\dfrac{b^2}{4a} + c$

The vertex of the graph is located at coordinates (h, k) and the graph is symmetrical about the vertical line x = h passing through the vertex.

Note: The expression for h, the x coordinate of the vertex is fairly straight forward and worth memorizing. The formula for k, the y coordinate of the vertex is much more complicated. So instead of memorizing the formula, remember that given an x value you can always calculate the corresponding y value by substituting the x value into the equation.

Example: Given the quadratic height function, H = $34 + 32t - 16t^2$ (where t is in seconds and H is in feet), find the coordinates of the vertex (h, k) and rewrite the equation in the a-h-k form. What significance does the vertex have?

Here a = -16 and b = 32 and c = 34. (Remember a is the coefficient of the squared term, which in this case happens to be the third term.) Recall from Chapter 10 that 34 represents the initial height in feet, 32 represents an initial upward velocity of 32 ft/sec, and -16 ft/sec^2 is (1/2) g or (1/2) of the acceleration due to gravity.

Using the formula for h. the horizontal coordinate of the vertex:

$$h = -\frac{b}{2a} \qquad \text{So } h = -\frac{32}{2(-16)} = \frac{-32}{-32} = 1$$

We can find k, the vertical coordinate of the vertex, by evaluating H = $34 + 32t - 16t^2$ when t = 1

Setting t = 1

$$\begin{aligned} H &= 34 + 32\,(1) - 16\,(1)^2 \\ &= 34 + 32 - 16 \\ &= 50 \end{aligned}$$

So k = 50.

We would also get the same value if we used the formula for k.

$$k = -\frac{b^2}{4a} + c \qquad \text{So } k = -\frac{32^2}{4(-16)} + 34 = \frac{-1024}{-64} + 34 = 16 + 34 = 50$$

The vertex lies at the point (1, 50). The equation rewritten in the a-h-k form would be

$$H = -16\,(t-1)^2 + 50$$

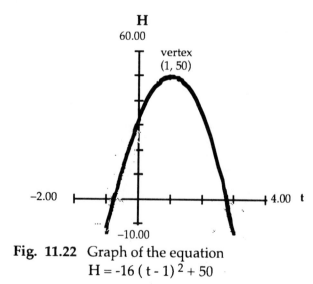

Note that only the solid part of the curve has meaning in our height model. Values of negative time or negative height are meaningless.

The vertex represents the point at which the object reaches a maximum height. So at 1 second, the object reaches a maximum height of 50 feet.

Fig. 11.22 Graph of the equation
$H = -16\,(t-1)^2 + 50$

Algebra Aerobics:

1. Express in the form $y = ax^2 + bx + c$

   a) $y = 2\left(x - \dfrac{1}{2}\right)^2 + 5$

   b) $y = -\dfrac{1}{3}(x+2)^2 + 4$

2. Express in the form $y = a(x - h)^2 + k$

   a) $y = x^2 + 6x + 7$

   b) $y = 2x^2 + 4x - 11$

3. Find the coordinates of the vertex, the vertical intercept, and graph:

   a) $y = x^2 + 8x + 11$

   b) $y = 3x^2 + 4x - 2$

4. Find the coordinates of the vertex, the vertical intercept, and graph:

   a) $y = 0.1(x + 5)^2 - 11$

   b) $y = -2(x - 1)^2 + 4$

Answers

1. a) $y = 2x^2 - 2x + \dfrac{11}{2}$ b) $y = -\dfrac{1}{3}x^2 - \dfrac{4}{3}x + \dfrac{8}{3}$

2. a) $y = (x + 3)^2 - 2$ b) $y = 2(x + 1)^2 - 13$

3. a) vertex: (-4,-5)   y-intercept:(0,11)   b) vertex:(-2/3, -10/3)   y-intercept: (0,-2)

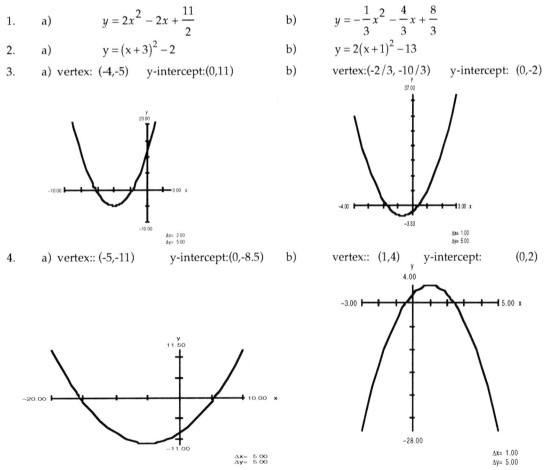

4. a) vertex:: (-5,-11)   y-intercept:(0,-8.5)   b) vertex:: (1,4)   y-intercept: (0,2)

11.4 Constructing Graphs and Equations of Quadratic Functions

---

The Characteristics of a Quadratic Function

A quadratic function of the form $y = ax^2 + bx + c$
where a, b and c are constants and $a \neq 0$:
- opens up if $a > 0$, and down if $a < 0$

- has a vertex at (h, k) where $h = -\dfrac{b}{2a}$ and $k = -\dfrac{b^2}{4a} + c$

- has a y–intercept at the point with coordinates (0,c)
- zero, one or two x-intercepts
  whose x coordinates can be found using the quadratic formula

The function can be rewritten in the a-h-k format as
$$y = a(x - h)^2 + k$$

Example 1: Sketch a graph of the quadratic function $y = x^2 - 2x - 3$

Here $a = 1$, $b = -2$ and $c = -3$.
a) Does the curve open up or down?
   The curve opens up, since a, the coefficient of x^2, is positive

b) Where is the vertex?

First find the x coordinate of the vertex using $x = -\dfrac{b}{2a}$.

$$-\frac{b}{2a} = -\frac{(-2)}{(2)(1)} = \frac{2}{2} = 1$$

Now find the y coordinate of the vertex by evaluating the function for x = 1.
$$\begin{aligned} y &= 1^2 - (2)(1) - 3 \\ &= 1 - 2 - 3 \\ &= -4 \end{aligned}$$

Hence the coordinates of the vertex are (1 , –4). We can now write the function in the a-h-k form as: $y = (x - 1)^2 - 4$ and do a rough sketch .

c) What is the y-intercept?
   From the original form of the equation we know the y-intercept is –3.

d) What are the x-intercepts (or zeros)?
   The zeros, occur when $y = 0$, that is when
$$0 = x^2 - 2x - 3$$
   The polynomial on the right hand side factors easily so we have
$$0 = (x + 1)(x - 3).$$

Applying the Zero Product Rule we have

$$(x+1) = 0 \quad \text{or} \quad (x-3) = 0$$

$$x = -1 \qquad\qquad x = 3.$$

Hence the function crosses the x-axis at –1 and at 3.

Figure 11.23 shows a sketch of the graph.

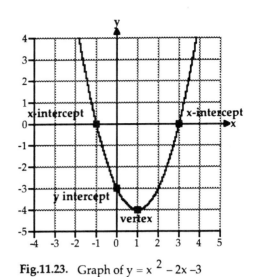

Fig.11.23. Graph of $y = x^2 - 2x - 3$

Example 2: Sketch a graph of $y = -x^2 - 6x - 10$

Here $a = -1$, $b = -6$ and $c = -10$

a) Does the curve open up or down?

The curve opens down, since a is negative.

b) Where is the vertex?

First find the x coordinate of the vertex using $x = -\dfrac{b}{2a}$.

$$-\frac{b}{2a} = -\frac{(-6)}{(2)(-1)} = \frac{6}{-2} = -3$$

Now find the y coordinate of the vertex by evaluating the function for $x = -3$.

$$y = -(-3)^2 - 6(-3) - 10 = -9 + 18 - 10 = -1$$

Hence the coordinates of the vertex are (-3, -1). The equation in the a-h-k format is:

$$y = -(x+3)^2 - 1$$

c) What is the y-intercept?

The y-intercept is – 10.

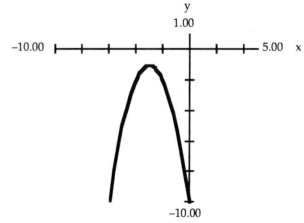

Figure 11.24 Graph of $y = -x^2 - 6x - 10$

d) What are the x-intercepts or zeros?

Since the vertex lies below the x-axis and the curve opens down, there are no x-intercepts. If we tried to evaluate the quadratic formula, we would get:

$$\frac{-b \pm \sqrt{b^2 - 4ac}}{2a} = \frac{-(-6) \pm \sqrt{(-6)^2 - 4(-1)(-10)}}{(2)(-1)}$$

$$= \frac{6 \pm \sqrt{36 - 40}}{-2}$$

$$= \frac{6 \pm \sqrt{-4}}{-2}$$

Since $\sqrt{-4}$ is not a real number, the quadratic formula produces no real roots. The parabola is sketched in Figure 11.24

Example 3: Find the equation of the parabola in Figure 11.25

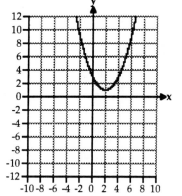

Fig. 11.25 A mystery parabola

The vertex of the graph appears to be at (2, 1) and the graph opens upward, so the equation must be of the form
$$y = a(x-2)^2 + 1 \qquad (1)$$
where a is positive. How can we find a value for a? If we can identify values for any other point (x, y) that lies on the parabola, we can substitute these values into (1) to find a. The y-intercept, (0, 3), is a convenient point to pick. So setting $x = 0$ and $y = 3$ we get:
$$3 = a(0-2)^2 + 1$$
$$3 = 4a + 1$$
$$2 = 4a$$
$$a = 1/2$$

So the equation in the a-h-k form is: $\qquad y = (1/2)(x-2)^2 - 1$

If we wanted it in the equivalent a-b-c form, we could multiply out and collect like terms to get: $y = (1/2)x^2 - 2x + 3$

Algebra Aerobics: (answers on bottom of next page)

1. Sketch the graph, and identify the vertex and the intercepts of the following.

   a) $y = x^2 + 3x + 2$
   b) $y = -\frac{1}{2}(x-2)^2 + \frac{7}{2}$
   c) $y = x^2 + 3$

2. Write an equation for the following parabola.

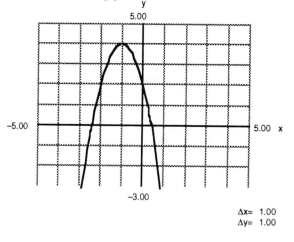

Δx= 1.00
Δy= 1.00

Using the discriminant to predict the number of zeros

If you know the vertex and which way the parabola opens, you can predict the number of zeros a quadratic function will have. You can also use the part of the quadratic formula under the radical, $b^2 - 4ac$, called the discriminant, to predict the number of zeros. It is called the discriminant because it allows us to discriminate whether a quadratic function has two, one, or no zeros. You don't need to memorize the following cases, since they will fall out naturally when you evaluate the quadratic formula. But you should recognize each of the three possibilities. If $b^2 - 4ac$ is negative, then $\sqrt{b^2 - 4ac}$ has no real value, so there will be no real zero. If $b^2 - 4ac$ is zero, then the quadratic formula will yield only one zero, namely $x = \dfrac{-b}{2a}$. If $b^2 - 4ac$ is positive, then the quadratic formula will determine two zeros.

Answers (for "Algebra Aerobics" from previous page)

1. a) vertex: (-3/2, -1/4); y-intercept: (0,2)   b) vertex: (2,7/2) y-intercept: (0,3/2)
        x-intercepts: (-2,0) and (-1,0)   x-intercepts:(-0.65,0) and (4.65,0)

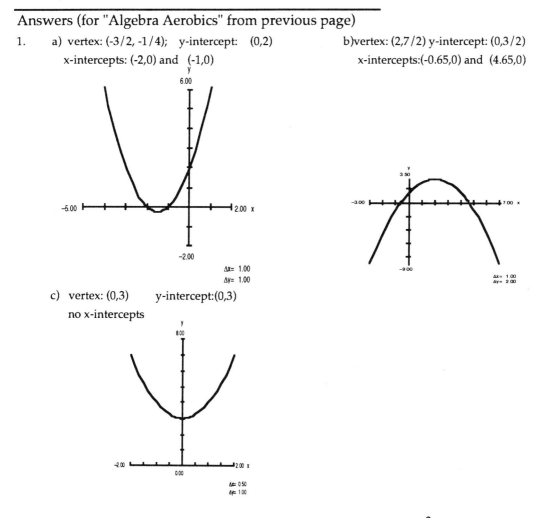

   c) vertex: (0,3)   y-intercept:(0,3)
       no x-intercepts

2.   vertex: (-1,4), y-intercept: (0,2) So the function is of the form $y = a(x- (-1))^2 + 4$
      Substituting in (0,2), we get $2 = a(0+1)^2 + 4$ $\Rightarrow$ $-2 = a$ So we get: $y = -2(x + 1)^2 + 4$

For any quadratic function of the form $y = ax^2 + bx + c$,
the expression b^2-4ac is called the *discriminant*, which
can be used to determine the number of zeros.

If $b^2 - 4ac > 0$, the function has two real zeros.
If $b^2 - 4ac = 0$, the function has one real zero.
If $b^2 - 4ac < 0$, the function has no real zeros.

Example 1: Let $y = x^2 + 3.4x + 2.1$. Predict the number of zeros.

Solution: Here $a = 1.0$, $b = 3.4$, and $c = 2.1$. So the discriminant:
$$b^2 - 4ac = 3.4^2 - 4(1)(2.1)$$
$$= 11.56 - 8.4$$
$$= 3.16$$

Since the discriminant, 3.16, is positive, the function has two distinct zeros:

$$x = \frac{-3.4 + \sqrt{3.16}}{2(1.0)} \quad \text{and} \quad x = \frac{-3.4 - \sqrt{3.16}}{2(1.0)}$$
$$\approx \frac{-3.4 + 1.78}{2(1.0)} \qquad\qquad \approx \frac{-3.4 - 1.78}{2(1.0)}$$
$$\approx \frac{-1.62}{2.0} \qquad\qquad\qquad \approx \frac{-5.18}{2.0}$$
$$\approx -0.81 \qquad\qquad\qquad \approx -2.59$$

So the function crosses the x-axis at approximately $(-0.81, 0)$ and $(-2.59, 0)$.

Example 2: Let $y = 2.89x^2 + 3.40x + 1.00$. Predict the number of times the function crosses
the x-axis.

Solution: Here $a = 2.89$, $b = 3.40$, $c = 1.00$. Evaluating the discriminant gives us:
$$b^2 - 4ac = 3.40^2 - 4(2.89)(1.00)$$
$$= 11.56 - 11.56$$
$$= 0$$

Since the discriminant is 0, the function has only a single real zero which occurs at:
$$x = (-3.4 + \sqrt{0}) / 2(2.89)$$
$$= -3.4 / 5.78$$
$$\approx -0.41$$

The graph touches the x-axis once. at the vertex $(-0.41, 0)$.

Example 3: Let $y = 4.6x^2 + 3.4x + 2.1$. How many times does the function touch the x-axis?

Solution: $a = 4.6$, $b = 3.4$, and $c = 2.1$. The discriminant is::
$$
\begin{aligned}
b^2 - 4ac &= 3.4^2 - 4(4.6)(2.1) \\
&= 11.56 - 38.64 \\
&= -27.08
\end{aligned}
$$

The discriminant is -27.08, a negative number. Since $\sqrt{-27.08}$ has no real value, the quadratic formula fails to yield a real zero. Therefore the function never touches the x-axis.

Algebra Aerobics: [10]

Use the discriminant to determine the number of x-intercepts for each of the following parabolas. Describe the graph.

1. $y = 3x^2 + 2x + 11$
2. $y = -x^2 + 8x - 16$
3. $y = \dfrac{1}{2}x^2 - 3x + 2$

---

[10] **Answers**

1. $d = 4 - 4(3)(11) = -128 < 0$ $\Rightarrow$ no x-intercepts
   The graph is narrow and opens up.
2. $d = 64 - 4(-1)(-16) = 0$ $\Rightarrow$ one x-intercept
   The graph opens down.
3. $d = 9 - 4(1/2)(2) = 5 > 0$ $\Rightarrow$ two x-intercepts
   The graph opens up and is broad.

11.5 The Average Rate of Change of Quadratic Functions (optional)

In Chapter 10 we calculated the average rate of change of distance with respect to time for a freely falling object. In this context, the average rate of change represented an average velocity. When we graphed these velocities as a function of time, the graph was linear. So while the velocity was not constant, it was changing at a constant rate. Is the average rate of change of a quadratic function is always linear, or is this a peculiarity of quadratic models for falling objects?

Let's examine another quadratic, $y = 3x^2 - 8x - 23$. Table 11.6 shows values for selected x's and corresponding y's in columns 1 and 2. and Figure 11.26 (a) shows the parabolic graph of the data. Column 3 contains the average rates of change of y with respect to x (between adjacent points). Now a linear relationship clearly appears– not between x and y, but between x and the average rate of change of y with respect to x. As the value of x increases by 1, the value for the average rate of change increases by 6. So the average rate of change, while not constant, is increasing at a constant rate of 6. Figure 11.26 (b) shows the linear graph (with a slope of 6) of the average rate of change as a function of x.

Table 11.6		
x	y	Average rate of change
−10	357.00	n.a.
−9	292.00	−65
−8	233.00	−59
−7	180.00	−53
−6	133.00	−47
−5	92.00	−41
−4	57.00	−35
−3	28.00	−29
−2	5.00	−23
−1	−12.00	−17
0	−23.00	−11
1	−28.00	−5
2	−27.00	1
3	−20.00	7
4	−7.00	13
5	12.00	19
6	37.00	25
7	68.00	31
8	105.00	37
9	148.00	43
10	197.00	49

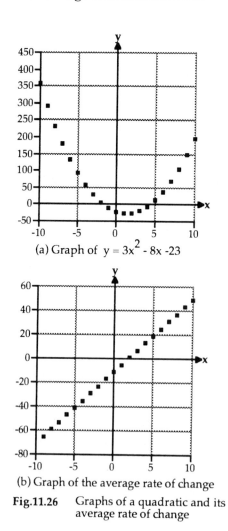

(a) Graph of $y = 3x^2 - 8x - 23$

(b) Graph of the average rate of change

Fig.11.26   Graphs of a quadratic and its average rate of change

We have seen numerically in two examples that the average rate of change of a quadratic function is a linear function. We now demonstrate algebraically that this is true for *every* quadratic function.

Suppose that y is a quadratic function of x, written using functional notation as $y = f(x) = a x^2 + b x + c$. In the previous examples we fixed an interval size over which to calculate the average rate of change: $1/60$ of a second for the free fall experiment and 1 in Table 11.6. Now we pick a constant interval size r, and for each position x, compute the average rate of change of y (or f(x)) between x and $x + r$.

It is simple to compute $f(x + r)$:

$$f(x+r) = a(x+r)^2 + b(x+r) + c$$

$$= a(x^2 + 2rx + r^2) + bx + br + c$$

$$= ax^2 + bx + c + (2ax + b)r + ar^2. \quad \text{(regrouping terms)}$$

Then the average rate of change of f(x) between x and $x + r$ is:

$$\frac{\text{change in } f(x)}{\text{change in } x} = \frac{f(x+r) - f(x)}{(x+r) - r}$$

$$= \frac{(ax^2 + bx + c + (2ax+b)r + ar^2) - (ax^2 + bx + c)}{r}$$

$$= \frac{(2ax+b)r + ar^2}{r}$$

$$= 2ax + (b+ar). \qquad (1)$$

This is a linear function of x, with

$$\text{slope} = 2a \quad \text{y-intercept} = b + ar.$$

Note that the slope depends only on the original equation (in particular only on the value of *a*.) The y-intercept depends not only on *a* and *b* in the original equation, but on *r*, the interval size over which we calculate the average rate of change. For example, the quadratic function in the example above, $y = 3x^2 - 8x - 23$, has $a = 3$, $b = -8$, and $c = -23$. Table 11.6 was generated using an interval size of $r = 1$. Substituting these values, we find that the average rate of change from x to x+1 is given by $6x + (-8 + 3) = 6x - 5$. For example, when x = 2, the average rate of change from 2 to 3 is $6(2) - 5 = 7$. (This value is recorded in Table 11.6 in the line with x = 3, since column 3 contains the average rate of change between the preceding value of x and the current one.)

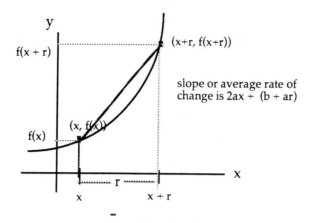

Fig.11.27   The slope of the line segment connecting two points
on the parabola separated by a horizontal distance
of r, is $2ax + (b + ar)$

What if we had chosen a different value for r? What if we had calculated the average rate of
change for interval sizes of $1/3$ instead of 1? Then using equation (1) the average rate of
change would be $2(3)x + (-8 + (3)(1/3)) = 6x - 7$. Again the slope is 6, but now the y-
intercept is -7.

If we took smaller and smaller values for r, then the term ar would get closer and closer to 0,
and hence the y-intercept $b + ar$ would get closer and closer to b. For very small r's the
average rate of change would get closer and closer to the linear expression $2ax + b$.

Given a quadratic function

$$y = ax^2 + bx + c \quad \text{(where a, b and c are constants and } a \neq 0 \text{),}$$

the relationship between x and the average rate of change of y with respect to x is linear.

The average rate of change over very small intervals approaches $2ax + b$

Chapter 11 Summary

A *quadratic function* or *polynomial function of degree two;*, has the form $y = ax^2 + bx + c$
where a, b and c are constants and $a \neq 0$. The graph of a quadratic function is a *parabola.*
and has the following properties:

- is in the shape of a symmetric "bowl"

- opens up if a > 0, and down if a < 0

- as $|a|$ (the absolute value or magnitude of *a*) increases, the parabola becomes narrower.

- has a vertex at (h, k) where $h = -\dfrac{b}{2a}$

  and $k = -\dfrac{b^2}{4a} + c$

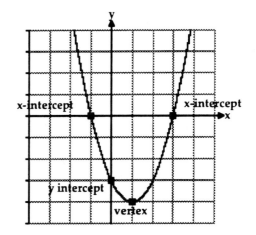

Fig.11.28. Graph of a quadratic with two x-intercepts

- has a *y-intercept;* at the point with coordinates (0,c)

- has zero, one or two *x-intercepts* whose x coordinates (called *zeros*) can be found using the quadratic formula

The Quadratic Formula says that the *zeros* of any quadratic function of the form
$y = ax^2 + bx + c$ are given by:

$$x = \frac{-b \pm \sqrt{b^2 - 4ac}}{2a}$$

The expression $b^2{-}4ac$ under the radical sign is called the *discriminant,* and can be used to predict the number of zeros. If $b^2 - 4ac > 0$, the function has two real zeros; if $b^2 - 4ac = 0$, the function has one real zero; If $b^2 - 4ac < 0$, the function has no real zeros.

Any quadratic function can be rewritten in the a-h-k format as
$$y = a(x - h)^2 + k$$
where (h,k) is the vertex.

If y is of the form $y = ax^2 + bx + c$, then the relationship between x and the average rate of change of y with respect to x is linear. The average rate of change over very small intervals approaches 2ax + b.

EXERCISES

1. On the same graph, sketch the plots of the following functions and label each with its equation.

$$y = 2x^2$$
$$y = -2x^2$$
$$y = -2x^2 + 3$$
$$y = -2x^2 - 3$$

2. For each part do a rough sketch of a graph of any function of the type:

$$y = ax^2 + bx + c$$

    a) where $a > 0$, $c > 0$, and the function has no real zeros

    b) where $a < 0$, $c > 0$, and the function has 2 real zeros

    c) where $a > 0$ and the function has one real zero

3. Write each of the following quadratic equations in function form (i.e., solve for y in terms of x). Then find the vertex and the zeros using any method. Finally, using these points, do a rough hand sketch of the quadratic function.

    a) $\qquad\qquad y + 12 = x(x + 1)$
    b) $\qquad 2x^2 + 6x + 14.4 - 2y = 0$
    c) $\qquad\quad y + x^2 - 5x = -6.25$
    d) $\qquad\qquad y - 8x = x^2 + 15$
    e) $\qquad\qquad y + 1 = (x-2)(x+5)$

4.   Match the following graphs with an equation. Explain your reasoning for each of your choices. Each of the grid lines is 2 units apart.

_____ $f(x) =\ 2x^2 - 8x - 2$
_____ $g(x) =\ 2x^2 - 8x + 3$
_____ $h(x) =\ 0.5x^2 - 2x + 3$
_____ $i(x) =\ -2x^2 - x + 2$
_____ $k(x) =\ 2x - 5$
_____ $j(x) =\ -0.5x^2 - 2x + 3$

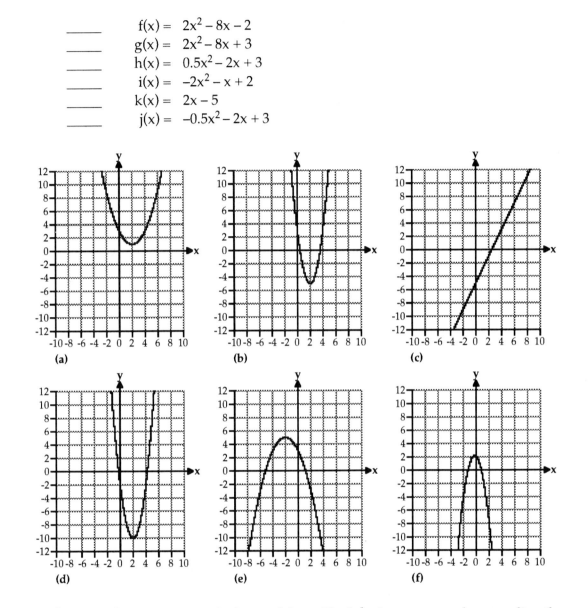

5.   These quadratics are given in factored form. Find the intercepts and vertex directly, without multiplying out and applying the quadratic formula. Sketch the graph.

a) $y = (x+2)(x+1)$
b) $y = 3(1-2x)(x+3)$

6.   Construct two quadratic functions that each have a y-intercept of 10 and a graph that turns down. Do a rough sketch for each function and identify how many zeros each function has.

7.   Construct two quadratic functions that each have a y-intercept of –3 and a graph that turns down. Do a rough sketch for each function and identify how many zeros each function has.

8.   Solve the following quadratic equations by using the quadratic formula.
      a) $6 t^2 - 7t = 5$
      b) $3x(3x - 4) = -4$
      c) $(z + 1)(3z - 2) = 2z + 7$
      d) $(x + 2)(x + 4) = 1$

9.   Parabolic mirrors have the useful property that there exists a special point called the focus lying on the axis of symmetry inside the parabolic curve. When any ray of light, originating from the focus, strikes the mirrored surface, it is reflected parallel to the axis of symmetry. Conversely, any rays coming into the mirror parallel to the axis of symmetry are reflected to the focus. Open up the Quadratic Graphing package and click on Parabolic Reflector for a demonstration. These properties are used in parabolic mirrors for telescopes and in parabolic radio antennae. We know already that the coordinates of the vertex are $(-b/2a, -b^2/4a + c)$. The focus is $1/(4a)$ units above the vertex (if the graph opens up) or below the vertex (if the graph opens down). Find an algebraic expression in terms of a, b, and c for the coordinates of the focus.

10.   A certain baseball, hit straight up in the air, is at height

$$y = 4 + 50t - 16t^2$$

      feet above the ground level at time, t, seconds after being hit. This formula is valid for $t > 0$, until the ball hits the ground.

      a)   How high is it at $t = 0$?
      b)   When does it hit the ground?
      c)   When, if ever, is it 30 feet high? 90 feet high?
      d)   What is the maximum height that it reaches? When does it reach that height?

11.   As we have seen, the distance that a freely falling object falls can be modeled by the quadratic function $d = 16t^2$ where t is measured in seconds, and d in feet. There is a closely related function $v = 32t$ which gives the velocity, v, in feet per second at time, t, for the same freely falling body.

      a)   Fill in the missing values in the following table.

Time (seconds)	Distance (feet)	Velocity (feet/sec)
1		
1.5		
2		
		80
	56	

b) Sketch of both functions, distance vs. time and velocity vs. time on two different graphs but on the same page. Label each of the above points on the curves.

12. (Requires graphing calculator or computer)

The following data show average growth of the human embryo prior to birth.

embryo age weeks	weight grams	length centimeter
8	3	2.5
12	36	9
20	330	25
28	1000	35
36	2400	45
40	3200	50

Source: Reprinted with permission from Prentice Hall, Kimber et al., *Textbook of Anatomy and Physiology*, copyright 1955, "Embryo Age, Weight and Height.", p. 785.

a) Plot the age vs. weight data and find the closest quadratic model that you can to approximate the data.

b) According to your model what would a 32 week embryo weigh?

c) Comment on the range for which your formula is reliable.

d) Plot the length vs. age data then make a mathematical model for the length vs. age of an embryo from 20 to 40 weeks.

e) Using your model compute the age at which an embryo would be 42.5 cm. long.

13. Open up the FAM1000 course software file for linear regression and examine mean personal wages vs. age for all men, and then for all women. Note the linear regression formulae for these two cases.

a) Given the shape of the data, it is also reasonable to consider quadratic models for wages in thousands of dollars vs. age for women and men:

$$\text{women wage} = -0.026 \, (\text{age})^2 + 2.26 \, (\text{age}) - 28$$
$$\text{men wage} = -0.042 \, (\text{age})^2 + 3.91 \, (\text{age}) - 52$$

At what age does the quadratic model for women predict the same wages as the linear regression model? Calculate your answer and sketch a graph showing the answer.

b) Is there an age at which both quadratic models predict the same income for men and women? If so, what is it?

c) Comment on the reliability zones of the linear and the quadratic models in this case.

14. (Requires computer with course software.)
    Practicing skills with quadratics in the form $y = ax^2 + bx + c$

    i) Open "Q7: Find coefficients a,b,c" in *Quadratic Graphs*". The software will pick a random quadratic function and display its graph. You predict the corresponding value of a, b, and c, graph your predicted quadratic, and see if your guess is correct. Repeat using integer and real values, until you are consistently right.

    ii) Open "Q6: Find (x,y) coordinates: abc" in *Quadratic Graphs*". The software will display a quadratic equation and ask you to click on any three points that lie on the graph. By clicking on "graph curve." the computer will graph the given equation and the quadratic to see if they are the same. Continue the process until you are 100% accurate.

    Practicing skills with quadratics in the form $y = a(x - h)^2 + k$

    iii) Open "Q15: Find (x,y) coordinates: ahk". The software will display a quadratic equation and ask you to click on any three points that lie on the graph. By clicking on "graph curve." the computer will graph the given equation and the quadratic to see if they are the same. Continue the process until you are 100% accurate.

15. We dealt previously with systems of lines and ways to determine the coordinates of points where lines intersect. Once you know the quadratic formula, it's possible to determine where a line and a parabola, or two parabolas, intersect. As with two straight lines, at the point where the graphs of two functions intersect (*if* they intersect) they both have the same x value and the same y value.

    a) Find the intersection of the parabola $y = 2.0x^2 - 3.0x + 5.1$ and
       the line $y = -4.3x + 10.0$.
    b) Plot both functions, labelling any intersection point(s)

16. Complete the table below for the function $y = 3 - x - x^2$

x	y	average rate of change $= \dfrac{\Delta y}{\Delta x}$	rate of change of the average rate of change
-3	-3	N/A	N/A
-2	1	$(1 - -3)/(-2 - -3) = 4/1 = 4$	N/A
-1	3	$(3 - 1)/(-1 - -2) = 2/1 = 2$	$(2 - 4)/1 = -2$
0	3	$(3-3)/(0--1) = 0/1 = 0$	$(0 - 2)/1 =$
1	1		
2			
3			

Graph the average rate of change on the vertical axis, x on the horizontal axis.
What type of function does the graph represent? What is its slope (the rate of change of the average rate of change)?

EXPLORATION 11.1
Properties of Quadratic Functions

Objective
- Part I: to explore the effects of a, b and c on the graph of quadratic equations in the a-b-c form: $y = ax^2 + bx + c$ (where a, b, c are real numbers and $a \neq 0$)
- Part II: to explore the effects of a, k and k on the graph of quadratic equations in the h-k form: $y = a(x - h)^2 + k$ (where a, h, k are real numbers and $a \neq 0$)

Materials/Equipment
- graphing calculator or several programs in *Quadratic Graphs* from the course software
- graph paper

Procedure

Part I: Exploring quadratics in the form $y = ax^2 + bx + c$

We start by choosing values for the coefficients a, b and c and graphing the resulting equations by hand. From these graphs we make predictions about the effects of a, b and c on the graphs of the equations. To check your predictions, you can use a graphing calculator or if you have access to a computer, you can use one of the quadratic programs. Be sure to record your predictions and observations, so you can share them with the class. Work in pairs and discuss your findings with your partner.

1. *Making Predictions*
    Start with the simplest case, where $b = 0$ and $c = 0$. The equation will have the form:
    $$y = ax^2$$

   a) Make a data table and sketch the graph of $y = x^2$ using both positive and negative values for x. (Note that $a = 1$ in this example.) On the same graph sketch the graph of $y = 4x^2$ (here $a = 4$) Make a prediction of what the graph of $y = 2x^2$ will look like. Check your work and predictions with your partner.

   b) How would you describe your graphs? Are they symmetrical? When are the functions increasing? Decreasing? Do they have a maximum or a minimum value? What is this value? Which graph shows y changing the fastest compared to x? How do two graphs compare if one has a larger value for a?

   c) Make a data table, and sketch the graph of $y = -x^2$. Make predictions for what the graphs $y = -4x^2$ and $y = -2x^2$ will look like. Check your predictions with your partner.

   d) How would you describe these graphs? Are they symmetrical? When are these functions increasing? Decreasing? Do they have a maximum or a minimum value?

What is this value? Which graph shows y changing the fastest compared to x ? How do two graphs compare if one has a larger absolute value for *a*?

2.   *Testing Your Predictions*

Use a graphing calculator or "Q1: Quadratic graphs with sliders" in the course software package *Quadratic Graphs* to check your predictions.

a)   What effect does *a* have?
     Make predictions based on the graphs you constructed by hand.

   - If using a graphing calculator, graph several equations where b=0 , c=0, and the value of *a* is different for each equation. Do your predictions about the effect of *a* hold true? Now keep $b = 0$, but c≠0., and vary *a*. What is the effect? Do your predictions about the effect of *a* still hold true if both $b \neq 0$ and $c \neq 0$?

   - If using software, *Quadratic Graphs,* first explore what happens when you set $b = 0$ and $c = 0$, and vary *a*. Dragging the line in the *a* column changes the value of *a*, which changes your equation and the corresponding graph on the screen. Did your predictions about *a* hold true? Now keep $b = 0$, but set c≠0., and vary *a*. What is the effect? Do your predictions about the effect of *a* still hold true if both $b \neq 0$ and $c \neq 0$?

Summarize how changing *a* changes the graph? What happens to the graph if you change the sign of *a* ? What happens if you keep the sign of *a* but increase the size of *a*? Is there any point on the graph that doesn't change as you increase or decrease the size of *a* ?

b)   What effect does *c* have?
     Using the method described above, explore how the coefficient *c* affects the graph.

c)   What effect does *b* have?

3.   *Class discussion*
     Summarize in your own words your conclusions. Does the rest of the class agree?

Exploration Linked Homework

1. Write a 60 second summary of your conclusions.
2. Open "Q7: Find coefficients a,b,c" in *Quadratic Graphs"*. The software will pick a random quadratic function and display its graph. You predict the corresponding value of *a, b,* and *c* using the sliders, graph your predicted quadratic, and see if your guess is correct. Repeat using integer and real values, until you are consistently right.

Part II: Exploring quadratics in the form $y = a(x - h)^2 + k$

> Work with a partner, recording your results as you go. Again you'll start graphing by hand and then use the computer to confirm your predictions.

1. *Making Predictions*

   a) Make a data table and sketch of the graph $y = 2x^2$ using both positive and negative values for x. On the same graph sketch the graph $y = 2(x - 1)^2$ and $y = 2(x + 1)^2$. Make a prediction of what the graphs $y = 2(x - 3)^2$ and $y = 2(x + 3)^2$ will look like. Check your work with your partner.

   b) What effect does replacing x with $(x - 1)$ have? What has happened to the vertex? The shape of the curve? What if you replace x with $(x + 1)$? With $(x - 3)$? With $(x + 3)$? In general how does the graph of $y = 2x^2$ compare with the graph of $y = 2(x - h)^2$ if h is positive? If h is negative? Do you and your partner agree?

   c) Using what you know from Part I about the effect of adding a constant term to a quadratic, on a new graph sketch $y = 2x^2$ and $y = 2x^2 + 4$ and $y = 2x^2 - 5$. In general how does the graph of $y = 2x^2$ compare with the graph of $y = 2(x)^2 + k$ if k is positive? If k is negative? Check your predictions with your partner.

   d) Now (without generating a data table) do a rough sketch of the graphs of the quadratic functions $y = 2(x - 1)^2 + 3$ and $y = 2(x + 1)^2 - 5$. Do you and your partner agree?

2. *Testing Your Predictions*

   Use a graphing calculator or the program "Q14: Quadratic ahk Form" in the course software package *Quadratic Graphs* to test your predictions.

   a) What effect does *a* have?
      Choose $h = 0$ and $k = 0$. What is the form of the function? What is the effect of varying *a*? How does the role of *a* here compare to the role of *a* in equations of the form $y = ax^2 + bx + c$

   b) What effect does *h* have?
      Make a prediction based on the graphs you constructed by hand. Choose a non-zero value for *a*, $k = 0$ and vary *h*.

      - If using a graphing calculator, compare your graphs to the graph of the equation when $h = 0$.

        Let $y_1 = ax^2$ $k = 0$ and $h = 0$
        Use y_2, y_3, y_4, and so on for functions when $k = 0$ and $h \neq 0$

- If using the software program, "Q14: Quadratic ahk Form", you'll get two parabolas on the screen: the red one is the graph of $y = ax^2$ and the white one is the graph of $y = a(x - h)^2$. What is the effect of varying h?

c) What effect does k have?

Make a prediction. Then choose a non-zero value for a, $h = 0$ and vary k.
Use method described above in (b) but this time vary k.

d) What effect do h and k together have?

What do you think will be the effect of changing both of them?

- If using a graphing calculator, choose a non-zero value for a. Now vary both h and k. Compare your graphs to the graph of the equation when $h = 0$ and $k=0$.

Let $y_1 = ax^2$	$k = 0$ and $h=0$
Use y_2, y_3, y_4, and so on for functions when	$k \neq 0$ and $h \neq 0$

- If using the software program, "Q14: Quadratic ahk Form", choose a non-zero value for a. Now vary both h and k. The screen may look rather confusing. It will display four graphs:

red:	$y = y = ax^2$
green:	$y = a(x - h)^2$
blue:	$y = a(x)^2 + k$
white:	$y = a(x - h)^2 + k$

Focus first on comparing the red and the white graphs. Can you predict the effects of varying a, h and k? Now, can you describe the relationship among the four graphs?

3.   *Class discussion*

Summarize your conclusions in your own words. Does the rest of the class agree?

Exploration Linked Homework

Write a 60 second summary of your conclusions.

EXPLORATION 11.2.
Sketching Quadratic Functions

Objective
- to find vertices, x and y-intercepts of several quadratic functions, and sketch their graphs

Materials/Equipment
- graphing calculator or function graphing program with root finding capabilities
- graph paper

Procedure

1. Find the vertex and y-intercept and then sketch by hand a graph of the following functions:

$$y_1 = 2x^2 - 3x - 20 \qquad\qquad y_3 = 3x^2 + 6x + 3$$

$$y_2 = -2(x-1)^2 - 3 \qquad\qquad y_4 = -(2x+4)(x-3)$$

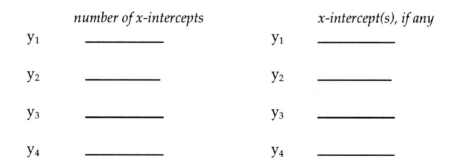

	vertex		*y-intercept*
y_1	—————	y_1	—————
y_2	—————	y_2	—————
y_3	—————	y_3	—————
y_4	—————	y_4	—————

2. Describe how many x-intercepts each function has, then calculate the coordinates of any x-intercepts.

	number of x-intercepts		*x-intercept(s), if any*
y_1	—————	y_1	—————
y_2	—————	y_2	—————
y_3	—————	y_3	—————
y_4	—————	y_4	—————

3. If you have access to a graphing calculator or function graphing program, use it to graph your equations. Enter y_1 and set a domain and range of an appropriate size to display the vertex any x-intercepts clearly. You might start by setting $-10 < x < 10$, and varying the windows size as needed. Does the graph agree with yours? Check the location of the vertex and y-intercept. Now use the calculator's root finder or trace

function to find estimates for any x-intercepts. Remember computers and graphing calculators give answers rounded to several decimal places, which are not necessarily exact answers. Record your estimates of the x-intercepts using technology.

estimated value of any x-intercepts

y_1 _____

y_2 _____

y_3 _____

y_4 _____

4.   Now write each function in both the a-b-c and a-h-k forms. In which form is it easier to find the intercepts? In which the vertex?

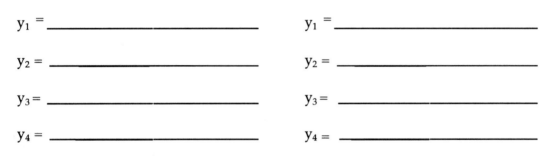

a-b-c form	*a-h-k form*
$y_1 =$ _____	$y_1 =$ _____
$y_2 =$ _____	$y_2 =$ _____
$y_3 =$ _____	$y_3 =$ _____
$y_4 =$ _____	$y_4 =$ _____

EXPLORATION 11. 3.
Parabolic Bounce

Objective
- to generate a parabolic bounce image for which a quadratic formula typical of ballistic formulas can be found.

Materials/Equipment
- manual control camera with 'bulb' shutter feature & tripod
- for same class processing: Polaroid Polagraph high contrast instant develop slide film, 12 or 36 exposure depending on class size, with 35 mm Auto Processor. For next class processing: any high contrast black and white slide film such as Kodalith
- room which can be completely darkened
- white or black ball which bounces well, color to contrast with wall of room (an ordinary 2.5 inch diameter red rubber ball can be darkened with a black magic marker, or a white ping pong ball works well)
- tall stepladder
- strobe light (this can be the sort used in dance halls, does not have to be scientific quality)
- slide projector
- roll of white tracing paper, 24" or 30" wide (sold in stores with drafting supplies)
- yard or meter sticks, T-squares or drafting triangles, drafting tape

Related Software:
"Q1: Quadratic graph with sliders" and "Q2: Find coefficients a,b,c" in *Quadratic Graphs*

Procedure
Before class:
Before class, have equipment set up. Set up ladder close to blank wall. Practice dropping ball so that it moves slightly sideways as it bounces. Set up and focus camera so that the general bounce region is visible through the lens. Place strobe light so it illuminates bouncing ball from below, but does not obstruct view of bounce. Turn on the strobe, turn out the lights, and watch through the lens as someone does some practice drop bounces. It is a good idea to take some shots the week before this is done in class, to be certain that the camera settings provide legible slides when developed.

In class:
Students work in groups of 2 or 3. Each group practices a sideways moving drop before taking a picture.

1. *Taking the picture*

   Student is poised on ladder ready to drop ball at height of about 8 feet, lights turned off, strobe turned on. Camera operator sets shutter open, says "Go", student drops ball, camera operator closes shutter after 1 or 2 bounces are complete. Each team keeps track of which number shots are theirs, or a numbered card can be placed in view near strobe to identify whose it is. Take a couple of shots per team if there is sufficient film. Have the film developed, or develop it with the Polaroid instant development box.

2.   *Using the slides*
     Project each slide on a sheet of tracing paper taped to a light wall. Adjust the projection until the bounce is about full size. Students draw around a circle around each ball image for one or two complete bounces.

Analysis of Results

1.   *Finding a formula for the bounce*

     Draw a spot in the middle of each ball image. Carefully draw in a smooth curve through the ball center spots. Because the curve is drawn on transparent paper it can be folded along the axis of symmetry of the curve so that both sides of the curve can be seen to lie on top of each other. The resulting fold line is the orientation of the y direction, and the perpendicular direction is the x orientation. Draw in an x and a y-axis: the x-axis should be about where the floor line is judged to be, and the y-axis could be the fold line, or a vertical line to the right or the left of the curve, going through the curve at some point. It is instructive if different teams choose different places for the y-axis.

     Fit a quadratic formula to the curve by finding the parameters a, b, c.
     * estimating a: find the vertex, measure horizontally 1 unit then vertically down,(negative a), or up, (positive a), to the curve. The vertical distance to the curve is close to a, depending on how accurately the curve is drawn.
     * estimating b: b is approximately the slope of the tangent line where the curve passes through the y-axis.
     * finding c: this is the y value where the curve crosses the y-axis.

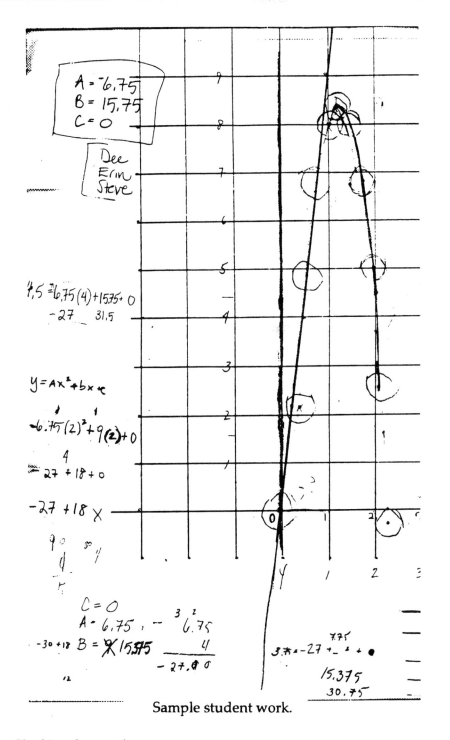

Sample student work.

2. *Checking the equation*

Pick a random point on the curve. Substitute its x value in the equation and see if the theoretical y value produced by the equation is equal to the actual y value for this point. If it is not quite right, then tinker with *a* and *b* until a closer value for y is found. Then check the formula with another random point on the curve. A final check can be done with the "Q9: Quadratic Curve Fitting" program in *Quadratic Graphs* or with a function graphing program.

EXPLORATION 11.4
Getting to the Root of Things

Objective
- to study the profiles of several quadratic functions that describe a physical phenomenon.

Materials/Equipment
- graphing calculator or function graphing program(optional)
- enclosed Function Profile Worksheets

Related software
- "Q12: Parabolic Reflector" in *Quadratic Graphs*

Procedure
Work in pairs and discuss your findings. Use the quadratic functions on the enclosed worksheets and for each function: Record your observations in your notebook.

1.   Describe the mathematical properties
         Does the parabola open up or down?
         What are the values of the coordinates of the y-intercept?
         What are the values of the coordinates of vertex

2.   Make a sketch of the graph by hand.

3.   Find the values for the coordinates of the x-intercepts by hand calculations, using the quadratic formula or other methods as appropriate.

4.   (Optional) Use a graphing calculator or function graphing program to plot the function and find estimates of the intercepts. Check these results with your hand calculations and graph.

Analysis of Results
Interpret your findings in terms of the physical phenomenon represented by each function.

FUNCTION PROFILE #1 $h = -4.84 \, t^2 + 381.00$

This function describes the height h from the ground at time t for an object dropped from the top of the Empire State Building. h gives the height in meters and t measures the time in seconds.

MATHEMATICAL PROPERTIES

Opens up or down?_____

Coordinates of h-intercept _____

Coordinates of vertex_____

Make a "quick and dirty" sketch by hand.

ZEROS OF THE FUNCTION:

From your sketch, how many zeros do you expect to find?

If appropriate, use one of the short cuts on the original equation to find the zeros(s) if any. Show your work.

Use the quadratic formula to find the exact value of the zeros.

List the coordinates of any t-intercepts:_____

If you have access to a graphing calculator or function graphing program, plot the function. Choose an appropriate range and domain that will show any zeros. Use it to give estimates for the t-intercepts and check with your values. Either print out and attach the graph and intercept estimates, or write down these estimates:

INTERPRETING IN CONTEXT

How tall is the Empire State Building?

How far away from the ground is the object after:

1 second ? _____ 3 seconds?_____10 seconds ?_____

Label these points by hand on your graph.

How many seconds does it take before it hits the ground?_____

For what values of **t** does the physical interpretation make sense in this problem?

FUNCTION PROFILE #2 $y = 16 x^2 + 40 x$

This function describes the distance traveled by an object that is thrown downward with an initial velocity of 40 ft/sec. **y** gives the distance in feet and **x** measures the time in seconds.

MATHEMATICAL PROPERTIES

Opens up or down?_____

Coordinates of y-intercept _____

Coordinates of vertex_____

Make a "quick and dirty" sketch by hand.

ZEROS OF THE FUNCTION

From your sketch, how many zeros do you expect to find?

If appropriate, use one of the short cuts on the original equation to find the zero(s) if any. Show your work.

Use the quadratic formula to find the exact value of the zeros.

List the coordinates of any x-intercepts _____

If you have access to a graphing calculator or function graphing program, plot the function. Choose an appropriate range and domain that will show any zeros. Use it to give estimates for the zeros and check with your values. Either print out and attach the graph and x-intercept estimates, or record these estimates:

INTERPRETING IN CONTEXT

How many feet has the object traveled in:

0 seconds?_____2 seconds ?_____ 8 seconds ? _____

Label these points by hand on your graph.

For what domain does the function have physical meaning in this problem?

FUNCTION PROFILE #3 $r = s^2 - 14s + 49$

The graph of this function describes a cross-section of a parabolic reflector, a mirror whose shape you can think of as being generated by rotating a parabola about the vertical line of symmetry that runs through its vertex. This cross section is a vertical slice that contains the axis of symmetry of the parabola.

MATHEMATICAL PROPERTIES
        Opens up or down?_____

        Coordinates of r-intercept _____

        Coordinates of vertex_____

        Make a "quick and dirty" sketch of the graph by hand.

ZEROS OF THE FUNCTION
        From your sketch, how many zeros do you expect to find?

        If appropriate, use one of the short cuts on the original equation to find the zeros if any. Show your work.

        Use the quadratic formula to find the exact value of the zeros.

        List the coordinates of any s-intercepts_____

        If you have access to a graphing calculator or function graphing program, plot the function. Choose an appropriate range and domain that will show any zeros. Use it to give estimates for the zeros and check with your values. Either print out and attach the graph and s-intercept estimates, or write down these estimates:

INTERPRETING IN CONTEXT
Parabolic mirrors have the unusual property that there exists a special point inside the parabolic shape called the focus. When any ray of light, originating from the focus, strikes the mirrored surface, it is reflected parallel to the axis of the reflector (the axis in this case being the y-axis). Conversely, any rays coming into the mirror parallel to the axis are reflected to the focus. Open up the *Quadratic Graphs* and click on "Parabolic Reflector" for a demonstration. These properties are used in parabolic mirrors for telescopes and in parabolic radio antennae.
        The coordinates for the focus for the general parabola $y = ax^2 + bx + c$ are given by $(-b/2a, (1 - b^2)/4a + c)$. Find the focus of our function example above. Show your work. Plot the point on your graph.

Generate a sketch of what would happen to at least two rays of light emitted from this focus that hit the parabolic surface.

Part III

Exploring On Your Own

Overview

Now you are on your own. Part III consists of an eclectic collection of independent explorations from the social and physical sciences. In Parts I and II, explorations appeared at the end of the chapter, where the title implied what function would be needed. Now *you* will have to decide what function models are appropriate. Small groups could all work on the same exploration or on different ones. The work could be done in or outside of class as a final project.

Some explorations involve exploring existing data sets. The Nations Data Set, for example, contains information about 114 different countries and the Classic Experiment with Sugars provides the original data from an experiment on the conversion rates of sucrose. Other explorations entail collecting data in a scientific experiment to study light or the cooling liquids. The approach varies from exploration to exploration. Some are quite open ended, some fairly structured. In each case you use the algebraic and technological tools developed throughout the course to search for patterns in the data and for ways to describe and analyze those patterns mathematically.

In each exploration you need to record, analyze and summarize your results. Your analysis should answer questions such as: What are the algebraic properties of the function(s) you are using? Is it an appropriate choice in this situation? In what ways does your model simplify reality? What are the limitations of your model and your data? What are the strengths of your argument? What are the weaknesses? Each exploration asks you for a written or verbal report of your results.

Exploring On Your Own

Using Algebra in Independent Explorations

A collection of explorations that includes political science data, data from a chemistry experiment and experiments in biology, physics, and chemistry where you collect you own data.

Measuring Light
Nations Data Set
A Classic Experiment with Sugars
The Case of Drink Q
Cooling of Liquid

These explorations are prototypes that we are trying out this year.

Measuring Light

Introduction:

In this laboratory, you will measure the relationship between the brightness of light versus the distance from the light source. To do this experiment we must have some way to measure illumination levels. Eyesight varies a lot among individuals; different people are comfortable working or reading at different light levels. Although what is "too bright" or "dark" is subjective if people attempt to describe light levels, there are fairly accurate meters which can be used to get objective measurement of light.

Photographers use light meters which read out in units designed to control the amount of light getting to the film. The scientific unit of illumination is the lumen/meter^2, but we will use the British footcandle (about 10.7 lumen/meter^2) which is in common use by illuminating engineers. The footcandle is the amount of illumination shed on a surface 1 foot away from a "standard candle" light source. Originally the standard candle was made to certain dimensions from whale blubber, now it is defined as an international unit called the candle, equal to the luminous intensity of 5 square millimeters of platinum heated to 1773.5 Celsius.

To get a practical idea of what a footcandle is, 2 FC is barely enough light to see anything. A difficult task, like sewing with black thread on black fabric, probably requires closer to 200 FC. A softly daylit room with no direct sun is around 20 FC. Hospital operating tables are at around 2000 FC.

Objective

To investigate the relationship between surface illumination and distance from light source.

Procedure

Our experiment will use an illumination meter that reads in footcandles from a light sensor that can be swiveled to face a light source directly. The Sekonic model 246 light meter reads from 0-100 footcandles on the red scale, and from 0-500 FC on the black scale. The light sensor has two plastic dome covers; to use the red scale mount the cover with the white dot. With the black metal shield (stored on the back pocket of the meter) slipped into the slot on the light meter, the meter will give a reading 10 times the black scale. For this experiment you probably will not need the black shield. Note that the red scale has major graduations marked at every 20 FC, so each small graduation is 2 FC apart; and the black scale has major graduations every 100 FC, so the small graduations are 10 FC apart..

The manufacturer claims the data from this meter will be ± 10% accurate: this means that if you get a reading of 45 FC, it could be somewhere between 45±4.5 or between 40.5 –49.5 FC. Other sources of error can come from putting the light sensor in your own shadow while trying to take a reading, reading the graduations wrongly, or positioning the light sensor inaccurately.

We will use a clear bulb light source, so that the filament which gives off the light is easy to see. CAUTION: DO NOT TOUCH BURNING BULB! Measurements of distance from the source need to be made to the filament, rather than to the glass bulb.

1.   Find a room which can be completely darkened, preferably over 16 feet long. Get a lamp socket which does not obstruct the light from the bulb, several clear bulbs of different wattage's, a rolling metal tape carpenters ruler at least 16 feet long, some duct tape and the footcandle meter.

2.   Tape down the metal rule with the zero end level with the filament, then stretch the rule 16 feet out from the source and tape down the end. This operation can be done on the floor or several tables can be pushed together to get a long work surface.

3   Using one source as the only light in the darkened room, take footcandle readings perpendicular to the source at .5, .75, 1, 1.5, 2, 3, 4, 5, 6, 7, 8, 9 feet. If there is enough room, and there is enough light to read the meter, it is good to take readings up to 16 feet. It is important to hold the meter with the light sensor swiveled so it is facing perpendicular to the source in a direct horizontal line from the filament. It may be helpful to prop the meter on some object that is the same height off the work table as the filament. Record the data, noting bulb wattage. Keep a record of your measurements.

4   Different teams of students can repeat the experiment using different wattage sources, to see how source intensity affects results.

Lab Report

Include the following information in your lab report along with your data tables and graphs.

Objective

Describe the purpose of the experiment. What were you trying to find out?

Methods

Describe the procedures that you followed. For example, how did you obtain your measurements for FC? What type of light source did you use? Did you vary the intensity of the light source?

Results

This section should include all of the data you obtained.
Create a graph of your data plotting distance from light source (d) on the horizontal axis and illumination (i) on the vertical axis. Be sure to include your units.

Analysis of results

1.   Use an available curve fitting program to find a reasonable best fit function for the variables you have chosen to study. Record your best fit function, carefully identifying the units. Give some indication of your level of confidence in it. (Note: You may find a function which you haven't studied yet.)

2.   Specify the general domain and range for the function, and the restricted domain and range for which this function models your data. List and interpret at least three specific points that lie on the graph of your function (not the original data)

3.   Describe what happens to your function as it models your data, starting at your "initial" value and moving through increasing values of the independent (horizontal) variable. Describe in algebraic terms what happens to:

        the dependent variable
           when is it positive, negative or zero ?
           when is it increasing, decreasing or staying the same?
           is there a maximum or minimum value?

        the average rates of change (or slope)
           when is the slope positive, negative or zero?
           when does the slope appear to be increasing, decreasing or flat?

Reporting your results

Summarize your results. Describe in words what happened to your measure of illumination as your increased the distance from the light source. Compare your results with the results of other groups in the class. Does the intensity of light source have an effect?
Write up a summary of your results. Include in your summary any graphs that you have created.

Nations Data Set

Objective

An open ended analysis of social and demographic data from 114 nations

Materials/Equipment

Attached printout of "Nations Data Set"

> If using a computer
> Spreadsheet and function graphing program
> "Nations Data" file which is on data disk

> If using graphing calculators
> Graphing calculators with best fit function capabilities
> Graph link file with "Nations Data Set" which is enclosed on data disk.

Procedure

The attached data set contains the following information about 114 different nations: Gross Domestic Product (GDP) in U.S. dollars, Gross Domestic Product per capita in U.S. Dollars, infant mortality rate, life expectancy in years, number of persons per hospital bed, number of persons per TV set, number of persons per physician and percent of population able to read and write. The information is compiled from various agencies throughout the world that routinely collect social and demographic data. A data set like this allows social scientists to examine relationships between variables and to explore the answer to questions, such as how does the per capita GDP affect a nation?

1. Exploring the raw data.

    Look through the data set looking for possible relationships between variables. Try testing out various hypotheses. Does the Gross Domestic Product affect life expectancy? What happens when the per capita Gross Domestic Product is high? Does it appear to relate to the infant mortality rate? What variables do not appear to be related at all? When you plot two variables that appear related what is the general pattern? Does it look as if you could fit a function to it? Perhaps you might need to break the data into sections and fit a different function to each piece. You can do this by looking at the poor countries vs the rich countries (or those with low life expectancy vs. high life expectancy, ect.) Or you may decide to compare the means of various data sets. (You will need to create two more columns, one with the values for the independent variable and the other with the means.)

2.   Generating a working hypothesis

Formulate what seems like a reasonable conjecture from your data about the relationship between at least one pair of variables. Indicate which you are thinking of as the independent variable (to be graphed on horizontal axis) and which the dependent variable (to be graphed on the vertical axis). Why would this conjecture make sense?

Analysis of results

1.   Use an available curve fitting program to find a reasonable best fit function for the variables you have chosen to study. Record your best fit function, carefully identifying the units. Give some indication of your level of confidence in it. You may find a function which you haven't studied yet.

2.   Specify the general domain and range for the function, and the restricted domain and range for which this function models your data. List and interpret at least three specific points that lie on the graph of your function (not the original data)

3.   Describe what happens to your function as it models your data, starting at your "initial" value and moving through increasing values of the independent (horizontal) variable. Describe in algebraic terms what happens to:

   •   the dependent variable

          when is it positive, negative or zero ?
          when is it increasing, decreasing or staying the same?
          is there a maximum or minimum value?

   •   the average rates of change (or slope)

          when is the slope positive, negative or zero?
          when does the slope appear to be increasing, decreasing or flat?

4.   How well does this function describe the data? What are some limitations in using this function to describe the data? What questions have been raised?

5.   What are next steps in examining your original conjecture(from #2 above) or in examining new conjectures?

Reporting your results

Write a research report that need only be three paragraphs long that contains the following information. Include your graphs in your report.

Data and Methods

State your working hypothesis or question in an opening topic sentence. Describe the data set and the type of data analysis you used.

Analysis of results

What is your algebraic evidence? Does it support your argument? Be sure to show your scatter plot and all the information about your best fit function specified above. Did you find a regression line, or another type of function? Did you use means to find it or the whole data set? Include any additional tables and graphs that support your argument.

Limitations of your evidence

Think about the variables being used. How are the variables defined and how are they measured? What information do you have about how the data were actually collected. What are the limitations on the mathematical model you constructed? How is it different from the data? What other information would be valuable in making a better analysis. What factors have not been taken into considerations? What questions have been raised?

Conclusions

What can you conclude from your work? What are the strengths and weaknesses of your argument? What are the implications of your conclusions? What questions have been raised? What suggestions can you offer for future research on this topic?

Country	GDP	GDPpc	Person/TV	LifexpM	LifexpF	Person/Bed	person/dr	infmor	Literacy%
Afghanistan	3	200	181	46	44	2054			
Albania			13	70	77	173	585	30	0.92
Algeria	42	1570	13	67	69	393	1062	52	0.52
Andorra	0.76	14000	8.6						0.99
Angola	5.1	950	210	44	48	845	15136	145	0.4
Argentina	112	3400	4.6	68	75	205	326	29	0.95
Armenia			5	69	76	246	117	35	
Australia	293.5	16700	2.2	74	81	199	438	7	0.99
Austria	141.3	18000	2.9	73	80	105	327	7	0.99
The Bahamas	2.6	10200	4.4	68	75				34
Bangladesh	23.8	200	316	55	55	3218	5264	107	0.47
Barbados	1.8	7000	3.7	71	77	121	1042	20	0.99
Belgium	177.9	17800	2.4	74	80	103	298	7	0.98
Belize	0.373	1635		66	70	332	2021	36	0.93
Brazil	369	2350	5	57	67	270	.848	60	0.81
Bulgaria			2.9	70	77	97	312	12	0.98
Burma	1.23	205		38	42	515	3177	114	0.5
Cambodia	2	280	129	48	51	632	27000	111	0.5
Cameroon	11.6	1010		55	59	377	12540	77	0.65
Canada	537	19600	1.6	75	82	143	449	7	0.99
Central African Republ	1.3	440		41	44	695	16447	137	0.27
Chad	1	190		40	42			132	0.3
Chile	34.7	2550	6.7	72	78	303	895	15	0.92
China			32	67	69	427	648	52	0.7
Colombia	51	1500	6.1	69	75		1061	28	0.8
Congo	2.5	1070	317	46	49	456	3873	111	0.57
Costa Rica	6.4	2000	9.2	76	80	442	981	11	0.93
Cuba	14.9	1370	4.3	75	79	141	303	10	0.99
Cyprus	5.3	7585	3.2	74	79	168	476	9	0.95
Czech	75.3	7300	3.2	69	77	98	319	9	0.99
Denmark	94.2	18200	2.1	73	79	184	360	7	0.99
Dominica	0.174	2100		74	80	247	1947	10	0.9
Dominican Republic	8.4	1120	10	66	71	508	934	52	0.83
Ecuador	11.8	1100	12	67	73	598	1039	39	0.88
Egypt	41.2	730	11	59	63	504	1698	76	0.44
EL Salvador	5.9	1060	11	64	70	973	1322	41	0.75
Ethiopia	6.6	130	541	51	54	3873	36660	106	0.18
Fiji	1.4	1900	73	63	68	413	2438	18	0.85
Finland	79.4	15900	2.6	72	80	80	406	5	0.99
France	1080	18900	2	74	82	81	374	7	0.99
Germany	1398	17400	2.6	73	80	121	319	7	0.99
Ghana	6.6	410	61	54	58		22452	83	0.6
Greece	82.9	8200	4.5	75	80	199	303	9	0.925

Country	GDP	GDPpc	Person/TV	LifexpM	LifexpF	Person/Bed	person/dr	infmor	Literacy%
Haiti	2.2	340	271	43	47	1258	6083	109	0.53
Honduras	5.5	1090	31	65	70	818	1586	45	0.73
Hungary	55.4	5380	2.4	67	76	100	262	13	0.98
Iceland	4.5	17400	3.4	77	81	58	355	4	0.99
India	240	270	44	58	59	1324	2337	78	0.48
Indonesia	133	680	17	59	63	688	2685	67	0.85
Iran			26	65	67	688	2685	60	0.54
Iraq			19	65	67	568	1922	67	0.6
Israel	57.4	12100	3.5	76	80	177	345	9	0.92
Italy	1.01	17500	3.4	74	81	153	228	8	0.98
Japan	2470	19800	1.2	76	82	74	588	4	0.99
Jordan	3.6	1100	15	70	74	571	813	32	0.71
Kenya	8.3	320	104	51	55	734	7313	74	0.5
North Korea			87	67	73	74	370	28	0.99
South Korea			5	67	74	429	1007	22	0.96
Kuwait	15.3	11000	1.5	73	77	347	515	13	0.71
Laos	0.9	200		50	53	402	3555	102	0.45
Lebanon	4.8	1400	2.5	67	72	263	771	40	0.75
Libya	26.1	5800	8.9	62	66		•	63	0.64
Luxmbourg	8.5	21700	3.9	73	81	87	496	7	1
Malaysia	54.5	2960	9.3	66	72	457	2638	26	0.8
Mexico	328	3600	6.7	69	77	1367	885	27	0.9
Monaco	0.475	16000	1.5					7	0.99
Mongolia	1.8	800	18	64	69	83	340	43	0.89
Morocco	28.1	1060	21	66	70	959	4415	50	0.5
Mozambique	1.75	115	428	47	50	1227	43536	129	0.14
Nepal	3.4	165		52	53	4015	16007	84	0.29
Netherlands	259.8	17200	2.5	75	81	170	400	6	0.99
New Zealand	49.8	14900	3.2	73	80	114	359	9	0.99
Nicaragua	1.7	425	20	61	67	804	1882	53	0.57
Niger	2.3	290	330	43	46			111	0.28
Nigeria	35	300	21	54	57	1160	6573	75	0.51
Norway	76.1	17700	2.9	74	81	177	305	6	0.99
Oman	10.2	6670	1.1	66	70	380	1078	37	0.2
Pakistan			63	57	58	1706	2364	102	0.35
Panama	6	2400	12	72	78	333	871	17	0.87
Papua New Guinea	3.4	850	383	56	57	234	9953	63	0.52
Paraguay	7.3	1500	13	72	75	816	1470	25	0.9
Peru	25	1100	11	63	68	625	997	54	0.85
Philippines			9.1	63	68	683	1062	51	0.88
Poland	167.6	4400	3.8	69	77	118	464	13	0.98
Portugal	93.7	9000	5.6	72	79	221	348	10	0.83
Romania	63.4	2700	5.8	69	75	106	549	20	0.96
Russia			2.7	64	74	74	226	27	0.99
Rwanda	2.35	310		39	41	9000	177	119	0.5

Country	GDP	GDPpc	Person/TV	LifexpM	LifexpF	Person/Bed	person/dr	infmor	Literacy%
Saudi Arabia	111	6500	3.7	66	70	359	523	52	0.62
Singapore	45.9	16500	5.1	73	79	282	757	6	0.87
Solomom Islands	0.2	600		68	73			28	0.6
South Africa	115	2800	11	62	68		1271	47	0.99
Spain	514.9	13200	2.3	74	81	234	257	7	0.97
Sri Landka	7.75	440	25	69	74	356	5823	22	0.9
Sudan	5.2	184	120	53	55	1222	9439	80	0.27
Sweden	145.6	16900	2.3	75	81	175	395	6	0.99
Switzerland	152.3	22300	3.2	75	82		311	7	0.99
Syria	30	2300	19	65	68	891	1037	43	0.64
Taiwan			3.1	72	79	221	868	6	0.9
Thailand	103	1800	17	65	72	604	4377	37	0.89
Togo	1.5	400	161	55	59	752	12992	89	0.45
Turkey	219	3670	5.6	69	73	465	1108	49	0.81
Uganda	6	300	150	37	38	817	20000	112	0.52
United Arab Emirates	34.9	13800	12	70	74	292	618	22	0.68
UK	920.6	15900	2.9	74	80	146	611	7	0.99
US	5950	23400	1.2	72	79	211	406	10	0.97
Uruguay	9.8	3100	5.2	71	77	127	341	17	0.96
Vietnam			31	63	68	298	2843	46	0.88
Yemen	8	775	34	49	51	995	5531	113	0.38
Yugoslavia			3.6	71	76	163	490	21	0.9
Zaire	6.6	180	1870	46	49	476	23193	111	0.72
Zambia	4.7	380	41	44	45	349	8437	85	0.54
Zimbabwe	6.2	545	72	40	44		6951	74	0.67

Data Dictionary for "Nations Data Set"

Country	Name of Country
CDP	Gross Domestic Product in billions of US dollars
GDPpc	Gross Domestic Product per capita
LifexpM	Life Expectancy Male in years
LifexpF	Life Expectancy Female in years
Persons/Beds	Number of Persons per hospital bed
Persons/TV	Number of Persons per TV set
Persons/dr	Number of Persons per physician
Infmor	Infant Mortality per 1000 live births
Literacy %	% of population able to read and write

A Classic Experiment with Sugars

Related Reading

Kenneth Kustin, *Crystals, Light, and Sweet-Tooth Satisfaction* in the Anthology of Readings.

Introduction

An important aspect of chemical reactions is the <u>rate</u> at which they occur. In a classic experiment (published in the Journal of the American Chemical Society in 1922) Fales and Morrell carefully studied the rate at which *sucrose* is converted to *fructose* and *glucose* under various conditions. Sucrose is the form of sugar occurring in products such as cane sugar, corn syrup, and honey. Fructose occurs naturally in many fruits, and glucose is the form of sugar most easily used in human metabolism. When we eat sucrose, our bodies convert it to fructose and glucose in a chemical reaction similar to the one studied by Fales and Morrell. The process is called the "inversion" of sucrose. Table 1 gives the data from one of the experimental runs of Fales and Morrell.

Table 1

t = time elapsed (seconds)	y = fraction of sucrose remaining
0	1.0000
69	0.9924
1035	0.9418
3113	0.8401
4857	0.7557
9231	0.5875
12834	0.4791
18520	0.3473
26320	0.2203
32640	0.1558
76969	0.0075

The "time" column is straightforward; 69 seconds is just over a minute after the beginning of the experiment. The "fraction remaining" column shows what fraction of the original concentration of sucrose remains unconverted at the given time. Initially (at time 0) it is all there - the fraction remaining is 1.0. After 3113 seconds, about 84% remains. After 76969 seconds, less that 1% of the original sucrose remains unconverted to fructose and glucose, and the experiment is considered over.

1.   How many hours did the experiment reported in Table 1 run?

2.   Make a scatter plot of the data in Table 1. What type of function seems to model this?

3.   Determine a specific function to model the data. Plot the function and the scatter plot together to demonstrate how well your function fits.

4.   In the actual experiment, Fales and Morrell reported the fraction <u>converted</u>, not the fraction remaining. They used the letter x to stand for the fraction converted.
a)   What is the relation between x and the fraction y in Table 1?
b)   Sketch the graph of x versus t.
c)   What happens to x as t increases from 0 to 76969? Interpret this in the context of the experiment.

5.   The speed of the reaction depends on the nature of the solution in which the sucrose is dissolved. Table 2 shows the data reported when this solution was different from that for Table 1.

a)   Is this reaction slower, or faster, than the one in Table 1? Explain your reasoning.
b)   Graph the data in Table 2, and determine a function modeling the data.
c)   How does the choice of time units (seconds, minutes, hours) affect the shape of the graph?

Table 2

Velocity of Sucrose	
t=time elapsed (min)	y=fraction of sucrose remaining
0	
360.5	0.8880
1075.87	0.7123
1752.87	0.5759
3258.87	0.3501
3872.87	0.2838
4624.37	0.2248
7161.47	0.1010
8307.87	0.0294
12673.87	0.0131

Measuring The Concentration Of Sucrose.

You might wonder how it is possible to determine the concentration of sucrose in a reacting mixture at a given time, and still let the reaction go on undisturbed. Chemists have discovered a wonderful phenomenon to exploit for this purpose, involving light that has been *polarized*. Polarized light consists of electromagnetic waves that oscillate in just one particular direction; this direction can be determined by passing the light through a special lense. When polarized light passes through a mixture of water containing compounds such as sucrose, glucose, and fructose, the direction of polarization is changed by a certain angle. Let's use the letter A to stand for this angle. As the reaction proceeds, the angle changes; we denote the angle at time t by A_t and the angle when the reaction is complete by A_∞. It turns out that the concentration of sucrose is directly proportional to the difference $A_t - A_\infty$. To express this algebraically, we need a symbol for the concentration of sucrose at time t. The symbol for this is $[S]_t$. So there is a constant c such that

$$[S]_t = c(A_t - A_\infty) \qquad (*)$$

Problem 1. Show that the *fraction of sucrose remaining* is given by the formula:

$$[S]_t/[S]_o = (A_t - A_\infty)/(A_o - A_\infty) \quad \text{(Use formula (*).)}$$

The right hand side of this formula is the ratio that is used to determine the fraction of sucrose remaining.

Equation (*) depends on the combined effects of the three compounds. We will find some algebraic relations which are inherent in the physical situation, and combine these to explain equation (*).

Each compound causes a change in the angle of polarization, proportional to its concentration. For sucrose, there is a constant of proportionality, call it s, such that

$$A_s = s[S]$$

Similarly, for fructose and glucose, there are constants f and g with

$$A_f = f[F]$$
and$$\quad A_g = g[G]$$

The total change in angle is the sum of these, $A = s[S] + f[F] + g[G]$. We put a subscript t on each concentration to indicate the concentration at time t. So

$$A_t = s[S]_t + f[F]_t + g[G]_t. \quad (1)$$

(There are no subscripts on the constants s, f, g, since these do not change with time.)

When t = 0, we have the original concentration of sucrose, $[S]_o$, and no fructose or glucose. So $[F]_o = 0$ and $[G]_o = 0$, and equation (1) with t = 0 is just

$$A_o = s[S]_o \quad (2)$$

When the experiment ends ("t = oo") there is no sucrose left. So $[S]_\infty = 0$, and then equation (1) is just

$$A_\infty = f[F]_\infty + g[G]_\infty \quad (3)$$

One final relation comes from the nature of the reaction. Each molecule of sucrose which is converted produces one molecule of fructose and one of glucose. So

$$[S]_o - [S]_t = [F]_t = [G]_t \quad (4)$$

From these equations you can deduce a few others, including the equation (*) which is used in the Fales-Morrell experiment.

Problem 2. The text states a reason for equation (4). Imagine someone still didn't see exactly why (4) is true; try to explain it to them more fully.

Problem 3. By combining various equations (1), (2), (3), and (4), derive the following relations:

a) $[F]_\infty = [G]_\infty = [S]_0$. Interpret this in the context of the experiment - what has happened when it is over?

b) $A_t - A_\infty = (s\text{-}f\text{-}g)[S]_t$.

Finally, how is the constant c in equation (*) related to s, f, and g?

The Case of Drink Q

Developed by Lowell M. Schwartz. Chemistry Department,
University of Massachusetts/Boston

This exercise will simulate a procedure very often done in the sciences. Frequently, a quantitative measurement must be made with an instrument which does not directly output the desired quantity. Several circumstances might lead to this situation: the original scale markings on the instrument might be inaccurate or simply missing; or the instrument's output variable might be functionally related to, but not the same as, the desired variable. In either situation, the procedure is to *calibrate* the instrument's output in terms of the desired variable. This involves reading the output of the instrument for a set of samples for which the desired quantity is <u>known</u>. A graph of the output readings versus the known quantities thereby relates the instrument output to the desired quantity. This graph is called a 'calibration curve' or a 'standard curve'. With the calibration curve in hand, any subsequent output reading by the instrument can be translated to the desired quantity by looking up the functional correspondence on the graph. In this exercise, we will calibrate an instrument called a 'spectrophotometer' in terms of a desired quantity: the strength of a beverage.

An enterprising group of youngsters have set up a stand by the roadside and are selling cold lemonade and similar soft drinks to people passing by. The group purchases the beverages in powder form and simply adds water to make the drinks. A thirsty consumer stops, buys a glass of cherry-flavored drink and is disappointed by the flavor. Perhaps the artificial flavor is bad or perhaps the youngsters added too much (or too little) water to a given amount of powder. Determined to figure out the reason, she saves some of her drink for testing. (We will call this the 'questionable' drink or just 'drink Q' for short.) She buys some of the same powdered beverage at a supermarket and, following the manufacturer's directions, makes up a glassful with the correct proportions of water and powder. She compares the taste of Q with her own drink and thinks that there is a slight difference. She also compares the colors of the two and again thinks she sees a slight difference. Neither of these qualitative tests is conclusive and so she decides to do a quantitative test to compare the strengths of the two; we have agreed to do the testing for her.

Our quantitative test will be based on color. Why are the cherry drinks red? Because white light (coming from the sun or from a light bulb) is a mixture of all colors and when white light passes through the drink, relatively little red is absorbed compared to other colors. Thus, light having passed through the cherry drink is long on red and short on other colors; the eye sees this as red. The stronger or more concentrated the drink, the more light is absorbed, the darker is the red color.

Experimental stategy

We have a spectrophotometer which generates a small beam of colored light and measures the loss of intensity of this beam passing through an object placed in the light path. Having chosen to use yellow light in this experiment, we use the spectrophotometer to measure the intensities of a yellow beam passing through several drinks made up to known strengths by different dilutions with water. (Call these drinks 'standards' and label them S1, S2, etc.) From this data we can develop a function, for the standard curve, which describes how the absorption of yellow light depends on the degree of dilution. After measuring the yellow light absorbed by the 'questionable' drink, we use the function to predict how dilute the youngsters made their product. We compare this result to the manufacturer's specified dilution.

To make up the standards we begin by dissolving a known amount of cherry-flavored powder in a known volume of water in a mixing vessel. This is the first standard S1. We withdraw a small sample, place it in the spectrophotometer and measure the absorption of yellow light (the 'absorbance') passing through. After returning the first sample to the mixing vessel, we add another measured volume of water, stir, and this becomes the second standard S2. Again we measure the absorbance of a small sample in the spectrophotometer and return the sample to the mixing vessel. Notice that by returning each test sample to the mixing vessel, we lose no liquid volume and so can keep track of the total volume of water in the successive standards. The standard curve that we want is absorbance as the dependent variable versus total water volume as the independent variable.

To add measured volumes of water we use a glass tube called a 'graduated cylinder'. Like a measuring cup, this cylinder is marked with lines indicating the volume contained up to each line. The volumes are in units of milliliters, abbreviation 'mL'. (A one-liter bottle of soda contains 1000 mL and is roughly a quart.) Our independent variable is the total volume of water used to make a standard. We will use the symbol V for this variable and will record these data in milliliter units.

beaker

graduated cylinder

The spectrophotometer detects the absorption of yellow light by the red-colored samples in terms of an 'absorbance' scale. This scale has the following property: When pure water is placed in the spectrophotometer so that 100% of the yellow light passes through (none is absorbed), the absorbance reading is zero. When 10% of the yellow passes through (90% is absorbed), the absorbance is one. When only 1% passes through (99% is absorbed), the absorbance is 2. And when none passes through, the absorbance is infinite. Absorbance is a dimensionless quantity, it has no units. We will use the symbol A to stand for absorbance.

flask

In the laboratory.

☐ Identify the following at your work station:
- cherry beverage powder in a plastic container (The powder has a mass of 7.0 grams, about a quarter of an ounce by weight.)
- beaker, 600 mL capacity, in which you will prepare the standards
- graduated cylinders, 50 and 100 mL capacity, for measuring volumes of water
- flask to keep a supply of water at your work station
- Spectronic 70 spectrophotometer
- tube, called a cuvet (pronounced *kew-vet*), used to hold a sample in the spectrophotometer
- plastic dropper used for mixing in the beaker and for filling the cuvet
- paper towels for wiping down the outside surface of the cuvet

cuvet

☐ Locate the dispenser containing Drink Q.

☐ Fill the flask with deionized water (from the DW tap) to use at your work station.

□ Adjust the spectrophotometer absorbance scale to read zero when the sample is just water.
- Pour water into the cuvet up to the horizontal line.
- Wipe down the cuvet, place it in the spectrophotometer and close the cover.
- Turn the zeroing knob until the meter reads zero absorbance (the same as 100% transmittance)
- Discard the water in the cuvet.

□ Set up a data table in your notebook that has the following form

Standard number	Added water volume in mL	Total water volume V in mL	Absorbance A	Concentration C in g per mL
S1				
S2				

□ Prepare the first standard S1 as follows:
- Fill a graduated cylinder with water to the 50 mL line.
- Empty the cherry powder into the 600 mL mixing beaker.
- Using the water in the graduated cylinder, wash any remaining powder into the beaker.
- Pour the rest of the 50 mL of water from the graduated cylinder into the beaker.
- Stir until all the powder has dissolved. The mixing beaker now contains S1.
Record the data: for S1 both "Added water" and "Total water" are 50 mL.

□ Measure the absorbance of S1: (TAKE CARE NOT TO MOVE THE ZEROING KNOB.)
- Rinse the cuvet with S1 by filling the cuvet with S1 and pouring this back into the mixing beaker.
- Again add S1 to the cuvet until the level is at or a bit above the horizontal filling line.
- Wipe down the cuvet, place it in the spectrophotometer and close the cover.
- Read and record the ABSORBANCE. This is the red scale printed below the mirror.
- Pour the sample back into the mixing beaker.

□ Prepare several more standards S2, S3, etc. by successively diluting the contents of the mixing beaker and measuring the absorbance after each dilution. Here are guidelines:
- You may choose your own series of dilutions. Prepare at least eight standards; the more you prepare, the better your data will define the standard curve. You must stop when the mixing beaker is full, about 500 mL. The data points will spread out more evenly if you gradually increase the volumes of added water as you progress from S2 upward. Start with 20 or 30 mL for S2 and increase up to 100 mL for your final standard.
- Be sure to stir the contents of the mixing beaker thoroughly after each addition.
- Fill the cuvet and measure the absorbance exactly as you did for S1.
- Be careful not to discard any liquid from a cuvet. We are keeping track of the total volume of each standard as the sum of the "Added volumes". This works only if you don't lose any
- Don't forget to record both "Added water" and the Absorbance data for each standard.

□ Take your empty cuvet to the Drink Q dispenser, rinse the cuvet with Drink Q and discard the rinse. Fill to the horizontal line with Drink Q, measure the absorbance in the spectrophotometer and record this in your notebook.

□ Rinse out all glassware that contained cherry solutions.

Data analysis

Before constructing the standard curve, you must sum up the accumulated water additions (listed under "Added water") to find the total volume V of each standard solution and write these in the column headed "Total water volume" for all your standards.

In addition to regarding the total water volume V as the independent variable, we will be interested in examining an alternative which is the 'concentration' (symbol C) of each standard. Concentration (or strength) indicates the relative proportion of powder to water. And although scientists express concentration in several different ways, the most convenient for our experiment is as 'grams of powder per milliliter of water'. Thus the concentrations to be entered in the last column of your data table are calculated by dividing the constant 7.0 grams of powder in each standard by the variable total volume V (in mL) of that standard.

Examine the data in the last three columns of your data table. Do you understand the trends? Figure out which answer in each set of [] brackets is correct and write these in your notebook.

As the independent variable V increases in the standards from S1 to S2 to S3, etc.,
- the corresponding dependent variable A [increases / decreases].

Why? Because as more water is added to a fixed quantity of cherry powder,

- the total volume of water [increases / decreases],
- the drinks become [more / less] dilute,
- the concentration of powder becomes [greater / less],
- the red color becomes [more / less] intense,
- the [more / less] yellow light is absorbed by the liquid in the spectrophotometer,
- the absorbance [increases / decreases].

In your notebook, construct a graph plotting the absorbance A of the standards on the vertical axis (the ordinate) versus the total water volume V on the horizontal axis (the abscissa). The plotted points drawn as circles here should trace out a curved path, absorbance decreasing from left to right.

According to theory, the function A vs. V should be a smooth function. Thus, if the data points do not fall exactly on a smooth curve, this is due to a variety of small experimental errors: inexact water volume additions, extraneous absorbances by finger prints or droplets on the sample tube, or inexact or fluctuating spectrophotometer readings. We are not interested in these experimental errors, but rather would like to have the smooth curve representing the function as if there were no errors. Therefore, draw a smooth curve passing as close as possible to as many data points as possible. Do not connect the dots unless you can do so with a perfectly smooth curve.

Use the best fit function program to find an algebraic equation which represents the A vs. V data. If the measurements were made carefully and the spectrophotometer was set up and operating properly, this function can fit the data using a single adjustable parameter. But an equation with two adjustable parameters is just as useful.

Write the algebraic equation that you find in your notebook.

Generate another plot of your algebraic equation curve superimposed on a graph of the data points.

Now we are in a position to compare the strength of drink Q with the strength of the manufacturer's recommended formulation. Find the position of the absorbance A_Q of drink Q on the vertical axis and use a ruler or straight-edge to draw a horizontal straight line from that axis and intersecting the smooth curve. Draw a straight vertical line from this intersection down to the horizontal axis. Record the total water volume at this point on the horizontal axis. This is the water volume used by the youngsters to dilute each 7.0 grams of powder in the cherry beverage that they sold. Let's use V_Q for this volume.

Use your algebraic A vs. V function to solve for the unknown V_Q at the measured value of A_Q for drink Q. This algebraic V_Q should be very close to the V_Q found from your graph.

In order to compare this with the manufacturer's formulation, we could make up some liquid beverage, measure its absorbance as we did that of Q and find the corresponding V value. However, it is simpler to calculate these quantities from the manufacturer's specification. Here is the problem: The purchased package contains 19 ounces of powder and this is supposed to be diluted with 8 quarts of water. If instead we use only 7.0 grams of powder, how many milliliters of water would we use to produce a drink of the same strength? V_m stands for this volume. Calculate V_m by yourself if you can. Otherwise follow these step-by-step instructions, entering the results in your notebook:

(Conversions are given below Step 5.)

1. Calculate the mass of powder in grams which is equivalent to 19 ounces.
2. Calculate the volume of water in liters equivalent to 8 quarts.
3. Calculate the volume of water in milliliters equivalent to 8 quarts.
4. Divide the volume of water in milliliters (step 3) by the mass of powder (step 1).
5. Multiply this quotient (step 4) by 7.0 grams of powder to obtain V_m.

(There are 28.3 grams per ounce and 1.06 quarts per liter.)

Compare the youngsters' V_Q to the manufacturer's V_m. Are the two the same? Probably not exactly.

• Which is more dilute, the youngster's drink or the manufacturer's specification?

• In your opinion, are the two dilutions close enough to be considered equivalent? (This may be a difficult question to answer. A quantitative study of the experimental uncertainties is often necessary to answer definitively.)

An alternative standard curve

The theory that predicts how the absorbance of most liquid solutions vary with dilution is called Beer's law. However, this theory is formulated with concentration as the independent variable rather than dilution volume. Beer's law states that the absorbance A is proportional to the concentration C, i.e. a plot of A vs. C is a straight line. Let us write the equation of the straight line for our purposes as $A = A_0 + kC$, where k is the slope of the line and A_0 is a constant which, in principle, is zero. However, in practice for A_0 to be zero, the spectrophotometer must be properly adjusted to read zero absorbance for a sample of pure water (for which C is zero).

Plot your experimental data on A vs. C axes. Use a straight-edge to draw a perfectly straight line which comes as close as possible to as many data points as possible. DO NOT connect the dots. The straight line which you have drawn is an alternative standard curve to the A vs. V standard curve developed previously.

Find the equation of your straight line by reading the intercept A_0 from your graph and calculating the slope k of your line. Write your straight line equation here by filling in the two blanks under A_0 and k and then copying the results to your notebook.

$$
\begin{array}{cc}
A_0 & k \\
\end{array}
$$

$$A = \underline{\hspace{2cm}} + \underline{\hspace{2cm}} \ C$$

Use this equation and A_Q to find the concentration C_Q of powder in the drink Q.

Calculate the concentration C_m of the drink as specified by the manufacturer.

Compare the two concentrations C_Q and C_m. Does the analysis using a straight line standard curve lead to the same conclusion about how the youngsters made up their product as does the A vs. V standard curve?

Use your straight line equation and algebra to derive the equation of the A vs. V function.

Prelab Exercises

The data that you gather in this experiment will conform to the expected functional form only if you properly read and record both the water volumes and the absorbances. Here are some practice exercises to help you get ready to read the volumes and absorbances.

Read the liquid levels in these graduate cylinders. If your glassware is clean, the surface will curve upwards to the cylinder wall. Read to the bottom of this curve (called the *meniscus*). Recording the levels to the nearest milliliter will be OK but you will obtain a smoother curve if you can estimate to the nearest tenth of a mL.

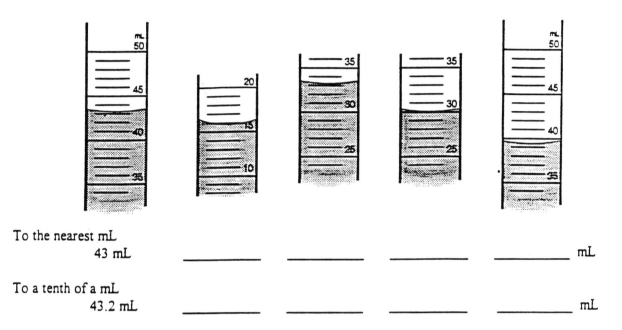

To the nearest mL
     43 mL   _____   _____   _____   _____ mL

To a tenth of a mL
     43.2 mL   _____   _____   _____   _____ mL

Read the absorbance data indicated by these pointers. Be careful - the ABSORBANCE scale runs from right to left and is not uniform.

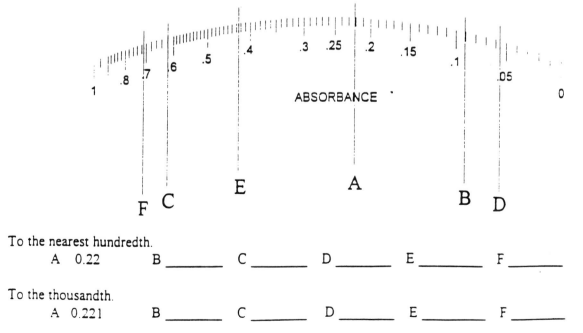

To the nearest hundredth.
    A 0.22   B _____   C _____   D _____   E _____   F _____

To the thousandth.
    A 0.221   B _____   C _____   D _____   E _____   F _____

Cooling of Liquid

Anyone who drinks hot liquids is familiar with the process of liquid cooling. When coffee or tea is just poured it is often too hot to drink without burning your tongue, and if you let it cool for ten minutes it probably will seem not hot enough. Different cups transmit heat at different rates; a metal cup permits hot liquids to cool much faster than a ceramic or insulated cup. To investigate the mechanism of cooling liquids, all that is required is a clock with a seconds readings, a cup of nearly boiling liquid, and a laboratory or cooking thermometer that can measure up to 212F or 1000C.

Procedure

1. Read the room temperature. Try to conduct the experiment at constant room temperature, out of drafts or other changing heat conditions. It does not matter whether you use Fahrenheit or Celsius, but use the same system for all measurements.
2. Boil water and fill a cup with a thermometer standing in it; or, place a thermometer in a cup of coffee you have purchased and are willing to sacrifice or drink cold in the interest of scientific enlightenment.
3. Immediately watch the thermometer rise and record the highest reading. Choose this time as t=0.
4. Every minute during the first period of rapid cooling take temperature readings. After the cooling slows down you can take readings every 2 minutes, or at longer intervals.
5. Check the room temperature periodically.
6. Different groups can use different cups to see how much cooling is affected by the container. An insulated plastic commuter mug, a tin camping cup, a paper cup and an ordinary ceramic mug are good ones to try.

Results

Since the liquid will eventually work its way down close to room temperature, the degrees above room temperature are more significant than the actual temperature reading.

1. Compute the Temp. Liquid -Temp. room differences for each time of your readings.

2. Create graph of your data, plotting Temp. liquid- Temp. of room (on vertical axis) vs. time (on horizontal axis).

Analysis of results

1. Examine this graph to find a possible generic curve to model it. Use a function graphing program or graphing calculator to find a best fit function. Adjust the formula to give actual temperature vs. time, rather than temperature difference vs. time.

2.   Specify the general domain and range for the function, and the restricted domain and range for which this function models your data. List and interpret at least three specific points that lie on the graph of your function (not the original data)

3.   Describe what happens to your function as it models your data, starting at your "initial" value and moving through increasing values of the independent (horizontal) variable. Describe in algebraic terms what happens to:

     the dependent variable
        when is it positive, negative or zero ?
        when is it increasing, decreasing or staying the same?
        is there a maximum or minimum value?

     the average rates of change (or slope)
        when is the slope positive, negative or zero?
        when does the slope appear to be increasing, decreasing or flat?

4.   Calculate the average rate of change of cooling for fixed time intervals and graph your results.

Reporting your results

Summarize your results. Compare your results with the results of other groups in the class. Do the different containers have an effect?
Write up a summary of your results. Include in your summary any graphs that you have created.

Explorations in College Algebra

Anthology of Readings

Anthology of Readings

TABLE OF CONTENTS

PART I: ALGEBRA IN THE SOCIAL SCIENCES

PAGE

An Op Ed article on cancer, with a letter in response

Background reading for class presentations

Readings related to Chapter 3: *Rates of Change*

Readings related to Chapter 4: *When Rates of Change Are Constant*

Readings related to Chapter 5: *Looking for Links*

Articles on federal and private data collecting

Articles on education and income

Articles on Regression Lines

Reading related to Chapter 6: *When Lines Meet*

Part II: Algebra In The Physical And Life Sciences

Readings related to Chapter 8: *Deep Time and Deep Space*

A poetic reflection on the Big Bang

Articles related to the relative sizes and ages of things in the universe

Reading related to Chapter 10: *The Mathematics of Motion*

An article related to the free fall experiment

Part III: Exploring on Your Own

Background reading for "A Classic Experiment With Sugars"

Science Musings

CHET RAYMO

True nature of math remains, in sum, a mystery

IF YOU ARE AFFLICTED WITH MATH ANXIETY, YOU may not like what I'm about to say.

Our most certain knowledge of the world is mathematical. Nonmathematical knowledge is held suspect by scientists. Dependence upon mathematics as the arbiter of truth defines the modern, Western way of knowing. For better or worse, it is the source of our power, our wealth and our physical well-being.

There it is. Take it or leave it. Math offers our best crack at figuring out what the world is all about.

And the big mystery is – no one knows what mathematics is, or why it works so unreasonably well.

Not even mathematicians.

Or the scientists who use mathematics as their language of discovery.

There are, I suppose, three possible explanations:

1. Mathematics is an invention of the human mind, like the English language or the game of chess, that has proven particularly useful for expressing patterns we observe in nature.

2. Nature is itself mathematical. We learn mathematics by observing the way the world works.

3. Mathematics exists as a kind of Platonic ideal, beyond and outside of nature. All reality, including the human mind, participates in some mysterious way in this preexisting order.

All of which is a wordy way of saying:

1. We invent math.

2. We discover math.

3. We are math.

John D. Barrow, professor of astronomy at the University of Sussex in Britain, takes up the mysterious effectiveness of mathematics in his recent book, "Pi in the Sky: Counting, Thinking and Being." He takes us through the history of mathematics, from counting on fingers to the high-blown abstractions of theoretical physicists who use mathematics to investigate the Big Bang moment of creation.

Math offers our best crack at figuring out what the world is all about.

He's after the answers: What is math? Why does it work? Why is science up to its ears in numbers – nay, to the very tip-top of its head? Barrow's answer: Uhhh... well... I guess... maybe... dunno.

The mystery remains.

And that, says Barrow, is the point.

His claim: "We have found that at the roots of the

scientific image of the world lies a mathematical foundation that is itself ultimately religious."

I think he's right.

And if I were a theologian, that is exactly where I would start constructing a concept of God that is relevant to our time – with mathematics.

At first blush, the idea sounds preposterous, even blasphemous. But wait. Isn't mathematics the most effective medium of exchange between the human mind and the cosmos? Through math we have come to know the Big Bang, the universe of galaxies, the unfolding infinities of space and time. We have plunged into the heart of matter and explored mysteries of life and mind. Nothing else has taken us to there. Math, only math.

And we haven't the foggiest idea why it works.

By and large, theologians still offer us the God of Michelangelo's Sistine Chapel ceiling – a gray-bearded version of ourselves ensconced in some paltry heaven up in the sky. "Oh, don't be silly," they'll protest, "the gray-bearded man is just a metaphor for something we cannot fully express." Exactly. So why stick with a medieval metaphor that opens a gulf between science and religion, between knowing and feeling?

Why not adopt a metaphor that embraces the full richness of the cosmos, a metaphor that links our minds to all that exists?

Namely, mathematics.

The human mind has evolved in response to the world in which we live, in the same way as did our senses of sight, sound, and smell. If we are mathematical creatures,

> **And maybe, just maybe, we are math incarnate – to the core of our souls.**

it is because the world is in some deeply mysterious sense mathematical. Call it, if you will, the mind of God. The phrase is metaphorical, but so is every other definition of God. The mathematical metaphor links us profoundly, deeply, to all of creation in a way that is consistent with the spirit and the substance of modern science.

That's where I'd start, if I were a theologian.

But I'm not a theologian. I don't even have a religion. I just know a ripping good mystery when I see one.

Of the three possible explanations for the mysterious effectiveness of mathematics, maybe all are true. Maybe we invent math. Maybe we discover it. And maybe, just maybe, we are math incarnate – to the core of our souls. Maybe, just maybe, mathematics is our glimpse of the eternal, omnipresent, creative foundation of the world. Why else does it work so astonishingly well?

Barrow may have got it exactly right: "Our ability to create and apprehend mathematical structures in the world is . . . a consequence of our oneness with the world. We are the children as well as the mothers of invention."

Chet Raymo is a professor of physics at Stonehill College and the author of several books on science.

U.S. Government Definitions of Census Terms

by Anthony Roman (1994), Center for Survey Research, University of Massachusetts, Boston.

Housing Unit

The government defines *housing unit* as a "house, an apartment, or a single room occupied as separate living quarters." The exact definition becomes more complex when one considers what constitutes separate living quarters. Perhaps the easiest way to define a housing unit is to describe what it is not. The following are <u>not</u> housing units and therefore the people living in them do <u>not</u> make up households: 1) most units in rooming or boarding houses where people share kitchen facilities, 2) units in transient hotels or motels, 3) college dormitories, 4) bunk houses, 5) group quarters living arrangements such as military housing, convents, prisons, and mental institutions, and 6) units within other housing units which do not have direct access from a hallway or outside. Most other living arrangements should come under the definition of a housing unit and therefore a household.

Household & Family

A *household* includes all persons occupying a housing unit. A *family* consists of all people in a household who are related by blood, marriage, or adoption. By strict definition, a family needs to have a minimum of two people, a householder and at least one other person. A householder is defined as the person in whose name the housing unit is owned or rented. If more than one such person exists, or if none exist, than any adult household member can be designated as a householder.

A household may contain no families. This, in fact, is not uncommon. Although they may consider themselves one, a group of single unrelated people or an unmarried couple living together are not considered a family. A household may also contain more than one family, although this is rare. In most cases in which multiple families share a single housing unit, there is at least one member of one family who is related in some way to a member of another family. By

definition then this group becomes one large family. For example, if a husband and wife rent a house and the wife's cousin and her two children come to live with them, this constitutes one large family since all members are related by blood, marriage, or adoption.

Size of Household

The *size of a household* is the number of persons who are residing in the household at the time of interview and who do not usually live elsewhere. A visitor staying temporarily at someone's house is <u>not</u> part of the household. A person who is away on vacation or in a hospital is a member of the household if he or she usually lives there. The status of college students may be the hardest to determine. If they usually live away from the household, they are <u>not</u> part of the household.

Household vs. Family Income

Household income is the sum of all incomes earned by all who live in the household. If a husband and wife each earn $30,000, their child earns $5,000, and an unrelated boarder living in an extra bedroom earns $15,000, then the household income is $80,000. All four members of the household are considered to be living in a household with an income of $80,000, even though all household members may not have access to that entire amount.

For the purposes of income reporting, *family income* is considered to be the sum of all incomes earned by an individual and all other family members. In the previous example, the husband, wife, and child would all be considered as part of a family whose income is $65,000. Although by strict definition, a family of one cannot exist, for the purposes of the distribution of total household income, the boarder would be considered a family of one whose income is $15,000.

Employment Status

A person's *employment status* falls into one of three distinct categories: employed, unemployed, or not in the labor force. An employed person is anyone at least 16 years old who worked <u>last week</u> for pay or for his or her own family's business, regardless of the number of hours worked. Persons with a job but who did not work last week due to illness, vacation, or other reasons, are also considered employed. An unemployed person is one who did not work at all last week, but who was available to accept a job and has looked for work during the last 4 weeks. A person is not in the labor force if he or she did not work at all last week and either hasn't looked for work during the last four weeks or did not want a job. Retired persons, housewives, and students are the most common examples of persons not in the labor force. A special class of persons called "discouraged workers" has recently been added to those who are not in the labor force. Discouraged workers are people who did not work last week, and have been out of work for a long time. They may have looked for work in the past, but have not looked in the last four weeks.

What constitutes "looking for work?" Interviewing for a job, sending resumes to companies, or answering newspaper ads are all considered looking for work. Reading the newspaper "help wanted" ads without following up is not considered looking for work.

Race

The Census Bureau classifies people into one of five distinct *racial categories*: 1) White, 2) Black, 3) American Indian, Eskimo or Aleut, 4) Asian or Pacific Islander, and 5) Other. Included among whites are those who claim their race is Canadian, German, Near Eastern Arab, or Polish among many others. Included among blacks are those who claim their race is Jamaican, Haitian, or Nigerian among many others. Included among Asian and Pacific Islanders are people who claim their race is Chinese, Filipino, Vietnamese, Hawaiian or Samoan among many others. Included among the "other" race category, are people who consider themselves to be multiracial and many people of Hispanic origin.

Ethnicity

An individual's *ethnicity* refers to what the person considers to be his or her origin. A person may consider him or herself Polish, Irish, African, Hispanic, English or one of many other ethnic origins. Origin means different things to different people. For example, origin can be interpreted as a person's ancestry, nationality group, lineage, or country of birth of the individual, his or her parents or ancestors. A person can have dual ethnicity, but may identify more closely with one and therefore consider themselves to be of that ethnicity.

Ethnicity is distinct from race. People of a given ethnicity may be of any race. A person can be black and English, or white and Greek. A person of Hispanic ethnicity may also consider her or himself to be black, white or neither of the two. The Census Bureau asks only one specific question on ethnicity which is, "Do you consider yourself to be of Hispanic origin?" This question is asked of all respondents along with questions of race. This produces a more accurate count of people of Hispanic origin than asking the single question, "Are you white, black, Hispanic, or something else?" Many people of Hispanic origin will choose white or black to this single questions.

There is no single list of ethnic categories now in use. Many people use race, Hispanic origin, and a question pertaining to origin to infer ethnicity. This way Poles can either be separated from Czechs or combined as Eastern Europeans.

HEALTH MEASUREMENTS

Medical professionals collect various data in order to measure the healthiness of individuals. Sometimes comparisons are made between individual measurements and healthy normal standards: e.g. blood pressure measurements give a direct indication of hypertension, temperature readings measure fever. Some information by itself does not give an indication of state of health, but can be useful when compared with other personal data: e.g. height by itself is not a health indicator, but height vs. weight does give medically helpful data.

Although many medical tests require invasive procedures, very useful data can be obtained without x-rays, blood tests or surgery. The following measurements, taken in class, are easily done and typical of data used in general or sports medicine.

Pulse

Pulse rate measures the number of heartbeats per minute. This varies considerably among individuals, but is typically lower when resting than when doing strenuous exercise. People who exercise regularly or are athletes have generally lower pulse rates than sedentary folk. There are several places on the body where a pulse can be felt; an easy one can be found by placing the first two fingers of one hand on the inside of the other wrist and experimenting with exact position until the pulse is found. The count is taken for 1 minute, under relaxed conditions, for a resting pulse rate. Typical results are between 60 and 100, with 72 being an average resting rate.

Blood Pressure

A blood pressure cuff wrapped around your upper arm at the height of your heart is the most common way of obtaining readings of systolic and diastolic blood pressure. These two readings correspond to different parts of the heart pumping cycle. With every heart beat blood pressure rises and falls in a cycle; the systolic pressure is the maximum and the diastolic is the minimum. The readings are given in millimeters of mercury, (mm Hg is the abbreviation), the same units by which atmospheric pressure is measured in a barometer.

High blood pressure is a medical condition called hypotension; it increases the risk of heart attacks and strokes, and may cause eye and kidney problems. For students in their late teens of twenties a systolic pressure over 140 or a diastolic over 90 would be considered overly high and reason to see a doctor. Hypertension or low pressure exists when the systolic pressure is below 90 mm of mercury. This may be normal or may represent a pathologic state depending on other symptoms such as lightheadedness, faintness or presence of fever. If possible a low or high blood pressure measurement should be compared to previous blood pressure measurements. A trained medical professional can use a blood pressure cuff, or readings can be obtained from home digital devices. An average healthy measurement might be 120/80 which is read as "120 over 80" and denotes the systolic, and then the diastolic reading.

Respirations

Respirations (rate of breathing, number of inhalations and expirations per minute) are an important indicator of essential body functions and consist primarily of carbon dioxide and oxygen exchange in the lungs. Without adequate functioning of the respiratory system none of the other body systems can perform effectively in maintaining a healthy body. The respiratory system includes many body organs and structures including the nose and throat, trachea and bronchi, lungs, diaphragm, intercostal muscles, brain and nerves. It depends heavily on the circulatory system for support of all functions.

Normally an individual is not aware of breathing during the usual activities of daily living and only notices increased respirations during strenuous exercise. This increase in rate and sometimes depth may be accomplished by mouth breathing, which is uncommon in healthy adults but normal for children under one year of age.

Individual differences in normal respiratory rates and capacity are influenced by body size and age. Generally breathing becomes slower and deeper as one ages through middle adulthood. Normal ranges are 20 to 30 respirations/minute for toddlers, 16 to 20 respirations/minute for adolescents and 12 to 20 respirations/minute for young and middle aged adults. Because of physiological changes due to aging, normal respiratory rates in the elderly may slightly decrease from their normal adult rate.

Assessing Respirations

To assess respirations count the rise (number of inspirations) of the chest for thirty seconds and multiply by two which is reported as respirations per minute. This is usually done as you finish assessing the pulse with fingers remaining in place on the wrist. The reason for this is to prevent the individual from consciously altering the rate of respirations which is sometimes done when he/she is aware that respirations are being counted.

Height/weight

Overweight or underweight for a given height and frame can be symptomatic of a variety of health problems. Insurance companies base premiums on extensive height/weight data; a majority of Americans are overweight by medical standards and they pay higher premiums because they are at greater risk, particularly for heart disease, the number one cause of death in the U.S..

Diet books and articles in the popular media discuss weight problems extensively, both from a cosmetic and a health perspective; although some promise painless weight loss with bizarre diets, others are more realistic and give accurate information about the health risks associated with weight gain or loss. Most books by physicians who specialize in weight control recommend low-fat eating habits combined with a regular program of safe exercise. Since muscle weighs more than fat, a previously sedentary person who starts an exercise program may gain weight initially as heavier muscle begins to replace lighter fat.

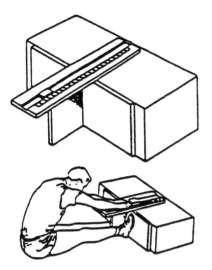

Standards for the sit and reach flexibility test.	
	SIT AND REACH
Excellent	22 in. or greater
Good	19-21 in.
Average	14-18 in.
Fair	12-13 in.
Poor	11 in. or less

From *Health Improvement Program*, National Athletic Health Institute, Inglewood, CA.

Flexibility

Lower back pain is a common complaint among adults. This problem is often related to lack of flexibility in the hips, lower back and ham strings (back of the legs). A flexibility measurement requires a simple device constructed by mounting a yardstick on a foot high box so that the 0 to 14 inch part of the yardstick hangs off the edge of the box. The subject sits with legs flat to the floor underneath the yardstick, and feet flat against the edge of the box. The measurement is taken by reaching forward as far as possible, without bending the knees, and reading the farthest inch mark at the tip of the fingers. For typical college age students an average flexibility measurement is 14-18 inches.

Eye chart

Visual ability varies immensely among different people, and even in one person each eye may have very different vision. Younger people tend to have more flexibility in their range of focus; as they get older they become less able to accommodate their vision to different distances and start to wear bi-focal or tri-focal lenses. Some diseases of the eye are more commonly seen in the elderly, and as age increases the likelihood of needing glasses also increases. For many students, college is the place where they first discover a need for corrective lenses.

A rough measurement of visual ability in each eye can be made with a small eye chart held 14 inches from the eyes. The text on the chart is in progressively smaller sizes; as with a wall mounted eye chart you read down as far as you can go. The eye rating, such as 20/20 or 20/40, is shown at the right of each paragraph. 20/20 means that you can see at 20 feet what a normal healthy eye can see at 20 feet; 20/40 means you see at 20 feet what normal eyes see at 40 feet, in other words, you can't see as far. A person with 20/20 vision is supposed to be able to read characters 1/3 inch high twenty feet away.

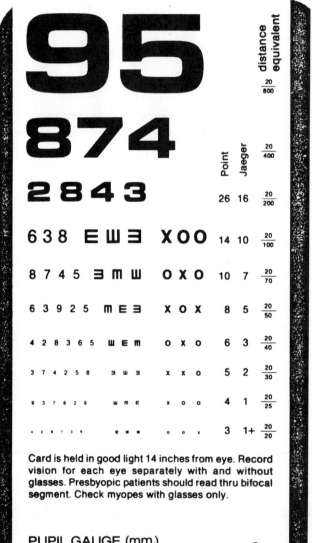

95

874

2843

638 EШЗ XOO

8745 ЗПШ OXO

63925 ПЕЗ XOX

428365 ШЕП OXO

374258 ЗШЗ X X O

937826 ШПЕ X O O

Point Jaeger distance equivalent

Point	Jaeger	distance equivalent
		20/800
		20/400
26	16	20/200
14	10	20/100
10	7	20/70
8	5	20/50
6	3	20/40
5	2	20/30
4	1	20/25
3	1+	20/20

Card is held in good light 14 inches from eye. Record vision for each eye separately with and without glasses. Presbyopic patients should read thru bifocal segment. Check myopes with glasses only.

PUPIL GAUGE (mm.)

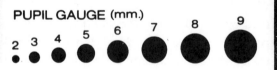

2 3 4 5 6 7 8 9

·UNITS & MEASUREMENT

Notes by Meg Hickey
© 1996

MEASUREMENT

In order to measure something, one must first be able to count, or quantify in some way. Human animals are unique in the sophistication of their ability to count, and thus make measurements. As the joke about mathematicians says: there are 3 kinds of mathematicians, those that can count and those that can't.

You can imagine early humans noticing that each of their arms and legs had a fixed quantity of similar items sticking off the end; with the aid of fingers and toes you can see how a system like our decimal numbers, based on 10, would be likely to evolve. If we were all born with 7 fingers on each hand, and 4 toes on each of 3 feet, no doubt things would have been different. As it is, though there are other counting systems in use by the scientific community, the decimal system is used and understood all over the world.

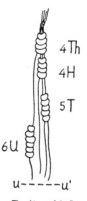

The *chimpu* of the Peruvian and Bolivian Indians, a descendant of the *quipu*. This one shows the number 4456.

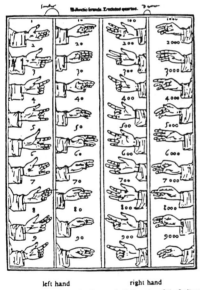

left hand   right hand

Finger counting from the *Summa de Arithmetica* of the Italian mathematician Luca Pacioli, the first important mathematical work to be printed, which was published in Venice in 1494.

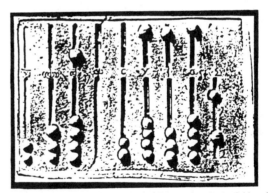

Roman hand abacus. Plaster cast of the specimen in the Cabinet des Médailles, Paris. Almost full size. Between the two rows of grooves are the Roman number symbols for

1 million 10 1   1

Integers   Unciae

From Karl Menninger's "Number Words and Number Symbols," published by the M.I.T. Press, Cambridge, MA

The counting system based on 2, called binary, is also not a surprising development since the quantity 2 is heavily featured in the human body. Since 2 can represent the dual states of a switch, either on or off, it is extremely convenient for use in computers, which operate by turning thousands of miniscule microchip switches on or off. Systems based on multiples of 2 like 8, octal, or 16, hexadecimal, have also been used in computers.

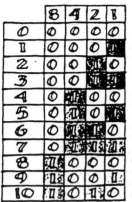

BINARY NUMBERS

Computers and calculators use binary code because it is the simplest number system. Its two digits — 0 and 1 — compare with our ten digits (0 to 9).

The table on the left shows how binary numbers relate to decimal numbers. Reading from right to left, the ones and zeros in each column indicate whether or not the number contains 1, 2, 4, 8 and so on, doubling each time. 0101, for example, is $0 \times 8 + 1 \times 4 + 0 \times 2 + 1 \times 1 = 5$. Each binary digit (0 or 1) is called a bit.

Most students have at some point been exposed to a letter grading system. Letters A to F may be given based on a student's numerical score, e.g. A for 100 to 91, B for 90 to 81, etc. Or, letter grades may be assigned by a subjective opinion of the teacher, e.g. a student's writing is lacking in originality but grammer and spelling have both improved, class participation is average and attendance is perfect: looks like a B+ grade.

A letter grade is a ranking system; it is looser than a number system for making judgements of quality or quantity. Similarly the egg grading system, "medium", "large", "extra large", and "jumbo" doesn't tell you the exact volume or weight of an egg, but it gives you a general idea of the size of the egg relative to other possible eggs. The clothing industry has size scales for shoes, socks, shirts, hats, gloves, dresses and trousers; the sizing may be fairly loosely determined, like unisex XS, S, M, L, XL; or it may be quite specific like $7^1/_2$AAA shoes which must fit closely in both length and width.

So, the first requirement for measurement is to have a quantity system to use as a scale; it could be 0-100 on the decimal number system, it could be X-rated to PG for movies, or it could be 1 to 8 on the Richter scale for earthquakes.

The second requirement for measurement is to know what it is you are measuring. Defining what is being measured could be as trivial as "the number of people in a room", or as sociologically complicated as "how well qualified a student is to enter college". Admissions counselors use a lot of different methods to measure preparedness for college: they may require high school grades, interviews, references from several people, a personal statement, essays, various national tests, and financial information. They must boil all this down to ranked categories like "early admit, admit, wait list, and reject", then they must make loan and scholarship decisions about the admitted students.

Measurement systems must be agreed upon by all of their users in order to accurately transmit the information obtained by measuring. Recent attempts to standardize systems internationally have resulted in the U.S. population becoming familiar with metric units like liters, meters, grams and degrees Celsius.

Before units of measurement were invented the passage of time by day and by moon could be told, courtesy of our solar system. The thousands of other units we use have been designed and refined by humans over time. Early units of length were based upon body parts such as the foot, the palm, the elbow to fingertip (cubit) and the stride or pace. Clearly these had the problem of not being standard.

The mediaeval rood measurement was defined as the combined length of the left feet of the first 16 men out of church in a given district. This was an attempt to base a unit on an average foot rather than a particular foot. The demand for standardization grew as international trade and scientific communication increased. Later, length was standardized on a platinum-irridium bar kept at a standard temperature in a vault in Sèvre, France. Currently the foot is based on the meter which is standardized in terms of the distance light travels in a certain time. This method gives the degree of precision necessary for modern physics, but is far more than the accuracy required for measuring the length of a room.

Queuing-up to arrive at a "right and lawful" rood in the 16th century. This authentic old depiction of the process laid down by Master Koebel does indeed place 16 assorted individuals toe-to-heel, old and young, obscure and important, just as they happen to come out of the door of the church at right. The three observing personages in the background are very likely the local commissioners for weights and measures, overseeing this averaging-out operation. (COURTESY OF ZEITLIN AND VER BRUGGE, BOOKS AND PRINTS, OF LOS ANGELES)

From H.A. Klein, "The Science of Measurement/
A Historical Survey" © 1974, Dover Publications, Inc.,
New York, N.Y. Reprinted with permission.

The increasing precision of measuring units has been made possible by advances in science and technology. Many things that are measured require a measuring tool or device, designed to compare what you are measuring with some established unit. The tool could be as simple as a meter stick, or as sophisticated as an astronomical observatory used for planetary measurements.

The science of measurement cuts across many fields, but it can be broadly divided into quantitative or qualitative measurements. Quantity measurements require a counting system, a definition of the units which are being counted, and sometimes a measurement tool.

Qualitative categorization of data organizes facts into named ("nominal") groups, or ordered ("ordinal") groups. Nominal data is something like "male or female" or "red, green, blue or purple"; these facts imply no value or particular order, they are simply observable qualities of the items being documented.

Ordinal, ranked measurements, on the other hand, imply order; either loosely for a group, like "small, medium or large", or quite specifically ranked for each individual, like 1st, 2nd, 3rd, 4th," etc. in a race. Ordinal measurements do not give you a specific unit measurement, like 30" waist or 2hrs 10min 32.5 sec running time in a race, but they do indicate where one item stands relative to others being measured. Sometimes quantity measurements are converted to ranked ordinal measurements.

Statisticians use all kinds of data. Our society uses statistics in a multitude of ways; we are bombarded with them at work, at school, in the news media, in all forms of advertising, and in the operation of our local, state, and national government. Each of us is the subject of considerable statistical measurement by the IRS, the US Census, the credit rating services, mail order catalog businesses, the medical establishment, insurance companies, schools,... to name a few. Here are some of the measurements they make on us, from highly quantified to loosely categorized:

Quantitative data

The IRS wants to know your total income, (say $23,671.75), measured in $ units This is a continuous quantitative measurement on a divisible scale.

The Registry of Motor Vehicles wants to know how many cars you have, (2), measured in whole cars only; they are not interested in fractions of cars you may have rusting away in your back yard. This is a discrete quantitative measurement using only integer numbers.

Qualitative data

Your college wants to know what position your grade point average was in your high school graduating class, (perhaps 7th from the top in a class of 22): this is an individually ranked ordinal measurement.

The U.S. census wants to know what level of education you have completed, grade school, middle school, high school, undergraduate, graduate or post graduate, (high school); this is a group ranked ordinal measurement.

The blood bank wants to know whether you have had hepatitus, measles, mumps; (no, no, yes); this is nominal measurement.

We have come a long way from the days in which the shop keeper would bite the coin you offered in order to decide whether you could be trusted. The vast sea of swirling computer data surrounding us all is available to the modern merchant who wants to check your credit card limit. The problem is that the data is only as reliable as the accuracy and appropriateness of how it is defined, collected, recorded, stored and interpreted. Despite the development of amazingly accurate computers, human errors in judgement, computation or data entry are compounded by mechanical failures of data processing or transmission equipment. Nonetheless, the pressures of doing business in a global economy with over 5 billion people have made cur society dependent on data and its machinery.

Financial data affects your ability to buy a house. Handgun data used effectively could prevent murders. AIDS data could save lives. Ecological data could save entire planetary species, including us. It is now a matter of global responsibility and human survival to be able to critique, understand, interpret and react to data measurements which are clues to future life on the planet.

UNITS

The history of measurement units has progressed from early attempts to measure quantities using things found in nature, such as parts of the body or seeds, to current scientific techniques which are accurate enough for sub-atomic particle physics, and often require a room full of equipment. For most of our daily measuring tasks we are able to use fairly simple devices; among the simplest are rulers for length, and measuring cups for volume. The clocks and scales we use to measure time and weight in the home are more complex, but still able to be built compactly.

Cesium Atomic Frequency
Standard No. NBS-6
at the National Bureau of Standards,
Boulder, Colorado:
the primary standard for
the unit of time in the U.S.

From The National Institute
of Standards and Technology

Over time, units have been defined and used with increasing precision. Many trades and professions have "rules of thumb" by which measurement decisions are made loosely; for instance the depth in inches of a concrete floor slab should be roughly 0.4 times the spanning distance in feet.

Some of the early attempts at defining measurement units we now find amusing. The density of beer used to be measured by an official tester who would pour some on a wooden bench, then, in his leather pants, sit in the puddle, swish it around, and feel how sticky the beer was when he tried to stand up.[1] This was a crude way to estimate how much the beer had been watered down. Some old units still in use today originated from very loose definitions. The candle power unit, used to measure the luminous intensity of a light source, was originally based on "the standard candle" which was a specified size and made of whale oil. The modern unit for luminous intensity is the candela, but photographers and lighting engineers still use the candle power, which is now defined very precisely using modern technology.

Many of the units now in use, in spite of attempts to replace them with the metric system, originated in the British Isles. The Magna Charta signed by the unpopular King John of England in 1215 had as one of its reforms a guarantee of uniform weight and measure standards. Queen Elizabeth I in the 1590's made another attempt at setting fair and uniform measures by law. By this time liquids like oil or honey were bought in units which doubled as they increased in volume, forming what is known as a geometric series, where each quantity is multiplied by the same number (in this case two) to get the next quantity.

---

[1] This and many other fascinating measurement units are described in "The Science of Measurement" by Herbert Arthur Klein.

The smallest unit was the mouthful (about 1/2 ounce U.S. weight). Double this was the jigger (this name is still in use in bars today, but it is now about $1^1/_2$ oz.), sometimes called the handful. Following this, each unit twice the amount of the preceding one, was the jack or jackpot, the gill or jill, the cup, the pint, the quart, the pottle, the gallon, the peck, the half bushel, the bushel, the cask, the barrel (32 gallons), the hogshead, the pipe or butt, and the tun. Variations on this basic doubling scheme exist today for volume or liquid capacity measures.

Increased royal taxation on volume units even as small as the jack and the jill was a cause of great ferment in the reign of Charles I. When the official size of all capacity units was also reduced, the poor were devastated; they had to pay much more to get a lot less. This familiar old nursery rhyme originated as their political protest against devalued units:

> Jack and Jill went up the hill
> To fetch a pail of water.
> Jack fell down and broke his crown,
> And Jill came tumbling after.

The oppressed people saw their rhymed prediction come true: King Charles "broke his crown" when he was beheaded in 1649.[2]

The definition and application of units can have broad political, sociological, economic and even public health consequences. In the early days of working with radioactive materials their dangers were not understood; even after it became known that serious sickness was caused by radiation, government decisions about exposure units were often made for economic convenience in nuclear weapons research rather than for the utmost safety of workers and surrounding citizens exposed to the radioactivity. In your working life you may never participate in the scientific definition of units, but as a consumer and a citizen you need to understand the implications of those that affect your life.

The world scientific community now uses the International System of Units (Systeme International des Unites) which is based on 8 primary units, 6 of them from the metric system: the meter for length, kilogram for mass, second for time, ampere for electric current, kelvin for temperature and candela for light intensity. The two other basic units, used for measuring plane and solid angles, are the radian and the steradian. From these eight basic units many other units derive, e.g. the liter is a unit of volume which is length x length x length = length^3, and it is defined from the meter.

The metric system length measurements, like the old capacity units, form a geometric series, but in this case each unit quantity is multiplied by ten to get the next unit.

> 10 millimeters is a centimeter,
> 10 centimeters is a decimeter,
> 10 decimeters is a meter,
> 10 meters is a dekameter,
> 10 dekameters is a hectometer,
> 10 hectometers is a kilometer.

Systems that multiply by 10 are called "decimal"; those that multiply by 2 are "binary".

---

[2] This historical side light from "The Science of Measurement" by Herbert Arthur Klein.

Some measurements are composed of several units, for instance speed of a car is measured in miles per hour, or in Canada and Europe kilometers per hour; in either case the basic units are length per time. By combining base properties such as length, time and mass, many other phenomena can be measured.

$$\text{Speed, distance traveled per time} = \frac{\text{length}}{\text{time}}$$

$$\text{Acceleration, speed change per time} = \frac{\text{length}}{\text{time}} \,/\, \text{time} = \frac{\text{length}}{\text{time}^2}$$

$$\text{Force} = \text{mass} \times \text{acceleration} = \frac{\text{mass} \times \text{length}}{\text{time}^2}$$

$$\text{Work} = \text{force} \times \text{distance} = \frac{\text{mass} \times \text{length} \times \text{length}}{\text{time}^2} = \frac{\text{mass} \times \text{length}^2}{\text{time}^2}$$

$$\text{Power} = \text{work per time} = \frac{\text{mass} \times \text{length}^2}{\text{time}^2} \,/\, \text{time} = \frac{\text{mass} \times \text{length}^2}{\text{time}^3}$$

Energy expended is work done, so all forms of energy or work can be measured in the same units. Since power is energy expended per time, power x time = work. Machines convert electrical energy into mechanical work, or vice versa, so that one form of energy can be transformed into another. Although the same units could be used for all forms of energy, each scientific discipline tends to use units convenient for them. Our electrical energy use is measured and paid for in kilowatthours, KWH. Heating engineers use British Thermal Units, BTU. Dieticians use Calories. The metric system uses joules.

Conserving energy resources is a global concern, yet most consumers of energy don't know much about energy units or how they relate to each other. The form of energy most people are familiar with is food. Any diet book will tell you that adults, depending on size and activity, need somewhere between 700 and 2500 Calories each day. The Calorie with a capital C, sometimes called a kilocalorie, is 1000 calories with a small c. An average chocolate chip cookie has about 250 Calories.

1 Calorie = 1 kilocalorie = 1000 calories = 4,185.5 joules.

The British Thermal Unit is roughly a quarter kilocalorie. 1 BTU = 1,054.5 joules.

Some air conditioners are rated in "tons;" this does not refer to the weight of the appliance. An air conditioning "ton" is based on the amount of cooling provided in one day by the melting of a 1 ton block of ice, hardly a convenient thing to test. A "ton" is about 12000 BTU, or roughly 3000 Calories, or 12 chocolate chip cookies.

Another form of energy unit the bill payer of a household becomes familiar with is the KWH. A 2 -bedroom apartment might typically use 200 KWH a month. The cost of a kilowatthour varies around the country; in some urban areas it is as high as 10 cents/KWH.

1 KWH = 3,600,000 joules = 3414 BTU = 860 Calories = about $3^1/_2$ choc. chip cookies.

Conversions from one unit to another are encountered in many professions, and as the metric system is used increasingly we are all forced to become familiar with new units. Anyone who travels abroad has to deal with conversions of monetary units, and, unlike other units where the conversion factor is defined as a constant, monetary exchange rates vary daily. Conversions are only possible for units having the same base properties: any length units are interchangeable, or any time units can be converted to each other; but you can not convert time units into velocity units, or work units into power units.

Questions:

1. At one time in the history of M.I.T. the shortest freshman in the DKE fraternity was selected as a unit of measurement. The unfortunate Mr. Smoot, who was about 5' 4", was used as a ruler to measure the nearby Harvard bridge, which turned out to be 340 Smoots long. Estimate the width of the river in feet, then convert to miles. There are 1760 yards in a mile, and 3 feet in a yard.

2. Research an ancient or unusual unit, from any country, and write a paragraph explaining it; or, invent a new unit that you think is appropriate to modern times. (e.g. an hours length of phone calls is actually called an Erlong, after the phone company engineer who invented it.)

3. You figure you can get by on £30 a day on a trip to England. After air fare you have $700 to spend on your vacation. If the current rate of exchange is £1 = $1.50, how many days can you afford? If the rate changes to £1 = $1.70, how long can you stay?

References:
Fundamentals of Physics, 3rd. edition, Halliday & Resnick, J. Wiley & Sons.
The Science of Measurement, A Historical Survey, Herbert Arthur Klein, Dover Publications.
Simplified Engineering for Architects and Builders, 5th. edition, Harry Parker, J. Wiley & Sons.
Number words and number symbols, Karl Menninger, M.I.T. press, Cambridge, MA.
The Universal Almanac, 1993, John W. Wright.
The VNR Concise Encyclopedia of Mathematics , Van Nostrand Reinhold, New York
The Science of Measurement, Herbert Arthur Klein, Dover Publications, N.Y.
Mathematics in Western Culture, Morris Kline, Oxford University Press.

VISUALIZATION

Notes by Meg Hickey
© 1996

Facts about our world are transmitted to us through all of our senses, but most of us acquire information primarily by seeing and hearing. Hearing is transitory, each sound occurs once in time then dies. With a recording you can hear sounds again, so you get another chance to assimilate what you heard the first time. Similarly, things we see in motion are transitory, and unless some record is made, video, painting, sculpture, photograph, chart, graph, or cartoon, the sight decays in memory.

Some visual records are made with the intention of giving the most accurate record possible of the subject; others, such as abstract paintings, political cartoons, and statistical charts, are often made with the deliberate intention of abbreviating, slanting, simplifying, changing, flattering, exaggerating or even misrepresenting the original subject. Regardless of their degree of faithfulness to the original, visual recordings have the advantage that they can be made to sit still in time while you study them; also, since they cover area or volume, a minimum of 2 or 3 dimensions can be seen in relation to each other simultaneously.

Visualization is an essential tool for writers who try to build a picture in your mind with words. A written description of a scene or person brings different images to mind for different readers, but a written description accompanied by an illustration or photograph pins down your vision to a particular image. Similarly, written lists of scientific data do not give an exact mental image of their meaning, but a chart, graph, or other visible picture of the data offers a much clearer interpretation of the data. Visualization is a vital tool for revealing trends otherwise hard to see in mathematical and scientific data. You can debate about the ratio, but it is generally fair to say that "a picture *is* worth a thousand words."

Mathematicians, scientists and statisticians from every discipline use tables, charts, graphs and maps for looking at data in a visual way rather than as a string of words and/or numbers and symbols. News media present statistical data to us with techniques designed to have more popular appeal, such as political cartoons, animated weather maps, or illustrated graphs. National debt statistics are a classic example; this is such a serious and vital topic that attempts to explain it are made throughout society by politicians, news media, teachers, and even comedians.

Just like credit card users whose debt is so great they can never get to paying off the principal because they must pay off the ever increasing interest first, the U.S. must devote more and more of its resources to paying off interest as the national debt grows faster and faster. Laurie Anderson tackles this issue by projecting a huge map of the United States on the screen in one of her performances in 1990. She then points out that the people on the east coast side of the Amtrak line running through Boston, New York and Washington are paying the interest on the national debt, and the people on the west side are paying the principal. Then she tells the audience that within 10 years, at the current rate of growth, those on the east of the Mississippi will be paying the interest, those on the west, the principal. The horrified audience reaction demonstrates that a theater full of voters understood the gravity of the issue from this very effective method of visualization.

The British national debt chart shown here was done in 1786 by William Playfair, who invented charts for the analysis and presentation of numerical data. Yet another treatment of the debt issue is the 1936 Herbert Johnson classic cartoon showing the overwhelming effect of government spending on the individual taxpayer.

239. "NONSENSE, IF IT GETS TOO DEEP, YOU CAN EASILY PULL ME OUT."
BY HERBERT JOHNSON, 1936.

From "Cartoons by Herbert Johnson"
by Herbert Johnson, published 1936

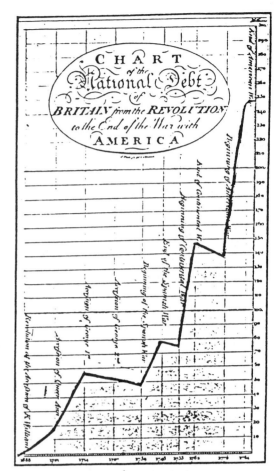

The Divisions at the Bottom are Years, & those on the Right hand Money.

From the Commercial and Political Atlas © 1786, London

The basic 2D visual techniques of presenting data are: tables, bar charts, pie charts, graphs, trees and maps. Following is a selection of examples, plain and fancy, from both technical and non-technical sources.

Tables:

Tables are a clear visual way of organizing data facts. A classic form of table, seen by children everywhere, is the multiplication table; as they grow older they are likely to encounter the duty roster; by the time they are old enough to buy a used car in this country they will probably consult the frequency of repair tables in Consumer Reports; and, when they drive the car they can look up the mileage between 2 cities on a mileage table. Tables of scientific or statistical data are often the first step in producing another form of visualization, such as a graph or bar chart.

From the New England Road Map by the American Map Corporation, © 1984, Maspeth, N.Y.

APPROXIMATE MILEAGES	Albany, N.Y.	Augusta, Me	Boston, Mass	Concord, N.H.	Hartford, Conn.	Montpelier, Vt	Montreal, P.Q	New York, N.Y.	Providence, R.I.
Albany, N.Y.		790	165	150	100	160	215	140	160
Augusta, Me	790		160	140	250	175	750	355	205
Bangor, Me	355	65	225	205	320	240	310	425	275
Bar Harbor, Me	790	105	260	240	355	280	355	460	305
Bennington, Vt	40	750	140	110	105	170	215	175	145
Boston, Mass	165	160		70	95	180	310	205	45
Bridgeport, Conn	175	305	150	195	50	245	345	55	120
Burlington, Vt	145	210	220	145	220	35	100	285	235
Calais, Me	450	160	320	300	410	335	360	520	365
Concord, N.H	150	140	70		140	110	240	250	110
Danbury, Conn	115	305	155	195	55	240	375	60	130
Greenfield, Mass	80	225	95	90	60	135	260	165	100
Hartford, Conn	100	250	95	140		195	320	105	70
Keene, N.H	95	195	80	55	90	115	240	200	100
Laconia, N.H	170	145	95	25	170	95	230	270	135
Lowell, Mass	155	150	25	45	100	150	285	210	65
Manchester, N.H	150	140	55	15	125	125	255	230	95
Montpelier, Vt	160	175	185	110	195		135	795	210
Montreal, P.Q	215	750	310	240	320	135		355	335
New Bedford, Mass	185	215	55	175	100	230	365	200	30
New Haven, Conn	115	285	135	180	35	230	345	75	105
New London, Conn	145	265	100	155	45	240	365	170	55
Newport R.I	185	225	65	135	85	240	365	175	30
Newport, Vt	215	165	270	150	245	60	105	350	260
New York, N.Y	140	355	205	250	105	795	355		175
Pittsfield, Mass	40	270	135	135	70	155	245	140	130
Portland, Me	235	55	105	85	195	155	270	300	150
Portsmouth, N.H	190	105	55	45	145	145	780	255	100
Presque Isle, Me	510	270	380	360	475	395	395	580	410
Providence, R.I	160	205	45	110	70	210	335	175	
Provincetown, Mass	260	280	115	190	185	795	430	785	115
Quebec, P.Q	365	220	385	320	420	215	150	510	420
Rutland, Vt	95	275	155	95	150	65	165	230	170
Springfield, Mass	83	240	90	120	25	170	290	130	80
White Mtns, N.H (Twin Mts)	230	110	155	85	275	70	175	330	195
Worcester, Mass	130	190	40	80	60	170	795	165	40

○ = Much better than average

○ = Better than average

○ = Average

◉ = Worse than average

● = Much worse than average

Trouble Spots	Toyota Corolla (except Tercel) 76 77 78 79 80 81	Volkswagen Rabbit (diesel) 76 77 78 79 80 81	Volvo 240 series 76 77 78 79 80 81
Air-conditioning	○ ○ ○ ○ ○ ○	● ● ○ ● ○	○ ○ ○ ○ ● ○
Body exterior (paint)	○ ○ ○ ○ ○ ○	○ ○ ○ ○ ○	○ ○ ○ ○ ○ ○
Body exterior (rust)	○ ○ ○ ○ ○ ○	○ ○ ○ ○ ○	○ ○ ○ ○ ○ ○
Body hardware	○ ○ ○ ○ ○ ○	○ ○ ○ ○ ○	○ ○ ○ ○ ○ ○
Body integrity	○ ○ ○ ○ ○ ○	○ ○ ○ ○ ○	○ ○ ○ ○ ○ ○
Brakes	○ ○ ○ ○ ◉ ○	○ ○ ○ ○ ○	○ ○ ○ ○ ○ ○
Clutch	○ ○ ○ ○ ○ ○	○ ○ ○ ○ ○	○ ○ ○ ○ ○ ○
Driveline	○ ○ ○ ○ ○ ○	○ ○ ○ ○ ○	○ ○ ○ ○ ○ ○
Electrical system (chassis)	○ ○ ○ ○ ○ ○	○ ○ ○ ○ ○	○ ○ ○ ○ ○ ○
Engine cooling	○ ○ ○ ○ ○ ○	○ ● ● ○ ○	○ ○ ○ ○ ○ ○
Engine mechanical	○ ○ ○ ○ ○ ○	○ ○ ○ ○ ○	○ ○ ○ ○ ○ ○
Exhaust system	○ ○ ○ ○ ○ ○	○ ● ● ● ○	◉ ● ● ○ ○ ○
Fuel system	○ ○ ○ ○ ○ ○	○ ○ ○ ○ ○	○ ○ ○ ○ ○ ○
Ignition system	○ ○ ○ ○ ○ ○	○ ○ ○ ○ ○	○ ○ ○ ○ ○ ○
Suspension	○ ○ ○ ○ ○ ○	○ ○ ○ ○ ○	○ ○ ○ ○ ○ ○
Transmission (manual)	○ ○ ○ ○ ○ ○	○ ○ ○ ○ ○	○ ○ ○ ○ ○ ○
Transmission (automatic)	○ ○ ○ ○ ○ ○		○ ○ ○ ○ ○ ○
Trouble Index	○ ○ ○ ○ ○ ○	○ ○ ○ ○ ○	○ ○ ○ ○ ○ ○
Cost Index	○ ○ ○ ○ ○	○ ○ ○ ○	● ○ ○ ○ ○

Graphs:

Graphs originated with René Descartes (1596-1650) in his study of geometry. In order to specify where a point was on a surface he used 2 crossed lines with measurement scales marked on them: he could locate any point by moving over a horizontal measurement, then up or down a vertical measurement.

This is like compass directions that say "Go 12 feet west then 7 feet north"; if we added "then dig down 5 feet" we would be changing from a 2-dimensional to a 3-dimensional system. In algebra it is customary to call the horizontal scaled line the x axis (or "abscissa"), and the vertical scaled line the y axis (or "ordinate"). The x and y values for each point are called its "coordinates," and the 4 quarters of the graph are called "quadrants". Since x and y can vary up and down the scaled axes, they are called "variables".

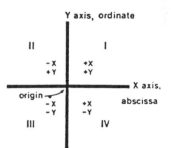

From A.J. MacGregor, "Graphics Simplified" ©, University of Toronto Press, 1979, Buffalo, N.Y.

Any 2 quantities related to each other can be plotted, such as height versus age of a person, or a formula like $y = 2x-1$. If your height was measured on each birthday you can plot a point for every year of your age; each point is the intersection of the age value on the x axis with the height value on the y axis. If you connect up the points with short lines you then have a graph to visualise your growth. With a formula like $y = 2x-1$ you can produce a continuous graph because you can calculate a y value for any x value. Some variables are completely discontinuous, like the number of umbrellas sold in New York each day; others are in fact a continuous fluctuation, like air temperature, but if measured as a daily average give jumpy looking data.

Charts:

The word "chart" is used loosely to describe several kinds of comparative data displays. A graph is sometimes called a line chart or "fever chart" after a hospital patient's temperature graph. Line charts are very versatile; several lines can be plotted on the same coordinates, such as the height vs. age for each of 3 children; or several lines can add up to a total line, such as number of students enrolled each year in a college, showing male and female sections of the total. Bar charts show horizontal or vertical bars whose lengths represent quantities. Bar charts that show data vertically are sometimes called column charts. Column charts are seen very frequently in company annual reports where annual income for the past years can be shown as vertical bars scaled to income. Sometimes the bars are illustrated, for income it would be typical to use a pile of coins to show bar height.

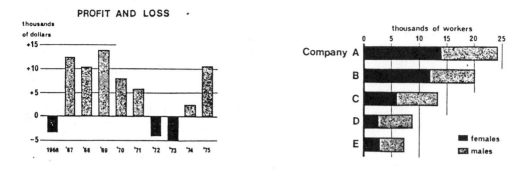

PROFIT AND LOSS

From A.J. MacGregor, "Graphics Simplified" ©, University of Toronto Press, 1979, Buffalo, N.Y.

Bars in charts show the quantity by changing in length, the width stays the same for each bar. It is misleading to use a column illustration that increases in both width and length to represent bigger numbers (see barrels). The problem is that your eyes compare the increase in area rather than the increase in height; if the data has only doubled but the bar is shown doubled in height and width, then the area of the bar has quadrupled. The branch of science that deals with dimensional distortion from models is called dimensional analysis.

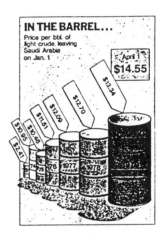

© 1979 Time Inc. Reprinted with permission.

Some bar charts show a spread of data, like the daily high and low of the stock market index. Another sort of chart is called a "timeline;" it marks events in history or in a persons life, as they occur in time. Timelines are also used for construction scheduling; bars show different phases of the work, including waiting time for materials to arrive or concrete to set.

Histograms are bar charts which show frequency distribution, or how many of your data population fit in graduated categories of a continuous variable. For example, if you measured the height of everyone in your class, probably you would get measurements between 58" and 84", but the majority would fall between 62" and 74", and the average, also called the "mean", might be 68".The human histogram shown uses the actual students at each height to represent the data count columns. The longest column in a histogram is called the "mode", if there are two columns of the largest size then the data is said to be "bi-modal". The mode, which is the most frequently occurring data category, is not necessarily the mean. If the data is arranged in ascending order of magnitude, then the middle piece of data is the "median".

Brian L. Joiner, "Living Histograms,"
International Statistical Review, 43 (1975), 339-340

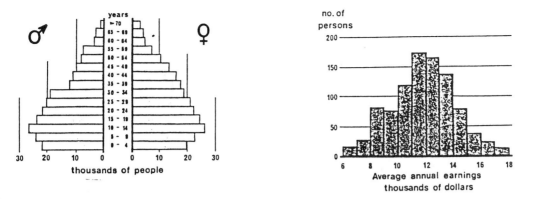

thousands of people

Average annual earnings
thousands of dollars

From A.J. MacGregor, "Graphics Simplified" ©, University of Toronto Press, 1979, Buffalo, N.Y.

A pie chart represents a total quantity as a circular area; it is divided up into wedge shaped pieces each proportional to parts of the total. Here is an example: A will leaves money to 4 survivors in proportion to their ages which are 56, 45, 32 and 17; show each person's share using a pie chart.

The first step is to find the age data total:

$56 + 45 + 32 + 17 = 150$

Each person gets age/total ages as a proportion.

The oldest gets $56/150 = 0.37 = 37\%$

Each proportion is converted to the number of degrees of a 360° circle: for the oldest this is $0.37(360°) = 126°$

If you do not have software that draws the pie chart, a protractor is needed for measuring angles on the chart.

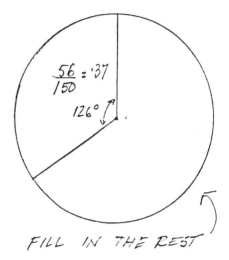

FILL IN THE REST

Trees:

Family trees are familiar to most people as a way of visualising who is related to whom. Trees have also been used to show the development of different languages from root languages, or to show organizational structure branching out from a chief executive. Decision trees are diagrams of what choices branch out from each decision; these can get so complicated when some of the later decisions feed back into earlier choices that the "tree" looks more like overgrown jungle. Critical Path Method diagrams are logical decision trees used in industry for making the most efficient choices to optimise cost or production time.

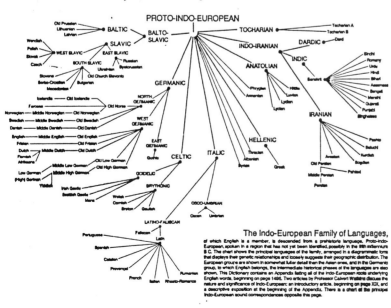

The Indo-European Family of Languages,

Maps:

Globes, road maps, house plans and orchestra seating plans are all maps; they show the physical relationship of each item or place with respect to the others. Anyone who has ever tried to wrap a spherical object in a piece of paper realises that representing a spherical planet to scale on a planar piece of paper, without wrinkles, is impossible. Continental maps must gradually expand from the equator to the poles, so that the northern part of Canada looks much bigger than it is relative to places near the equator. Some maps attempt to use a proportional measurement scale, others may be distorted, like a subway map, where you don't need to know the exact distance, but you do need to know which order the stations are in and where they connect.

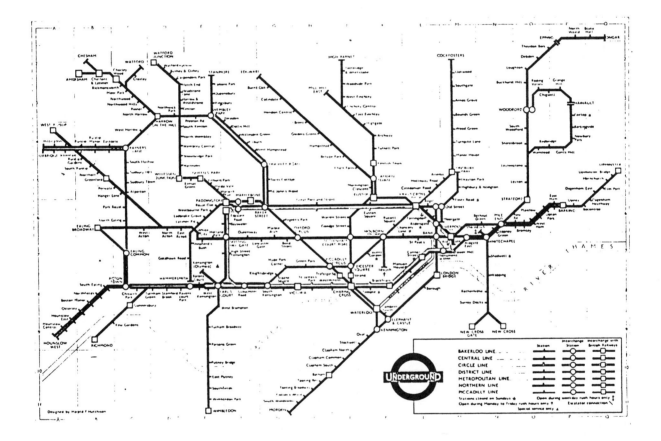

London Transport Museum copyright © 1983, LRT registered user number 95/E/676. Reprinted with permission.

A classic map/chart is Charles Minard's visualization of the decimation of Napoleon's army as it advanced to, and retreated from, Moscow in the ill-starred Russian campaign of 1812 and 1813. The width of the lighter band represents 422,000 men starting at the Polish border and, after suffering appalling losses, finally arriving with 100,000 men in Moscow. The return of these unfortunate troops in sub-zero weather is shown by the steadily diminishing darker band with an accompanying temperature graph; only 10,000 made it back, many of their companions had frozen to death. Minard's brilliant technique shows army size related to route, significant geographic locations where battles or obstacles were encountered, and temperatures related to the times and places of the retreat.

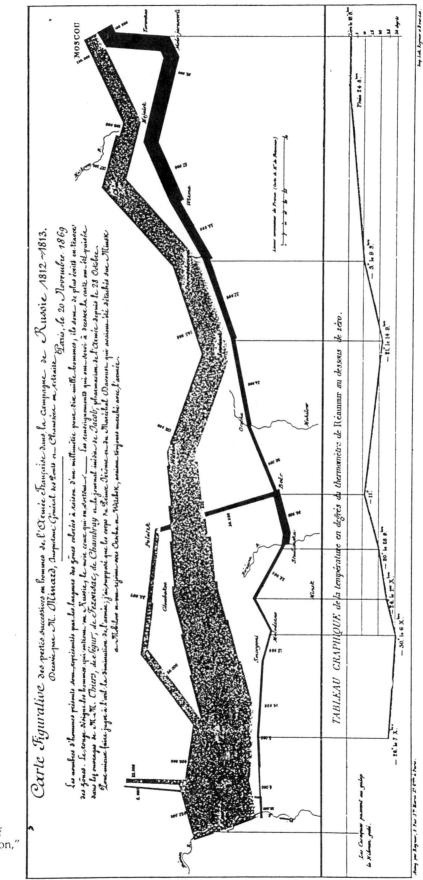

Edward R. Tufte,
"The Visual Display of
Quantitative Information,"
(Cheshire, Connecticut:
Graphics Press, 1983).

Another classic is this map·of cholera cases in London in 1854. Dr. John Snow plotted a spot on the map showing the house of every person who died of Asiatic cholera in a virulent epidemic centered around Broad Street. He suspected the water from the public pump at Broad St. was contaminated, but for a while he was confused by the fact that some of the deaths occurred in locations closer to other pumps. Further investigation revealed that people closer to other pumps were making special trips to the Broad St. pump because its contaminated water tasted better! Inactivation of the Broad Street pump finally stopped the epidemic.

Dr. Snow had a very hard time convincing the authorities that his medical detective work was true, but now he is recognised as a founder of the science of epidemiology, which uses statistics to uncover reasons for the spread of disease.

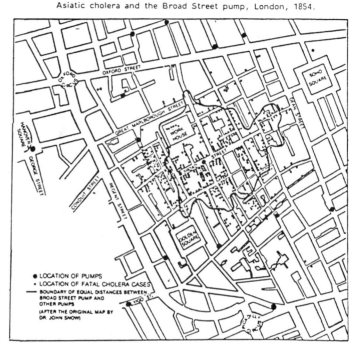

Asiatic cholera and the Broad Street pump, London, 1854.

● LOCATION OF PUMPS
• LOCATION OF FATAL CHOLERA CASES
— BOUNDARY OF EQUAL DISTANCES BETWEEN
BROAD STREET PUMP AND
OTHER PUMPS
(AFTER THE ORIGINAL MAP BY
DR. JOHN SNOW)

From "Snow on Cholera" by J. Snow. Reprinted in "Principles of Epidemiology" by L.H. Roht et al. by permission of the Harvard University Press.

Regardless of what combination of techniques is used to visualise data, the important questions to ask are:

Does the data make sense? Is it biased by special conditions or interest groups? Was it obtained by consistent and well considered methods? Are there any obvious omissions?

What does the data mean? Can the data be projected for the future? What implications does it have in our lives?

Who needs to know this data? What are effective ways to communicate the data to these people?

All of these questions require common sense, ethics, and enough mathematics to know when to be suspicious of what you are seeing.

References:
The Graphic Display of Quantitative Information, and, Envisioning Information, Edward R. Tufte
American Heritage Dictionary, Houghton Mifflin Co., Boston, MA
A History of American Graphic Humor (1865-1938), William Murrell, Cooper Square Publishers, Inc., N.Y., 1967
Graphics Simplified, A.J. MacGregor, University of Toronto Press, Toronto
Designer's Guide to Creating Charts & Diagrams, Nigel Holmes, Watson Guptill Publications, N.Y.
Graphis Diagrams, Walter Herdeg, Graphis Press Corp., Zurich, Switzerland, 1983
Innumeracy, and, Beyond Numeracy, J.A. Paulos, Vintage Books, N.Y.
Principles of Epidemiology, Roht, Selwyn, Holguin & Christensen, Academic Press/ Harcourt, Brace Jovanovich Publishers

ESSAY STEPHEN JAY GOULD

The Median Isn't the Message

My life has recently intersected, in a most personal way, two of Mark Twain's famous quips. One I shall defer to the end of this essay. The other (sometimes attributed to Disraeli), identifies three species of mendacity, each worse than the one before—lies, damned lies, and statistics.

Consider the standard example of stretching truth with numbers—a case quite relevant to my story. Statistics recognizes different measures of an "average," or central tendency. The *mean* is our usual concept of an overall average—add up the items and divide them by the number of sharers (100 candy bars collected for five kids next Halloween will yield 20 for each in a just world). The *median*, a different measure of central tendency, is the half-way point. If I line up five kids by height, the median child is shorter than two and taller than the other two (who might have trouble getting their mean share of the candy). A politician in power might say with pride, "The mean income of our citizens is $15,000 per year." The leader of the opposition might retort, "But half our citizens make less than $10,000 per year." Both are right, but neither cites a statistic with impassive objectivity. The first invokes a mean, the second a median. (Means are higher than medians in such cases because one millionaire may outweigh hundreds of poor people in setting a mean; but he can balance only one mendicant in calculating a median).

The larger issue that creates a common distrust or contempt for statistics is more troubling. Many people make an unfortunate and invalid separation between heart and mind, or feeling and intellect. In some contemporary traditions, abetted by attitudes stereotypically centered upon Southern California, feelings are exalted as more "real" and the only proper basis for action—if it feels good, do it—while intellect gets short shrift as a hang-up of outmoded elitism. Statistics, in this absurd dichotomy, often become the symbol of the enemy. As Hilaire Belloc wrote, "Statistics are the triumph of the quantitative method, and the quantitative method is the victory of sterility and death."

This is a personal story of statistics, properly interpreted, as profoundly nurturant and life-giving. It declares holy war on the downgrading of intellect by telling a small story about the utility of dry, academic knowledge about science. Heart and head are focal points of one body, one personality.

In July 1982, I learned that I was suffering from abdominal mesothelioma, a rare and serious cancer usually associated with exposure to asbestos. When I revived after surgery, I asked my first question of my doctor and chemotherapist: "What is the best technical literature about mesothelioma?" She replied, with a touch of diplomacy (the only departure she has ever made from direct frankness), that the medical literature contained nothing really worth reading.

Of course, trying to keep an intellectual away from literature works about as well as recommending chastity to *Homo sapiens*, the sexiest primate of all. As soon as I could walk, I made a beeline for Harvard's Countway medical library and punched mesothelioma into the computer's bibliographic search program. An hour later, surrounded by the latest literature on abdominal mesothelioma, I realized with a gulp why my doctor had offered that humane advice. The literature couldn't have been more brutally clear: mesothelioma is incurable, with a median mortality of only eight months after discovery. I sat stunned for about fifteen minutes, then smiled and said to myself: so that's why they didn't give me anything to read. Then my mind started to work again, thank goodness.

If a little learning could ever be a dangerous thing, I had encountered a classic example. Attitude clearly matters in fighting cancer. We don't know why (from my old-style materialistic perspective, I suspect that mental states feed back upon the immune system). But match people with the same cancer for age, class, health, socioeconomic status, and, in general, those with positive attitudes, with a strong will and purpose for living, with commitment to struggle, with an active response to aiding their own treatment and not just a passive acceptance of anything doctors say, tend to live longer. A few months later I asked Sir Peter Medawar, my personal scientific guru and a Nobelist in immunology, what the best prescription for success against cancer might be. "A sanguine personality," he replied. Fortunately (since one can't reconstruct oneself at short notice and for a definite purpose), I am, if anything, even-tempered and confident in just this manner.

Hence the dilemma for humane doctors: since attitude matters so critically, should such a sombre conclusion be advertised, especially since few people have sufficient understanding of statistics to evalu-

---

Stephen Jay Gould teaches biology, geology, and the history of science at Harvard.

ate what the statements really mean? From years of experience with the small-scale evolution of Bahamian land snails treated quantitatively, I have developed this technical knowledge—and I am convinced that it played a major role in saving my life. Knowledge is indeed power, in Bacon's proverb.

The problem may be briefly stated: What does "median mortality of eight months" signify in our vernacular? I suspect that most people, without training in statistics, would read such a statement as "I will probably be dead in eight months"—the very conclusion that must be avoided, since it isn't so, and since attitude matters so much.

I was not, of course, overjoyed, but I didn't read the statement in this vernacular way either. My technical training enjoined a different perspective on "eight months median mortality." The point is a subtle one, but profound—for it embodies the distinctive way of thinking in my own field of evolutionary biology and natural history.

We still carry the historical baggage of a Platonic heritage that seeks sharp essences and definite boundaries. (Thus we hope to find an unambiguous "beginning of life" or "definition of death," although nature often comes to us as irreducible continua.) This Platonic heritage, with its emphasis on clear distinctions and separated immutable entities, leads us to view statistical measures of central tendency wrongly, indeed opposite to the appropriate interpretation in our actual world of variation, shadings, and continua. In short, we view means and medians as the hard

"realities," and the variation that permits their calculation as a set of transient and imperfect measurements of this hidden essence. If the median is the reality and variation around the median just a device for its calculation, the "I will probably be dead in eight months" may pass as a reasonable interpretation.

But all evolutionary biologists know that variation itself is nature's only irreducible essence. Variation is the hard reality, not a set of imperfect measures for a central tendency. Means and medians are the abstractions. Therefore, I looked at the mesothelioma statistics quite differently—and not only because I am an optimist who tends to see the doughnut instead of the hole, but primarily because I know that variation itself is the reality. I had to place myself amidst the variation.

When I learned about the eight-month median, my first intellectual reaction was: fine, half the people will live longer; now what are my chances of being in that half. I read for a furious and nervous hour and concluded, with relief: damned good. I possessed every one of the characteristics conferring a probability of longer life: I was young; my disease had been recognized in a relatively early stage; I would receive the nation's best medical treatment; I had the world to live for; I knew how to read the data properly and not despair.

Another technical point then added even more solace. I immediately recognized that the distribution of variation about the eight-month median would almost surely be what statisticians call "right skewed." (In a symmetrical distribution, the profile of variation to the left of the central tendency is a mirror image of variation to the right. In skewed distributions, variation to one side of the central tendency is more stretched out—left skewed if extended to the the left, right skewed if stretched out to the right.) The distribution of variation had to be right skewed, I reasoned. After all, the left of the distribution contains an irrevocable lower boundary of zero (since mesothelioma can only be identified at death or before). Thus there isn't much room for the distribution's lower (or left) half—it must be scrunched up between zero and eight months. But the upper (or right) half can extend out for years and years, even if nobody ultimately survives. The distribution must be right skewed, and I

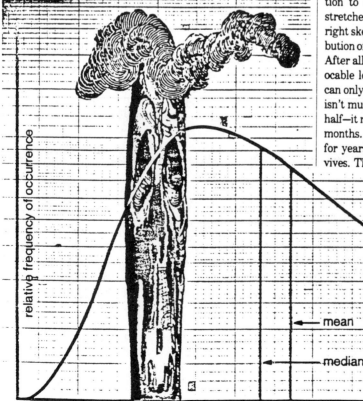

relative frequency of occurrence

← mean

← median

needed to know how long the extended tail ran—for I had already concluded that my favorable profile made me a good candidate for that part of the curve.

The distribution was, indeed, strongly right skewed, with a long tail (however small) that extended for several years above the eight month median. I saw no reason why I shouldn't be in that small tail, and I breathed a very long sigh of relief. My technical knowledge had helped. I had read the graph correctly. I had asked the right question and found the answers. I had obtained, in all probability, that most precious of all possible gifts in the circumstances—substantial time. I didn't have to stop and immediately follow Isaiah's injunction to Hezekiah—set thine house in order: for thou shalt die, and not live. I would have time to think, to plan, and to fight.

One final point about statistical distributions. They apply only to a prescribed set of circumstances—in this case to survival with mesothelioma under conventional modes of treatment. If circumstances change, the distribution may alter. I was placed on an experimental protocol of treatment and, if fortune holds, will be in the first cohort of a new distribution with high median and a right tail extending to death by natural causes at advanced old age.

It has become, in my view, a bit too trendy to regard the acceptance of death as something tantamount to intrinsic dignity. Of course I agree with the preacher of Ecclesiastes that there is a time to love and a time to die—and when my skein runs out I hope to face the end calmly and in my own way. For most situations, however, I prefer the more martial view that death is the ultimate enemy—and I find nothing reproachable in those who rage mightily against the dying of the light.

The swords of battle are numerous, and none more effective than humor. My death was announced at a meeting of my colleagues in Scotland, and I almost experienced the delicious pleasure of reading my obituary penned by one of my best friends (the so-and-so got suspicious and checked; he too is a statistician, and didn't expect to find me so far out on the left tail). Still, the incident provided my first good laugh after the diagnosis. Just think, I almost got to repeat Mark Twain's most famous line of all: the reports of my death are greatly exaggerated. ▣

> The problem may be briefly stated: What does "median mortality of eight months" signify in our vernacular? I suspect that most people, without training in statistics, would read such a statement as "I will probably be dead in eight months"—the very conclusion that must be avoided, since it isn't so.

CHANCE News

Prepared by J. Laurie Snell, with help from William Peterson, Fuxing Hou and Ma. Katrina Munoz Dy, as part of the CHANCE Course Project supported by the National Science Foundation.

Please send comments and suggestions for articles to:

jlsnell@dartmouth.edu

Back issues of Chance News and other materials for teaching a CHANCE course are available from the Chance Web Data Base.

http://www.geom.umn.edu/locate/chance

==
" Data, data everywhere, but not a thought to think"
                                        Jesse Shera's paraphrase of Coleridge.
==

We found this quote in John Paulos' new book *A Mathematician Reads the Newspaper*.

FROM OUR READERS

Jerry Johnson sent us the following excerpts from a discussion on a journalism listserve group.

I teach statistics at the University of Texas at Arlington. Two weeks ago I read in the science section of a local paper an article defining the difference between the median, the mean, and the average. Everything was fine until he defined the mean as the average of the largest and smallest numbers in a set of data. I have always used and taught that the (arithmetic) mean is the same as the average of the numbers. When I talked to him about this, he indicated that this was the definition given in an Associated Press list of definitions. Is this the definition used by journalists?

Thanks for everyone who answered my question concerning the mean. I have contacted the Associated Press and hope to change their definition. One of the problems seems to come from dictionaries that define the mean as midway between extremes. I contacted Merriam-Webster and got one editor there to agree that midway between extremes is in a philosophical sense and not a mathematical sense. Webster's New World Dictionary has a more specific definition as "a middle or intermediate position as to place, time, quantity, kind, value,..."

After discussion with Norm Goldstein, Director of APN Special Projects, the Associated Press Style Book will be modified to indicate that the calculation of the mean is identical to that of the average.

A Fragmented War on Cancer

By Hamilton Jordan

ATLANTA

It has been 25 years since President Richard Nixon declared war on cancer. Having had two different cancers, I am a survivor of that war and a grateful beneficiary. My first, an aggressive lymphoma, was treated with an experimental therapy developed at the National Cancer Institute. Ten years later, my prostate cancer was detected early by the simple P.S.A. blood test, a diagnostic tool supported by Federal grants.

But I am also a symbol of the limited success of that war. The treatments I received were merely updated versions of the methods used 25 years ago. A powerful cocktail of chemicals killed my lymphoma while ravaging my body. A surgeon, using an elegant procedure with no permanent side effects, cut out my prostate.

Scientists are still looking for both the "magic bullet" that kills only

Hamilton Jordan, who was President Jimmy Carter's chief of staff, is a board member of Capcure, a nonprofit organization that finances prostate cancer research.

cancer cells and the genetic switch that turns off random cancer growth or prevents genetic flaws from causing cancer. While significant progress has been made, twice as many people will be diagnosed with cancer this year as in 1971, and twice as many will die. One in three women and one in two men will have cancer in their lifetimes. The raw data suggest we are on the verge of an epidemic. What happened to the "war"?

Groups compete for a shrinking pie.

● Our rhetoric exceeded our commitment. Dr. Donald Coffey, a cancer researcher, says we promised a war but financed only a few skirmishes. The Federal budget expresses our national priorities: The Federal Aviation Administration, for example, will spend $8.92 billion to make air travel safe. The chances of dying in an airline accident are one in two million. But the National Cancer Institute will spend only $2.2 billion — one-tenth of a cent of every Federal tax dollar — to find a cure for the disease that kills

more Americans in a month than have died in all commercial aviation accidents in our history.

● We created expectations not based on scientific reality. Our political leaders failed to appreciate the simple reality that there is not just one cancer but more than 100 cancers that have all defied a single solution. At the same time, the numerous organizations representing those different cancers have fought among themselves for bigger slices of a shrinking pie instead of forging a consensus on behalf of a larger pie.

● Huge successes with some cancers have been offset by rises in others. In addition, mortality from other diseases has declined, leaving an aging, cancer-prone population.

With adequate financing, breakthroughs in cancer prevention and treatment are likely over the next decade. Yet promising research that would have been automatically financed a decade ago is rejected today because of belt-tightening, discouraging brilliant young investigators from entering cancer research in the first place.

Is one-tenth of one cent enough to find a cure for a disease that will strike 40 percent of Americans? You will not think so when cancer strikes you or your loved ones. □

The New York Times Editorials/ Letters, Saturday June 1, 1996

Promote Cancer Treatment, Not Cancer Phobia

To the Editor:

In "A Fragmented War on Cancer" (Op-Ed, May 29), Hamilton Jordan says, "The raw data suggest we are on the verge of an epidemic."

His assertions that cancer will be diagnosed in twice as many people and that twice as many will die of it this year than in 1971 are misleading.

More people will be diagnosed with and die of cancer this year than in 1971 because the population is now much larger. Moreover, cancer is much more common for older

age groups than younger age groups, and the proportion of Americans who are in older age groups is larger this year than in 1971.

After you take into account population size and the number of people in different age groups, recent national cancer data suggest a slight downward trend in overall cancer mortality during the current decade.

A reported increase in the diagnosis of cancer usually reflects changes in diagnostic and reporting practices rather than an actual increase in cancer incidents.

It was also misleading for Mr. Jordan to present as evidence of an emerging cancer epidemic the often quoted

figures, "One in three women and one in two men will have cancer in their lifetimes." Such high lifetime risks figures actually mean many people will live long enough lives to be likely to develop cancer. These figures will mask much lower risk of people of any given age developing cancer in the next 10, 20 or 30 years.

While I support Mr. Jordan's continued efforts to promote progress in cancer prevention and treatment, I encourage him to be more careful lest he unwittingly promote cancer phobia.

William M. London Dir. of Public Health, American Council on Science and Health New York, May 29, 1996

Presentation of Mathematical Papers at Joint Mathematics Meetings and Mathfests

Robert M. Fossum
Secretary
American Mathematical Society

Kenneth A. Ross
Associate Secretary
Mathematical Association of America

Visual Aids at Joint Mathematics Meetings

Preparation of Transparencies for the Overhead Projector

The most frequent complaint heard at meetings is that transparencies are unreadable.

Although transparencies for the overhead projector can be written on while the talk is in progress, you are strongly encouraged to prepare them in advance and test them out with a projector. Writing as you lecture is not a good idea. For one thing, it invites recording too many details. In addition, if you make an error, you will either fail to notice it or else will cross it out and correct it. Either way the audience is distracted.

Transparency copy may be prepared by word processing on a computer or by hand. Typewritten copy is not recommended as the characters are too small to read. To prepare transparencies in advance you are advised as follows:

Preparation of Text by Hand
If you write by hand, use either a #2 lead pencil or a thick felt pen (such as Flair # 844-01 or Pentel Permanent Point Bullet Marker MM50) on sheets of ordinary white 8 1/2" x 11" paper.

Never write in script; it is too hard to read. If you prepare a transparency by hand, you should print. (This does not mean all caps.)

Prepare all copies on a hard-surfaced desk or table, not on a cloth-covered table, blotter, or other resilient surface.

There should be no smudges, erasures, or corrections on a prepared transparency.

General Instructions
1. **Leave at least a 1" margin** on all four sides of the text.

2. **Use characters not less than 1/2" in height.** If you use word processing software, use bold type, 14-point or larger. Please note that the projected image is distorted in such a way that the upper part of each page is considerably larger than the lower part. To balance the image, make the characters on the bottom of the sheet larger and farther apart than those at the top.

3. **Use no more than 12 lines per sheet, and leave ample space between lines.**

4. **Limit each transparency to one topic.** Complicated problems may, however, be simplified in presentation by the use of overlays, which consist of several acetate sheets hinged together like the pages of a book. A complex image can be built up from simple components added to the picture, one at a time, by turning the pages. This mode of presentation can be very effective but calls for careful preparation.

Please Note: The speaker who needs to **illustrate while lecturing** will be provided with a supply of blank transparency sheets and an overhead projector pen. Use only black, blue, green, or red overhead projector pens. Do not use pink, yellow, orange, or any pastel colors.

Use of the Overhead Projector

Overhead projection equipment is relatively easy to use; however, **a speaker unfamiliar with the overhead projector should practice with it prior to the lecture** in order to become familiar with its features and feel comfortable with it during the lecture.

For maximum effectiveness in the use of the projector, please note the following suggestions:

1. **Keep your shoulder out of the way!** If it is lighted by the projector, it is blocking the screen.

2. **Avoid distracting the audience by continually turning around to look at the screen.** Be sure, however, to glance at the screen when placing a new transparency on the projector to ensure its proper placement.

3. Keep in mind that **the projector's lamp can be turned on or off** to direct the audience's attention to the speaker or to the screen as desired.

Additional Guidelines for Speakers

1. **Speak loudly enough** to be heard in all areas of the room.

2. **Practice your speech**, timing yourself to ensure that important points are not rushed and ample time is left for a summary.

3. Define key terms briefly.

SLOPES

Notes by Meg Hickey
© *1996*

As children we learn about slopes as something hard to walk up and fun to slide down. Because humans find it much easier to walk on flat surfaces, we are physically aware of the increase in effort required to climb hilly or sloped ground. Some sloped ground has recreational possibilities, such as skiing or sledding, or dramatic interest in a marathon, such as Heartbreak Hill in the Boston Marathon. Though we experience slopes in different ways, we all know that once you have expended energy to go up them, they offer the potential to roll or slide down, and, the steeper the slope, the faster the movement downwards.

There are a lot of practical uses of slopes that involve their potential for moving people or things downwards, by the force of gravity upon them. In buildings, sloped roofs are designed to shed water and snow to the ground. Streets are sloped from the middle down to the gutters so that rain can run along the gutters into the catch basins. Planted areas and paved parking areas need to be sloped at least 1 foot down for every 100 feet for the surface to drain well. Sloped pipes also use gravity to move drain water and sewage down into the main sanitary sewer in the street. A slope as small as 1/8" down for every foot of horizontal length is sufficient for plumbing pipes to drain well: note that a slope of 1/8" per foot is equivalent to 1/8" per 12" or 1 unit down for every 96 horizontal units, roughly the same as for surface drainage.

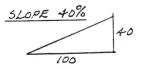

1 FOOT = 12"

ACCEPTABLE DRAINAGE SLOPE FOR PLUMBING PIPES

There are a number of ways to specify how steep a slope is:

With roofs slope is often referred to as "pitch" and might be specified as 5:12 which means 5 feet up for every 12 horizontal feet. Sometimes carpenters refer to slopes as "rise over run" meaning the ratio of vertical to horizontal. Slopes can be laid out with a carpenter's framing square.

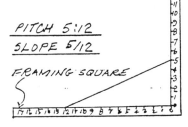

PITCH 5:12

SLOPE 5/12

FRAMING SQUARE

When a slope is specified as a percentage, such as 40%, then the vertical rise is 40 feet for a horizontal run of 100 feet. You can use any other units like meters or yards and you will get the same slope.

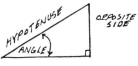

SLOPE 40%

Slopes can also be specified by the angle at which they rise from the horizontal. The angle can be found by trigonometry very easily if you know the rise and run of a slope, since the *tangent* of an angle is the rise/run. On a triangle showing the rise, run, and angle of a slope, the tangent is sometimes called the "opposite side over the adjacent side" referring to the sides of the triangle relative to the angle.

HYPOTENUSE   OPPOSITE SIDE   ANGLE   ADJACENT SIDE

The angle of a slope is the *arctangent* of the slope ratio. Many calculators have an ATAN function which gives angles from tangents; if your calculator doesn't have it you can look it up in math books with trigonometric tables.

For a triangle with 40 rise and 100 run, the tangent is:

40/100 = 0.40

ATAN(0.40) = 20.8°

What is the slope, from the horizontal, of this italic type?

Roof plan A is the view looking down on the roof of the Victorian house shown. The arrows show the down direction of drainage, and the dotted path shows the path of a raindrop starting on the top of the tower and rolling down to the ground. Notice the small section of ridged roof behind the chimney: it is called a "cricket"; if it was not there the water and snow would drain into the back of the chimney and cause a leakage problem.

Now look at roof plan B, below, which shows three proposed additions to an L shaped building. You should be able to identify three potential leakage problems with this plan. Where will the snow build up?

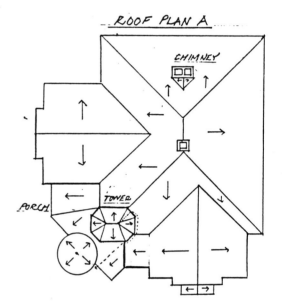

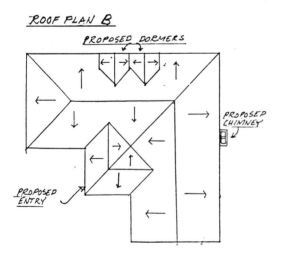

If an earth slope is too steep it may cause erosion. Roots of trees and shrubs can help to prevent erosion of planted banks, but if the slope is greater than 50% retaining walls are required to stop the earth from sliding. Because slope is critical to ground water drainage and building and planting potential, architects start with a contour map of a site, usually available from the U. S. Geological Survey map service. From these you can draw a sideways cut-through view, called a "section" showing the slope of the land.

One of the obvious problems in placing a building on sloped land is that you need to avoid having all the surface water runoff running into your cellar- as in Section A. What is done is to slope the ground away from the house, all the way around it, as in Section B.

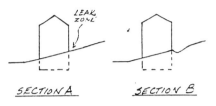

If you look at the contour map of Amesbury you can compute the slope of the ski runs on the northeast side of Powwow hill.

The scale is one centimeter represents 200 meters, and the contour lines are 3 meter differences in height.

The shorter run starts on the 69m. contour line and ends down at the 27m. line for a rise of 69 - 27 = 45m.; it covers about 1 centimeter on the map, so the horizontal distance is 200m.

The average slope then is 45/200 = .225 or 22.5%.

Disregarding the slight inconvenience of ending up in the lake, calculate the slope % on the opposite, southwest side of the hill, from the 60 meter contour line to the base.

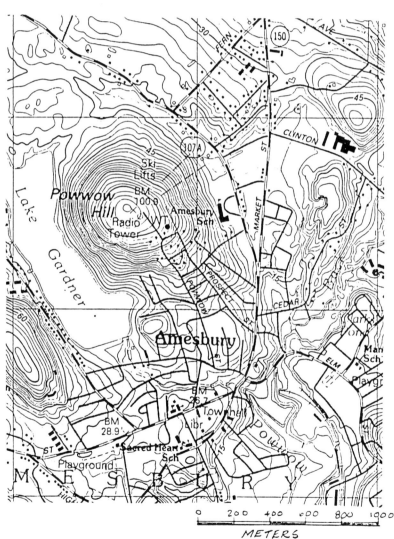

From Massachusetts Highway Department & the U.S. Geological Survey

After the VietNam war many young veterans confined to wheelchairs successfully lobbied for much needed improvements in the architectural standards for wheelchair access to public buildings and housing. As a result many intersections now have curb cuts where a section of curb has been replaced with a sloped ramp from sidewalk to street area, and the steepness of ramps for rising into buildings has been changed from 1 foot of rise for every 10 feet of horizontal length to 1 foot of rise for every 12 feet of ramp length. This gentler slope makes it easier for people with less strength to feel more in control of the downwards roll of the chair, and more able to roll the chair up the ramp. Curb cuts, because they are shorter, are allowed to be steeper but should be kept under 15%. If wheelchair ramps are specified 1:12 and the front door is 3 feet above the ground, then the ramp must be 12x3' or 36 feet long.

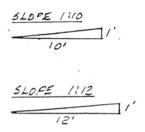

Ramps that rise around a central space have been used in two famous buildings: the Guggenheim art museum in New York city has a spiralling ramped gallery 5 stories tall, and the Boston Aquarium has a series of ramps winding around a huge fish tank about 3 stories tall. The museum was designed by Frank Lloyd Wright, and the aquarium by the Cambridge Seven Associates. Both of these places are worth a visit.

GUGGENHEIM MUSEUM

Another practical use of slopes is in making 3D objects that have angled surfaces rather than just horizontal and vertical planes. People who sew clothing are familiar with producing sloped surfaces in skirts or trousers where the fabric must be sloped in from the larger hip to the smaller waist. For example, if a 40" hip must be reduced to a 30" waist, 10" must be gradually removed over the 7" distance from hip to waist. If the entire 10" was removed only at the side seams the garment would fit badly over the butt and stomach. These parts can be fitted with sloped "darts", which taper the fabric in gently to the waist. Darts are made by folding the fabric and stitching at an angle down to the edge of the fold. The tapering resulting from a dart is twice the width of the top of the dart.

For the skirt shown: if the front and back are each 20" wide with a 1" taper on each side from the hip to the top, there remains 18" at the top of each piece; this is still too big since half of the 30" waist is 15". The front and back pieces each have to be reduced from 18" to 15" with darts. If the front has 2 darts, one each side, each dart can take up half of the 3" difference required, or 1.5". The top of the dart when folded is 1.5"/2 = .75".

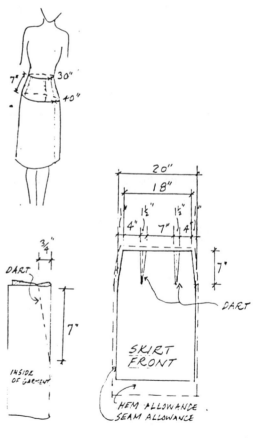

Incidentally, the old fashioned word for a basic clothing pattern is "sloper", deriving from the old English "slupan" -to slip. This word is not only the root of clothing to slip over your body, but you can see how words like slip and slope are related to the same root.

Another example of slopes in clothing is the high heel. If cowboy boots are sold with a 2" heel in both men's and women's sizes, ranging from a women's size 5, 7.5" long, to a men's size 13, 12.5" long, you can calculate the difference in slope for each case.
The small size has a slope of 2/7.5 = .29 = 29%; the large size has a slope of 2/12.5 = .16 = 16%.
What angles are the 2 slopes?
What size heel would a woman with a size 5 foot have to buy in order to have a slope of 16%?

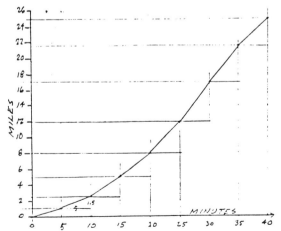

All of the examples of slopes cited so far have been physical, and measurable in space as one dimension divided by another, in effect, a ratio.
The word "rate", which comes from the same root as "ratio", is often used for slopes on graphs.
We talk about "birth rate" (16.2 live births per 1000 population in 1991 in U.S.A.), "tax rate" (5 cents per dollar = 5%), or "rate of travel" (miles/hr or mph).

Any 2 related quantities which can be graphed, such as US population versus time, or position of mercury in a thermometer versus temperature, can produce a plotted line which has one or more slopes along it. If the line is not straight you can find the slope for short segments.

For example, if you set your car trip meter to 0, then read the mileage every 5 minutes for a 25 mile trip from the center of one town to the suburbs of another, then plot distance versus time, you can find the average speed during any 5 minute interval, or you can find the overall average speed of the whole trip.

minutes	miles
0	0
5	1.0
10	2.5
15	5.0
20	8.0
25	12.0
30	17.0
35	21.5
40	25.0

The average speed for the whole trip is:

25 miles/40 minutes = 5/8 mile per minute

or, to get it in mph, multiply by 60 minutes per hour:

60 min/hr x 5/8 mile/min = 37.5 mph

You know that at some times you were travelling faster or slower than this average. You can get a more accurate picture of the variations in speed by looking at individual time segments.

During the first 5 minutes 1 mile was covered. The average speed is 1 mile per 5 minutes, or 1/5 mile per minute: this can be read as the slope of the first line segment. At the same rate for an hour, since there are 60 minutes in an hour, you would go 60 x 1/5 mile = 12 miles. The speed is then 12 miles per hour (mph).

During the next segment 2.5 - 1 = 1.5 miles was covered in 5 minutes. The speed was:
1.5/5 miles/min x 60 min/hr = 18 mph.

Calculate the average mph for the segment from 25 to 30 minutes.

Obviously, if you took mileage data at smaller time intervals you could get a more accurate reading of the speed as it changes. The branch of mathematics called differential calculus teaches how to find slopes for infinitesmal segments of a graphed line or curve.

Once you know the slope of a line you have a ratio you can apply to similar right angled triangles of the same slope but different size.
The law of Pythagoras, AxA + BxB = CxC, and the proportional rules of similar triangles, A/a = B/b = C/c, can be used to find unknown dimensions. Here is one worked out example, and one for you to try:

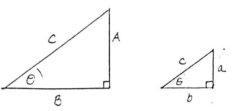

For the house with entry overhang shown, how long is the total rafter, R + r ?

Using Pythagoras: AxA + BxB = RxR
9x9 + 12x12 = RxR = 81 + 144 = 225
R = √225 = 15

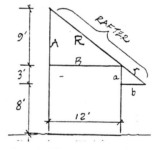

SIDE VIEW OF HOUSE WITH ENTRY OVERHANG

Using similar triangles, because triangles A B R and a b r have the same slope:

a/A = r/R
3/9 = r/15
15 * 3/9 = r = 5
So: Total Rafter = R + r = 15 + 5 = 20

For the house section with skylight, how long is the total rafter, R + r, and what is the height of the skylight, a + a?

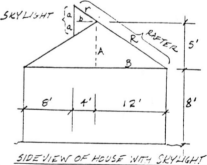

SIDE VIEW OF HOUSE WITH SKYLIGHT

References:
Architectural Graphic Standards, Ramsey & Sleeper, J. Wiley & Sons, N.Y.
Vogue Pattern Book, Vogue Pattern Service.
Webster's Unabridged Dictionary, Merriam-Webster Inc.
United States Coastal & Geodetic Survey.
Frank Lloyd Wright, Solomon R. Guggenheim Museum.
Daughters of Painted Ladies, Pomada, Larsen, Keister; E.P. Dutton, N.Y.

Sketches by Meg Hickey, building illustrations by Myrna Kustin and Juanita Jones.

THE UNIVERSITY OF SOUTHERN MISSISSIPPI

FACTBOOK
1995-1996

The University of Southern Mississippi

10-YEAR TREND:
ENROLLMENT BY ETHNIC GROUP

Hattiesburg Campus
Fall 1986-95

FALL SEMESTER	CAUCASIAN		ASIAN		BLACK		AMERICAN INDIAN		HISPANIC		TOTAL
1986	9,164	82.3%	297	2.7%	1,544	13.9%	24	0.2%	100	0.9%	11,129
1987	8,716	82.2%	283	2.7%	1,506	14.2%	25	0.2%	72	0.7%	10,602
1988	9,125	82.6%	270	2.4%	1,575	14.3%	27	0.2%	54	0.5%	11,051
1989	9,576	83.0%	248	2.1%	1,650	14.3%	21	0.2%	49	0.4%	11,544
1990	9,872	82.9%	260	2.2%	1,702	14.3%	20	0.2%	58	0.5%	11,912
1991	10,146	82.2%	307	2.5%	1,811	14.7%	24	0.2%	60	0.5%	12,348
1992	9,468	81.1%	329	2.8%	1,791	15.3%	26	0.2%	66	0.6%	11,680
1993	9,187	80.0%	311	2.7%	1,874	16.3%	33	0.3%	82	0.7%	11,487
1994	9,202	79.4%	306	2.6%	1,953	16.9%	23	0.2%	103	0.9%	11,587
1995	9,454	78.0%	291	2.4%	2,218	18.3%	38	0.3%	112	0.9%	12,113

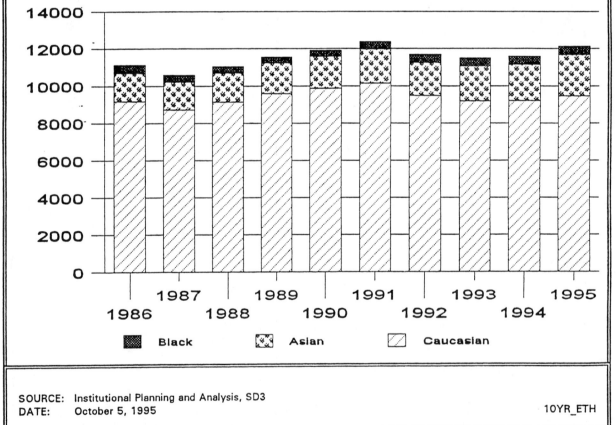

SOURCE: Institutional Planning and Analysis, SD3
DATE: October 5, 1995

10YR_ETH

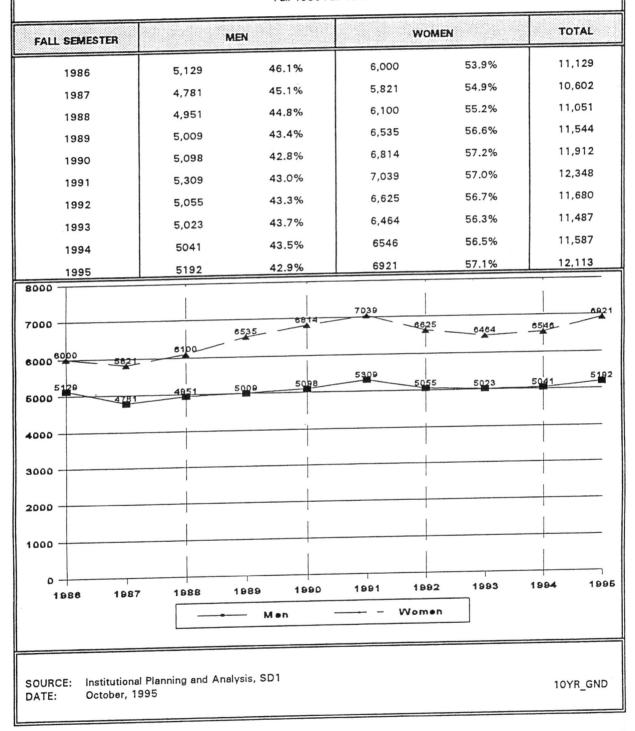

The University of Southern Mississippi

**10-YEAR TREND:
ENROLLMENT BY GENDER**

Hattiesburg Campus
Fall 1986-Fall 1995

FALL SEMESTER	MEN		WOMEN		TOTAL
1986	5,129	46.1%	6,000	53.9%	11,129
1987	4,781	45.1%	5,821	54.9%	10,602
1988	4,951	44.8%	6,100	55.2%	11,051
1989	5,009	43.4%	6,535	56.6%	11,544
1990	5,098	42.8%	6,814	57.2%	11,912
1991	5,309	43.0%	7,039	57.0%	12,348
1992	5,055	43.3%	6,625	56.7%	11,680
1993	5,023	43.7%	6,464	56.3%	11,487
1994	5041	43.5%	6546	56.5%	11,587
1995	5192	42.9%	6921	57.1%	12,113

SOURCE: Institutional Planning and Analysis, SD1
DATE: October, 1995

10YR_GND

The University of Southern Mississippi

RETENTION OF FIRST-TIME ENTERING FRESHMEN
(All First-Time Freshmen)

Hattiesburg Campus
Fall 1983-1994

	ALL FIRST-TIME FRESHMEN	AVERAGE ACT	CUMULATIVE GRADUATION AND CONTINUATION RATES							
			AFTER 1ST YEAR	AFTER 2ND YEAR	AFTER 4TH YEAR		AFTER 6TH YEAR		AFTER 8TH YEAR	
			CONTINUED	CONTINUED	GRADUATED	CONTINUED	GRADUATED	CONTINUED	GRADUATED	CONTINUED
1983	1,179	19.7	72.3%	61.7%	21.3%	29.3%	37.9%	9.7%	42.7%	5.0%
1984	1,151	19.8	72.7%	59.3%	18.1%	31.8%	35.6%	10.6%	40.7%	6.0%
1985	1,093	19.7	71.6%	57.9%	20.8%	29.4%	38.0%	10.9%	43.5%	8.6%
1986	1,015	20.1	72.0%	60.6%	18.9%	31.4%	35.0%	13.1%	42.3%	6.3%
1987	1,033	20.3	75.7%	65.0%	19.9%	34.7%	39.9%	12.7%	47.0%	6.5%
1988	1,183	20.0	74.0%	61.5%	19.5%	34.9%	39.5%	10.0%	44.2%	5.2%
1989	1,204	20.4	76.5%	63.0%	22.2%	32.1%	41.2%	11.8%	47.3%	6.3%
1990	1,084	21.6	78.9%	63.1%	21.7%	32.2%	39.5%	13.1%		
1991	1,124	21.8	71.3%	58.4%	16.9%	33.3%				
1992	1,204	21.8	71.9%	55.6%						
1993	1,183	21.7	72.9%	61.5%						
1994	1,084	21.7	73.9%							
1995	1,063									
1996	1,017									
1997	1,004									

SOURCE: Institutional Planning and Analysis, Retention Report
DATE: January, 1996

RETEN.ALL

The University of Southern Mississippi

10-YEAR TREND:
ENROLLMENT BY CLASSIFICATION

Hattiesburg Campus
Fall 1986-95

FALL SEMESTER	FRESHMEN	SOPHOMORE	JUNIOR	SENIOR	GRADUATE	TOTAL	% CHANGE
1986	1,913	1,539	2,393	3,709	1,575	11,129	0.44%
1987	1,786	1,458	2,250	3,487	1,621	10,602	-4.74%
1988	1,931	1,563	2,461	3,424	1,672	11,051	4.24%
1989	1,950	1,659	2,571	3,597	1,767	11,544	4.46%
1990	1,773	1,722	2,725	3,898	1,794	11,912	3.19%
1991	1,818	1,605	2,886	4,148	1,891	12,348	3.66%
1992	1,550	1,470	2,522	4,299	1,839	11,680	-5.41%
1993	1,580	1,385	2,468	4,172	1,882	11,487	-1.65%
1994	1,685	1,405	2,410	3,956	2,131	11,587	0.87%
1995	1,797	1,472	2,575	4,024	2,245	12,113	4.54%

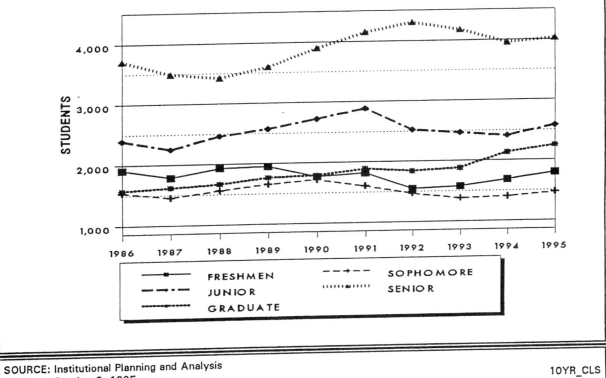

SOURCE: Institutional Planning and Analysis
DATE: October 3, 1995

10YR_CLS

The University of Southern Mississippi

DEGREES AWARDED:
5-YEAR PROFILE BY TYPE OF DEGREE

Fiscal Year 1990-91 through 1994-95

	1990-91	1991-92	1992-93	1993-94	1994-95
UNDERGRADUATE DEGREES	**2,225**	**2,425**	**2,581**	**2,383**	**2,274**
Bachelor of Arts	100	129	150	176	193
Bachelor of Fine Arts	23	18	21	33	31
Bachelor of Music	12	16	23	16	17
Bachelor of Music Education	17	18	16	15	19
Bachelor of Science	1,679	1,709	1,708	1,570	1,513
BS in Business Administration	394	419	480	394	326
Bachelor of Science in Nursing	0	98	153	138	145
Bachelor of Social Work	0	18	30	41	30
MASTER'S DEGREES	**696**	**714**	**776**	**744**	**704**
Master of Arts	30	41	53	36	39
Master of Art Education	3	2	9	4	2
MA in the Teaching of Languages	0	0	5	10	17
Master of Business Administration	76	81	56	51	28
Master of Education	226	247	239	191	216
Master of Fine Arts	4	5	2	6	4
Master of Library Science	30	19	37	41	29
Master of Music	4	6	7	8	11
Master of Music Education	7	6	17	12	11
Master of Professional Accountancy	6	11	11	13	20
Master of Public Health	0	0	8	9	13
Master of Science	268	227	275	282	257
Master of Science in Nursing	0	5	7	38	20
Master of Social Work	42	64	50	43	37
SPECIALIST DEGREES	**18**	**19**	**22**	**9**	**14**
DOCTORAL DEGREES	**106**	**95**	**93**	**95**	**103**
Doctor of Education	27	18	10	11	10
Doctor of Music Education	0	1	0	1	1
Doctor of Musical Arts	0	1	4	3	3
Doctor of Philosophy	79	75	79	80	89
TOTAL DEGREES AWARDED	**3,045**	**3,253**	**3,472**	**3,231**	**3,095**

SOURCE: Institutional Planning and Analysis, Degrees Awarded by Major, STU0510
DATE: January, 1996

DEG_TYPE

The University of Southern Mississippi

DEGREES AWARDED:
5-YEAR PROFILE BY COLLEGE AND TYPE OF DEGREE

Fiscal Year 1990-91 through 1994-95

	1990-91	1991-92	1992-93	1993-94	1994-95
THE ARTS	58	56	67	69	72
Bachelor of Arts	6	4	7	5	5
Bachelor of Fine Arts	23	18	21	33	31
Bachelor of Music	12	16	23	16	17
Bachelor of Music Education	17	18	16	15	19
BUSINESS ADMINISTRATION	470	526	586	459	360
BS in Business Administration	394	419	480	394	326
Bachelor of Science	76	107	106	65	34
EDUCATION AND PSYCHOLOGY	563	647	638	579	533
Bachelor of Science	534	630	617	564	510
Bachelor of Arts	29	17	21	15	23
HEALTH AND HUMAN SCIENCES	355	353	429	423	437
Bachelor of Science	355	237	246	244	262
Bachelor of Science in Nursing	0	98	153	138	145
Bachelor of Science in Social Work	0	18	30	41	30
LIBERAL ARTS	453	490	532	529	521
Bachelor of Science	390	384	411	373	356
Bachelor of Arts	63	106	121	156	165
SCIENCE AND TECHNOLOGY	326	353	329	324	351
Bachelor of Science	324	351	328	324	351
Bachelor of Arts	2	2	1	0	

SOURCE: Institutional Planning and Analysis, Degrees Awarded by Major, STU0510
DATE: January, 1996

Page 1 of 2
DEG_COLL

The University of Southern Mississippi

DEGREES AWARDED:
5-YEAR PROFILE BY COLLEGE AND TYPE OF DEGREE

Fiscal Year 1990-91 through 1994-95

	1990-91	1991-92	1992-93	1993-94	1994-95
THE GRADUATE SCHOOL	**820**	**828**	**891**	**848**	**821**
Master of Arts	30	41	53	36	39
Master of Art Education	3	2	9	4	2
MA in the Teaching of Languages	0	0	5	10	17
Master of Business Administration	76	81	56	51	28
Master of Education	226	247	239	191	216
Master of Fine Arts	4	5	2	6	4
Master of Library Science	30	19	37	41	29
Master of Music	4	6	7	8	11
Master of Music Education	7	6	17	12	11
Master of Professional Accountancy	6	11	11	13	20
Master of Public Health	0	0	8	9	13
Master of Science	268	227	275	282	257
Master of Science in Nursing	0	5	7	38	20
Master of Social Work	42	64	50	43	37
Specialist in Education	16	19	22	9	14
Specialist in English	2	0	0	0	0
Specialist in History	0	0	0	0	0
Doctor of Education	27	18	10	11	10
Doctor of Music Education	0	1	0	1	1
Doctor of Musical Arts	0	1	4	3	3
Doctor of Philosophy	79	75	79	80	89
TOTAL DEGREES AWARDED	**3,045**	**3,253**	**3,472**	**3,231**	**3,095**

SOURCE: Insitutional Planning and Analysis, Degrees Awarded by Major, STU0510
DATE: January, 1996

Page 2 of 2
DEG_COLL

STATISTICAL
PORTRAIT
1995

Office of Institutional Research

Table 1

Undergraduate Admissions Summary: Fall 1987 - 1995

FRESHMEN

	1987	1988	1989	1990	1991	1992	1993	1994	1995
Applied	3349	3274	3272	2775	2487	2378	2494	2356	2439
Decision Ready	2925	2932	2804	2420	2181	2035	2237	1963	2110
Admitted	1877	1601	1648	1516	1388	1366	1496	1320	1363
Admit Rate	64.2%	54.6%	58.8%	62.6%	63.6%	67.1%	66.9%	67.2%	64.6%
Enrolled	1033	818	823	751	745	734	800	662	691
Admitted but Deferred	94	41	53	68	70	53	34	41	23
Yield Rate	55.0%	51.1%	49.9%	49.5%	53.7%	53.7%	53.5%	50.2%	50.7%
Not Admitted:									
Denied	1048	1331	1156	904	793	669	741	643	747
Incomplete Applications	424	342	468	355	306	343	257	393	329

TRANSFERS

	1987	1988	1989	1990	1991	1992	1993	1994	1995
Applied	3468	3683	4118	3653	3189	3375	3382	2875	2761
Decision Ready	3037	3012	3398	3149	2808	2912	3018	2480	2374
Admitted	2574	2224	2623	2610	2455	2555	2741	2249	2081
Admit Rate	84.8%	73.8%	77.2%	82.9%	87.4%	87.7%	90.8%	90.7%	87.7%
Enrolled	1668	1319	1619	1552	1442	1505	1666	1397	1225
Admitted but Deferred	239	170	182	218	166	170	110	93	79
Yield Rate	64.8%	59.3%	61.7%	59.5%	58.7%	58.9%	60.8%	62.1%	58.9%
Not Admitted:									
Denied	463	788	775	539	353	357	277	231	293
Incomplete Applications	431	671	720	504	381	463	364	395	387

TOTAL UNDERGRADUATES

	1987	1988	1989	1990	1991	1992	1993	1994	1995
Applied	6817	6957	7390	6428	5676	5753	5876	5231	5200
Decision Ready	5962	5944	6202	5569	4989	4947	5255	4443	4484
Admitted	4451	3825	4271	4126	3843	3921	4237	3569	3444
Admit Rate	74.7%	64.4%	68.9%	74.1%	77.0%	79.3%	80.6%	80.3%	76.8%
Enrolled	2701	2137	2442	2303	2187	2239	2466	2059	1916
Admitted but Deferred	333	211	235	286	236	223	144	134	102
Yield Rate	60.7%	55.9%	57.2%	55.8%	56.9%	57.1%	58.2%	57.7%	55.6%
Not Admitted:									
Denied	1511	2119	1931	1443	1146	1026	1018	874	1040
Incomplete Applications	855	1013	1188	859	687	806	621	788	716

Table 13

Enrollment Trends at UMass Boston (State Funded Enrollment), Fall Semesters 1986-1995

	1986	1987	1988	1989	1990	1991	1992	1993	1994	1995
Total Enrollment	12919	13374	12451	12584	12478	11606	11775	12136	12142	11602
Full-Time Enrollment (Hct)	6969	7448	7007	7002	7002	6556	6561	6657	6532	6064
FTE Enrollment	8983	9526	8921	8921	8863	8300	8439	8607	8552	8095
Matriculated Undergraduate	9065	9615	9283	9514	9216	8589	8693	8972	8556	8007
% Full-Time	66.8%	68.0%	67.9%	66.6%	68.0%	67.9%	66.9%	65.1%	66.2%	66.6%
Matriculated Graduate	1010	1474	1678	1756	1802	1890	1897	1958	2035	2258
% Full-Time	33.4%	30.9%	27.0%	24.0%	26.4%	27.4%	30.6%	32.9%	32.5%	30.8%
Non-degree Students	2844	2485	1490	1314	1460	1127	1185	1206	1551	1337
% Undergraduate	82.3%	79.9%	75.0%	77.5%	76.7%	76.5%	68.9%	69.3%	67.0%	74.0%
Total Undergraduate										
HCT Enrollment	11406	11601	10399	10532	10336	9451	9509	9808	9595	8997
FTE Enrollment	8005	8252	7605	7613	7489	6891	6933	7035	6849	6312
% Female	54.6%	55.8%	56.9%	56.5%	55.6%	55.0%	53.5%	53.0%	53.3%	53.7%
Mean Age	27	27	27	27	27	28	28	28	28	29
Median Age	24	24	24	24	25	25	25	25	25	25
Total Graduate										
HCT Enrollment	1513	1973	2052	2052	2142	2155	2266	2328	2547	2605
FTE Enrollment	978	1274	1316	1308	1374	1409	1506	1572	1703	1783
% Female	58.6%	60.4%	62.6%	62.0%	64.0%	64.7%	64.2%	63.6%	63.3%	62.3%
Mean Age	34	34	34	35	34	35	34	34	34	35
Median Age	33	33	32	33	32	33	32	32	32	32

Table 5

SAT Scores of New Freshmen by College/Program, Ten-Year Trend
(Excluding the DSP Program, Learning Disabled and Foreign Students)

		1986	1987	1988	1989	1990	1991	1992	1993	1994	1995
College of	SATVerbal	453	453	464	464	449	447	432	431	433	430
Arts &	SATMath	468	475	479	488	471	473	477	464	483	474
Sciences	Combined	921	928	943	952	920	920	909	895	916	904
	[N]	[429]	[444]	[330]	[299]	[294]	[262]	[240]	[256]	[211]	[251]
College of	SATVerbal	448	453	449	409	436	428	423	418	412	413
Management	SATMath	512	529	513	507	520	502	518	500	490	515
	Combined	960	982	962	916	956	930	941	918	902	928
	[N]	[108]	[93]	[68]	[54]	[47]	[36]	[32]	[52]	[31]	[32]

Table 27

Degrees Conferred by Level, Academic Year 1984-85 to 1994-94

	AY 83-84	AY 84-85	AY 85-86	AY 86-87	AY 87-88	AY 88-89	AY 89-90	AY 90-91	AY 91-92	AY 92-93	AY 93-94	AY 94-95
UMass Boston												
Certificate	24	24	52	38	31	53	41	44	59	62	82	55
Bachelor's	1010	1251	1149	1232	1319	1401	1535	1630	1636	1579	1579	1467
Graduate Certificates										8	25	23
Master's	89	97	140	192	252	358	421	427	474	508	502	478
CAGS	6	8	11	14	12	25	19	26	32	30	31	22
Doctorate					1	2	1	2	2	3	4	8
Total	1129	1380	1352	1476	1615	1839	2017	2129	2203	2190	2223	2053
Public Institutions												
Bachelor's	12574	12769	12849	13599	13211	13714	14235	14728	14678	n.a.	n.a.	n.a.
Statewide												
Bachelor's	40135	39780	39779	40748	40308	41613	42825	43520	44487	n.a.	n.a.	n.a.

Enrollment Trends
Fall Semesters 1982-1995

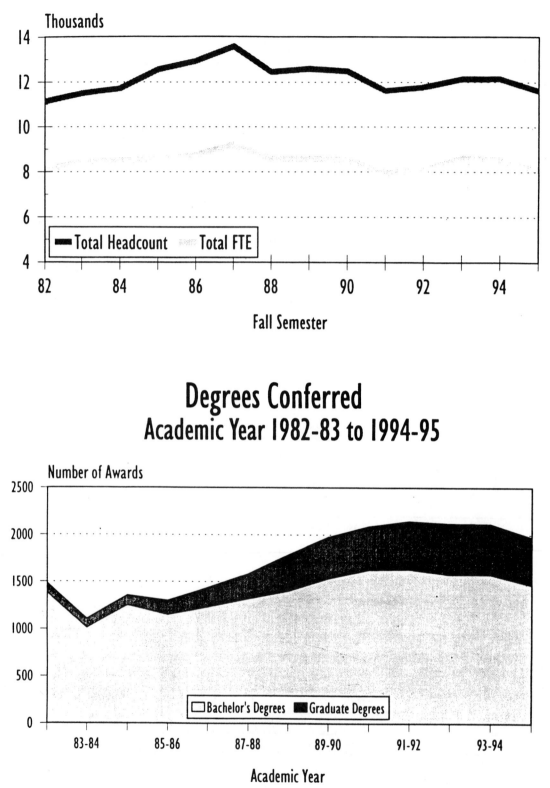

Degrees Conferred
Academic Year 1982-83 to 1994-95

Graduate awards include master's, CAGS and doctoral degrees.

Why International Statistical Comparisons Don't Work

By SYLVIA NASAR

LIVING standards in the United States have fallen behind those in Germany, Japan, even Finland. The pay and productivity of once-proud American workers trail those of a dozen or so other nations. The short-sighted United States invests half as much as far-sighted Japan.

These dispiriting and widely cited statistics have one thing in common: They are wrong.

In fact, contrary to today's conventional wisdom, the United States remains the richest and most productive economy in the world. That's true even though other industrial nations have been growing faster — and narrowing the gap — practically since the United States became No. 1 in the first place, in about 1910.

Certainly, Americans can't afford to be complacent. If some of the worrisome trends in the United States economy — from increasingly ill-prepared workers to skimpy savings — continue long enough, today's dubious statistic could turn into tomorrow's hard fact.

But part of what's unnecessarily stoking Americans' anxiety is a handful of false comparisons based on faulty ways of calculating the economic scorecard. Much of the trouble arises in converting other nations' output, consumption or investment from yen, marks or lire into dollars of comparable buying power.

The widespread practice — in place everywhere from the lofty World Bank to the nightly news — of using currency market exchange rates seriously distorts comparisons because it does not take account of

---

The U.S. doesn't really invest only half as much as Japan or have less per capita income.

---

different price levels — higher rents in Tokyo than Toledo, cheaper cherries in Bologna than Bristol, the fact that maids are affordable for most middle-class families in Mexico but for few families in Minneapolis.

Put another way, not taking account of different prices in different countries makes as little sense as comparing your pay today with your salary in 1978 without factoring in the doubling of prices in that interval.

Besides, exchange rates gyrate with wars, business cycles and urges by foreign investors to buy American bonds. None of these reflect real changes in economic fundamentals.

Economists rely on so-called purchasing power parity exchange rates. "If you're making comparisons using PPP's, you come much closer to the truth with a capital T, " said Robert Summers, an economist at the University of Pennsylvania who, together with his colleague Allan Heston, helped introduce the whole notion.

So what do the facts, correctly measured, say? Start with that most basic yardstick of wealth, economic development and well-being, a country's gross domestic product — the sum of what it produces inside its own borders — divided by the population. The United States is still on top, 17 percent ahead of Germany and a surprising 22 percent ahead of Japan.

What a nation produces, however, includes a great many things — from missiles to private security guards to new factories — that don't add immediately to the material well-being of its citizens. What about consumption, public and private — which reflects how well people are living? According to Mr. Summers and Mr. Heston,

---

Americans live better and can loaf more because they're more productive.

---

the gap between America's consumer society and Germany's and Japan's, which devote more to investment, is even wider.

At this point someone is apt to point out that there's more to life than material goods — namely, the time to enjoy them. As it turns out, busy Japanese workers work about 260 hours more a year than Americans — 2,052 hours in total — to achieve their lower living standard. Germans, on the other hand, may have living standards 15 percent below those of Americans, but they also work 10 percent fewer hours. In many peoples' eyes, these leisure-loving people — who get six-week vacations — may be very close to having it all.

What lets Americans live better

while loafing more than their Japanese rivals is higher — yes, higher — productivity. Contrary to the widespread view that the Japanese economy is vastly more efficient, every comparison shows that output per employee is 40 percent greater in the United States. America also has the most efficient service industries and agriculture.

In some sectors of manufacturing, Japan and Germany are much closer to Yankee efficiency. Japanese productivity is about 80 percent that of America's, while Germany's hovers at about 75 percent. Canada's is the closest. What's more, factory productivity growth abroad slowed in the 1980's, just as it spurted in the United States.

Hard to believe? Japan is hyperefficient as a maker of cars, VCRs and industrial and office equipment. But not enough to offset being relatively inefficient at turning out pharmaceuticals, aircraft, textiles, construction material, processed food and raw materials.

$21.53 an Hour

American politicians have lately been griping that American factory workers trail their counterparts in Europe badly when it comes to pay. Pay here is $14.77 an hour, they say, lower than in at least 12 other countries and positively paltry compared with Germany's princely $21.53 an hour. But adjusting for Germany's higher cost of living shrinks the German advantage to a few pfennig.

What matters for future living standards, of course, is how much a nation invests in new technology, plants and equipment. A rallying cry of late has been that Japan spends twice as much as the United States. That turns out to be a bit of a tall tale based on ignoring the high cost of building in Japan. Correctly measured, Japan invests about 28 percent of G.D.P., versus 19 percent for the United States — nowhere near twice as much. Surprisingly, the American rate is not far below that of Germany, France and Italy, which plow just 21 or 22 percent of their G.D.P. into the next generation of machines, factories and office buildings.

The United States, not Japan and certainly not Finland, is still the richest and most competitive economy. But for the last 15 years, the United States has consumed more, invested less and grown more slowly than any other industrial nation. Will it remain in the lead? Simple extrapolation often fails because it ignores peoples' ability to change course. Which road is taken will depend on what happens from here on, not on the past.

THE NEW YORK TIMES, MONDAY, FEBRUARY 3, 1992

North Dakota, Math Country

By Daniel Patrick Moynihan

WASHINGTON

In his State of the Union Message, the President reaffirmed his commitment to making our country "the world leader in education," adding that to do so, "We must revolutionize America's schools."

He didn't say how. But he asked for help. *And help is at hand!*

I have discovered the formula. In a flash of insight. Like Leo Szilard in London waiting to cross on a green light, thinking up nuclear fission, or James Powell thinking up magnetic levitation while waiting for a toll ticket at the Bronx-Whitestone Bridge.

I am a little old for that sort of thing, but still it happened. I was allotted two minutes in a gathering of Democrats last week to explain, yet again, that there is simply no significant connection between school expenditure and pupil achievement.

I had a chart. Utah at the bottom, under $3,000 average expenditure per pupil per year. New York at the top, over $6,000. Now what of average scores on the national eighth-grade exam in 1990? Well, No. 1 was North Dakota, eighth from the bottom in spending; No. 9 was Idaho, third from the bottom. (As for Utah, the test was never given there but the state has the highest high school graduation rate.) New York ranked 20th.

Uh huh, nodded the audience. Same old stuff. Then it came to me. "Fellow countrymen!" I exclaimed. "If you would improve your state's math scores, move your state closer to the Canadian border!"

The whole room got it!

On to the Congressional Budget Office! Please, I asked, get me the correlation between math scores and distance of state capitals from the

Daniel Patrick Moynihan is Democratic Senator from New York.

Canadian border. Back came the answer. A negative 0.522 — which may be the strongest correlation known to education, and which means that the further a capital is from the border, the lower its test score. By contrast, the correlation between expenditures and math tests is a paltry 0.203.

Not coincidentally, these findings may provide the first empirical support for the "theory of climates" of the 16th-century French philoso-

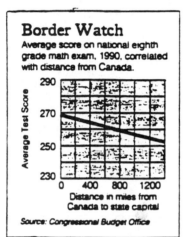

Border Watch

Average score on national eighth grade math exam, 1990, correlated with distance from Canada.

Average Test Score (vertical axis: 230, 250, 270, 290)

Distance in miles from Canada to state capital (horizontal axis: 0, 400, 800, 1200)

Source: Congressional Budget Office

The New York Times

pher Jean Bodin, who much influenced Montesquieu, so important to U.S. constitutional history. Bodin found "northerners unequaled in wars and industry, and southerners ... unequaled in the contemplative sciences, but the habitants of the median region ... particularly fit ... for the blossoming of arts and letters." That France should occupy the median region is perhaps no accident.

Even so, the indicated course for American education is obvious. Disadvantaged states should establish summer capitals in the Thousand Islands of the St. Lawrence River, which happens to include New York State territory bordering on French-speaking Quebec.

This study, commissioned under the Civil Rights Act of 1964, determined that after a point there is precious little association between school resources and school achievement. The resources that matter are those the student brings to the school, including community traditions that value education. Or don't.

The President was right enough in stating that "the major cause of problems of the cities is the dissolution of the family." But it is not just slum families: affluent families, too — the self-destructive poor and the self-indulgent affluent.

Furthermore, those long summer evenings in St. Lawrence skiffs might be given over in part to reading the literature on educational achievement that begins with the 1966 report on Equality of Educational Opportunity, known as the Coleman Report.

More spending on schools doesn't improve scores.

The plain fact is that in a rush of findings in the 1960's, social science disabused us of most of what we used to "know" about social issues. Generally speaking, social policy has not been able to accept this. In this climate, conservatives have fared best because of a general inclination to leave things alone anyway. (Last year's 1990 State of the Union goal — that Americans will be "first in the world in math and science achievement" by the year 2000 — is pure fantasy. But the Administration knows it, and knows also that it will be long out of office by the time the country finds out.)

Liberalism, by contrast, is reduced to a mantra. That's fine if you are a fakir; if a child, not so good. □

QUANTITATIVE EXAMPLES

Edited by John G. Truxal

VERBAL SAT SCORES

One of the major arguments in federal education circles concerns whether to publish national test results on a state-by-state or even district-by-district basis. The National Assessment of Educational Progress (NAEP) has historically released data only on a regional basis so particular states or districts are not criticized. The SAT scores for 1989 are available by state as shown below.

SAT Scores 1989

Mean SAT Verbal Score

1.	Iowa	512	27.	Washington	448
2.	N. Dakota	500	28.	N. Hampshire	447
3.	Utah	499	29.	Alaska	443
4.	S. Dakota	498	30.	Oregon	443
5.	Kansas	495	31.	Nevada	439
6.	Nebraska	487	32.	Connecticut	435
7.	Tennessee	486	33.	Delaware	435
8.	New Mexico	483	34.	Maryland	434
9.	Alabama	482	35.	Maine	431
10.	Oklahoma	479	36.	Massachusetts	432
11.	Wisconsin	477	37.	Vermont	435
12.	Kentucky	477	38.	Virginia	430
13.	Minnesota	477	39.	Rhode Island	429
14.	Louisiana	473	40.	New Jersey	423
15.	Mississippi	472	41.	Pennsylvania	423
16.	Missouri	471	42.	California	422
17.	Arkansas	471	43.	Florida	420
18.	Montana	469	44.	New York	419
19.	Idaho	465	45.	Texas	415
20.	Wyoming	462	46.	Indiana	412
21.	Illinois	462	47.	D.C.	407
22.	Colorado	458	48.	Hawaii	406
23.	Michigan	458	49.	Georgia	402
24.	Arizona	452	50.	South Carolina	399
25.	Ohio	451	51.	North Carolina	397
26.	W. Virginia	448			

Before we jump to the conclusion that Iowa is the optimum place to educate children and North Carolina is the worst, we might also want to consider the following data.

Percent Taking SAT | Mean SAT Verbal Score

1.	Iowa	5	27.	Washington	39
2.	N. Dakota	5	28.	N. Hampshire	66
3.	Utah	5	29.	Alaska	42
4.	S. Dakota	6	30.	Oregon	63
5.	Kansas	10	31.	Nevada	23
6.	Nebraska	10	32.	Connecticut	75
7.	Tennessee	12	33.	Delaware	60
8.	New Mexico	11	34.	Maryland	60
9.	Alabama	8	35.	Maine	59
10.	Oklahoma	8	36.	Massachusetts	72
11.	Wisconsin	12	37.	Vermont	63
12.	Kentucky	10	38.	Virginia	59
13.	Minnesota	15	39.	Rhode Island	63
14.	Louisiana	9	40.	New Jersey	67
15.	Mississippi	4	41.	Pennsylvania	63
16.	Missouri	13	42.	California	44
17.	Arkansas	7	43.	Florida	47
18.	Montana	20	44.	New York	69
19.	Idaho	17	45.	Texas	43
20.	Wyoming	14	46.	Indiana	55
21.	Illinois	17	47.	D.C.	67
22.	Colorado	29	48.	Hawaii	52
23.	Michigan	12	49.	Georgia	59
24.	Arizona	23	50.	South Carolina	55
25.	Ohio	23	51.	North Carolina	57
26.	W. Virginia	15			

The relation between these two data sets is displayed clearly in the following graph.

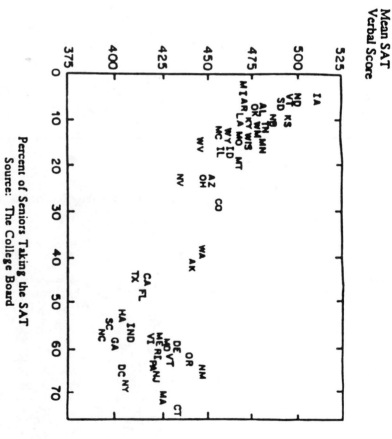

Percent of Seniors Taking the SAT
Source: The College Board

When we look at test scores from *past* years we can clearly see that the Mean SAT Average goes down as the percentage of students taking the test rises. For example, partial data from two times four years apart show changes in average Math scores:

	Old		New	
	%	Average	%	Average
Montana	9	547	20	523
Arizona	11	512	23	500
Colorado	17	520	29	508

Who Collects Data and Why?

Computerized databases detailing the lives of U.S. residents are compiled on an enormous scale by government agencies and private firms of every description.

Government Agencies

The most politically charged, and therefore the most discussed, databases have in the past been those maintained by the Federal government: e.g., those of the Census Bureau, the Social Security Administration, all branches of the military, the FBI and other security agencies, and myriad lesser-known Federal agencies which inevitably compile extensive databases of use to someone.

Less known to the public, but equally important, are various obscure computer files established by the 50 state governments and thousands of local government entities. Unlike the well known Federal databases, many state and local data files are in the public domain -- available for free and a for a fee to any interested party. The compilation and sale of such files is a growing business for database companies.

Important state and local data archives -- often computerized and readily accessible -- include:

    - Public registries of births, deaths, marriage and divorce, real estate property sales and assessments, and so on.

    - State department of motor vehicles listings of automobile ownership and accidents.

    - Local tax records (such as property tax payments), lists of residents and voter registration lists.

Banks and Insurance Firms

Commercial banks, insurance companies of all kinds, and health care insurers in particular all assemble massive amounts of data on the finances, health status, and other personal business of nearly all Americans. While this data is not public information as such, it is very often available for a fee to outside parties.

Even the most confidential kinds of information are often available with little difficulty to those who know how to gain access: for example, certain health insurance firms will provide employers with detailed monthly reports on the health care expenditures and diagnoses of their employees (e.g., John Jones saw a psychiatrist four times for depression last month).

Other commonly used criteria include what are called the "recency, frequency and amount" of each list member's purchases during a certain period. For example, you could purchase a mailing list whose members had each bought at least $50 worth of books on at least three different occasions during the past six months.

Various databases firms specialize in compiling and continuously updating lists based on certain public sources. For example, one firm tracks every house sold in the nation, using local property tax and courthouse records as raw material. The firm compiles not only the buyer's and seller's identity but the sales price , location, and other data. This data is valuable to many other firms -- for example, to carpet and furniture retailers for whom recent homebuyers are prime customers.

For a sense of how these firms sell their data, see the attached advertisement by Database America Companies, which recently appeared in the weekly "Direct Marketing News."

How is This Data Used?

As indicated above, all of this data is compiled , sold, and used for many commercial purposes. The most obvious of these purposes is direct marketing -- especially the sale of retail consumer goods by mail and by telephone.

As computer memories becomes ever-larger and the speed and complexity of computer capabilities grow, the cross-tabulation and synthesis and the many types of database mentioned above is becoming remarkably sophisticated.

One example of such sophistication is the Claritas Corporation, a database and marketing firm which has analyzed the U.S. population one the basis of zip code and sub-area classifications. Each small geographical zone -- down to the neighborhood level -- is given one of about 40 tag names, such as "Shotguns and Pick-up trucks," "Brains and Money," and so on. These categories are based not only on income, but other variables such as occupation, family size, property values in the area, spending habits, and so on. Claritas continuously updates and sells this data not only to direct marketers but to many types of businesses deciding where to initiate new ventures.

Besides direct marketing, other uses of data include the following:

- Deciding where to locate new stores, malls, distribution hubs, office buildings, residential developments and other enterprises.

By Bob Martin

U.S. Government Data Collection by Anthony Roman, 1994.

The primary data collection arm of the Federal government is the Bureau of the Census. Although most people are familiar with the Bureau of the Census through the decennial Census, many people are unaware that its data collection activities continue throughout each and every year. The U. S. Government routinely collects information about many aspects of American life, both personal and corporate, through surveys.

One major aspect of the Census Bureau's work involves economic surveys of American businesses. The Bureau tracks manufacturing firms, importing and exporting firms, wholesale and retail trade businesses and virtually any kind of business. Among the data collected from these surveys are the types and amounts of products being manufactured, sold, and exported. The number of workers being employed within each type of industry is also estimated. Economic forecasters and legislators rely on these data when setting polities designed to keep the U.S. economy stable. On many days, the business section of U.S. daily newspapers will contain articles describing how a recent government report on "manufactured durable goods" or "exports" or some other economic variable, affected the stock market or had direct or indirect impact on our economic situation.

The Bureau of the Census collects many other types of data through its demographic surveys conducted with randomly selected U.S. residents. The National Health Interview Survey collects information about American health and the proliferation of various types of diseases. National crime trends are examined using data from the National Crime Survey which measures victimization and exposures to crime. The Consumer Expenditures Surveys collect information about U.S. residents' purchasing habits, from houses and cars to pickles and bread. The Survey of Income and Program Participation examines the component of American's income and to what extent people are or are not availing themselves of government programs. This survey, for example, is used to estimate the number of families receiving food stamps and the number who are eligible to receive them but don't.

The largest, ongoing demographic survey is the Current Population Survey (CPS). The CPS collects data from approximately 60,000 households each month in order to estimate employment, unemployment, and other characteristics of the general labor force. Each month, network new broadcasts , local news broadcasts, and newspapers will detail how the national or statewide unemployment rate went up or down the previous month. An increase of 1 tenth of 1 percent in the national unemployment rate may mean that about 60,000 more Americans were unemployed last month as compared to the

month before. Needless to say, the numbers are carefully watched by legislators, policy makers, and business leaders.

The origin of the Current Population Survey corresponds to the origin of government data collection in general. During the Great Depression of the 1930's, it became painfully clear that government leaders lacked sufficient information to develop strategies for dealing with the economic plight of U.S. citizens. How many Americans were out of work? Was the South in worse shape than the Midwest? Were farmers in worse shape than laborers? Was Ohio better off than Oklahoma? Which industries were strong and which were about to fail? Leaders had no answers to any of these questions and no centralized database had been kept for this type of information. The Enumerative Check Census of 1937 was the first attempt by government to conduct a nationwide probability survey and measure unemployment. The experience led to a great deal of research on methods of conducting large-scale probability surveys and methods of estimating unemployment and other important national statistics. The Sample Survey of Unemployment began in March 1940 and has been conducted monthly ever since. The Census Bureau took over conducting the survey in August, 1942 which became known as the Current Population Survey (CPS).

Since its inception as a national survey of approximately 8,000 households, the CPS has grown in size and complexity. It has become useful as a tool for the equitable distribution of federal funds and is used by researchers to study many attributes related to employment and unemployment in the U.S.

In general, the data collection conducted by the Census Bureau for the federal government has provided a wealth of information for legislators and researchers to help understand the effects of national policies and the relationships among earnings, industry and job classification, and a host of demographic characteristics.

Current Population Reports   Special Studies Series P23-189

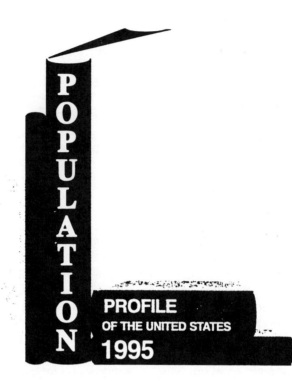

POPULATION

PROFILE
OF THE UNITED STATES
1995

Issued July 1995

U.S. Department of Commerce
Ronald H. Brown, Secretary
David J. Barram, Deputy Secretary

Economics and Statistics Administration
Everett M. Ehrlich, Under Secretary
for Economic Affairs

BUREAU OF THE CENSUS
Martha Farnsworth Riche, Director

Educational Attainment

ANDREA ADAMS

The Nation's educational level has risen dramatically in the past 50 years.

Since the Bureau of the Census first collected data on educational attainment in the 1940 census, educational attainment among the American people has risen substantially. In 1940, one-fourth (24.5 percent) of all persons 25 years old and over had completed high school (or more education), and 1 in 20 (4.6 percent) had completed 4 or more years of college. By 1993, over four-fifths (80.2 percent) had completed 4 years of high school or more, and over one-fifth (21.9 percent) had completed 4 or more years of college.

The increase in educational attainment over the past half century is primarily due to the higher educational attainment of young adults, combined with the attrition of older adults who typically had less formal education. For example, the proportion of persons 25 to 29 years old who were high school graduates rose from 38.1 percent in 1940 to 86.7 percent in 1993, while for persons 65 years old and over, it increased from 13.1 to 60.3 percent.

There is no difference in the educational attainment of young men and women.

Differences in educational attainment between men and women have historically been attributed to differences in attainment at the college level. In 1940, the percentages of men and women 25 to 29 years old completing 4 or more years of college were close to equal, but at a very low level (6.9 percent compared with 4.9 percent). Between 1940 and 1970, both sexes increased their college attainment, but men's gains were significantly greater. The college completion rates for men and women 25 to 29 years old in 1970 were 20.0 and 12.9 percent, respectively. Since 1970, however, the college gains of young adult women have outstripped those of young adult men, until by 1993, there was no statistical difference in the proportions of men and women 25 to 29 years old with 4 or more years of college (23.4 and 23.9 percent, respectively).

Educational attainment levels continue to rise for race and Hispanic groups.

Blacks have made substantial progress in narrowing the educational attainment gap relative to Whites. In 1940, only 7.7 percent of Blacks 25 years old and over had completed high school, compared with 26.1 percent of Whites. In 1965, the corresponding figures were 27.2 and 51.3 percent, respectively. By 1993, 70.4 percent of Blacks 25 years old and over had completed high school, compared with 81.5 percent of Whites. Hence, the

Percent of Persons Who Have Completed High School or College: Selected Years 1940 to 1993

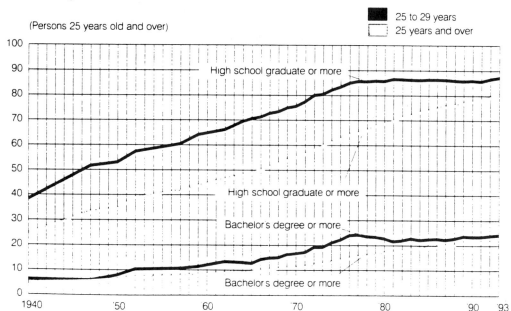

difference between the Black and White rates was smaller in 1993 than in the earlier years.

Among persons 25 to 29 years old in 1940, only 10.6 percent of Black and Other-races men had completed 4 years of high school, compared with 38.9 percent of White men. By 1993, there was no statistical difference in the proportions of Black men and White men who had completed high school: 85.0 and 86.0 percent, respectively. Similar gains were made by young Black women but they remained different from White women in 1993, when 80.9 percent of Black women 25 to 29 years old had completed 4 years of high school, compared with 88.5 percent of White women. In 1940, the proportions were 13.8 percent of Black and Other-races women and 43.4 percent of White women.

Although the proportion of Blacks 25 years old and over who have completed college has increased since 1940, it is about one-half the proportion of their White counterparts (12.2 percent compared with 22.6 percent in 1993). Among young adults 25 to 29 years old in 1993, Blacks were more than half as likely as Whites (13.2 percent compared with 24.7 percent) to have completed 4 or more years of college. Data for persons of Hispanic origin[1] have not been collected for as long a period as for race groups, but the patterns also indicate some improvement in educational attainment over time. Among Hispanics 25 years old and over in 1993, 53.1 percent had completed high school, up from 36.5 percent in 1974. Completion of college stood at 9.0 percent, a significant increase from the level of 5.5 percent in 1974.

[1]Persons of Hispanic origin may be of any race. These data do not include the population of Puerto Rico.

For Further Information

See: Current Population Reports, Series P20-476, *Educational Attainment in the United States: March 1993 and 1992.*

Contact: Rosalind R. Bruno or Andrea Adams Education and Social Stratification Branch 301-457-2464

Educational Attainment of Persons 25 Years Old and Over, by Sex, Race, Hispanic Origin, and Age: March 1993

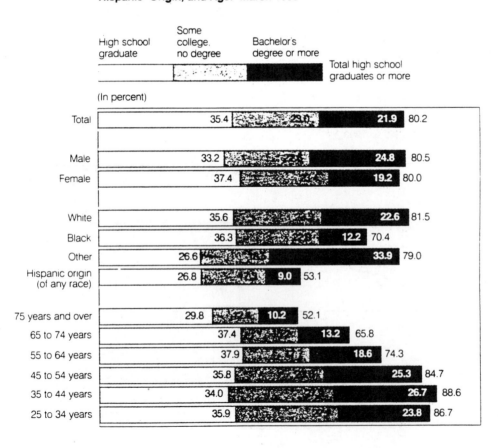

Money Income

WILFRED T. MASUMURA

Household income declined between 1989 and 1993.

Real median household income (in 1993 dollars) in the United States fell 7.0 percent from $33,585 in 1989 to $31,241 in 1993.[1] Most recently, real median household income fell 1.0 percent between 1992 and 1993. This continuing decline occurred during and after the 1990-91 recession.

Household income varied by household composition.

In 1993, the median income of married-couple households was $43,129. The median income was much less for households with a female householder, no husband present ($18,545), and for nonfamily households, mainly one-person households ($18,880).[2]

Households with the oldest householders and the youngest householders had the lowest median incomes in 1993. Households with householders 65 years old and over had a median income of only $17,751. Somewhat higher was the median income of households with householders 15 to 24 years old, $19,340. In contrast, households with householders 45 to 54 years old had the highest median income, $46,207.

Household income varied by race and ethnic origin.

In 1993, Asian and Pacific Islander households had the highest median income ($38,347); whereas, Black households had the lowest ($19,532). The 1993 median income was $32,960 for White households and $22,886 for Hispanic households.[3]

Household income varied by the number of earners in a household.

Households with no earners had the lowest median income in 1993 ($11,807); whereas, households with two or more earners had the highest median income ($49,430). Households with one earner had a median income of $25,560.

Household income varied by the householder's job status.

In 1993, the median household income of householders who were employed year-round, full-time was $44,834. Among householders who were part-time workers, the median household income was $21,608. Householders who did not work had a median household income of only $14,787.

Median earnings of year-round, full-time workers varied by gender.

In 1993, the median earnings of year-round, full-time workers was $30,407 for men and $21,747 for women. The

[1] Changes in "real" income refer to comparisons after adjusting for inflation based on changes in the Consumer Price Index.

[2] The difference between $18,880 and $18,545 is not statistically significant

[3] Persons of Hispanic origin may be of any race. These data do not include the population of Puerto Rico. Due to the small number of American Indian, Eskimo, and Aleut households, a median income figure for them would be statistically unreliable.

Median Household Income, 1989, 1992, and 1993

(In 1993 dollars)

1989	1992	1993
$33,585	$31,553	$31,241

Percent change   1992-93 = 1 0 percent decline
1989-93 = 7 0 percent decline

female-to-male earnings ratio for year-round, full-time workers was 0.72, comparable to the all-time high reached in 1990.[4]

Median earnings of year-round, full-time workers varied by occupation.

In 1993, among male year-round, full-time workers, the median earnings was $42,722 for executives and managers; $32,327 for sales workers; and $27,653 for precision production, craft, and repair workers. For women, the figures were $28,876, $18,743, and $21,357, respectively.

Median earnings of year-round, full-time workers varied by educational attainment.

In 1993, the median earnings of male year-round, full-time workers 25 years old and over with a college degree was $45,987. In comparison, the median earnings for those

[4]The earnings data and female-to-male earnings ratio for 1989 and 1990 were modified based on the inclusion of data on members of the Armed Forces.

with only a high school diploma was $26,820, and the median earnings for those with some high school education but no diploma was $21,402. For female year-round, full-time workers, the comparable figures were $32,291, $19,168, and $14,700, respectively.

The distribution of income has become somewhat more unequal over time.

The household income distribution changed over the past 25 years. In 1993, those at the bottom 20 percent of the income distribution received less of the Nation's income than previously, while those at the top 20 percent received more.

In 1968, the poorest 20 percent of households received 4.2 percent of the aggregate household income. By 1993, their share declined to just 3.6 percent. In contrast, the highest 20 percent of households received 42.8 percent of the aggregate household income in 1968. By 1993, their share had increased to 48.2 percent.

Those in the middle of the income distribution also received proportionally less of the Nation's income in 1993 than previously. The middle 60 percent of households received 53.0 percent of the aggregate household income in 1968. By 1993, their share had declined to 48.2 percent.

For Further Information

See: Current Population Reports, Series P60-188, *Income, Poverty, and Valuation of Noncash Benefits: 1993.*

Contact: Wilfred T. Masumura
Income Statistics Branch
301-763-8576

Share of Aggregate Household Income, by Quintile: 1968 to 1993

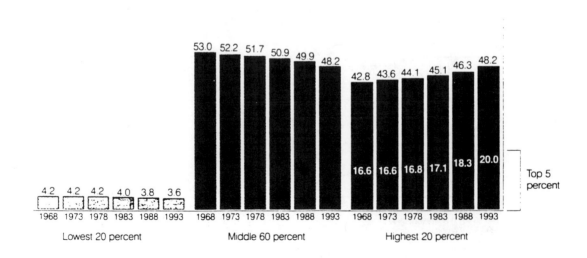

Poverty

BERNADETTE D. PROCTOR

The number of poor rose significantly between 1992 and 1993.

There were 39.3 million persons below the official poverty level[1] in 1993, significantly different from the 38.0 million poor in 1992. However, the proportion of the population with income below the poverty level, 15.1 percent in 1993, was not significantly different from the 1992 poverty rate of 14.8 percent.

Poverty estimates based on the current definition date back to the early 1960's. The number of persons in poverty as well as the poverty rate fell dramatically during the 1960's, from 40 million (22.2 percent) in 1960 to 24 million (12.1 percent) in 1969. The next decade experienced relatively small changes, with the number of poor fluctuating

[1]The poverty definition used by the Federal Government for statistical purposes is based on a set of money income thresholds which vary by family size and composition and do not take into account noncash benefits. The average poverty threshold for a family of four was $14,763 in 1993. A four-person family with cash income below their threshold would be counted as poor.

from 23 to 26 million, and poverty rates varying between 11.1 and 12.6 percent. This was followed by a rapid increase in poverty during the 1978-83 period as the number of persons in poverty increased by 11 million to a total of 35.3 million. The poverty rate reached 15.2 percent in 1983, unsurpassed since 1965. While the poverty rate in 1993 (15.1 percent) was not statistically different from this recent peak of 15.2 percent in 1983, it remains well above the 1978 level of 11.4 percent, a recent low point.

The poverty rate varied across race and ethnic groups.

In 1993, the poverty rate was 12.2 percent for Whites, 33.1 percent for Blacks, and 30.6 percent for persons of Hispanic origin.[2] For Asians and Pacific Islanders, the largest component of the remaining race groups, the poverty rate was 15.3 percent in 1993, not significantly different from

[2]Persons of Hispanic origin may be of any race. These data do not include the population of Puerto Rico.

the 1992 rate of 12.7 percent. Blacks showed no significant increase in poverty between 1992 and 1993. Whites and persons of Hispanic origin showed an increase only in the number living in poverty, not in the rate.

Even though the poverty rate for Whites was lower than that for the other racial and ethnic groups, the majority of poor persons in 1993 were White (66.8 percent). Blacks constituted 27.7 percent of all persons below the poverty level, whereas Asians and Pacific Islanders represented 2.9 percent of the Nation's poor. Persons of Hispanic origin comprised 20.7 percent of the poor in 1993.

About one-third of families maintained by women with no spouse present had income below the poverty level.

While 12.3 percent of all families had incomes below the poverty level in 1993, 35.6 percent of families maintained by female householders with no spouse present were poor. In contrast, only 6.5 percent of married-couple families lived

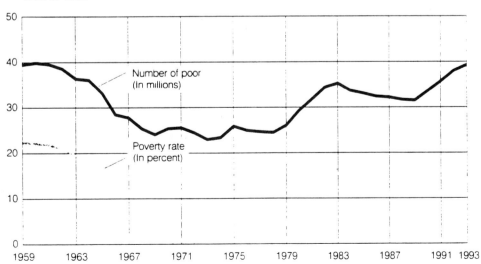

Persons Below the Poverty Level: 1959 to 1993

in poverty. The proportion of female-householder families in poverty was substantially higher for Blacks and Hispanics than for Whites. White families with a female householder, no spouse present, had a poverty rate of 29.2 percent. The corresponding rates for Blacks and Hispanics (49.9 and 51.6 percent, respectively) were not significantly different.

In 1993. 41.5 percent of all poor families were maintained by a married couple: whereas. 52.7 percent were headed by a female householder. no spouse present.

One-half of the poverty population consisted of the elderly and children.

Half of the Nation's poor in 1993 were either children under 18 years old (40.1 percent) or persons 65 years old and over (9.6 percent). The poverty rate for children was 22.7 percent, higher than that for any other age group. The proportion of the elderly living in poverty was 12.2 percent, 2.9 percentage points below the poverty rate for all persons. However, a higher proportion of elderly than non-elderly were concentrated just over their respective poverty threshold, that is, between

100 and 125 percent of their threshold. Consequently. 18.5 percent of the Nation's 12.5 million "near poor" persons were elderly.

Half of poor family householders worked in 1993.

Poor family householders were much less likely to work than nonpoor family householders, regardless of family composition. Among poor family householders, 49.4 percent worked in 1993, and 15.3 percent worked year-round, full-time. In contrast, 80.2 percent of nonpoor family householders worked, and

61.9 percent worked year-round. full-time. In 68.1 percent of poor married-couple families. at least one person worked in 1993. and in 25.5 percent of these families both spouses worked. In poor families maintained by women with no spouse present, 42.7 percent of the householders worked, with only 9.2 percent working year-round, full-time. For the nonpoor householders in this category, 76.7 percent worked in 1993, and 54.9 percent worked year-round, full-time.

The number of poor persons varied considerably under alternative definitions of income.

Since much of means-tested assistance is in the form of noncash benefits, such as Medicaid and food stamps, experimental estimates were prepared by the Census Bureau to demonstrate the effects of including such benefits. Subtracting all government cash transfers from the official definition of income resulted in a poverty population of 60.6 million persons and a corresponding poverty rate of 23.4 percent in 1993. When taxes are subtracted from income and government cash transfers as well as noncash benefits (such as food stamps, housing, and Medicaid) are included in income, the number of persons below poverty was 31.5 million and the poverty rate was 12.1 percent.

For Further Information
See: Current Population Reports, Series P60-188, *Income, Poverty, and Valuation of Noncash Benefits: 1993.*

Contact: Bernadette D. Proctor
Poverty and Wealth Statistics Branch
301-763-8578

Poverty Rates for Persons and Families With Selected Characteristics: 1993

(In percent)

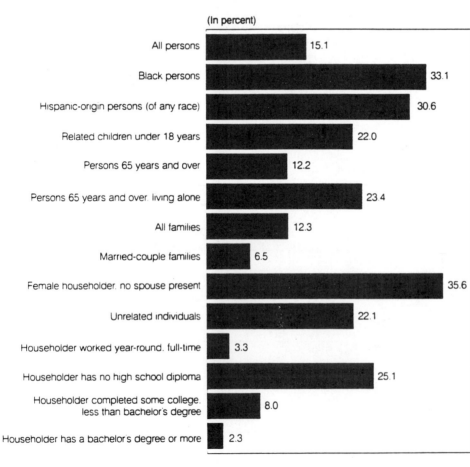

All persons	15.1
Black persons	33.1
Hispanic-origin persons (of any race)	30.6
Related children under 18 years	22.0
Persons 65 years and over	12.2
Persons 65 years and over. living alone	23.4
All families	12.3
Married-couple families	6.5
Female householder. no spouse present	35.6
Unrelated individuals	22.1
Householder worked year-round. full-time	3.3
Householder has no high school diploma	25.1
Householder completed some college. less than bachelor's degree	8.0
Householder has a bachelor's degree or more	2.3

Wealth, Income, and Poverty

Economists differentiate between wealth, a stock of value such as a house or stock, and income, a flow of value over time, such as a salary or interest payment. Data on wealth and income typically come from different sources, each with its own set of problems. There is relatively little information on the distribution of wealth in the United States. Because many items of wealth have not been sold recently, they do not have a readily identifiable price. A second reason for the scarcity is difficulty of measurement. One reason for the scarcity is the lack of data on wealth is its ownership by a relatively small group who are not eager to share information about how much wealth they own. As the controversies below illustrate, this second factor severely restricts how much we know about wealth. Unlike wealth, incomes leave a "paper trail" of readily measured dollar amounts. As a result, income is well documented in a variety of government and private-sector sources. But, as this chapter demonstrates, there is still serious disagreement about fundamental findings, including the trend in well-being for the typical family, and the extent of poverty in the United States.

DATA SOURCES

Wealth

Federal Reserve Survey The most comprehensive survey on wealth holdings is conducted by the U.S. Federal Reserve, a quasi-independent government body that acts as the country's central bank

THE DATA GAME

Where the Numbers Come from

Organizations	Data Sources	Key Publications
Board of Governors U.S. Federal Reserve System	Survey of Consumer Finances	*Federal Reserve Bulletin*
Bureau of the Census U.S. Department of Commerce	Survey of Income and Program Participation; U.S. Census of Population; Current Population Survey	*Census of Population; Current Population Reports* (Series P–60, P–70)
Bureau of Labor Statistics U.S. Department of Labor	Establishment Survey	*Employment and Earnings; Current Wage Developments*
Internal Revenue Service U.S. Department of the Treasury	Tax returns	*Statistics of Income*

Summary data in *Statistical Abstract of the U.S.*

(see chapter 11). At three-year intervals during the 1980s, the Fed conducted its own independent survey of several thousand households, asking 100 pages of questions about each family's financial status.

Data Sample: The 1986 Survey of Consumer Finances found that 19.3 percent of households owned stocks, with a median value of $6,000, and an average value of more than $80,000.

Survey of Income and Program Participation The second major U.S. wealth survey is conducted by the Census Bureau. These data were collected on a regular basis only between 1850 and 1890, and then not again until 1984 in the Survey of Income and Program Participation (SIPP).

Data Sample: In the 1984 survey, equity in homes comprised 59.7 percent of the wealth of households with incomes less than $10,000, but only 30.1 percent of the wealth of households with incomes over $48,000.

Indirect Estimates from Tax Records Historical data on wealth distribution are calculated by two major methods. The "estate-multiplier" method pioneered by Robert J. Lampman of the National Bureau for Economic Research uses Internal Revenue Service records for the 1 percent of estates subject to taxes (only holdings greater than $600,000 in 1987). A related method called "income capitalization" works backwards from tax reports of rent, dividends, and interest to estimate wealth from which these incomes are derived.

Data Sample: Lampman measured a decline in the share of wealth held by the top 1 percent from 36 percent in 1929 to 26 percent in 1956.

Direct Counts The extraordinarily rich usually are well-known individuals about whom information can be obtained from public sources. Based on stock-ownership records, media coverage, and independent investigation, *Forbes* and *Fortune* magazines each estimates wealth holdings for a select number of these individuals.

Income

U.S. Census Every decennial Census of Population since 1940 has included questions about income, providing researchers with a tremendously detailed data source, but one that is available only at ten-year intervals.

Data Sample: In the 1980 U.S. Census, McAllen-Edinburg-Mission, Texas, had 35.2 percent of the population below the poverty line, the highest rate for all U.S. metropolitan areas.

Current Population Survey By far the most frequently cited source of income data is the Census Bureau's Current Population Survey (CPS). Covering over 60,000 households, the CPS is the largest survey taken in between census years. (See chapter 9 on the origins of the CPS.)

Data Sample: According to the Current Population Survey, Manchester-Nashua, New Hampshire, had the fastest increasing per capita income between 1979 and 1983 of all U.S. metropolitan areas.

Bureau Labor Statistics' Establishment Survey Data collected in the BLS Establishment Survey are often quite accurate because the survey is based on employer records, rather than on respondent recall. But the Establishment Survey measures earnings for individual jobs, not the total income for a worker, who may have more than one employer, or for a family with several separate sources of income.

Data Sample: In the March 1985 Establishment Survey, the lowest-paying manufacturing industry was "Children's dresses and blouses," with average weekly earnings of $170.16 for nonsupervisory production workers.

Panel Study of Income Dynamics For comparisons over time, or longitudinal studies, researchers frequently turn to a private survey, the University of Michigan's Panel Study of Income Dynamics (PSID). Begun with 5,000 households in 1968, PSID followed 7,000 households in 1986, including many of the original sample, as well as those split off when children married or couples separated.

CONTROVERSIES

Are the Rich Getting Richer?

Are the rich getting richer? The answer is "we don't know." Attempts during the 1980s to measure wealth distribution generated considerable controversy, but ended in continued uncertainty whether wealth ownership is becoming more or less concentrated. Nonetheless, this research failure illustrates the problem of surveying wealth, or any other variable so unequally distributed that even large surveys are unlikely to include those who own a significant part of the total wealth.

In 1983 the U.S. Federal Reserve changed its usual format for measuring wealth by adding 438 individuals already known to be wealthy. In theory, this commonly used method of "enriching" the data sample should have increased our knowledge about wealth holdings. Indeed, the survey indicated that the share of wealth going to the top 0.5

percent increased to 35 percent in 1983 from 25 percent in 1962. But these startling results were not destined for much public notice. As usual, the Federal Reserve survey was published in the *Federal Reserve Bulletin*, from which the numbers made their way into academic studies and textbooks.

Then, in 1986, the Joint Economic Committee of Congress released its own interpretation of the Federal Reserve's study using catchy labels—"super rich," "very rich," "rich," and "everyone else"—to underscore the vast holdings of the wealthy that apparently increased since 1962. After a barrage of media reports on these results, the Reagan administration asked the Federal Reserve to reexamine the survey. Suddenly an error appeared: one of the 438 wealthy individuals was not as rich as he or she reported. Removing this individual caused the estimate for the share owned by the super rich to fall by almost 10 percent, wiping out the apparent increase since 1962.

The accuracy of the one data point is difficult to determine. A 1986 follow-up survey showing this person's wealth to be only $2.3 million led the Fed to conclude that the reported 1983 wealth of $200 million was a coding error. Critics responded that the individual, known to own Texas oil and gas wells, indeed may have suffered a financial setback reported between 1983 and 1986. Whatever the actual circumstances of this one Texas magnate, the dispute shows how difficult it is to measure wealth. A basic research principle is to include a survey sample that is large enough that errors for a single individual do not affect overall findings. Even the relatively large Federal Reserve survey—less than 4,000 households—was still too small to accurately measure changes in wealth holdings because so few individuals owned so much.

What Is Wealth?

For the less-than-extremely wealthy, Federal Reserve Board data are relatively accurate. Overall they measure total average net worth in 1986 at over $145,000 per household, while the median net worth was nearly $44,000 (see Box 8.1 on the difference between average and median). For black and Hispanic households, median net worth was only $11,000 in 1986, about one-fifth of median net worth for white households.

Most household wealth is in housing, an average value of $80,000 in 1986 for those who owned a home. For some research purposes, it is best to exclude ownership of homes and other consumer durable goods

in order to focus on ownership of wealth that is a source of economic power such as stocks and bonds. This "financial wealth" is even more unequally distributed than total wealth; the top 10 percent of households owned 86 percent of financial wealth in 1983. Some researchers argue that we should move in the other direction, expanding the definition of wealth to include the value of pensions and social security. Although most people do not consider these benefits to be wealth, they are similar to bonds and other arrangements that promise a source of future income. Former chief economic adviser in the Reagan administration, Martin Feldstein, advocated such an approach, by which measured inequality in U.S. wealth holdings would be reduced by nearly 30 percent.

Researchers need to choose carefully between these different measures of wealth. For analysis of the distribution of resources, total wealth measures are appropriate, perhaps also including the value of retirement funds. But if the research goal is to understand how wealth affects power, then it might be proper to use a narrower definition that excludes housing and retirement benefits.

Who Is the Richest of Them All?

In 1987, the wealthiest individuals in the *Fortune* and *Forbes* surveys were as follows. Both *Forbes* and *Fortune* openly discuss problems in

---

Box 8.1. Mean, Median, and Mode

The *median* is the data point for which one-half the observations are above and one-half below; the *mean* is the average, or in this case, total income divided by number in the sample; the *mode* is the data point with the most observations, a statistic rarely used in income analysis. Median income is the most common starting point for research on the "typical" household. A relatively small number of families with very high incomes skews the mean at about 15 percent higher than the median. Thus, mean annual family income in the United States was about $24,000 in 1980, while the median income in the United States was about $21,000. The major drawback to using the median is that it is unchanged by redistribution of income above and below it. For example, when the poor become poorer and the rich become richer, it is possible for the median to stay the same.

THE DATA GAME

Figure 8.1.

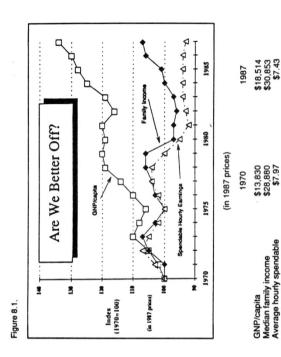

(in 1987 prices)

	1970	1987
GNP/capita	$13,830	$18,514
Median family income	$28,880	$30,853
Average hourly spendable earnings	$7.97	$7.43

Sources: Statistical Abstract of the United States, 1989 (Washington, D.C.: U.S. Government Printing Office, 1989), pp. 424, 444; Thomas E. Weisskopf, "Use of hourly earnings proposed to revive spendable earnings series," *Monthly Labor Review*, Nov. 1984, p. 40 (updated courtesy Thomas E. Weisskopf and David M. Gordon, Graduate Faculty, New School for Social Research).

WEALTH, INCOME, AND POVERTY

Fortune (worldwide)	*Forbes* (U.S. only)
Wealth (in billions)	
1. Sultan Hassanal Bolkiah $25.0	1. Sam Walton $8.5
2. King Fahd $20.0	2. John Werner Kluge $3
3. Samuel and Donald Newhouse $7.5	3. Henry Ross Perot $2.9

finding out about wealth; every year an individual will catapult to near the top when it is discovered that he or she is in fact quite wealthy. A previously uncounted Mars Candy Company benefactor, billionaire heiress, Jacqueline Mars Vogel, was belatedly "discovered" and added to the *Forbes* list in 1987. Lester Crown, owner of General Dynamics stock inherited from his father, as well as real estate and sports teams, weighed in at $5.7 billion in the 1987 *Fortune* list, but only $2.1 billion according to *Forbes*.

One source of such discrepancies is the lack of information on privately held wealth. The Mars Company, for example, is one of the small number of large firms still owned entirely by a few individuals. Unlike publicly held corporations, there is no stock price to estimate the value of the company (see chapter 10). Similarly, *Forbes* and *Fortune* must estimate the value of property that has not been bought or sold in recent years. Finally, family trusts complicate wealth measurement by dispersing fortunes among various descendants. As a result, *Forbes* publishes a separate wealth list for total family holdings, led by the DuPonts, Gettys, and Rockefellers.

Are We Better Off?

Is the typical American better off than twenty years ago? This simple question has provoked a broad spectrum of answers depending on the data source used. Between 1970 and 1987, Americans were somewhere between 10 percent *worse* off measured by average weekly earnings and more than 30 percent *better* off measured by income per capita. At stake in the interpretation of these data is the direction of U.S. economic policy. During the 1980s, supporters of the economic policies of presidents Reagan and Bush pointed to the optimistic income data, while critics emphasized data that showed an income decline.

Per Capita Income The rosiest view of economic progress during the 1980s, indicating 20 percent growth, is based on per capita personal income, a statistic from the U.S. Commerce Department national income accounts (see chapter 7). Per capita income is simply the total of all wages, interest, rents, and other incomes divided by the number of people in the country. Although this number is often cited as evidence of success in the Reagan-Bush economic program, critics charge that the statistic is misleading because the number of wage earners has increased relative to the number of dependents. Or to make the same point with a story: Assume a household of four people in 1980 with one person working outside the home earning $20,000 had a per capita income of $5,000. If in 1990, the original worker's income falls to $18,000, but a second family member now works outside the home, and earns $10,000, then per capita income increases to $7,000 ([$18,000 + $10,000] ÷ 4). In other words, total income increased, but only because more people were working, each of whom earned less.

Earnings What is the trend in earnings of the typical worker? This question changes the original question from one of *income* from all sources including rents, interests, and dividends, to one of *earnings*, including only wages and salaries. (Sometimes earnings is defined also to include self-employment income and farm income.) For many years the Bureau of Labor Statistics Establishment Survey measured "spendable weekly earnings for the typical family of four with one full time worker and three dependents." According to this measure, this statistically "average" family was better off in 1959 than in 1981! In 1982, the Reagan administration suspended publication of the series on the grounds that it was inflammatory and inaccurate.

Even critics of the administration agreed that this measure used outdated assumptions about tax rates and the composition of the workforce. But, even though correcting for these deficiencies was a relatively simple matter, no official replacement data were introduced. Consequently, University of Michigan economist Thomas E. Weisskopf attempted to revive the series on his own, using spendable *hourly* earnings data in order to avoid the problems with the discredited *weekly* earnings data. According to Weisskopf, this new statistical series still shows steady deterioration of earnings after 1972, falling by more than 13 percent by 1987. Thus, it appears that pay levels for U.S. workers indeed stagnated during the 1970s and 1980s (see Figure 8.1).

Family Income One way to combine the effect of additional workers with the effect of lower earnings is to look at the trend in income for household units. In fact, the most widely cited statistic on income is featured in the annual Current Population Survey publication, "Money Income of Households, Families and Persons." According to this source, income was nearly constant during most of the 1970s and 1980s at about $29,000 in 1987 buying power, and improved to only

Box 8.2. The Economic Consequences of Divorce

According to Lenore Weitzman's 1985 book, *The Divorce Revolution*, women suffer a 73 percent loss in economic status after divorce. Criticism published subsequently in the journal *Demography* suggested that this widely cited statistic is quite inaccurate. Weitzman's sample consisted of only 228 men and women who became divorced in Los Angeles County in 1977. Data from the far larger Panel Study of Income Dynamics measured an average of about 30 percent loss in economic status for women in the first year after divorce. The critics conclude that the economic well-being of men and women indeed diverges substantially in the years after divorce, but not nearly to the extent publicized by Weitzman's findings.

$30,853 in 1987. This statistic is often quoted, but usually without reference to serious shortcomings in the data.

One problem is that the U.S. Census carefully defines a "family" as two or more related individuals living together, thus leaving out more than 30 million individuals in 1990 who lived in nonfamily households, either as single-person households or with unrelated individuals. Median income for families is more than double the median income for nonfamily individuals, so looking only at families overstates the typical income of U.S. households. On the other hand, in recent years nonfamily household incomes increased faster than the average, so research only on families underestimates improvement in overall well-being.

A second problem with family income data is that they do not take into account the decline in family size from 3.58 in 1970 to 3.28 in 1980. According to one measure of family well-being, these smaller families are 20 percent better off because there are fewer children to support. However, other researchers object that these families might have fewer children precisely because of stagnating incomes. In this view, families are worse off because they could not afford as many children as they might have liked.

Because of these problems, researchers need to be careful in using the family income data. Overall, the limited improvement in the median family income level was a worrisome trend during the 1970s and 1980s. But used by itself it does not tell the whole story about changing living standards.

Good Jobs, Bad Jobs

Average well-being, whether measured by per capita income, earnings, or family income, is based on a single statistic that may hide important changes in the *distribution* of income. For example, it is possible for average income to remain unchanged at the same time that it is possible for the rich to become richer and the poor to become poorer. In fact, according to some researchers precisely this trend was underway during the 1980s, an assertion that provoked an often bitter debate about the actual distribution of pay for newly created jobs in the U.S. economy.

This controversy is a lesson for researchers about the potential pitfalls in using income data. Popular magazine articles titled "A Surge in Inequality," "The Shrinking Middle Class," "Low-Pay Jobs: The Big Lie," and "Chicken Little Income Statistics" confused readers with seemingly convincing statistics for opposite conclusions. Less frequently reported were the reasons for these differences. When the political controversy ended, it turned out that we had learned quite a bit about trends in U.S. income distribution.

It is well documented that pay levels decline during economic bad times, sending individuals down from middle to lower income levels. The debate about the disappearing middle class occurred just after the 1981–82 recession, so results depended critically on how those years were treated. For example, economists Barry Bluestone and Bennett Harrison received nationwide attention for their study showing that nearly one-half of net new jobs created between 1979 and 1985 paid less than $7,400 (in 1986 buying power). But researchers at the Bureau of Labor Statistics attributed the trend to changing economic conditions, in particular the 1981–82 economic downturn.

A second rejoinder to Bluestone and Harrison came from Brookings Institution economist Robert Lawrence, who attributed most of the growth in low-wage jobs to the sudden appearance of baby-boom youth in the labor market. As these workers gain in age and experience, Lawrence believed income distribution would move closer to its previous norm. Finally, other researchers attributed Bluestone and Harrison's results to incorrect use of inflation adjustments. The relationship between inflation measures and the measurement of income distribution is discussed in chapter 11. To summarize, by using a different inflation index, American Enterprise Institute researchers Mar-

vin Kosters and Murray Ross measured a *decline* in low-wage jobs for the same time period that Bluestone and Harrison measured their much-publicized increase in low-wage jobs.

Each of these researchers made quite different policy recommendations. Bluestone and Harrison advocated government-sponsored industrial policy to increase the number of high-paying jobs. Lawrence argued against industrial policy, calling instead for retraining and relocation assistance for low-paid, young workers. Kosters and Ross opposed all such government intervention in favor of continued reliance on private markets. The jury is still out on which policies will best help the U.S. economy. Nonetheless, it is possible to assess what we have learned about income distribution in this debate. First, there is widespread agreement that a shift has occurred toward low-pay jobs for some groups. Even Kosters and Ross, who gave the most optimistic appraisal of the U.S. economy, measured an increase in the proportion of low-paid jobs for full-time, year-round workers. In addition, the number of part-timers (who obviously earn less) has increased, although there is debate about how to measure the effect of the increase in part-time work on overall pay levels. Overall, there has been an increase in the proportion of low-income individuals, but the impact on households or families is a more complex picture because, as noted above, households may have higher earnings at the same time that each individual earns less.

Statistics for Every Theory?   At this point it may be tempting to question the usefulness of income statistics that tell such conflicting stories. If policy advisers from every political persuasion can find ample evidence to support any position, what good are income statistics? The answer is that each income statistic measures a slightly different concept; they must be used together, with careful distinction about what is being measured. For example, per capita income provides an indication of well-being, but it tells us nothing about the distribution of that well-being, which must be measured by other statistics. The confusing statistics on earnings and family income showed that major changes were occurring in the growth of income. Finally, in addition to an underlying slowdown in the growth of income, the good jobs–bad jobs debate illustrated what can go wrong when researchers and the popular media debate the findings do not pay careful attention to the limitations inherent in the data. Research by

Bluestone and Harrison faced considerable criticism when it was shown that alternative assumptions caused a reversal in their findings. The lesson for researchers is that it is necessary to anticipate objections by proving the robustness of results under different assumptions. Indeed, subsequent research confirmed Bluestone and Harrison's conclusion that low-pay jobs increased, but in a politically acrimonious climate opposed to their interventionist policy recommendations, these later findings received little attention.

What Is Poverty?

Research on poverty typically confronts a perplexing question: how to define poverty? The easy solution is to rely on the official U.S. Census Bureau poverty line, the most commonly used in social science research. But, as demonstrated by its origins, the Census Bureau poverty line is quite arbitrary, and therefore inadequate for some purposes.

The Official Poverty Line During the early 1960s, Molly Orshansky, a Social Security Administration staff economist, was asked to develop a definition of poverty for the War on Poverty. By her own admission, this "Orshansky" poverty line was a compromise between scientific justification and political expediency. She began with a U.S. Agriculture Department estimate for a nutritionally sound diet, and then estimated the total poverty budget based on the proportion of food expenditures in a typical total family budget. Orshansky was unhappy with these assumptions, in particular the food budget, which was intended as a stop-gap grocery list but required a sophistication of purchasing and food preparation that was not likely to be practiced by poor families. But Orshansky was under pressure from the administration to calculate a poverty line at about $3,000 for a family of four, low enough so that the War on Poverty could reasonably be expected to help all those designated as poor. She had more leeway in defining poverty for other-size families, so she set the poverty line on slightly more favorable terms for smaller-size households and families with more than four people.

Not surprisingly, these highly arbitrary poverty lines have been criticized on many grounds. But researchers disagree whether this measure under- or overestimates the actual rate of poverty.

The Poverty Line Is Too Low The official U.S. poverty line is an absolute standard, adjusted only for inflation. (In 1990 there was a renewed debate about how to adjust for inflation, including a Census Bureau proposal to use a new inflation adjustment that would substantially reduce the official poverty line; see chapter 11.) Over the years since 1965 the standard of living in the United States has fallen. (In 1965, it was just under one-half the U.S. median income; by 1986 the official poverty line was less than one-third of the median income. If the poverty line measured poverty at a constant *relative* rate, then measured poverty would have increased between 1965 and 1983, instead of falling, as it did in the official absolute standard.

The Poverty Line Is Too High Others have criticized the Orshansky poverty line for overestimating the poverty line. For example, economist Rose Friedman points out that poor families spend a higher proportion of their income on food than the ratio used by Orshansky. Friedman recalculates Orshansky's poverty level with new ratios, estimating U.S. poverty at one-half its official level.

A second correction that reduces the apparent level of poverty is to include the value of noncash government programs such as food stamps, school meals, housing subsidies, and medical care. One such effort measured government programs at their market value, that is, what it would cost the poor to buy services comparable to those provided by the government. By this estimate, the poverty rate falls by about one-third, which prompted President Reagan's chief economic adviser Martin Anderson to conclude that the War on Poverty had been "won." But many economists question the relevance of market values for adjusting the poverty rate. In particular, medical care is so expensive in the private sector that the market value of Medicare and Medicaid is sufficient by itself to lift—in theory—almost all the elderly poor out of poverty.

Implications The official "Orshansky" poverty line has withstood the test of time. It makes sense for many research projects because conclusions can be compared readily with other studies that also use this poverty line. Nonetheless, researchers should be aware of biases in the official number. Most notably, the absolute standard means that poverty today is measured by the same living standards used in 1965 (again, corrected only for inflation). The arbitrary assumptions made by Orshansky may be less of a problem. Any poverty line will be arbitrary, and the problems with the official one in part cancel out one another. Friedman's complaint about the food budget ratio is offset by the low food budget originally used, and the failure to include noncash benefits for the poor is offset by tax breaks and other government assistance available to the nonpoor.

One alternative research strategy is to adopt more than one poverty measure, including both absolute and relative standards. By using several poverty lines, it is possible to demonstrate the constancy, or robustness, of results under alternative assumptions. For example, the issue of how to count noncash benefits can be resolved by looking at alternative poverty measures published by the U.S. Census Bureau. Replacing the questionable market valuation for noncash benefits with more realistic assumptions reduces by about one-half the number of the poor who are raised above the poverty line by their noncash benefits.

Do the Poor Stay Poor?

The official poverty rate is based on a snapshot view. If our research interest is to find out if the same households are poor in different years, then we need longitudinal studies, that is, surveys that reinterview respondents over a long period of time. Surprising data from the University of Michigan's Panel Study of Income Dynamics (PSID) and the Census Bureau's Survey of Income and Program Participation (SIPP) show remarkable movement in and out of poverty. The PSID found that only 2.6 percent of the population was poor in eight out of ten years between 1969 and 1978, while over 24 percent were poor for at least one year. Similarly, in SIPP, only 6 percent of the population was poor in *every* month of 1984, but 26 percent were poor for at least one month.

For some conservative policy advisers, these data prove that poverty is less serious than CPS data indicate. In this view, most poor people require little government assistance because their poverty is temporary. Moreover, the small number who stay poor have become dependent on welfare programs and would gain by being forced to find regular employment. On the other hand, liberal social scientists often emphasize the surprisingly large proportion of the population experiencing tempo-

rary poverty. In this view, the welfare system is a much-needed security cushion for many families who are precariously close to the poverty line. Along similar lines, sociologist Mary Jo Bane used longitudinal data to show that most poverty occurred because of income or job changes, not because of family composition changes such as divorce or abandonment. Bane concludes: "the problem of poverty should be addressed by devoting attention to employment, wages and the development of skills necessary for productive participation in the labor force rather than hand-wringing about the decline of the family." This policy recommendation has not been tested, but it shows how one can use dynamic data available in the PSID or SIPP to go beyond the debate about the "actual" low or high rate of poverty.

SUMMARY

Without improved data on wealth, we will not understand its distribution in the United States. At present, however, the barriers to better data appear insurmountable. The uneven distribution of wealth means that even extremely large surveys will tell us nothing about the very wealthy; it is unlikely that any of the *Forbes* 400 will be interviewed in the extremely large-sample Current Population Survey, and even more improbable that they will be included in the Federal Reserve survey. For example, the enriched sample of 438 known-to-be-wealthy households in the 1983 Federal Reserve survey included none of the wealthiest Americans identified by business magazines.

Additional data on income also would be helpful, for example, replacing the spendable weekly earnings series and increasing funding to the Survey of Income and Program Participation. But in general, simply increasing the quantity of data will not be sufficient. As economist Isabel Sawhill concludes, "We are swamped with facts about people's incomes and about the number and composition of people who inhabit the lower tail, but we don't know very much about the process that generates these results." In other words, we need more data about income dynamics, that is, when and how income changes over time, and we need data about why incomes differ for different jobs.

Controversies about the distribution of income and the rate of poverty both demonstrate the need to test different assumptions in research projects. This kind of sensitivity analysis can preempt criticism about what may appear to be arbitrary choices such as years chosen for study, inflation indexes, or poverty lines. Such care can help researchers reach an audience beyond the already convinced.

Finally, despite the apparent existence of contradictory statistics, researchers have been able to reach important conclusions about income distribution and poverty. In other words, although it is possible to "lie with statistics," bending the numbers to agree with one's prejudice, it is also possible to use income statistics correctly if one understands the limitations of the underlying data.

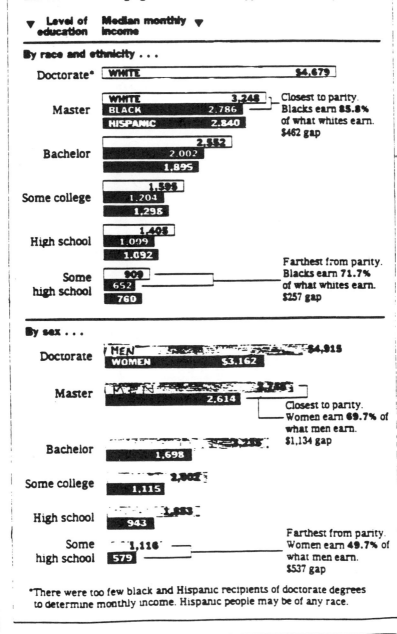

STATISTICAL PORTRAIT OF THE NATION

Education and Income

Educational levels and gross income according to figures from the 1990 Census released yesterday for those over age 18. Charts and maps that reflect the changing face of the nation appear occasionally.

▼ Level of education   Median monthly ▼ income

By race and ethnicity . . .

Doctorate* WHITE $4,679

Master
WHITE 3,248
BLACK 2,786
HISPANIC 2,840

Closest to parity. Blacks earn **85.8%** of what whites earn. $462 gap

Bachelor
2,552
2,002
1,895

Some college
1,595
1,204
1,298

High school
1,405
1,099
1,092

Some high school
909
652
760

Farthest from parity. Blacks earn 71.7% of what whites earn. $257 gap

By sex . . .

Doctorate
MEN $4,915
WOMEN $3,162

Master
MEN 3,748
2,614

Closest to parity. Women earn **69.7%** of what men earn. $1,134 gap

Bachelor
3,215
1,698

Some college
2,302
1,115

High school
1,883
943

Some high school
1,116
579

Farthest from parity. Women earn 49.7% of what men earn. $537 gap

*There were too few black and Hispanic recipients of doctorate degrees to determine monthly income. Hispanic people may be of any race.

LINEAR REGRESSION SUMMARY

A linear regression line is an attempt to find an average line equation to approximate a set of data points that looks as though it is headed roughly along a straightish path. If you want to know what's behind the scenes in calculating the linear regression line on the computer screen, here it is:

STEP 1
Start with a set of data, x and y points.

STEP 2
Find the average point $\bar{x}$, $\bar{y}$; (called "x bar" and "y bar", these symbols mean "average x" and "average y"). To do this add all the x values and divide by the number of points; then do the same for the y values.

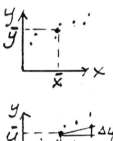

STEP 3
Find the vertical rise, Δy, and the horizontal run, Δx, for the line connecting each point to the average point.
On the data table this is $\Delta x = x - \bar{x}$ and $\Delta y = y - \bar{y}$ for each point.
S
STEP 4
Notice that if we are trying to find a slope for our average line, the lines connecting the average point to points farther away are likely to be closer to what we need. If we are computing an average slope we need to give greater weight in proportion to the distance from the average point.

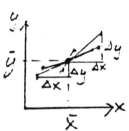

To give more importance to the slopes from the points farther away from the average point, $\bar{x}$, $\bar{y}$, we multiply each Δx and each Δy by the Δx for that point.
Now you have $\Delta x \cdot \Delta x$ and $\Delta x \cdot \Delta y$ for each point.
(Note that you can also multiply each Δx and each Δy by the Δy instead, this will give a differently weighted line.)

STEP 5
Now sum the $\Delta x \cdot \Delta y$ for all the points; this is $\Sigma \Delta x \cdot \Delta y$.
Then sum the $\Delta x \cdot \Delta x$ for all the points; this is $\Sigma \Delta x \cdot \Delta x$.
The weighted average slope for all the points is
$$m = \frac{\Sigma \Delta x \cdot \Delta y}{\Sigma \Delta x \cdot \Delta x}$$

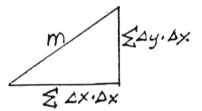

STEP 6
Find "b" by using $\bar{y} = m\bar{x} + b$ so $b = \bar{y} - m\bar{x}$

Now we know m and b we can use them to write the linear regression line equation y = mx + b.

Draw the line y = mx + b through the average point, $\bar{x}$, $\bar{y}$, using slope m and y-intercept b.

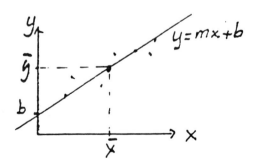

Linear Regression Calculation Example

7 babies between the age of 0 and 24 months are measured for length in inches. If we call the age x and the length y the data points are:

age, x	length, y
1	15
3	18
6	20
10	25
14	27
18	30
23	33

STEP 1: Plot points.

STEP 2: Calculate average point.

average x = $\bar{x}$ = $\underline{1 + 3 + 6 + 10 + 14 + 18 + 23}$ = 75/7 = 10.71
$\phantom{average x = \bar{x} = }$ 7

average y = $\bar{y}$ = $\underline{15 + 18 + 20 + 25 + 27 + 30 + 33}$ = 168/7 = 24.00

x	y	$\Delta x = $ $x - \bar{x}$	$\Delta y = $ $y - \bar{y}$	$\Delta x \cdot \Delta x$	$\Delta y \cdot \Delta y$	$\Delta x \cdot \Delta y$
1	15	-9.71	-9	94.28	81	87.39
3	18	-7.71	-6	59.44	36	46.26
6	20	-4.71	-4	22.18	16	18.84
10	25	-.71	-1	0.50	1	0.71
14	27	3.29	3	10.82	9	9.87
18	30	7.29	6	53.14	36	43.74
23	33	12.29	9	151.04	91	110.61
Sum 75	Sum 168			Sum= 391.40	Sum= 270	Sum= 317.42

STEP 3: Calculate $\Delta x = x - \bar{x}$ and Δy $y - \bar{y}$

STEP 4: Calculate $\Delta x \cdot \Delta x$ and $\Delta y \cdot \Delta y$ and $\Delta x \cdot \Delta y$

STEP 5: Calculate (for an x weighted line) m = $\dfrac{\sum \Delta x \cdot \Delta y}{\sum \Delta x \cdot \Delta x}$ = $\dfrac{317.42}{391.4}$ = 0.81

(For a y weighted line use m = $\dfrac{\sum \Delta x \cdot \Delta y}{\sum \Delta y \cdot \Delta y}$)

STEP 6: Calculate   b = y - mx = 24 - 0.81•10.71 = 15.32

So the linear regression line (x weighted) equation is:

$$y = 0.81 \, x + 15.32 \qquad \text{length}$$

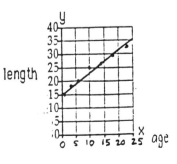

The Correlation Coefficient

Karen Callaghan, Ph.D
The University of Massachusetts-Boston

A correlation coefficient measures the strength and direction of a linear association between two variables. It ranges from -1 to +1. The closer the absolute value is to 1, the stronger the relationship. A correlation of zero indicates there is no linear relationship between the variables. The coefficient can be either negative or positive. The scatterplots below indicate linear association of the same strength but opposite directions.

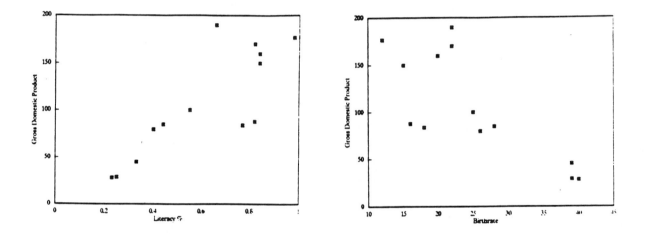

The graph on the left represents the relationship between literacy % and gross domestic product. As literacy % goes up, the gross domestic product goes up, so the coefficient is positive. The graph on the right represents the relationship between birthrate and gross domestic product. As birthrates go up, the gross domestic product goes down, so the coefficient is negative.

The scatterplots on the next page will give you an intuitive grasp of the correlation coefficient, labelled as cc. All the coefficients are either zero or positive. Figure a represents a perfect correlation of +1, all the points fall on a perfectly straight line with a positive slope, an unlikely occurrence in any social science data set. Figure b represents a strong correlation where the behavior of one variable is similar but not identical with the behavior of the other variable (e.g., like miles traveled versus gasoline used).

In Figure c, a correlation of .50 represents a moderately strong relationship. The relationship in Figure d is weak, so the coefficient is only .25.

The scatterplot in Figure e looks like a shotgun blast so the correlation is zero; the x and y variables are not so linearly related. You might be surprised to learn that the coefficient in Figure f is also zero. This is because the correlation coeficient measures a linear association, while the relationship in Figure f is curvilinear.

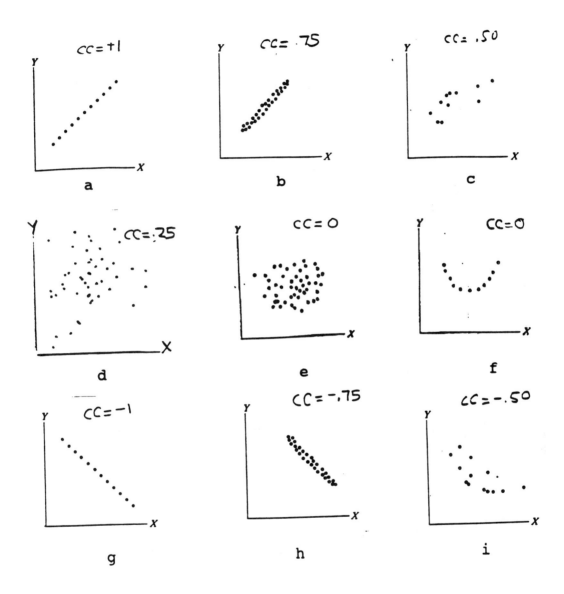

Figures **g, h, i** are mirror images of Figures **a, b, c.** All the correlation coefficients are negative. Figure **g** represents a perfect correlation of -1, all the points fall on a straight line with a negative slope, an unlikely occurence in any social science data set. Figure **h** represents a strong negative correlation. In figure **i** the correlation is moderately strong.

Computing the Correlation Coefficient (cc.)
(Optional Section)

While a perfect correlation is easy to decipher, it is difficult to guess the coefficient of weaker correlations. That is why Karl Pearson developed a precise mathematical measure of correlation known as Pearson's *r*, which is called cc. throughout the text. For those who are interested in knowing how the correlation coefficients are actually calculated, the steps are outlined below.

$$\text{Correlation Coefficient} = \frac{\sum (x - \bar{x}) * (y - \bar{y})}{\sqrt{\sum (x - \bar{x})^2 * (y - \bar{y})^2}}$$

The symbol $\sum$ is "sigma", the Greek letter S, which is a mathematical shorthand meaning "sum up." So if x takes on values 2, 3, and 6, then: $\sum x = 2 + 3 + 6 = 11$.

Literacy % and Life Expectancy for Six Selected Nations

	(x) Literacy %	(y) Life Expectancy
STEP 1: Start with a set of data, x and y points.Each data point is kept in a separate row.	.29	42
	.92	77
	.52	58
	.55	47
	.40	48
	.66	55
	3.34	327

STEP 2:
Find x ,the mean of x and y, the mean. of y. To do this add the values of x and divide by the number of points; then do the same for y.

$x = \sum x/n = 3.34/6 = .56$
$y = \sum y/n = 327/6 = 54.50$

STEP 3:
Subtract x from each value of x; subtract y from each value of y to get a new table of rows.

$x - \bar{x}$	$y - \bar{y}$
-.27	-12.50
.36	22.50
-.04	3.50
-.01	-7.50
-.16	-6.50
.50	.10

Products

STEP 4:
Take the products of each row in step 3 and sum them up.

```
-.27 x -12.50 =   3.38
 .36 x  22.50 =   8.10
-.04 x   3.50 =  -0.14
-.01 x  -7.50 =   0.07
-.16 x  -6.50 =   1.04
 .10 x   .50  =   0.05
```

Sum of products = 12.50

STEP 5:
Take each x value in step 3,
square it and sum all the points;
then do the same for y

x squared	y squared
$-.27^2 = .07$	$-12.50^2 = 156.25$
$.36^2 = .13$	$22.50^2 = 506.25$
$-.04^2 = .00$	$3.50^2 = 12.25$
$-.01^2 = .00$	$-7.50^2 = 56.25$
$-.16^2 = .03$	$-6.50^2 = 42.25$
$.10^2 = .01$	$.50^2 = .25$

Sums of Squares = .24   = 773.50

STEP 6:
Take the square root of the product of
the sum of the square in step 5.

$$\sqrt{.24 * 773.50} = 13.62$$

STEP 7:
Divide the sum in step 4 by
the value in step 6 to calculate
the correlation coefficient.

$$cc. = \frac{12.50}{13.62} = .92$$

So the correlation coefficient is .92 which says that literacy percent and life expectancy are strongly related.

Suppose we reverse the x and y axis. Now 'x' the independent variable is life expectancy and 'y' the dependent variable is literacy percent. You do not need to retrace steps 1 through 6 again. The correlation is still .92 since the formula does not make a distinction between the x and y variables. You cannot test whether literacy percent affects life expectancy, or life expectancy affects literacy percent. You can only test whether these two variables vary together in a linear relationship.

<u>An example of a negative correlation.</u>

News reports sometimes rate the educational systems by comparing the mean of scores of seniors in each state on college entrance exams. This method is misleading because the percent of high school seniors who take any particular college entrance test varies greatly from state to state. The figure below shows a scatterplot of the mean score on the Scholastic Aptitude Test (SAT) mathematics examination for high school seniors in each state compared with the percent of graduates in each state who took the test.

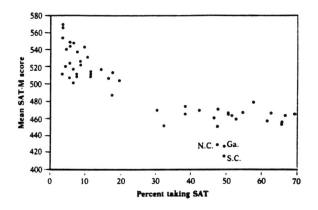

The negative association between the two variables is evident: SAT scores tend to be lower in states where the percent of students who take the test is higher. Only 3% of the seniors in the three highest scoring states took the SAT.

Performing Arts– The Economic Dilemma

by WILLIAM J. BAUMOL *and* WILLIAM G. BOWEN

A Study of Problems common to Theater, Opera, Music and Dance

APPENDIX X–1

Number of Performances and Audience Size

While the typical length of season is currently increasing, primarily in response to the demands of the performers, it is by no means obvious that the increased number of performances will bring with it a corresponding increase in the size of the total audience. There is some reason to expect that when more performances are provided, more tickets will be sold. If most of the orchestras that are lengthening their seasons are also developing new types of concerts, new series and different types of orchestral service, these may attract new persons to attend or they may lead others to attend more often. The new performances will probably take place on nights of the week or at times of the year which are convenient for some persons who would otherwise not have been able to come. But, other things being equal, we may well suspect that more frequent performance will serve, in part, just to spread the concert audience more thinly.

Since, as we saw in Chapter VIII, cost per concert does decrease up to a point as the number of concerts goes up, it is important to see whether the added concerts can be expected to bring an additional audience of average size along with them. Obviously, if attendance per concert declines sufficiently as the number of concerts grows, economies of scale will not help finances.

We compared average attendance per concert with the number of concerts per season for two of the major orchestras for which we had obtained statistically significant cost functions (see the appendix to Chapter VIII). Appendix Figure X–A shows what we found for the same orchestra whose unit cost curve is depicted in Figure VIII–4. It is evident from the diagram that average attendance does fall somewhat with the number of concerts.[1] Furthermore, the decline shown in the diagram is probably a considerable understatement of the rate of decrease involved in the underlying relationship. Many of the historical increases in number of con-

---

[1] Actually the decline is small in terms of the historical relationship shown in Appendix Figure X–A. An increase of 1 per cent in number of concerts yields at 1964 attendance levels a decrease of about 0.3 per cent in attendance per concert, a loss of about 8 persons per concert. With an average of slightly more than 100 concerts per season, this would reduce attendance by a total of about 800 admissions. Since, however, the new concert would on the average be attended by about 2,500 persons, it would bring in a net gain of some 1,700 paid admissions.

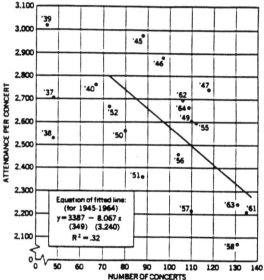

certs occurred only in response to popular demand when the potential audience was already available. If the increased number of performances had been undertaken haphazardly, without regard for autonomous demand changes, and if the new concert had been required, as it were, to hunt up its own audience, one may surmise that the resulting decrease in attendance per concert would have been far more marked. But as more orchestras strive for year-round operation, the number of concerts per season is bound to increase, whether or not they are accompanied by a considerable decline in attendance per concert.

While a larger supply of performances does increase the *total* audience, it undoubtedly yields diminishing returns in terms of audience per concert, and so it is not always an unmixed financial blessing.

PRINCETON · METRO

The Times

SUNDAY
MAY 8, 1994

His stats can oust a senator or price a Bordeaux

■ A Princeton economics professor uses statistics to uncover the mysteries of human behavior — from voting patterns that changed the shape of Pennsylvania's Senate, to the price of fine French wines.

By PETER ASELTINE
Staff Writer

PRINCETON BOROUGH — Orley Ashenfelter's days are numbered.

Behind the tragedy of discrimination and the bumbling of bureaucrats, the Princeton econometrician sees numbers.

Behind the bouquet of a fine wine — more numbers.

Ashenfelter's ability to see the telling patterns of the world and interpret them with numbers has led people to seek the Princeton University professor's expertise for a wide variety of very practical reasons.

Last month a federal judge in Philadelphia relied partly on Ashenfelter's analysis of voting patterns to give a Pennsylvania Senate seat to a losing Republican candidate. The judge found the Democratic winner had used fraud in soliciting absentee votes in the election in Philadelphia's Second District.

The case raised a difficult question, Ashenfelter says, because while the fraud justified unseating the Democrat, William Stinson, it was unclear whether the proper remedy was to seat the Republican, Bruce Marks, or hold a new election.

Marks had outpolled Stinson on the voting machines, 19,691 to 19,127. But in absentee ballots, Stinson received 1,391 votes to Marks' 366, winning the November 1993 special election by 461 votes.

U.S. District Judge Clarence C. Newcomer found that Stinson workers had improperly influenced absentee voters, mostly in Latino and African-American neighborhoods, in some cases marking the voters' ballots or forging their names.

In February, Newcomer threw out the absentee votes and ordered that Marks be seated. The following month, however, a federal appellate court ruled that while the judge was not wrong to unseat Stinson, he should not have seated Marks without first analyzing whether the fraud actually had swayed the election.

"It was a tricky business for the judge, because the issue really was whether it was reasonable to assume that Marks had actually won that election," Ashenfelter said. "The circuit court wanted some evidence that, in the absence of the fraud on the voters, there would have been a presumption that Marks could have been expected to get at least enough of the absentee votes so that he wouldn't be overwhelmed."

ASHENFELTER, a 51-year-old economics professor, was appointed as an expert for the new hearing by Newcomer, who had taken one of the statistics seminars that Ashenfelter regularly teaches for federal judges. Two other statistics experts testified, one hired by Marks and one hired by a group of voters opposed to him.

Ashenfelter has testified in court about a dozen times, mostly in dis-

• see NUMBERS, A18

Staff photo by Katherine Wyland

Princeton University's Orley Ashenfelter is a respected statistics expert.

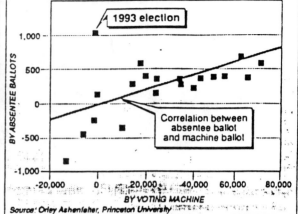

Princeton Professor Orley Ashenfelter used a scatter diagram to show that when the difference between Democratic and Republican machine votes is plotted against the difference in absentee votes for senatorial elections held in Philadelphia in the last decade, points representing the elections tend to fall around a line representing an ideal correlation between the machine and absentee tallies. The graph shows that the 1993 election falls well outside the shaded area where 95% of the results would be expected to fall. The probability that the deviation in the 1993 results was simply due to random changes in voting patterns is less that 1 percent, Ashenfelter calculated.

Times graphic by Laura Sommerville

Numbers

• continued from A1

crimination cases, which often rely on statistical analysis.

What Ashenfelter did in this case was quite simple, he says. He used a standard statistical technique called regression analysis to compare machine votes to absentee votes for each of the 22 senatorial elections held in Philadelphia in the last decade. What he found was that the difference between Democratic and Republican votes on the machines generally is a good indicator of the difference in absentee votes.

Ashenfelter used a scatter diagram to show that when the difference between Democratic and Republican machine votes is plotted against the difference in absentee votes, points representing the elections tend to fall around a line representing an ideal correlation between the machine and absentee tallies.

The graph shows that the 1993 election falls well outside the pattern of typical elections. Ashenfelter's analysis would predict a 133-vote advantage for Marks in absentee votes, given his 564-vote margin on the machines. Instead, Stinson had a 1,025-vote advantage.

Ashenfelter calculated that the probability was less than 1 percent that the deviation of 1,158 votes between his predicted result and the actual result was simply due to random changes in voting patterns. He calculated that the probability that Stinson could have received enough absentee votes to win was about 6 percent.

Ashenfelter said surveys of absentee voters conducted by The Philadelphia Inquirer and the Republicans provided information that was extrapolated in court to produced estimates of fraudulent votes that were surprisingly close to his deviation figure of 1,158.

Ashenfelter said there was an amusing moment in court when a lawyer for the Democratic-controlled board of elections tried to suggest that Ashenfelter's analysis

would not establish public confidence in a decision to seat Marks. The circuit court had called in its opinion for "evidence and an analysis ... worthy of the confidence of the electorate."

"He had asked me in my deposition whether I considered myself an expert on establishing the public confidence, and I had said, 'No.'" Ashenfelter said. "So in cross-examination, he got up, and he said, in the usual lawyer way, 'Isn't it true, Dr. Ashenfelter, that you are no more of an expert on what would establish the public confidence than I am?' He had been having his way with me, so at that point I just thought, if a guy asks you a question like that, you've got to let him have it. So I said, 'Well, I may be more of an expert than you are.' Well, the courtroom just cracked up."

NEWCOMER ULTIMATELY issued a new order to seat Marks, based on the testimony of the experts. The appeals court upheld the decision and Marks was sworn in on April 28, returning control of the Pennsylvania Senate to the Republicans.

Ashenfelter is modest about his role in the important case. He seems more excited about the prospect of using his court analysis as a teaching tool. He said many other professors have asked for it.

"It's very simple as these things go," he said. "The reason people want to use it for teaching is because it's hard to come up with such a simple example that's so telling for

an actual problem."

Ashenfelter spent only a week developing his analysis for the election case. He currently is working on a much bigger project for the state of New York, evaluating the state's Wicks Law, which requires that state construction projects be done without a prime contractor. By forcing the state to act as prime contractor, the law has probably cost New York taxpayers a billion dollars, Ashenfelter said.

"Expecting bureaucrats to manage a construction project is a pretty expensive proposition," he said.

Ashenfelter's greatest claim to fame, however, is a regression analysis that predicts the prices of Bordeaux wine vintages by charting rainfall and temperature during each growing season. His analysis has proved quite accurate and he publishes his predictions twice each year in a newsletter called "Liquid Assets."

The newsletter has nearly a thousand subscribers, including some wine-industry insiders in Bordeaux. But Ashenfelter said the industry still refuses to use the information in setting prices.

"Most people in the industry refuse to believe this," he said. "They will not change prices to reflect quality, and they have never had to do it in the past. So the whole system

is built on this bizarre sucker operation. The wine writers — I guess they're the ones who go along with it. It's a mystery to me."

Meanwhile, however, Ashenfelter is storing away some underpriced vintages. His days may be numbered in one sense, but he's looking forward to enjoying the fruits of his numerical labors for years to come.

Ideas & Trends

How a Flat Tax Would Work, For You and for Them

By David Cay Johnston

The big issue of the Presidential campaign in these first weeks of 1996 has been how Americans should tax themselves. And most of the talk is about a flat tax on incomes, with a large deduction for individuals so that lower-income people would pay no income tax.

There are about as many flat tax variants floating around as there are would-be occupants of the White House. But all derive from "The Flat Tax" (Hoover institution Press, 1995), by Robert Hall and Alvin Rabushka, both economists at the Hoover Institution at Stanford University, who devised the concept in 1981 and have been refining it ever since.

The two economists say their plan, with a 19 percent flat tax rate, would spur investment and growth by taxing income only once, eliminating taxes on capital gains and taxing individuals only on wages, salaries and pensions. It would allow no deduction for mortgage interest and charitable contributions. Interest and dividends would not be taxed directly, but would be affected by a business tax, also at 19 percent. (Businesses would no longer be able to deduct the cost of fringe benefits, except pensions, for their employees, however.)

So how would a flat tax affect the typical family? And how would it affect the taxes of the Presidential candidates who have pushed the issue to center stage?

The New York Times recalculated the taxes that some of the major candidates would have owned in 1994 had the Hall-Rabushka flat tax been in effect, consulting with the republican candidates' advisers and making minor adjustments to eliminate anomalies in the complex tax returns that are irrelevant to a flat tax.

The two candidates most closely identified with the flat tax - Steve Forbes, the magazine owner, and Patrick J. Buchanan, the television commentator - would no make their tax returns available. Mr. Forbes said he would never disclose his returns; Mr. Buchanan said he would make his public if he wins the Republican nomination.

President Clinton and three prominent Republican candidates - Senator Bob Dole and Phil Gramm and Lamar Alexander, the former Education Secretary - made available copies of their 1994 Federal income tax returns. (Mr. Clinton opposes the flat tax, as does Mr. Alexander. Mr. Dole has said he would go along with a flat tax. Mr. Gramm favors a 16 percent flat tax.)

All four candidates would enjoy considerable savings in personal income taxes under the Hall-Rabushka flat tax. A comparison also requires calculating the effect of the Hall-Rabushka business tax on business income, but even with both taxes, the taxes owned by each of the four candidates would decline - although probably none would save as much as Mr. Forbes, a fervent flat-tax advocate who is one of the wealthiest men in the United States.

Senator Dole and his wife, who had $607,000 in income in 1994, would save about $49,000; the Clintons whose income was $267,000, would save about $21,000.

Senator Gramm, who with his wife reported $305,000 in income, would save about $40,000; his aides emphasized that under his own proposal, the Senator's taxes would decline even more sharply.

Mr. Alexander's savings would be about $184,000 on an income of about $1 million. The savings would presumably be significantly greater for Mr. Forbes, who has an income of $1.4 million as chief executive officer of Forbes Inc. and a net worth that Fortune magazine last week estimated at $439 million.

The picture is not so bright for the Smiths, a hypothetical family of four used by the Tax Foundation, a conservative research group, in its analyses of the various flat-tax plans. The Smiths, who have $50,00 in total income, nearly all from wages, would see no change in their individual tax bill. And if interest rates decline as predicted, the Smiths, who collected $1,425 in interest in 1994, would come out a few dollars behind.

Sources: Tax Foundation (average family's return).
Those candidates that provided tax returns were consulted on redefining 1994 income into categories appropriate to the flat tax form.

If the Tax World Was Flat: Hypothetical Returns.

Recalculations of the 1994 tax returns of some of the Presidential candidates and that of a typical family. Some of the candidates would not release their tax returns..

Form 1 — **Individual Wage Tax** — **1994**

Your first name and initial (if joint return, also give spouse's name and initial): William J. & Hillary Rodham — Last name: Clinton — Your social security number

Present home address (number and street including apartment number or rural route): 1600 Pennsylvania Avenue, N.W. — Spouse's social security no.

City, town, or post office, state, and ZIP code: Washington, D.C. 20500

Your occupation ▶ U.S. President
Spouse's occupation ▶ Lawyer

1	Wages and salary	1	201,421
2	Pension and retirement benefits	2	
3	Total compensation (line 1 plus line 2)	3	201,421
4	Personal allowance		
	(a) ☐ $16,500 for married filing jointly	4a	16,500
	(b) ☐ $9,500 for single	4b	
	(c) ☐ $14,000 for single head of household	4c	
5	Number of dependents, not including spouse	5	1
6	Personal allowances for dependents (line 5 multiplied by $4,500)	6	4,500
7	Total personal allowances (line 4 plus line 6)	7	21,000
8	Taxable compensation (line 3 less line 7, if positive; otherwise zero)	8	180,421
9	Tax (19% of line 8)	9	34,280
10	Tax withheld by employer	10	
11	Tax due (line 9 less line 10, if ...)	11	
12	Refund due (line 10 le...	12	

Other Clinton Income — $0

Business income
Would not be taxed — $38,014

Capital gains

Income that might have changed with all businesses also paying a 19 percent flat tax
Dividends — $4,119
Interest — 17,140
Tax-exempt interest — 5,956
Rents — 258
Royalties — $27,474

Paid $55,313 under current laws

Saves $21,033

With no additional business taxes.

...lted on the flat tax form.

Form 1 — **Individual Wage Tax** — **1994**

Your first name and initial (if joint return, also give spouse's name and initial): Robert J. and Elizabeth H. — Last name: Dole — Your social security number

Present home address (number and street including apartment number or rural route): — Spouse's social security no.

Washington, D.C.

U.S. Senator
Pres. - Am. Red Cross

1		1	351,495
2		2	43,508
3		3	395,003
4a		4a	16,500
4b		4b	
4c		4c	
5		5	0
6		6	0
7		7	16,500
8		8	378,503
9		9	71,916
10		10	
11	Tax due (line 9 ...	11	
12	Refund due (...	12	

Other Dole Income

Business income — $145,256
Would not be taxed

Capital gains — $3,199

Income that might have changed with all businesses also paying a 19 percent flat tax
Dividends — $17,359
Interest — 7,150
Tax-exempt interest — 38,852
Rents — 0
Royalties — 0

$63,361

Paid $148,700 under current laws

Saves $49,185

After also paying a business tax of $27,599

Let's begin by admitting one thing that may be central to any thought we have about beginnings. Whether it is the mythic concept of divine power or the scientific concept of sheer vastness, we end up with something we cannot conceive.

We can put matters into words, numbers, expressions, formulas, but does that help? And if you cannot grasp it—

IMAGINE*
JAMES E. GUNN

IMAGINE ALL THE MATTER IN EXISTENCE GATHERED
TOGETHER
IN THE CENTER OF THE UNIVERSE—
ALL THE METEORS
COMETS
MOONS
PLANETS
STARS
NEBULAS
THE MYRIAD GALAXIES
ALL COMPACTED INTO ONE GIANT PRIMORDIAL ATOM,
ONE MONOBLOC, MASSIVE BEYOND COMPREHENSION,
DENSE BEYOND BELIEF . . .
THINK OF WHITE DWARFS, THINK OF NEUTRON STARS,
THEN MULTIPLY BY INFINITY. . . .

THE CENTER OF THE UNIVERSE? THE UNIVERSE ITSELF.
NO LIGHT, NO ENERGY COULD LEAVE, NONE ENTER.
PERHAPS TWO UNIVERSES, ONE WITHIN THE GIANT EGG
WITH EVERYTHING
AND ONE OUTSIDE WITH ALL NOTHING . . .
DISTINCT, UNTOUCHABLE . . .
IMAGINE!
ARE YOU IMAGINING?

ALL THE MATTER BROUGHT TOGETHER,
THE UNIVERSE A SINGLE, INCREDIBLE MONOBLOC,
SEETHING WITH INCOMPREHENSIBLE FORCES AND
POTENTIALS
FOR COUNTLESS EONS OR FOR INSTANTS
(WHO MEASURES TIME IN SUCH A UNIVERSE?),
AND THEN . . .
BANG!
EXPLOSION!
BEYOND EXPLOSION!
TEARING APART THE MONOBLOC, THE PRIMORDIAL
ATOM,
THE GIANT EGG HATCHING WITH FIRE AND FURY,
CREATING
THE GALAXIES
THE NEBULAS
THE SUNS
THE PLANETS
THE MOONS
THE COMETS
THE METEORS

SENDING THEM HURTLING IN ALL DIRECTIONS INTO
SPACE
CREATING SPACE
CREATING THE EXPANDING UNIVERSE
CREATING EVERYTHING . . .
IMAGINE!

YOU CAN'T IMAGINE?
WELL, THEN, IMAGINE A UNIVERSE POPULATED WITH
STARS AND
GALAXIES, A UNIVERSE FOREVER EXPANDING,
UNLIMITED,
WHERE GALAXIES FLEE FROM EACH OTHER,
THE MOST DISTANT SO RAPIDLY
THAT THEY REACH THE SPEED OF LIGHT
AND DISAPPEAR FROM OUR UNIVERSE
AND WE FROM THEIRS . . .

IMAGINE MATTER BEING CREATED CONTINUOUSLY,
A HYDROGEN ATOM POPPING INTO EXISTENCE
HERE AND THERE,
HERE
    AND
      THERE,
PERHAPS ONE ATOM OF HYDROGEN A YEAR
WITHIN A SPACE THE SIZE OF THE HOUSTON
ASTRODOME,
AND OUT OF THESE ATOMS,
PULLED TOGETHER BY THE UNIVERSAL FORCE OF
GRAVITATION,
NEW SUNS ARE BORN,
NEW GALAXIES TO REPLACE THOSE FLED BEYOND
OUR PERCEPTION,
THE EXPANDING UNIVERSE
WITHOUT END,
WITHOUT BEGINNING . . .

YOU CAN'T IMAGINE?
WELL, PERHAPS IT IS ALL A FANTASY . . .

* Excerpted from *The Listeners.*

The University of Kansas

Department of English

Feb. 2, 1995

Myrna Kustin
c/o CALC Lab
Massachusetts College of Art
621 Huntington Avenue
Boston, MA 02115-5882

Dear Myrna Kustin,

I am returning the signed permission form you sent to Martin Greenberg, who reprinted the piece of verse called "Imagine" in <u>Creations . . . .</u>

It may be of interest to you and the readers of <u>Explorations in Algebra</u> that "Imagine" was part of a novel titled <u>The Listeners</u>, published by Scribner's in 1972 and reprinted several times since then, that concerned a century-long project to pick up communications from extraterrestrials, a scientific project that now goes under the name of SETI. In the context of the novel, "Imagine" was the invention of the computer used to compile and analyze data from civilizations and languages on Earth as well as signals from the stars, and in this process the computer itself becomes self-aware and begins putting together information in various creative ways that its human programmers are not aware of. One of those creative ways is poetry, and "Imagine" is the culmination of that process; the poem has not title in the novel but is merely another part of information and musing upon it that is recorded in the interchapter sections that I called "Computer Run."

Sincerely,

James Gunn

POWERS OF TEN

A book about the relative size of things in the universe
and the effect of adding another zero

Philip and Phylis Morrison
and
The Office of Charles and Ray Eames

based on the film *Powers of Ten*
by The Office of Charles and Ray Eames

POWERS OF 10: HOW TO WRITE NUMBERS LARGE AND SMALL

This book uses a notation based on counting how many times 10 must be multiplied by itself to reach an intended number: For example. 10 x 10 equals 10^2, or 100; and 10 x 10 x 10 equals 10^3. or 1000. Multiplying a number by itself produces a *power* of that number: 10^3 is read out loud as "ten to the third power," and is another way to say one thousand. In this case, there is no great advantage. but it is much easier and clearer to write or say 10^{14} than 100,000,000,000,000 or one hundred trillion. After 10^{14}, we even run low on names. The number written above in smaller type—the 14 in the last example—is called an *exponent*. and the powers notation is often called *exponential notation*.

It is not hard to grasp the positive powers of ten—10^4, 10^7. 10^{19}—and how they work; but the negative powers—10^{-2} or 10^{-3}—are another matter. If the exponent tells how many times the 10 is to be self-multiplied. what can an exponent of –5 (negative five) mean? The system requires a negative exponent to signal division by 10 a certain number of times: 10^{-1} equals 1 divided by 10. or 0.1 (one-tenth); 10^{-2} equals 0.1 divided by 10, or 0.01 (one-hundredth). Because *adding* 1 to the exponent easily works out to be

the equivalent of multiplying by 10, it is self-consistent that *subtracting* 1 there works out to a division by 10. It is all a matter of placing zeros. Adding another terminal zero is simply to multiply by 10: 100 x 10 equals 1,000. Putting another zero next after the decimal point is to *divide* by 10: 0.01 ÷ 10 equals 0.001. The powers notation makes these operations even clearer.

But what of 10^0? That seems a strange number. However. notice it is equal to 10^1 (10) divided by 10 (or to 10^{-1} multiplied by 10). Although surprising. it is at least logical that 10^0 should be equal to 1.

Because you can make any power of ten ten times larger by adding 1 to its exponent (10^4 x 10 = 10^5), it follows that to multiply by 100 you add 2 to the exponent: 10^3 x 100 = 10^5 or 1000 x 100 = 100,000. In general. you can multiply one power of ten by another simply by adding their exponents: 10^6 x 10^3 = 10^9. Subtracting the exponents is the equivalent of division: 10^7 ÷ 10^5 = 10^2.

All numbers. not only numbers that are exact powers of ten. like 100 or 10,000, can be written with the help of exponential notation. The number 4000 is 4 x 10^3; 186,000 is 1.86 x 10^5. This convenient scheme is referred to as scientific notation.

All of this can be extended to basic multipliers other than ten: 2^4 = 2 x 2 x 2 x 2 (the fourth power of two); 12^2 = 12 x 12, and 8^{-1} = one-eighth. (But note that 2^0 = 1, 12^0 = 1, and 8^0 = 1.)

Logarithms arise from extensions of this scheme.

The symbol $\sim$ is mathematicians' shorthand for "approximately" or "about."

UNITS OF LENGTH

Parisians still consult this public meter displayed near the Palais du Luxembourg *1*. In London it is at Greenwich that you can confirm your yardstick *2*. At *3* is printed a real 10^{-1} meter, or 10 centimeters.

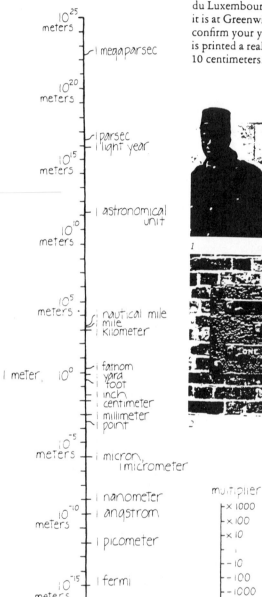

Grow your own food and build your own house, and no formal units of measurement much interest you; such is the gentle rule of thumb. But commerce has implied agreement on units of measurement. The legal yard has long been displayed for the use of Londoners, and the meter is still open to public comparison on a wall of a Paris building.

The system we call metric is the work of the savants of Revolutionary Paris in the 1790s. Even their determination to celebrate both novelty and reason met limits: Our modern second, minute, and hour remain resolutely nondecimal. That was no oversight—the metric day of ten hours, each of a hundred minutes with a hundred seconds to the minute, was formally adopted. But the scheme met fierce resistance. About the only costly mechanism every middle-class family then proudly owned was a clock or watch, not to be rendered at once useless by any mere claim of consistency! Practice won out over theory.

In much the same way, people who today frequently use units in a particular context are not always persuaded to sacrifice appropriateness to consistency. We list here a few nonstandard units of linear measurement that retain their utility even in these metric days, some even within the sciences.

COSMIC DISTANCES

parsec

The word is a coinage from *parallax* of one *second*. The *parsec* is in common use among astronomers because it hints at the surveyor's basic technique of measuring stellar distance by using triangulation. The standard parallax is the apparent shift in direction of a distant object at six-month intervals as the observer moves with the orbiting earth. It is defined so that the radius of the earth's orbit seen from a distance of one parsec spans an angle of one second of arc. The nearest known star to the sun is more than one parsec away.

light-year

This graspable interstellar unit rests on the relationship between cosmic distance and light travel over time.

The speed of light in space is 3.00×10^8 meters per second; in one year light thus moves 9.46×10^{15} meters, which is usually rounded off to 10^{16} meters, especially since only a few cosmic distances are so well known that the roundoff is any real loss of accuracy.

astronomical unit

The mean distance between sun and earth is a fine baseline for surveying the solar system; it is a typical length among orbits. $1 \text{ AU} = 1.50 \times 10^{11}$ meters. *Note*: 1 parsec = 3.26 light-years = 206,300 AU. The interstellar and the solar-system scales plainly differ; intergalactic distances run to megaparsecs.

TERRESTRIAL LENGTHS

miles, leagues, etc.

These are units suited for earth-bound travel, for distances at sea, or for road distances between cities. Nobody ever measured cloth by the mile, or train rides by the parsec.

yards, feet, meters

These rest on human scale, in folklore the length of some good king's arm. They suit well the sizes of rooms, people, trucks, boats. Textiles are yard goods. The meter was defined more universally, but clearly it was meant to supplant the yard and the foot. It was related to the size of the earth: One quadrant of the earth's circumference was defined as exactly 10^7 meters, or 10^4 kilometers. In 1981 the meter is defined with great precision in terms of the wavelength of a specific atomic spectral line. It is "1,650,763.73 wavelengths in vacuum of the radiation corresponding to the transitions between the levels $2p_{10}$ and $5d_5$ of the krypton-86 atom."

inches, centimeters, etc.

The same king's thumb? Human-scale units intended for the smaller artifacts of the hand: paper sizes, furniture, hats, or pies.

line, millimeter, point

Small units for fine work are relatively modern. The seventeenth-century French and English line was a couple of millimeters, and the printer's point measure is about 0.35 mm. Film, watches, and the like are commonly sized by millimeters. The pioneer microscopist Antony van Leeuwenhoek used sand grains as his length comparison, coarse and fine: He counted one hundred of the fine grains to the common inch of his place and time. Smaller measurement units are generally part of modern science, and thus usually metric.

ATOMIC DISTANCES

angstroms, fermis, etc.

Once atoms became the topic of meaningful measurement, new small units of length naturally came into specialized use. The Swedish physicist Anders Ångström a century ago pioneered wavelength measurements of the solar spectrum. He expressed his results in terms of a length unit just 10^{-10} meters long. It has remained in widespread informal use bearing his name: convenient because atoms measure a few angstroms (Å) across. The impulse for such useful jargon words is by no means ended; nuclear particles are often measured in fermis, after the Italian physicist Enrico Fermi. 1 fermi equals 10^{-15} meters.

ANGLES AND TIME

Angles are measured, especially in astronomy, by a nonmetric system that goes back to Babylon. A circle is 360 degrees; 1 degree = 60 minutes of arc; 1 minute = 60 arc-seconds. An arc-second is roughly the smallest angle that the image of a star occupies, smeared as it is by atmospheric motion. This page, viewed from about twenty-five miles away, would appear about one arc-second across.

Time measurement shares the cuneiform usage of powers of sixty. Note that a year of 365.25 days of 24 hours, each of 60 minutes with 60 seconds apiece, amounts to about 3.16×10^7 seconds.

THE DRAGONS OF EDEN

Speculations on the Evolution of Human Intelligence

Carl Sagan

BALLANTINE BOOKS · NEW YORK

What seest thou else
In the dark backward and abysm of
time?

WM. SHAKESPEARE
The Tempest

HE WORLD IS very old, and human beings are very young. Significant events in our personal lives are measured in years or less; our lifetimes in decades; our family genealogies in centuries; and all of recorded history in millennia. But we have been preceded by an awesome vista of time, extending for prodigious periods into the past, about which we know little—both because there are no written records and because we have real difficulty in grasping the immensity of the intervals involved.

Yet we are able to date events in the remote past. Geological stratification and radioactive dating provide information on archaeological, palenotological and geological events; and astrophysical theory provides data on the ages of planetary surfaces, stars, and the Milky Way Galaxy, as well as an estimate of the time that has elapsed since that extraordinary event called the Big Bang—an explosion that involved all of the matter and energy in the present universe. The Big Bang may be the beginning of the universe, or it may be a discontinuity in which information about the earlier history of the universe was destroyed. But it is certainly the earliest event about which we have any record.

The most instructive way I know to express this cosmic chronology is to imagine the fifteen-billion-year lifetime of the universe (or at least its present incarnation since the Big Bang) compressed into the span of a single year. Then every billion years of Earth history

The Cosmic Calendar

COSMIC CALENDAR

DECEMBER

SUNDAY	MONDAY	TUESDAY	WEDNESDAY	THURSDAY	FRIDAY	SATURDAY
	1 Significant oxygen atmosphere begins to develop on Earth.	**2**	**3**	**4**	**5** Extensive volcanism and channel formation on Mars.	**6**
7	**8**	**9**	**10**	**11**	**12**	**13**
14	**15**	**16** First worms.	**17** Precambrian ends. Paleozoic Era and Cambrian Period begin. Invertebrates flourish.	**18** First oceanic plankton. Trilobites flourish.	**19** Ordovician Period. First fish, first vertebrates.	**20** Silurian Period. First vascular plants. Plants begin colonization of land.
21 Devonian Period begins. First insects. Animals begin colonization of land.	**22** First amphibians. First winged insects.	**23** Carboniferous Period. First trees. First reptiles.	**24** Permian Period begins. First dinosaurs.	**25** Paleozoic Era ends. Mesozoic Era begins.	**26** Triassic Period. First mammals.	**27** Jurassic Period. First birds.
28 Cretaceous Period. First flowers. Dinosaurs become extinct.	**29** Mesozoic Era ends. Cenozoic Era and Tertiary Period begin. First cetaceans. First primates.	**30** Early evolution of frontal lobes in the brains of primates. First hominids. Giant mammals flourish.	**31** End of the Pliocene Period. Quaternary (Pleistocene and Holocene) Period. First humans.			

THE DRAGONS OF EDEN

would correspond to about twenty-four days of our cosmic year, and one second of that year to 475 real revolutions of the Earth about the sun. On pages 14 through 16 I present the cosmic chronology in three forms: a list of some representative pre-December dates; a calendar for the month of December; and a closer look at the late evening of New Year's Eve. On this scale, the events of our history books—even books that make significant efforts to deprovincialize the present—are so compressed that it is necessary to give a second-by-second recounting of the last seconds of the cosmic year. Even then, we find events listed as contemporary that we have been taught to consider as widely separated in time. In the history of life, an equally rich tapestry must have been woven in other periods—for example, between 10:02 and 10:03 on the morning of April 6th or September 16th. But we have detailed records only for the very end of the cosmic year.

The chronology corresponds to the best evidence now available. But some of it is rather shaky. No one would be astounded if, for example, it turns out that plants colonized the land in the Ordovician rather than the Silurian Period; or that segmented worms appeared earlier in the Precambrian Period than indicated. Also,

PRE-DECEMBER DATES	
Big Bang	January 1
Origin of the Milky Way Galaxy	May 1
Origin of the solar system	September 9
Formation of the Earth	September 14
Origin of life on Earth	~September 25
Formation of the oldest rocks known on Earth	October 2
Date of oldest fossils (bacteria and blue-green algae)	October 9
Invention of sex (by microorganisms)	~November 1
Oldest fossil photosynthetic plants	November 12
Eukaryotes (first cells with nuclei) flourish	November 15

~ = approximately

The Cosmic Calendar

The construction of such tables and calendars is inevitably humbling. It is disconcerting to find that in such a cosmic year the Earth does not condense out of interstellar matter until early September; dinosaurs emerge on Christmas Eve; flowers arise on December 28th; and men and women originate at 10:30 P.M. on New Year's Eve. All of recorded history occupies the last ten seconds of December 31; and the time from the waning of the Middle Ages to the present occupies little more than one second. But because I have arranged it that way, the first cosmic year has just ended. And despite the insignificance of the instant we have so far occupied in cosmic time, it is clear that what happens on and near Earth at the beginning of the second cosmic year will depend very much on the scientific wisdom and the distinctly human sensitivity of mankind.

DECEMBER 31	
Origin of Proconsul and Ramapithecus, probable ancestors of apes and men	~ 1:30 P.M.
First humans	~ 10:30 P.M.
Widespread use of stone tools	11:00 P.M.
Domestication of fire by Peking man	11:46 P.M.
Beginning of most recent glacial period	11:56 P.M.
Seafarers settle Australia	11:58 P.M.
Extensive cave painting in Europe	11:59 P.M.
Invention of agriculture	11:59:20 P.M.
Neolithic civilization; first cities	11:59:35 P.M.
First dynasties in Sumer, Ebla and Egypt; development of astronomy	11:59:50 P.M.
Invention of the alphabet; Akkadian Empire	11:59:51 P.M.
Hammurabic legal codes in Babylon; Middle Kingdom in Egypt	11:59:52 P.M.
Bronze metallurgy; Mycenaean culture; Trojan War; Olmec culture; invention of the compass	11:59:53 P.M.
Iron metallurgy; First Assyrian Empire; Kingdom of Israel; founding of Carthage by Phoenicia	11:59:54 P.M.
Asokan India; Ch'in Dynasty China; Periclean Athens; birth of Buddha	11:59:55 P.M.
Euclidean geometry; Archimedean physics; Ptolemaic astronomy; Roman Empire; birth of Christ	11:59:56 P.M.
Zero and decimals invented in Indian arithmetic; Rome falls; Moslem conquests	11:59:57 P.M.
Mayan civilization; Sung Dynasty China; Byzantine empire; Mongol invasion; Crusades	11:59:58 P.M.
Renaissance in Europe; voyages of discovery from Europe and from Ming Dynasty China; emergence of the experimental method in science	11:59:59 P.M.
Widespread development of science and technology; emergence of a global culture; acquisition of the means for self-destruction of the human species; first steps in spacecraft planetary exploration and the search for extraterrestrial intelligence	Now: The first second of New Year's Day

in the chronology of the last ten seconds of the cosmic year, it was obviously impossible for me to include all significant events; I hope I may be excused for not having explicitly mentioned advances in art, music and literature or the historically significant American, French, Russian and Chinese revolutions.

Elizabeth Cavicchi

Watching Galileo's Learning
Elizabeth Cavicchi

1 Introduction

By closely following Stillman Drake's biographies of Galileo, this essay interprets Galileo's free fall studies. Drake's biographies piece together how Galileo came to understand that the distance an object has fallen increases as the square of its descent time. He infers this story from calculations recorded among Galileo's working papers, including some that were not included in the definitive twenty volume set of Galileo's *Opera*, edited by Favaro in 1934.

Prior to Drake's studies, the interpretation of the historian Koyre prevailed among Galilean scholars. Koyre maintained that Galileo never made any observations of motion (except with the telescope); he says Galileo derived laws of motion through thought alone. For Koyre, Galileo's innovation in science lay in this introspective method of reasoning.

Koyre influenced the textbook treatment of Galileo's work. Many physics texts do not mention Galileo's free fall experiments. While some recent texts refer to his experiments, they mostly misrepresent Galileo's process of discovery and the mathematical and political context of the time. This misrepresentation is one of many ways textbooks misrepresent how we learn about nature through science. As a result, unable to imagine imitating Galileo by proceeding from not-knowing, students may doubt their own potential to learn through science.

By comparison with what we know today, the mathematical tools and physical picture available to Galileo seem very limited. The task of stripping away our twentieth century sophistication, to better approximate the outlook and thinking of Galileo, is formidable. As a biographer tracking Galilean documents Drake acquired this outlook. However, he does not carefully sort out clues that would assist a reader in entering Galileo's world view. My effort to understand Drake's argument seems parallel to Galileo's effort to understand motion. We begin from not-knowing. In watching Galileo's learning, I am repeatedly bewildered as my assumptions prevent me from seeing what Galileo did not know. I am also astonished by Galileo's creativity in using the tools he did have to find something new.

Accelerated motion is subtle: Galileo devoted most of a lifetime to its study. We can re-experience some of the subtlety and complexity Galileo encountered by observing our own students as they try to make sense of it. One student and I sorted out a partial understanding of free fall motion by carefully examining the evidence and our ideas. I was surprised when notions that seemed obvious, such as the periodicity of a timer, were not at all obvious to the student. As I tried to understand free fall from her perspective. I was repeatedly surprised by her creativity in overcoming apparent limitations.

Elizabeth Cavicchi

I prepared this essay to supplement readings in an algebra course. I did not expect that it would precipitate so much change in my understanding of free fall, science, and teaching.

2 Falling Objects

When younger than we can now remember, we played by dropping a toy and noisily goading an attentive older companion to retrieve it. While the game now seems a test of patience, seeing something fall was a new experience for us, which we observed without asking either why or how things fall. When, as an older child, we asked what makes things fall, the answer was "gravity". When we asked what gravity was, we were told "it makes things fall". Gravity was an empty word, devoid of explanation.

Better (or more complete, or less circular) answers to this question have evaded not only parents, but also people who explicitly study things and try to explain them. This simple question is too grand. Why does it take effort to get things to go where we want– lifting, dragging, carrying, bicycling, while they fall by themselves? No strings or other apparatus visibly pulls them down. Seeking a mechanism for falling is premature if we are not yet sure what is happening when something falls. Falling is so quick that it is not easy to say what is happening as something falls, even if it is watched carefully. Asking the question differently– "How do things fall?" - provides a clearer way to start towards an answer: by careful description of the phenomenon itself.

3 Early Explanations

The seeming simplicity of the "grand" question why things fall– motivated early historical attempts to talk about falling in a way that made sense to ordinary people. The Greeks said the weight of an object makes it go towards earth's center. Things that have lightness, like fire, move away from the center. Weightier things fall faster than lighter ones. Aristotle (384-322 B.C.) extended this explanation with logic, to make many other statements about motion and change. His mostly incorrect view was considered authoritative for about a millennium.

People who looked carefully found that some of these ideas just did not match with nature. Two sixteenth century Italian instructors dropped objects differing only in weight from a high window or tower. The objects struck the ground almost at once. These observations were augmented when Galileo Galilei (1564-1642. Figure 1. [7]) began investigating what happens when something falls.

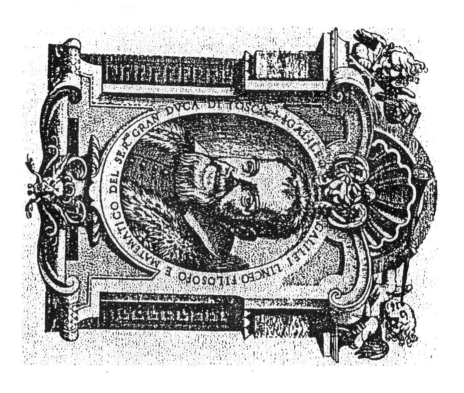

Figure 1: A portrait of Galileo, taken from his book on the sunspots, *Istoria e dimostrazioni intorno alle macchie solare*, published in Rome in 1613. By permission of the Houghton Library, Harvard University.

4 Galileo and the Inclined Plane

Galileo revised the question: he asked "how" things fall, instead of "why". This question suggested other questions that could be tested directly by experiment. By alternating questions and experiments, Galileo was able to identify details in motion no one had previously noticed or tried to observe. His questions about free fall suggested questions about other motions, such as motions Aristotle had classed as "natural". These motions start spontaneously without a push, such as a ball rolling down a slope, or a pendulum swinging. What he learned about one motion drove his study of the others.

In one experiment, performed near 1602, Galileo rolled balls down a ramp. He mounted lute strings crosswise to the motion, so that a distinct sound was produced each time a ball rolled over a string. He adjusted positions of the strings until time intervals between sounds were roughly equal. Perhaps Galileo hummed a tune, with beats about a half-second apart, to estimate constant time intervals.

Once he had equalized the rolling time between each pair of strings, he measured the distance between each pair of adjacent strings. The distances between strings were unequal: each consecutive pair of strings was further apart than the preceding pair of strings (Figure 2). When he divided all the distances between pairs of strings by the first distance, the quotients were the odd integers 1, 3, 5, 7, ...

This discovery was revolutionary; as the first evidence that motion on Earth was subject to mathematical laws. Until this time, people assumed mathematics governed only motions in the heavens; terrestrial motions were considered disorderly. Galileo did not simply accept this belief, but tried to test it with direct observation.

Galileo interpreted the separation between successive pairs of lute strings as a measure of the ball's speed while rolling between the two strings. The numerical pattern showed Galileo that the ball's speed increased, but it did not tell him how it increased. To understand this subtlety he again had to contradict physical theory of his contemporaries.

People already realized that one time or position measurement could change to another time or position by imperceptibly small, continuous increments. But speed was depicted as a motion that occurred in a perceptible time interval. While an object's speed could change between time intervals, they did not think speed could change instantaneously as we know it does today. Instantaneous change in speed was ill-defined for them, because they could not conceive of a speed without being able to see the motion directly. Galileo eventually understood and resolved this apparent contradiction, and concluded that speed changes continuously during free fall and natural motions. He wrote:

I suppose (and perhaps I shall be able to demonstrate this) that the naturally falling body goes continually increasing its *velocità* according as the distance increases from the point which it parted[6].

5 Mathematical Limitations

Galileo's ability to construct patterns from his measurements was limited by the mathematics available to him at the time. Although European mathematicians were then developing algebra beyond its Arabic and Hindu origins, and Galileo was aware of their work, he never used algebra in his physical studies. He never expressed any results using equations. His analysis of data and publications relied on a more classical training– popular at the time– in the logic, geometry, and numerical proportions of Euclid's *Elements* (fl. 300 B.C.).

One result of these mathematical limitations was that Galileo could not define speed the way we do today, as a distance divided by a time (for example miles per hour). In manipulating numbers according to the Euclidean theory of proportions, it was not legitimate to construct ratios of measurements with different units, such as dividing a distance by a time. Ratios could only be constructed by dividing one distance measurement by another like measurement (for example (20 meters)/(10 meters)), or one time by another time (for example (6 seconds)/(2 seconds)). The units of distance or time then canceled out, yielding a pure, dimensionless number.

Galileo's interpretation of speed as the separation between lute strings prevented him from seeing it as the distance traveled while the object rolled between the strings. If he had recognized that this one measurement could represent two different quantities (speed and distance) at once, he might have discovered the true relation between distance and time two years earlier. Instead he continued experimenting, rolling balls down ramps of different lengths and of different vertical heights. He incorporated results from all these experiments into his mature, formal understanding of motion, but at the time he performed them, they did not lead him to the relation he sought.

6 Pendulum Swings

After failing to discover the law for motion along the incline, in 1604 Galileo began looking for it in the motion of pendulums. He had already experimented extensively with pendulums. He had already been the first person to demonstrate that a pendulum swings through a small arc in very nearly the same time that it swings through a wide arc.

Galileo timed pendulum swings with a water clock he designed. A water pipe was fed from a large elevated reservoir and plugged with a finger. When it was turned on by removing his finger from the pipe, water spurted from the pipe into a bucket. He unplugged the pipe for the duration of a pendulum swing and weighed the water that flowed into the bucket. As long as the flow rate was

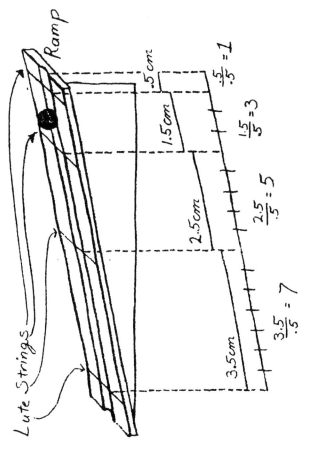

Figure 2: Whenever a ball rolling down a ramp bumps over a string, it makes a sound. The strings are positioned so the sounds mark out equal intervals in time. If we measure the distances between each pair of strings, the ratio of each distance to the first one makes a series of odd integers.

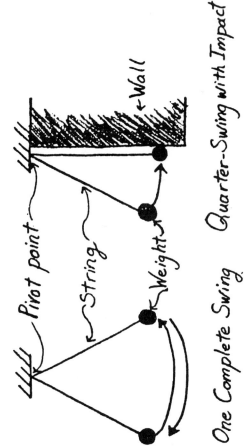

Elizabeth Cavicchi

One Complete Swing.   Quarter-Swing with Impact

Pivot point
String
Weight
Wall

Figure 4: A complete pendulum swing and one quarter pendulum swing. When the bob completes one quarter swing, it hits a wall, making a sound.

steady, weights of water were a measure of time; the longer the run, the more water, and hence the more weight (Figure 3). Galileo checked consistency of water flow through the pipe with a built-in timer– his own pulse. He used this clock to measure times for a wide variety of physical phenomena, such as free fall, balls rolling down ramps, and pendulum swings.

He measured one-quarter of a pendulum swing: from its moment of release to the moment when it was vertical. He mounted the pendulum so that when it hung vertically (like a plumb bob), the bob just touched a wall (Figure 4). He started his water clock when he released the pendulum and stopped it when he heard the bob bang against the wall. The sound helped him make measurements repeatably and reliably.

Since Galileo used the same clock for measurement of the pendulum and other natural motions, the times for all his measurements were expressed in the same 'units', which suggested that he could compare times for *different kinds* of natural motion, for example, pendulum swings and falling objects. Using a gear mechanism he designed, he tried to adjust the length of a pendulum string so that the time of its quarter swing exactly matched the time it took for an object to fall a given distance, by matching the weights of water that flowed during both motions. But his free fall measurement required too long a pendulum to be practical, so he tried matching the pendulum swing with half the free fall time he had measured, which corresponds to a fall of one meter. The length of

Elizabeth Cavicchi

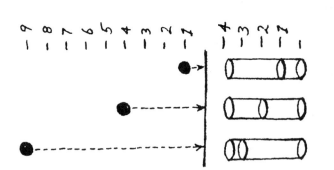

Figure 3: The water clock was started when a weight was released, and stopped when it hit the ground. More water was collected during long falls than during short falls. The amount of water is a measure of the elapsed time.

Elizabeth Cavicchi

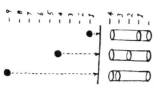

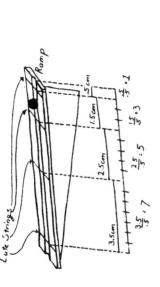

the resulting pendulum string was a little less than one meter.

Galileo then made a pendulum twice as long as this one, and timed its quarter swing. He decided to double the pendulum length, rather than to try shorter lengths, because the mathematics of the time did not condone the construction of fractional lengths such as 1/2 or 1/4 of a given length. He found that swing time was proportional to a 'geometrical mean' involving pendulum length, which he computed using a straightedge and compass on a geometric drawing. This geometrical computation is equivalent to taking the square root of pendulum length. Today we say that a pendulum's period is proportional to the square root of its length. Galileo used his version of this rule to correctly predict the swing time of a thirty foot pendulum which he may have hung in the courtyard of the University of Padua where he worked.

Galileo linked the pendulum swing to the free fall by performing the experiment backward. He used the water clock to time a weight as it fell through the same height as the length of one of the pendulums he had already timed. He expressed the ratio of the two times as a ratio of the whole number weights 942/850. While he did not reduce this ratio to the decimal form 1.108, he was quite close to the true ratio $\pi/2\sqrt{2} = 1.1107$.

For Stillman Drake, Galileo's biographer, this ratio is the closest analogy in Galileo's work to something like a physical constant. Physical constants are an artifact of algebra. They drop out of ratios. Thus, although we now commonly say that Galileo discovered the constant acceleration due to gravity, g, he did not and could not. The mathematics, which he so carefully applied, and which revealed so much to him, masked numerical constancy.

Galileo combined his method of relating a pendulum's length to its swing time with the ratio he found between fall and pendulum times. He thus calculated the height a weight would have to fall from in order to match the swing time of one of his pendulums. When he analyzed the ratio calculation, he found that the falling time appeared twice. Time was multiplied by time; time was *squared*. In following the mathematical rules for combining ratios, he had constructed a new physical law: falling distance is proportional to the square of falling time. He had also constructed the new physical quantity 'time squared'. Until Galileo discovered this pattern in his calculation, no one had ever squared a physical quantity other than length, whose square is usually area.

7 Extending the Law

Galileo wondered if the same rule applied to the data collected in the inclined plane/lute string experiments. There he had computed the sequence of odd integers 1, 3, 5, 7, ... from ratios of the first distance to each successive distance covered in equal times. This pattern suggests to us that the link between distance and time involves a square: sums of the series of odd integers produce the series of squared integers $(1, 1 + 3 = 4, 1 + 3 + 5 = 9, 1 + 3 + 5 + 7 = 16)$.

Figure 5: The total distance the ball had rolled after its release was proportional to the square of the integer identifying that equal time interval. The same law related distance and time in free fall motions.

Galileo did not identify this pattern when he originally performed the experiment, because he interpreted the lute string separations as speeds rather than distances.

Galileo recorded the integer squares (1, 4, 9, 16, ...) beside his original distance measurements on the notebook page containing data from the lute string experiment. He found that the product of the first distance and each successive integer square (1, 4, 9, 16, ...) yielded the same number that he had originally measured as the ball's cumulative distance traveled from the start of motion. These squared integers behaved like the squares of times in free fall (Figure 5). Balls rolling down an incline and falling weights exhibited the same relation: distance traveled is proportional to the square of time. We can also say that the pendulum motion is somehow similar: the square root of its length is proportional to its swing time. All these 'natural' motions were somehow similar.

Galileo commented, though not in these modern words, that if you plot the total distance traveled for several consecutive and equal time intervals, the dots fall along a parabola (Figure 6). But to Galileo this comment was purely

Elizabeth Cavicchi

geometric: it did not provide him with a definition of speed or a characterization of projectile motion. For some time he questioned whether an object's speed in free fall could be represented by the total distance it had fallen, or by the distance it had fallen in the most recent interval of time. If the first choice were correct, the speeds of successive time intervals would also fall along a parabola, just like the distance measurements, while the second choice would place the speeds on a straight line of increasing slope (Figure 7).

Galileo tried to measure speed of a falling object by observing the effects of impact, measured, for example, by how much of a depression the object makes in the dirt when it hits. These estimates were misleading. The depth of the impact crater increased like a parabola with the total distance fallen. At first Galileo assumed that the depth represented the object's speed when it hit the ground. The results seemed to confirm this, but eventually he guessed that impact depth is proportional to the square of speed. He later constructed a geometric argument confirming the second choice, showing that speed is proportional to the total time, rather than the distance, of free fall.

8   Formal Publication

Galileo conveyed his new understanding of the proportionality between distance and time squared only in personal correspondence and in teaching students. It was finally incorporated into Galileo's last book, *Two New Sciences* half a century after legend claims he first dropped weights from the Leaning Tower of Pisa.

By then, Galileo had been discovered by the Roman Inquisition, which had ruled that his work bordered on heresy by entertaining the possibility that the earth moved around the sun. He had been sentenced to life imprisonment (commuted to home incarceration) and prohibited from further publication. *Two New Sciences*, dedicated to a French ambassador who smuggled the manuscript out of Italy, was printed in the Protestant Netherlands by the Elsevier family in 1638.

But Galileo's long process of redesigning experiments and reworking data analyses was not suitable for description in a publication of the time. *Two New Sciences* had to present explanations formally derived from definitions and postulates. The work is a dialogue inspired by the writings of Plato, and includes discussions of long formal derivations by the characters. The proportionality between distance in free fall and the square of the falling time is proven geometrically as a theorem. The data through which Galileo first came to identify this law is not reproduced.

It is easier to understand the classical presentation of *Two New Sciences* than to understand the process by which Galileo derived these results. The former is a treatise similar to the work of Euclid and Plato, while the latter represents a whole new way of thinking about motion. Galileo's actual paths

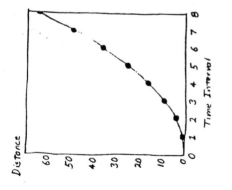

Figure 6: A plot of the total distance traveled in successive time intervals is a parabola.

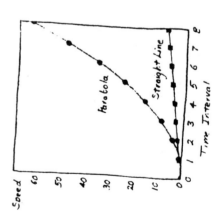

Figure 7: Is the plot of the average speed in successive time intervals a parabola or a straight line?

of learning included stops, new starts, interconnections among phenomena and mathematical analogy.

9 Conclusion

While Galileo had described "how" things fall, first by direct measurement and then abstractly with the time-squared rule, he did not aim to explain "why" things fall. Previous philosophers, in the tradition of Aristotle, regarded their task as finding causes for natural processes. Galileo diverged from this tradition in changing the question from "why do things fall" to "how do things fall". Galileo admits that he does not know the cause of free fall in an exchange between Simplicio (the Aristotelian) and Salviati (the speaker for Galileo) in the *Dialogue* which invoked the Inquisition's condemnation (Figure 8, [4]):

Simplicio: The cause of this effect is well known; everyone is aware that it is gravity.

Salviati: You are mistaken, Simplicio; what you ought to say is that everyone knows it is called 'gravity'. What I am asking for is not the name of the thing, but its essence, of which essence you know not a bit more than you know about the essence of whatever moves stars around... the name (gravity)... has become a familiar household word through the daily experience we have of it. But we do not really understand what principle or what force it is that moves stones downward, any more than we understand what moves them upward after they leave the thrower's hand, or what moves the moon around[9].

By revising experiments to make measurement practical, and analyzing those measurements with mathematics, Galileo composed the first accurate account of how things fall. He came to see numerical patterns in his data that he did not expect to see. These patterns were the first indication that terrestrial motions could be described by mathematics.

The outgrowth of that mathematical metaphor for nature is so tightly knitted into our contemporary physical understanding that it is almost impossible for us to see nature as freshly as he did. We translate physical quantities into algebraic abstractions; Galileo never did. By composing his measurements into ratios, he found a general relationship between distance and time. He worked with a few carefully made measurements; his work was not statistical. Galileo's method of learning from concrete, familiar examples, rather than from abstractions, may have something in common with that of students today.

Without knowing what he was going to find, or what methods might reveal it– and with the opposition of an authoritarian tradition that presumed to dictate how nature worked– Galileo extracted a new regularity from falling motions.

Figure 8: Participants in Galileo's *Dialogo sopra i due Massimi Sistemi del mondo Tolemaico e Copernicano* published in Florence in 1632. By permission of the Houghton Library, Harvard University.

Elizabeth Cavicchi

References

[1] *Aristotle's Physics*, W. D. Ross, trans., ed., (Oxford: Clarendon Press) 1966.

[2] L. N. H. Bunt, P. S. Jones, J. D. Bedient, *The Historical Roots of Elementary Mathematics*, (New York NY: Dover Pub., Inc.) 1988.

[3] I. Bernard Cohen, *The Birth of a New Physics*, (New York NY: W. W. Norton & Co.) 1985.

[4] Bern Dibner, *Heralds of Science*, (Cambridge MA: MIT Press) 1969.

[5] Stillman Drake, *Galileo*, (New York, NY: Hill and Wang) 1980.

[6] Stillman Drake, *Galileo: Pioneer Scientist*, (Toronto: University of Toronto Press) 1990.

[7] Stillman Drake, *Galileo at Work: His Scientific Biography*, (Chicago IL: University of Chicago Press) 1978.

[8] Howard Eves, *A Survey of Geometry*, (Boston MA: Allyn and Bacon, Inc.) 1972.

[9] Galileo Galilei, *Dialogue Concerning the Two Chief World Systems Ptolemaic & Copernican*, Stillman Drake, trans., Albert Einstein, forward, (Berkeley CA: University of California Press) 1967.

[10] Galileo Galilei, *Discoveries and Opinions*, Stillman Drake, trans., (New York, NY: Doubleday) 1957.

[11] Galileo Galilei, *Two New Sciences Including Centers of Gravity and Force of Percussion*, Stillman Drake, trans., (Madison WI: University of Wisconsin Press) 1974.

[12] Morris Kline, *Mathematical Thought from Ancient to Modern Times*, (new York NY: Oxford University Press) 1972.

[13] A. Rupert Hall, *The Scientific Revolution 1500-1800: The Formation of the Modern Scientific Attitude*, (Boston MA: Beacon Press) 1966.

[14] David C. Lindberg, *The Beginnings of Western Science: The European Scientific Tradition in Philosophical, Religious, and Institutional Context, 600BC to AD 1450*, (Chicago IL: University of Chicago Press) 1992.

[15] George Sarton, *Ancient Science Through the Golden Age of Greece*, (New York NY: Dover Pub., Inc.) 1980.

[16] Thomas Settle, "Galileo and Early Experimentation" in *Springs of Scientific Creativity: Essays on Founders of Modern Science*, R Aris, H. T. Davis, R. H. Stuewer, eds., (Minneapolis MN: University of Minnesota Press) 1983.

[17] Thomas Settle, "An Experiment in the History of Science", *Science*, Jan. 1961, v. 133, 19-23.

[18] Steven Weinberg, *Gravitation and Cosmology: Principles and Applications of the General Theory of Relativity*, (New York NY: John Wiley & Sons) 1972.

These are the texts which contributed to the reading on free fall. You might find others. The essay closely follows the accounts and approach of Stillman Drake. Because the essay is intended for undergraduate students, references are listed at the end rather than in citation.

Elizabeth Cavicchi

Crystals, Light, and Sweet-Tooth Satisfaction

Kenneth Kustin, Brandeis University

Two paths of scientific exploration converged in the first half of the nineteenth century to produce a connection between the shapes of molecules and the shapes of the crystals they form. Most of the many benefits of this discovery to human well-being are profound and widespread, but one benefit rewards neither the quest for good health nor that for betterment of the human condition. This benefit rewards out demand for sweet-tasting drinks, and for sweet-tasting foods with a stiff, crunchy quality that delights our manifold craving for sweetness, flavor, and texture. The availability of sugar, satisfying that craving and coincidentally providing employment and income to millions of people worldwide, was possible by scientific researchers motivated by curiosity about shining light through transparent substances.

Naturally occurring sugar is dissolved in water as juice. Solid sugar obtained from natural sources and its solutions are colorless. Sugar therefore appeared to lack distinctive properties that would allow sampling and rapid analysis. Without such a method of analysis, how could one test manufacturers' claims that the sugar they were offering was purified sucrose of a solution of purified sucrose in water, and not a mixture or a solution of other similar but less sweet-tasting substances with water? That's where light and transparency come into the picture.

From antiquity people were aware of colorless transparent rocks that were pretty and could be used as ornaments. Polishing accented their regular geometric forms, and mounting or stringing them together enhanced their beauty. It was also known that some of these crystals, as they were called, produced a double image when nearby object was viewed through them. About the same time that sugar, used for centuries as a medicine and as a coating to mask the often bitter taste of the medications in pills, became accepted as a food, scientists began to study how light was modified when it passed through a crystal. They discovered that crystals which yielded double images did so by splitting ordinary white light into two differently polarized components.

Experimentation with polarized light and crystals and crystals immersed in liquids yielded some additional surprises. Certain substances in pure liquid form could also polarize light; all such substances were said to be optically active. Then, in a novel experiment devised and performed by Jean Baptiste Biot in 1817, turpentine, a common liquid substance found to be optically active, was vaporized and analyzed with polarized light. The direction of polarized light that passed through a very long tube filled with turpentine vapor was changed or, more specifically, rotated when emerged at the other end. This experiment suggested that there must be some connection between the shapes of molecules, that is, their structures and the polarization of light, because a vapor is a gas consisting only of molecules separated by a vacuum. Louis Pasteur commenced his scientific career by finding that connection.

Like Biot, Pasteur started to experiment with crystals. He discovered a substance obtained from wine-making that crystallized in two otherwise completely identical forms and that differed only slightly in geometric appearance. The flat sides or faces of the crystals were so arranged that one type of crystal was the mirror image of the other. Solutions of one type rotated polarized light counterclockwise. When Pasteur made a solution of exactly equal amounts of the two types of crystals, the solution did not rotate polarized light, because there was perfect compensation of the two different directions of rotation. Pasteur's experiment, published in 1850, rightly suggested that the molecules forming the tow types of crystals must also differ in the same subtle fashion.

The optical rotation of sugar solutions had been discovered in 1818 by Biot, and the inversion of sucrose by the same investigator in 1835. The understanding that these effects were due to the geometries of the sucrose molecule and its products meant that a virtually universal method for detecting sucrose was available, and that the integrity of commercial sucrose preparations could be ascertained with certainty. The direction and magnitude of the rotation of sugar solution signify the presence of sucrose and are measures of how much is present. All one needs to do to make the analysis is take a small sample of a solid sucrose preparation, weigh it, dissolve it in water, and measure its optical rotation. From these data you can back calculate the percentage of sucrose in the solid; with honest manufacturers this is usually an astonishing 99.98%.

The research on optical activity by Biot, Pasteur and others was very timely. The anti-slavery movement provided an impetus to change the method of sugar production, and production of sugar from beets began to rival production from sugar cane. Naturally, people consuming sugar wanted to be reassured that they were getting the same substance, in the same quantity, with the same purity from beets that they were getting form sugar cane. Soon, as the price of sugar declined and its production sky-rocketed, people demanded more of the substance, and a general assurance of the honesty and quality of the product arose throughout the world.

Although other methods of sugar analysis such as measuring solution density have their place, optical rotation is still the chief method for analyzing sugar today. As experience with and confidence in the method increased, and its reliability and convenience became apparent, other uses for this technique were found. Sugar solutions can change by spoiling; sugar is changed in the human body when it is digested; sugar changes when it undergoes reaction to prepare other, useful substances. What better method for studying *how* sugar changes over time that a measurement of its optical rotation? Indeed, this technique plays an important part in studying how sugar reactions occur in many different contexts: biological, biochemical, and technological. Experiments like the one you have performed are carried out on sucrose and similar colorless optically active compounds as an important part of the process of finding out how chemical substances react in the human body.

TI-82 or TI-83
Graphing Calculator Instructions

Beverly K. Michael
Jeremiah V. Russell
Robert Lee

Contents

To the student:
Each Graphing Calculator Appendix Section corresponds to a chapter in the text. You are expected to read and work the examples in the Appendix <u>before</u> attempting the explorations or exercises at the end of each chapter. You should have a TI-82 or TI-83 graphing calculator in your hand as you read and work through each section in this guide.

---

[1] Hard copy of all the data files and instructions for using graph lnk files are in the Instructor's Manual.

The following table summarizes the Microsoft Excel spreadsheet and TI-82 and TI-83 graphing calculator program files contained on the course diskette "EXPLORATIONS DATA ":

	Spreadsheet or Program Name	Application or Data Set Description
1	DEFAULTS	Resets the graphing calculator to standard settings and clears all TI-82 and unnamed TI-83 statistical lists
2	USAGES	Data on the total population and ages of men and women as of July 1, 1991 (Ch.1, Expl. 1.4)
3	MEDICAL	Selected medical data for twenty one of the United States (Ch.1, ex. 21).
4	USPOP	Data on population of U.S. from 1790 to 1994 (Ch. 3, text, and Ch. 8, ex. 16)
5	MEDAGE	Data on median age of U.S. population from 1850 to 2050 (Ch 2, Sec 2.1, Ch. 3, ex. 10)
6	NEWS	Data on newspapers printed and number of newspapers published from 1915 to 1990 (Ch. 3, ex. 11)
7	TV	Data on number of on air TV stations from 1950 to 1990 (Ch. 3, ex. 11)
8	FEDDEBT	Data on accumulated gross federal debt from 1945 to 1995 (Ch. 3, ex. 16)
9	KALAMA	Data on mean height of Kalama children (Ch. 4, Sec. 4.4 and Ch. 5, ex. 9)
10	FAM1000	Data on mean personal wages for all individuals and for men and women from the March 1992 Current Population Survey (CPS)
11	FAM1000A	Data on mean personal total income for all individuals and for men and women from the FAM 1000 data set
12	FAM1000B	Data on mean personal total income for all nonwhite men, all white men, and all men from the FAM 1000 data set
13	FAM1000C	Data on mean personal total income for all nonwhite women, all white women, and all women from the FAM 1000 data set
14	COLLCOST	Data on mean annual cost for tuition and fees at public and private four year colleges from 1980 to 1994 (Ch. 5, ex. 7)
15	CALORIES	Data on calories burned by a 154 pound person at various pace speeds (Ch. 5, ex. 10)
16	SMOKERS	Data on smokers from 1965 to 1993 (Ch 5, ex. 11)
17	SKIJUMP	Data on U.S. ski jumping records for men from 1916 to 1978 (Ch. 5, ex. 12)
18	SHOTPUT	Data on Olympic shot put records for men from 1900 to 1992 (Ch. 5, ex. 14)
19	MARATHON	Data on winning times for women in the Boston marathon from 1972 to 1995 (Ch. 5, ex. 16)
20	WORLDPOP	Data on world population from 1800 to 1992 (Ch. 8, ex. 3)
21	REINDEER	Data on reindeer population on St. Paul Island from 1911 to 1938 (Ch. 8, ex. 14)
22	ECOLIB1	Data on the growth of E. coli bacteria in LB and SB cultures at 30 deg C and 37 deg C. Concentrations of bacteria are given in number of cells per milliliter of solution (Ch. 8, Expl. 8.1)
23	ECOLIB2	Data on the growth of E. coli bacteria in LB and SB cultures at 30 deg C and 37 deg C. Concentrations of bacteria are given in Klett readings (Ch. 8, Expl. 8.1)
24	FREEFALL	Data from a free fall experiment using a spark timer (Ch 10, Expl. 10.1)
25	NATIONS	Data from the "Nations Data Set" (Part III)

Appendix Section 0: Getting Started on the TI-82 or TI-83[1]

0.1 Turn the Calculator ON / OFF

Turn your calculator on by using the ON key, located in the lower left hand corner of the calculator. To turn the calculator off press 2nd OFF, located above the ON key.

0.2 Adjusting the Screen Contrast

Depending on the room lighting you may want to adjust the screen contrast.

Alternately press and release the 2nd then up arrow △ to darken the screen. Repeat as necessary. To lighten the screen contrast alternately press and release the 2nd down arrow ▽.

As the display contrast changes, a number in the upper right corner of the screen between 0 (lightest) and 9 (darkest) appears.

If you adjust the setting to 0, the display may become completely blank. If this happens, increase the contrast and the display will reappear. When contrast needs to be set at 8 or 9 all the time, it is probably time to change the batteries.

0.3 MODE Default Settings

The calculator should be set to the default mode settings. Press MODE to see the settings. Set your calculator to the settings in Figure 1 or 2 using your arrow keys and pressing ENTER to activate your choice.

Figure 1
The TI-82 default mode screen

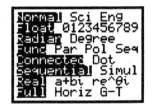

Figure 2
The TI-83 default mode screen

0.4 The Home Screen

The Home Screen is your calculation and execution of instruction screen. To return to the Home screen from any other screen, press 2nd QUIT.

The Home Screen is the primary screen of the TI-82 or TI-83.

If there is something displayed on the Home Screen, press the CLEAR key.

---

[1] The key stroking and menus for the TI-82 and TI-83 are nearly the same. Where they differ, screen images are presented side by side.

0.5 Calculating

The bottom six rows of keys on the graphing calculator behave like any scientific calculator, except that your entry is seen on an eight line computer screen. When you want the calculator to perform any calculation or instruction, press ENTER

> Note: The 2nd key will access the commands to the above left of any key. These commands are color coded with 2nd

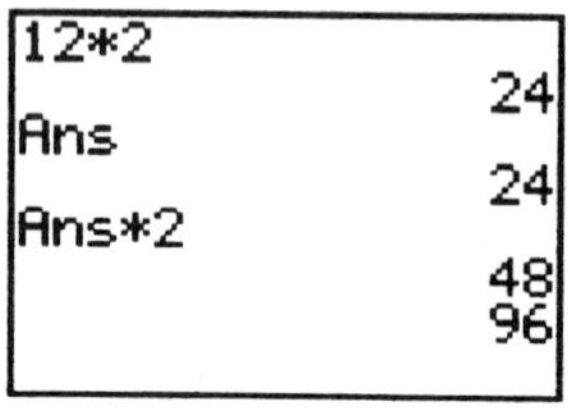

Example 1:

From the Home screen, do the following:

a. Type 12 X 2 then press ENTER .

24 is now displayed and *stored* as the answer. See Figure 3.

b. Press 2nd ANS and ENTER .

24 is again displayed

> The result of your last calculation is always stored in memory . To recall it press ANS .

c. Press the multiplication key X ,

then 2 and then ENTER . Pressing any operation key (+, -, x, ÷, x², x⁻¹ etc.) assumes that you want to operate on the stored answer. See Figure 3.

0.6 Iteration, Recalling a Process

Notice how ANS is also used.

Repeatedly press ENTER . Your screen should look like the bottom of Figure 3 and Figure 4.

This process is called iteration (repeating some process over and over again). The last operation (multiplying by 2) is repeated on the new answer.

```
12*2
            24
Ans
            24
Ans*2
            48
            96
```

Figure 3

The asterisk , * , is used for multiplication in place of the "times" sign to avoid confusion with the letter x.

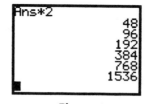

Figure 4

Example 2:
Interest compounded at 5% annually on an initial investment of $1000 could be represented by 1000*1.05. Or A = P(1+R) (Amount = original investment(1 + rate) Use iteration to determine the number of years for the amount of accumulated investment to be greater than $1300

Clear the Home screen, press CLEAR .

Enter 1000 followed by ENTER .

The number 1000 is now stored.

Press X 1.05 ENTER .

The number 1050 will now be displayed. By repeatedly pressing ENTER , you can see the growth of your initial $1000 investment year by year and determine how many iterations (years) are necessary for you to exceed $1300. See Figures 5 & 6.

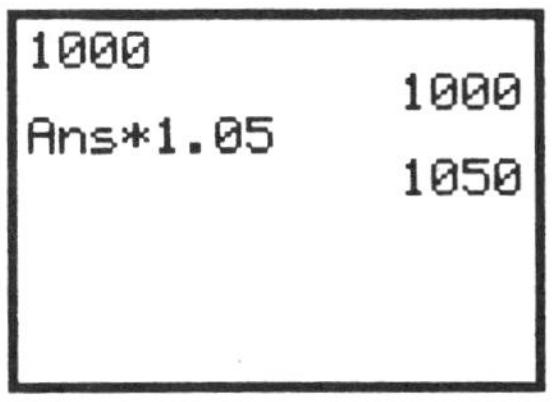

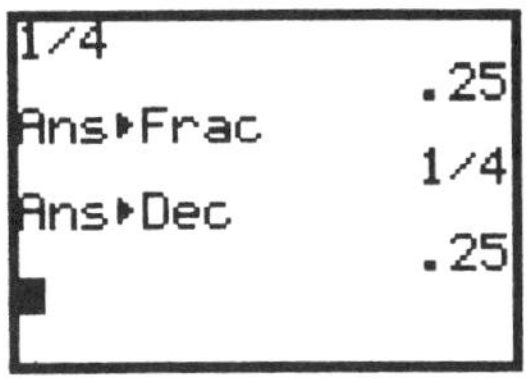

1000
1000
Ans*1.05
1050

Figure 5

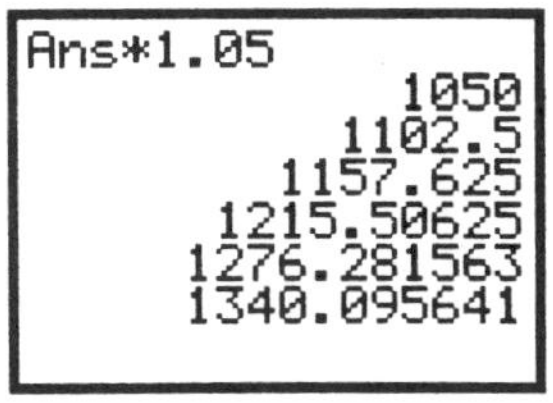

Ans*1.05
1050
1102.5
1157.625
1215.50625
1276.281563
1340.095641

Figure 6

Between year 5 and 6 the amount > 1300

0.7 Converting Decimals and Fractions

The TI-82 or TI-83 can be used to convert decimals and fractions.

Press 1 ÷ 4 ENTER . See Figure 7.

The decimal answer for this operation, 0.25, is displayed.

Now press MATH . You are in the MATH menu (Figure 8 or 9).

Press 1 or ENTER to select the highlighted option. This option [1:Frac] will change the decimal answer back into a fraction.

> Note: When the denominator of a fraction has more than four digits the answer is displayed as a decimal and will not return to a fraction

1/4
.25
Ans▸Frac
1/4

Figure 7

MATH NUM HYP PRB
1:▸Frac
2:▸Dec
3:³
4:³√
5:ˣ√
6:fMin(
7↓fMax(

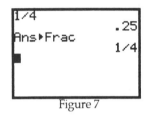

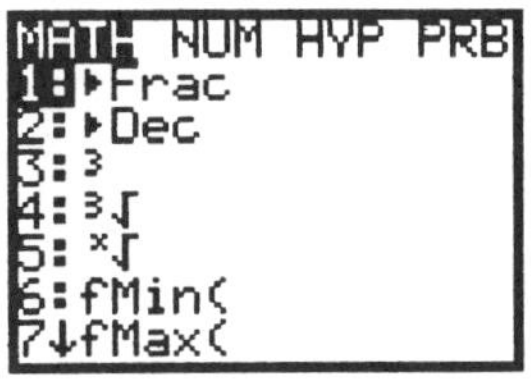

Figure 8 TI-82 MATH menu

MATH NUM CPX PRB
1:▸Frac
2:▸Dec
3:³
4:³√(
5:ˣ√
6:fMin(
7↓fMax(

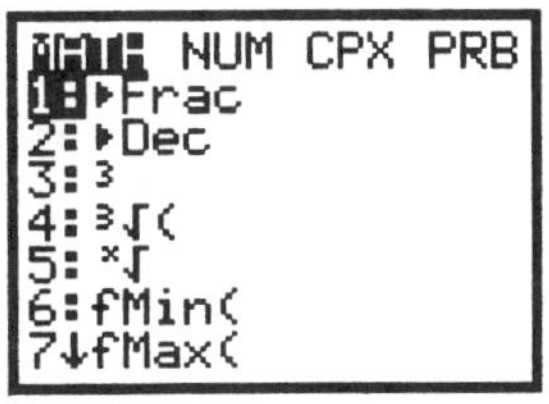

Figure 9 TI-83 MATH menu

0.8 Selecting Items from a Menu.

You can select an item from a menu by typing the number or by moving to that menu option with the down arrow key. You press ENTER to select your menu option.

Press MATH ▽ select [2: Dec] . Pressing ENTER changes the fraction back to a decimal. See Figure 10.

1/4
.25
Ans▸Frac
1/4
Ans▸Dec
.25

Figure 10

Example 3:
Type in the following fraction problems, then use the MATH menu to change the answers back to fractional form.

a. $\dfrac{1}{2} + \dfrac{1}{3}$

Press 1 $\boxed{\div}$ 2 $\boxed{+}$ 1 $\boxed{\div}$ 3 $\boxed{\text{ENTER}}$.

Press $\boxed{\text{MATH}}$ $\boxed{1}$ $\boxed{\text{ENTER}}$.

b. $3\dfrac{5}{9} + 5\dfrac{3}{7}$

Press (3 $\boxed{+}$ 5 $\boxed{\div}$ 9) $\boxed{+}$ (5 $\boxed{+}$ 3 $\boxed{\div}$ 7) $\boxed{\text{ENTER}}$.

Press $\boxed{\text{MATH}}$ $\boxed{1}$ $\boxed{\text{ENTER}}$.

See Figure 11.
The answer to part b is 566/63.

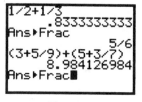

Figure 11

0.9 Raising a Number to a Power

The calculator can be used to raise bases to powers by using the power or exponent key $\boxed{\wedge}$

For 3^2 press 3 $\boxed{\wedge}$ 2 $\boxed{\text{ENTER}}$, or use a short cut, press 3 $\boxed{x^2}$ $\boxed{\text{ENTER}}$. This last method pastes the exponent above the 3.
See Figure 12.

Example 4:

Type the expression: $3^4 \times 2^5 \div 6^2$ $\boxed{\text{ENTER}}$

See Figure 12.

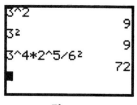

Figure 12

0.10 Order of Operations
The TI-82 or TI-83 uses an algebraic order of operations: inside parenthesis first, powers next, then multiply or divide from left to right and lastly add or subtract from left to right.

Example 5:
a.   Enter: $1 + 2(4 - 2)^2 + 6 \div 2$
See Figure 13.
The order of operations are performed algebraically in the following steps:

$1 + 2(4 - 2)^2 + 6 \div 2 =$
$1 + 2(2)^2 + 6 \div 2 =$ inside parenthesis
$1 + 2(4) + 6 \div 2 =$ raise to power
$1 + 8 + 6 \div 2 =$ multiply
$1 + 8 + 3 =$ divide
$9 + 3 =$ add
12 add

b. Enter: "One hundred fifths times two."
See Figure 14 for two methods.

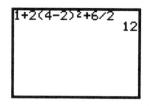

Figure 13

Figure 14

Note: For the TI-82 compare the difference in order of operations in Figures 14 & 15.
Parentheses in the denominator of a fraction are interpreted as a grouping .
For the TI-83 parenthesis are interpreted the same as the multiplication sign. See Figure 16.

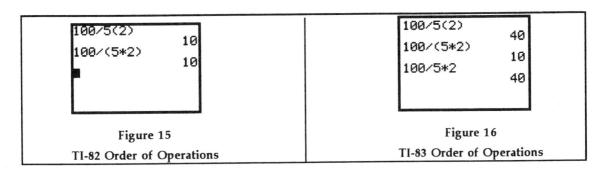

Figure 15	**Figure 16**
TI-82 Order of Operations	**TI-83 Order of Operations**

Example 5 continued:
c. Enter "Sixteen raised to the one half
power."
This is the same as the square root of 16.

Fractional exponents must always be
enclosed in parenthesis. See Figure 17.

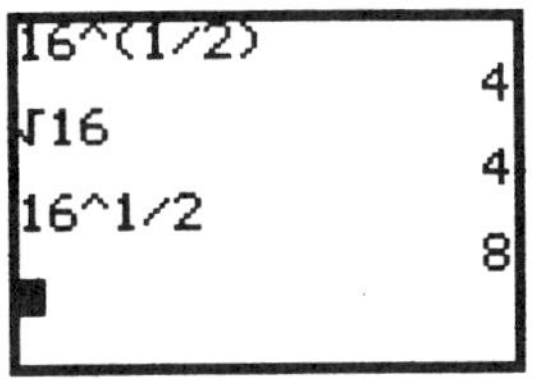

Figure 17

0.11   A Truth Test
The graphing calculator can be used to
evaluate the truth value of an expression.

To use this feature, you must use the 2nd

TEST menu. Figure 18 shows the TEST
LOGIC menu. This is where the equal and
inequality symbols are located.

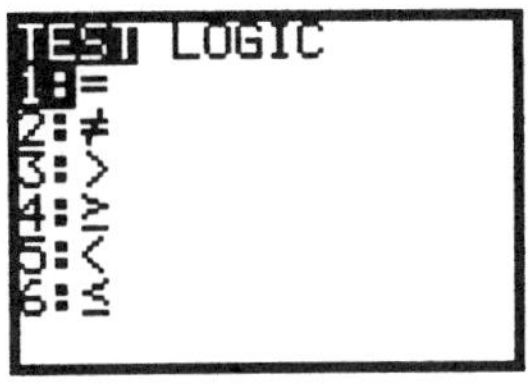

Figure 18

Example 6:
a.   Is $3 < 7$ true or false?

Press 3 2nd TEST ▽ .

Select [5:<.] ,press 7 ENTER See Fig. 18

When performing a TEST, remember that
0 means FALSE and 1 means TRUE.

b.   Does $3(4+5) = (3 \times 5)+5$?
This is a false statement, thus the answer is
zero (see Figure 19) because:
$$3(4+5) = 3 \times 4 + 3 \times 5$$

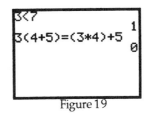

Figure 19

0.12 Deep Recall and Editing
Press CLEAR . To recover your last entry

press 2nd ENTRY . To evaluate press

ENTER .To edit an expression use the left

and right arrows to position the cursor for

editing and press delete DEL or insert 2nd

INS

Change the expression in example 6 b to :
$$3(4+5) = (3 \times 4) + (3 \times 5).$$

First recall the expression, 2nd ENTRY ,

then edit using arrows and 2nd INS .

See Figure 20. Now the expression is
evaluated as true   (i.e. 1).

Also try pressing 2nd ENTRY several
times and you will see some of the old
expressions you typed. This is called *deep
recall* and it is used to retrieve expressions
that were typed many steps ago

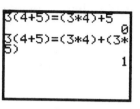

Figure 20

0.13 Storing Values to Variables

Recall Example 2 where we were finding
the amount of money accumulated using the
formula A = P(1+R), where P=$1000 and
R=.05. The calculator allows you to store
values to alphabetical letters A through Z.
You access the letters by first pressing the

ALPHA key and you store number values to

letters by using the store STO key.

Alphabetical letters are located to the
above right of keys and are color coded to
match the ALPHA key.

To store 1000 to P   press 1000 STO ALPHA

P ENTER .To store .05 to R press .05 STO

ALPHA R ENTER .See Figure 21.

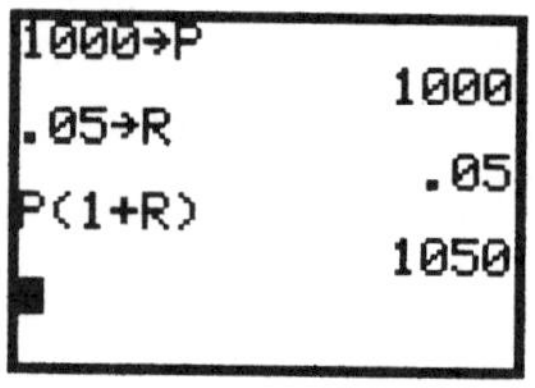

Type the expression P(1+R), remember to

press ALPHA before typing the letter.

The expression has been evaluated using
the stored values to P and R. These values
will remain the same until you store a new
value to R and P. See Figure 21.

Note: If your calculator is new or if the
memory has been cleared, the initial stored
value to all letters is zero.

```
1000→P
            1000
.05→R
             .05
P(1+R)
            1050
■
```

Figure 21

0.13.1 A special note about X and Y

Since the variables X and Y are used in
plotting graphs, their values are
constantly updated when you TRACE on a
graph. There are two ways to access the X
variable since it is usually the variable of

choice in algebra. Press ALPHA X or

use the handy key X T Θ See Figure 22.

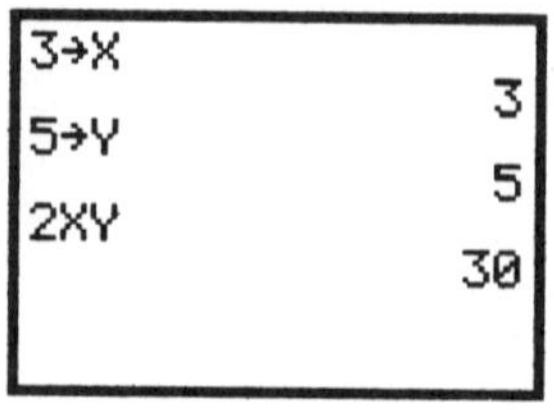

```
3→X
              3
5→Y
              5
2XY
             30
```

Figure 22

0.14 Subtraction and Negative

In algebra the minus sign is used two different ways:

1. as the operation sign between to numbers to mean "subtract", as in 5 - 3 , and
2. in front of a number to mean "the opposite of or negative of", as in - 7.

This calculator has two different keys for minus.

For 5 - 3 press 5 $\boxed{-}$ 3, for subtraction.

For -7 find the negative key $\boxed{(-)}$ located to the left of ENTER , now press $\boxed{(-)}$ 7.

See Figure 22.

> The negative sign is actually a little bit shorter and slightly raised compared to the subtraction symbol.

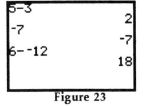

Figure 23

Example 6:

Type the following problems:

a. 6 - ⁻ 12
b. ⁻ 3 X ⁻ 9
c. (⁻ 5)²
d. ⁻ 5²

See Figures 23 and 24

Note that the values for **c** and **d** above are different. See Figure 24.

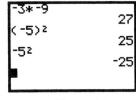

Figure 24

> To square a negative number you must put it in parenthesis.

Order of operations in **d** says: "Square five then take its opposite."

> **Trouble Shooting:**
> The most common calculator error is using the subtraction symbol instead of the negative symbol. See Figures 25 & 26.

0.15 The Error Message

Using the subtraction sign incorrectly produces an error message when you press $\boxed{\text{ENTER}}$.

Choose [2:Goto] to position the cursor to the place where the error occurred. Choose [1:Quit] to begin a new line on the Home Screen. See Figure 26

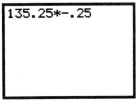

Figure 25

Figure 26

Appendix Section 1
Visualizing Data Using The TI-82 or TI-83[1] Statistics
Menus. Graphing Calculator Techniques for Chapter 1

1.1   Using Lists for Data Entry

To enter data into the calculator you use the statistics menu. You can store data in lista labeled L1 through L6.

> **Note:** The TI-83 can store data to aditional name lists by giving them a name. See the owners manual for more detail.

Press [STAT] . To clear any data stored in List 1, and List 2, select [4:ClrList] then press [2nd] [L1] [,] [2nd] [L2] [ENTER]
See Figures 1 & 2.
Now you are ready for data entry.
Press [STAT] , select [1:Edit]. See Figure 3.

Example 1:

Ten students were surveyed. Below is a table showing the number of hours each student worked last week:

0
0
4
10
10
16
16
18
20
20

Table 1

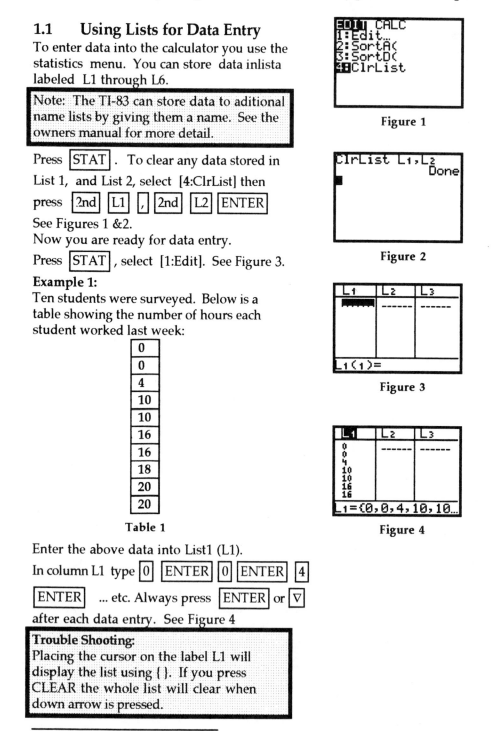

Figure 1

Figure 2

Figure 3

Figure 4

Enter the above data into List1 (L1).

In column L1 type [0] [ENTER] [0] [ENTER] [4]
[ENTER] ... etc. Always press [ENTER] or [▽]
after each data entry. See Figure 4

> **Trouble Shooting:**
> Placing the cursor on the label L1 will display the list using { }. If you press CLEAR the whole list will clear when down arrow is pressed.

---

[1] The keystroking and menus for the TI-82 and TI-83 are nearly the same. Where they differ, screen images are presented side by side.

1. 2   Histogram Setup

A histogram is a graph that helps you visualize data. To graph the above data you must first set up this statistical plot.

Press 2nd STAT PLOT , select [1: Plot1] as shown in Figure 5.

To set up the histogram correctly:

1. Select [ON] press ENTER .

2. Select the graph type ▽ ▷ to the histogram icon, then press ENTER .

3. Select the Xlist (for your horizontal axis) [L1] press ENTER .

For the TI-83 press 2nd L1 .

4. Select the frequency. In example 1 each item has a frequency of one so select [1] then press ENTER

Your plot should be set up as in Figure 6 or 7.

Figure 5

Figure 6 TI-82 Plot1 Menu

Figure 7 TI-83 Plot1 Menu

1.3   Selecting the Correct Window for the Histogram

Note: Before you plot a STAT PLOT graph, CLEAR or deselect all functions in Y =

1.3.1   Manual Window Sizing

To see your histogram you must set the graphing calculator window to the correct size. Manually set your window size based on the data in L1 . Since the L1 data from example 1 starts at 0 and ends at 20, your X minimum and X maximum values must be: $X \leq 0$ or $X \geq 20$. Y maximum should be slightly larger than the highest frequency for any one data entry.

Press |WINDOW| |∇| and enter the window settings as in Figure 8.

Press |GRAPH| then |TRACE| and use your right arrow until your screen looks like Figure 9.

When you TRACE on a histogram, the cursor moves to the top center of each column. The P1 in the upper right corner indicates you are tracing Plot 1. The **min** = 16, **max** =1 7 indicates X is on the interval 16 to 17 or, in this case, 16. The **n** value (frequency) is 2. So there are 2 people who worked 16 hours last week.

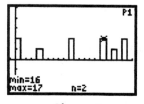

Figure 8

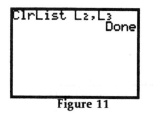

Figure 9

1.4   Working With More than One List

Example 2.

Suppose we have two data lists. List1 represents the sum of the numbers on a pair of dice. List2 represents the possible ways of getting each number by tossing a pair of dice (the frequency). See Table 2 below. Since we already have data in L1 , enter the following dice data into List 2 (L2) and List 3 (L3).

Press |STAT| .

To clear any data stored in List 2, and List 3, select [4:ClrList] then press |2nd| |L2| |,|

|2nd| |L3| |ENTER| . See Figures 10 & 11

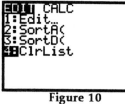

Figure 10

Figure 11

L2	L3
2	1
3	2
4	3
5	4
6	5
7	6
8	5
9	4
10	3
11	2
12	1

Table 2

Trouble Shooting
When using lists as frequency, the data must be whole numbers. Decimal values for frequency will produce the error message **ERR:STAT.**

1.4.1 Entering the data.

Press [STAT] , select [1:Edit]. See Figure 12.

In column L2 type [2] [ENTER] [3] [ENTER] [4]

[ENTER] ... etc., always pressing [ENTER] or

[∇] after each data entry. Press [▷] to get to

column L3. Type in all the L3 data pressing
ENTER after each number . See Figure 13

1.4.2 Turn OFF old Plots

Press [2nd] [STAT PLOT] , select [4: Plotsoff]

and [ENTER] as shown in Figures 14 & 15.

1.4.3 Setting up Plot2

Press [2nd] [STAT PLOT] , select [2: Plot2].

1. Select [ON] press [ENTER] .
2. Select the graph type (histogram icon)
   , press [ENTER] .
3. Select the Xlist (for your horizontal
   axis) [L2] press [ENTER] .
4. Select the frequency. In example 2 each
   item has a frequency located in L3 then
   press [ENTER]

Your plot should be set up as in Figure 16 or
17.

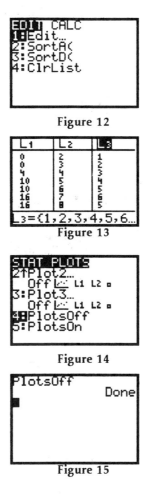

Figure 12

Figure 13

Figure 14

Figure 15

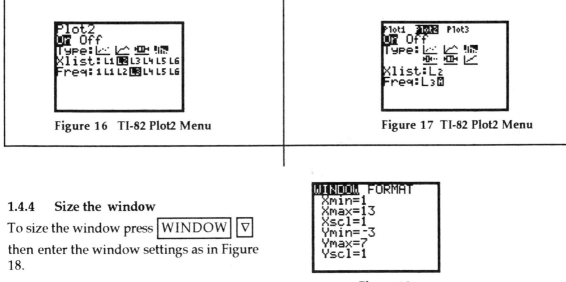

Figure 16 TI-82 Plot2 Menu

Figure 17 TI-82 Plot2 Menu

1.4.4 Size the window

To size the window press [WINDOW] [∇]

then enter the window settings as in Figure
18.

Figure 18

Automatic window sizing.
You can have the calculator automatically set your window by pressing [ZOOM] & selecting [9:ZoomStat]. See Figure 19. However you lose control of the max and min values and still may want to adjust the window.

ZOOM MEMORY
3↑Zoom Out
4:ZDecimal
5:ZSquare
6:ZStandard
7:ZTrig
8:ZInteger
9:ZoomStat

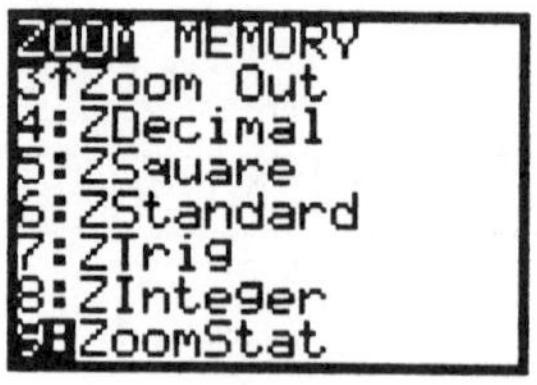

Figure 19

1.4.5 Tracing on the Histogram

Press [GRAPH] then [TRACE] and use your right arrow until your screen looks like Figure 20 or 21. When you TRACE on a histogram, the cursor moves to the top center of each interval column. Your histogram represents the possible ways of getting each outcome by tossing a pair of dice. The P2 tells that you are tracing on Plot 2. The min = 6, max = 7 indicates that you are tracing on the interval 6 to 7 or, in this case, 6. The **n** value (frequency) is 5. This means that the number 6 has 5 possible outcomes when a pair of dice is thrown.

P2
min=6
max=7
n=5

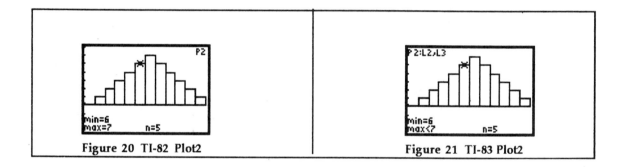

Figure 20 TI-82 Plot2

P2:L2,L3
min=6
max<7
n=5

Figure 21 TI-83 Plot2

Trouble Shooting
Dimension Mismatch. The error message shown in Figure 22 is telling you that your lists are not the same size. To graph two lists they must have the same number of elements. Either one list is longer than the other or you have set up your plot with the wrong lists.

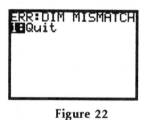

ERR:DIM MISMATCH
1:Quit

Figure 22

1.5 Changing the Interval Size.

Change your WINDOW settings to Figure 23. When you change the X scale, you are setting the width size of the interval. In this case the histogram will be 2 units wide, beginning at Xmin.

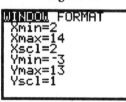

WINDOW FORMAT
Xmin=2
Xmax=14
Xscl=2
Ymin=-3
Ymax=13
Yscl=1

Figure 23

Figure 24 shows the new histogram with interval width of 2 units. You would interpret the TRACE point to mean there are 11 ways to get a 6 or 7 on the roll of two dice.

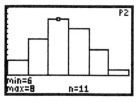

Figure 24

Change your WINDOW so that Xscl = 3 and Ymax = 16. Figure 25 shows an interval width of 3 units.

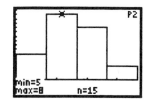

Figure 25

1.6 One Variable Statistics
Mean and Median

Example 3:
Below are SAT scores of 10 randomly selected students

SAT	600	640	430	500	510	530	550	370	500	530

Enter this data using the STAT menu. First CLEAR all values from L1 , L2 and L3 using the techniques learned in section 1.1 and 1.4. See Figure 26.

Enter the SAT scores in L1. Refer to section 1.1 if you need help. Press ENTER after each number. See Figure 27.

1.6.1 One Variable Statistics.
Let us look at the *one variable* statistics performed on list 1 (L1).

Press: STAT ▷ to < CALC>.

Select [1:1-Var Stats]. See Figure 28.

Press ENTER .

Figure 26

Figure 27

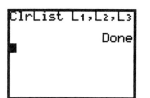

Figure 28

We see the information in Figure 29.

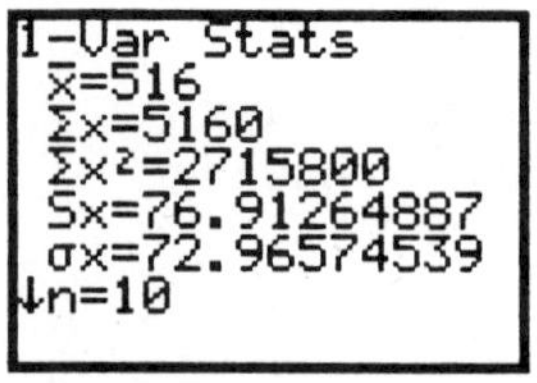

```
1-Var Stats
 x̄=516
 Σx=5160
 Σx²=2715800
 Sx=76.91264887
 σx=72.96574539
↓n=10
```

Figure 29

1.6.2 The Mean.

In Figure 29 the statistics we are most interested in are:

1. $\bar{x}$, the **mean** (average) represented by $\bar{x} = 516$ and
2. **n**, the number of elements in the list , n=10,

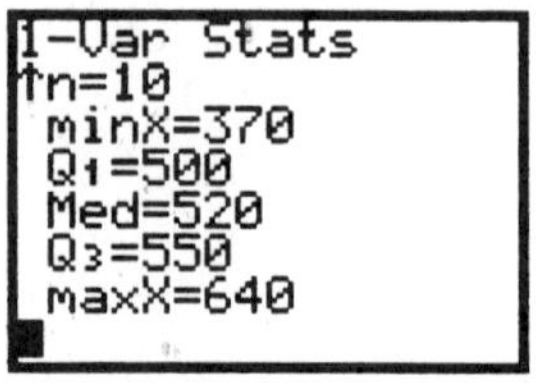

1.6.3 The Median.

Press $\boxed{\nabla}$ to see the information in Figure 30.

We are interested in **Med**, the **median** score , the middle score in rank order, represented by Med = 520.

```
1-Var Stats
↑n=10
 minX=370
 Q₁=500
 Med=520
 Q₃=550
 maxX=640
```

Figure 30

1.6.4 Sorting a List.

Sort L1 from low to high.

Press $\boxed{\text{STAT}}$, select [2:SortA].

See Figure 31.

Type the list you want to sort $\boxed{\text{2nd}}$ $\boxed{\text{L1}}$ $\boxed{)}$

$\boxed{\text{ENTER}}$. Figure 32 shows that we have sorted L1 .

```
EDIT CALC
1:Edit…
2:SortA(
3:SortD(
4:ClrList
```

Figure 31

Trouble Shooting:
Note: If you have two lists that you want to sort as an ordered pair, type:
SortA(L1,L2) $\boxed{\text{ENTER}}$.

```
SortA(L₁)
              Done
```

Figure 32

1.6.5 Looking for the Median.

Press $\boxed{\text{STAT}}$ select [1:Edit].

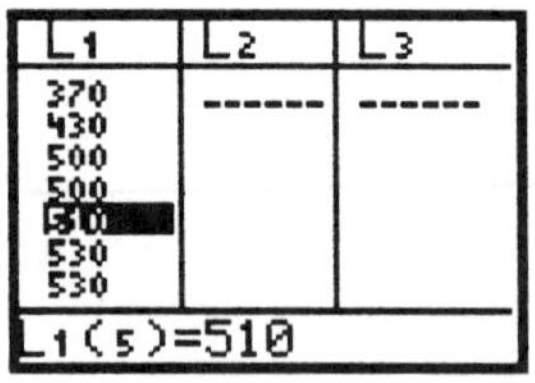

Look at your list now. See Figure 33 The **median** score is the middle score on the list, however, since there are 10 elements (an even number) the median is the score in between the 5th and 6th score, or

$$\frac{510 + 530}{2} = 520$$

You may think of the median of an even number of element as the average of the middle two scores.

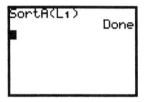

L₁	L₂	L₃
370	------	------
430		
500		
500		
510		
530		
530		

L₁(5)=510

Figure 33

1.7 Visualizing the Median
The Box and Whiskers Plot

Do a box and whisker plot. Select a plot.

Press [2nd] [STATPLOT], turn all plots OFF
then select [1:Plot1] .

1. Select: ON [ENTER] [∇] .

2. Select graph type: [▷] to the Boxplot
   icon [ENTER] .

TI-83 icons are slightly different

3. Select Xlist: L1 [ENTER] . See Figure 34.

Figure 34

To see the Boxplot: press [ZOOM] , select
[9:ZoomStat]. See Figure 35.

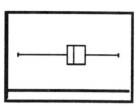

Figure 35

The boxplot show one variable statistics.
Press [TRACE] . The middle of the box is the
median (**Med**). See Figure 36. The whiskers
on the plot extend from the minimum list
value on the left to the first quartile (Q_1)
and from the third quartile (Q_3) to the
maximum list value. Use your left and
right arrows to see these scores.

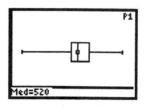

Figure 36

**Quartiles are 1/4 th or 25% of the list
when put in rank order.**

Figure 37

1.7.1 Show a Histogram of L1.
Make a histogram. Select the plot.

Press [2nd] [STATPLOT], select [1:Plot 1]

Select: ON [ENTER]

Type: (histogram) [ENTER]

Xlist: L1 [ENTER]

Press [WINDOW] [∇]

Select an appropriate range and interval
scale. See Figure 38.

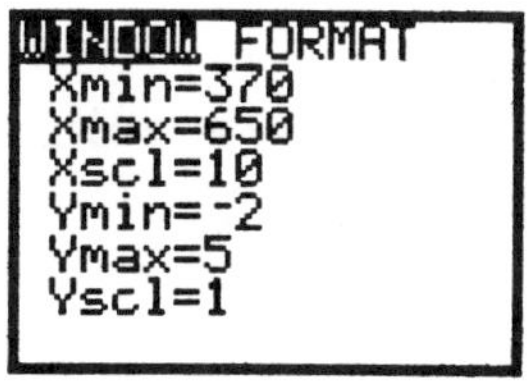

Figure 38

Press [GRAPH] .

Press [TRACE] to see the frequency, **n**, of
each SAT score.

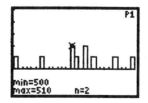

Figure 39

1.8   Other List Techniques
Example 4
The following chart gives the age of students in a mathematics class.

Age Interval	Frequency Count
15 - 19	2
20 - 24	8
25 - 29	4
30 - 34	3
35 - 39	2
40 - 44	1
45 - 49	1
Total	21

Since the age is given as an interval, calculate the median of each interval so that a single number can be entered into a list. Go to the home screen and type the lists below, remember to enclose the list in braces, { }. Store the values to L1 and L2 See Figure 40. Press:

{17,22,27,32,37,42,47} $\boxed{STO}$ $\boxed{2nd}$ $\boxed{L1}$ $\boxed{ENTER}$

{2,8,4,3,2,1,1,} $\boxed{STO}$ $\boxed{2nd}$ $\boxed{L2}$ $\boxed{ENTER}$.

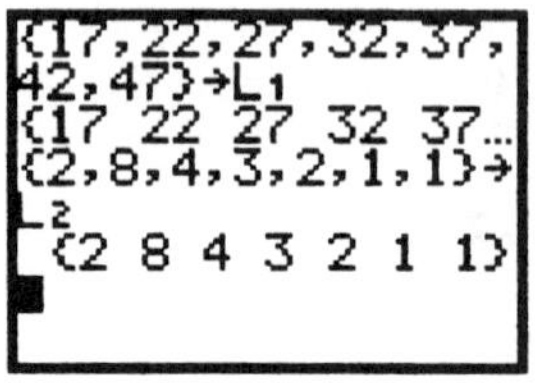

To calculate the mean you need to multiple the age times the frequency, sum all the ages then divide by the total frequency. Give it a try from the home screen.

1.8.1   Multiply L1 X L2 then store to L3.

$\boxed{2nd}$ $\boxed{L1}$ $\boxed{X}$ $\boxed{2nd}$ $\boxed{L2}$ $\boxed{STO}$ $\boxed{2nd}$ $\boxed{L3}$

See Figure 41.

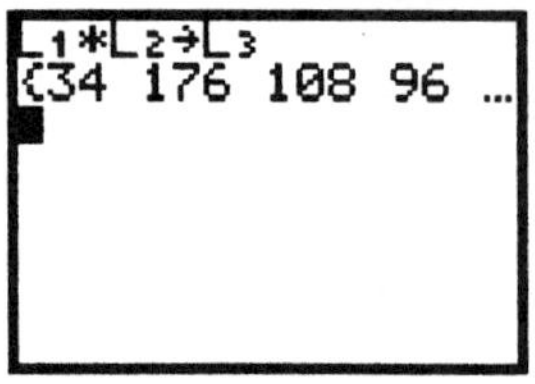

1.8.2   Sum List3.

$\boxed{2nd}$ $\boxed{LIST}$ $\boxed{\triangleright}$ to <MATH>, select [5:sum]

$\boxed{2nd}$ $\boxed{L3}$ $\boxed{ENTER}$. See Figure 42 & 43.

1.8.3   Divide by the Total.

$\boxed{\div}$ 21 . See Figure 44

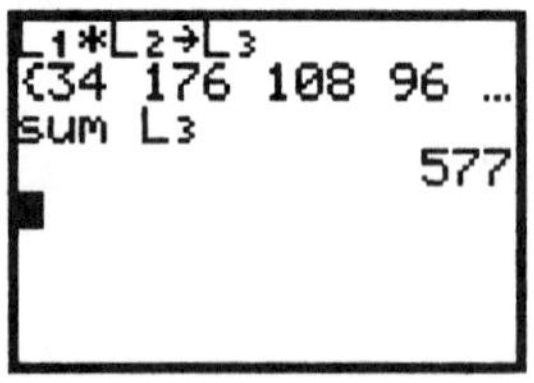

1.8.4   Use the Mean Command for Two Lists.

$\boxed{2nd}$ $\boxed{LIST}$ $\boxed{\triangleright}$ to <MATH>, select [3:mean],

$\boxed{2nd}$ $\boxed{L1}$ $\boxed{2nd}$ $\boxed{L2}$. See Figure 44.

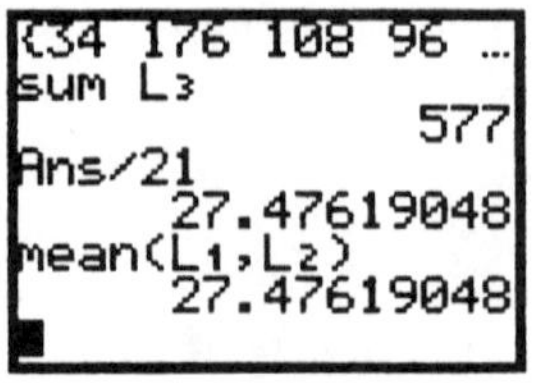

Either method gives the same answer.

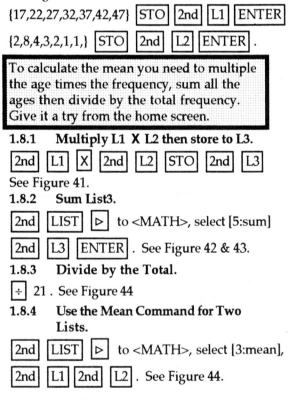

Figure 40

Figure 41

Figure 42

Figure 43

Figure 44

Appendix Section 2
Scatter Plots and Introduction to Graphing on the
TI-82 or TI-83[1] . Graphing calculator techniques for Chapter 2.

2.1   Scatter Plots

Relationships between two variable can be
visualized by graphing data as a scatter
plot. Think of the two list as ordered
pairs. An ordered pair (x,y) can represent a
point on a graph.

Example 1

Below in Table 1 are the SAT and Math
placement scores of ten randomly selected
freshmen students. Graph the data to see if
a relationship exits between the SAT scores
and the Math placement scores.

SAT	600	640	430	500	510	530	550	370	500	530
Math Place	25	29	14	12	11	8	17	16	9	26

Table 1

Refer back to Appendix Section 1.1-1.4 for
information on entering data into lists.

2.1.1   Enter the Data
CLEAR List1 and List2.
Enter the math placement scores into List1
(L1).
Enter the SAT scores into List2 (L2). See
Figure 1

2.1.2   Set up the Scatter Plot
Press 2nd   STATPLOT . Select [1:Plot1].
Set up the Plot1 as a Scatter Plot as in
Figure 2 or 3.

Figure 1

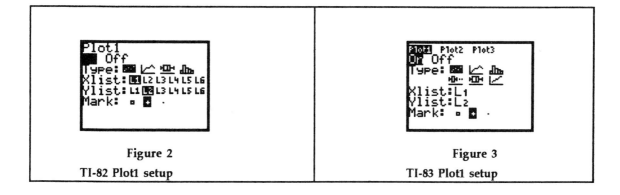

Figure 2	Figure 3
TI-82 Plot1 setup	TI-83 Plot1 setup

---

[1] The key stroking and menus for the TI-82 and TI-83 are nearly the same. Where they differ, screen images
are presented side by side.

Press WINDOW ▽ .

Set the appropriate window as in Figure 4. The Xlist is the math placement scores and the Ylist is the SAT scores, so the window is set just beyond the lowest and highest data values.

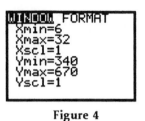

Figure 4

Press GRAPH . There appears to be an increasing relationship, i.e. as math placement score increases SAT increases, but there are a few exceptions.

2.1.3   Switch the x and y Variables.
What does the plot look like when L2, the SAT scores, is the independent variable? Set up the plot as in Figure 6 or 7.

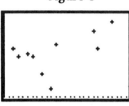

Figure 5

Math Placement on the X axis.

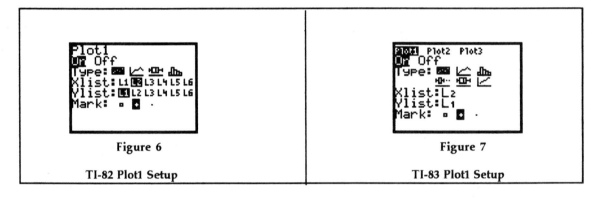

| **Figure 6** | **Figure 7** |
| TI-82 Plot1 Setup | TI-83 Plot1 Setup |

2.1.4   Sort List 2 and List1 together.
To sort List2 while keeping the pairing with List1, Press STAT , select [2:SortA]

2nd L2 , 2nd L1) ENTER .
See Figure 8.

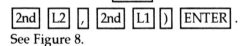

Figure 8

2.1.5   Setting the Window
Look at the sorted list. Press STAT select [1:Edit]. Use the sorted List2 to determine the window setting .

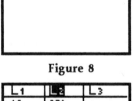

Figure 9

Press WINDOW ▽ .

Set the window as in Figure 10.

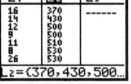

Figure 10

Press $\boxed{\text{GRAPH}}$. See Figure 11

Here the relationship is less clear. Perhaps more data is necessary to determine a trend or relationship.

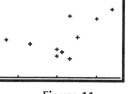

Figure 11

SAT on the X axis

Trouble Shooting

Before you press graph:

1. Clear $\boxed{Y=}$ or turn off all graphs.
2. Turn OFF all plots except the one you want to see.

ERR: INVALID DIM means your lists are not the same size, you have selected a list with no data in it or you have a plot turned on that you did not want and it has different size lists.

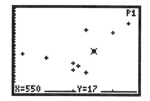

Figure 12

2.1.6   A Check List for Plotting

To plot statistical data in lists, follow these steps:

1. Clear old data in lists.
2. Store the statistical data in one or more lists.
3. Set up the STAT PLOT.
4. Turn Plots ON or OFF as appropriate.
5. Select or deselect Y= equations as appropriate.
6. Define the viewing WINDOW.
7. Explore the plot or graph by pressing TRACE.

2.2   Introduction to Graphing Functions

The graphing calculator can be used to graph equations that are functions. The top row of keys under the viewing screen contains all the graphing menus.

Before you begin graphing turn all plots OFF.

See Figure 13. Press $\boxed{Y=}$ and CLEAR all equations. See Figure 14.

2.2.1   The Standard WINDOW

Press $\boxed{\text{ZOOM}}$ select [6:ZStandard] and you see and xy- coordinate system in Figure 16.

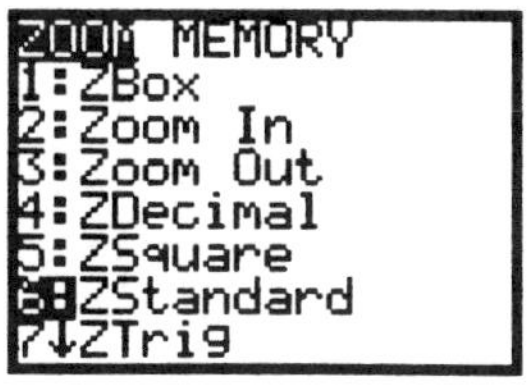

Figure 13

Figure 14

Figure 15

You see only a portion of the real number lines on the xy-coordinate plane. The size of the viewing window is determined by the window variables : Xmin, Xmax, Ymin, and Ymax . See Figure 17.

To see the current *Standard window* , press WINDOW . See Figure 18.

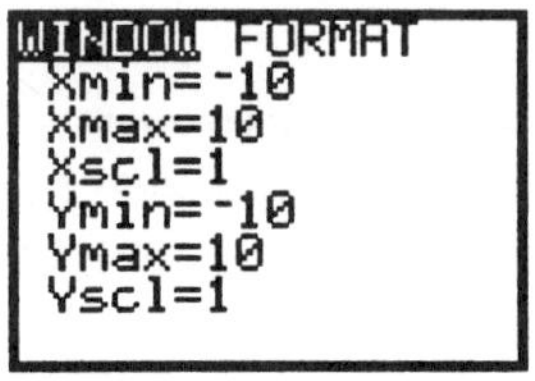

The distance along the X-axis goes from -10 to 10 or x:[-10,10]. The distance along the Y-axis goes from -10 to 10 or y:[-10,10]. The distance between the tic marks on the X axis is 1 unit (Xscl = 1) and the distance between the tic marks on the Y axis is 1 unit (Yscl = 1).

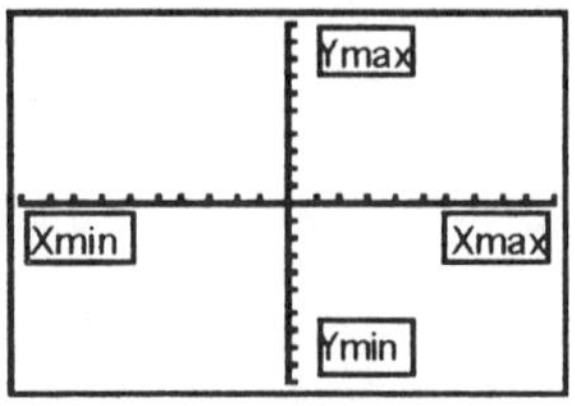

2.2.2 The Free Moving Cursor

Press GRAPH . When you press arrow keys the cursor can move anywhere on the graphing window. The cursor has changed to a cross and at the bottom of the screen we see the coordinates of the screen position, which change as we jump form pixel to pixel. A pixel is a point of light on the screen. See Figure 19.

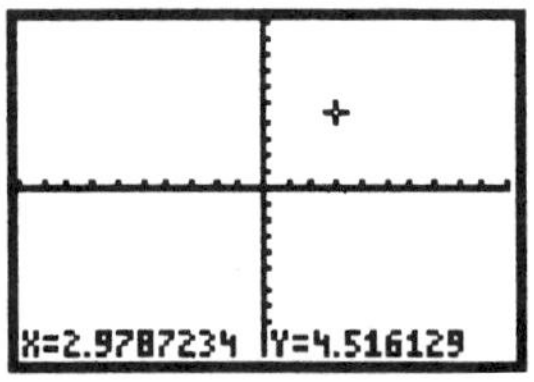

2.2.3 The Decimal Window

Notice that on the standard window we get rather "ugly" decimals when we press the arrow keys

Press ZOOM select [4:ZDecimal].

Now use the arrow keys. As you jump from pixel to pixel you increment by the decimal value .1 (1/10 th). See Figure 20.

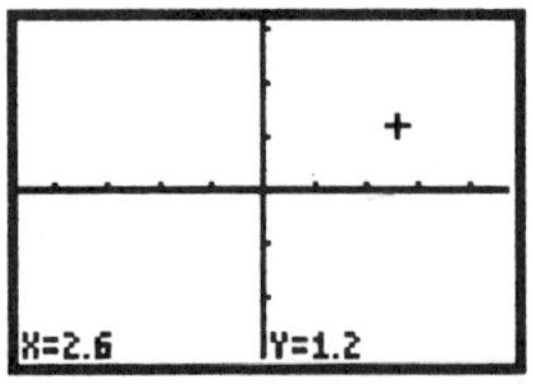

Press WINDOW . You see the decimal window settings for x and y. See Figure 21.

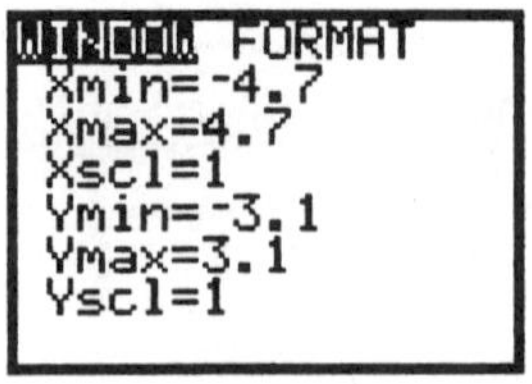

Example 2

To change Centigrade temperature to Fahrenheit use the formula F = (9/5)C +32. Enter this formula into the calculator and graph the function.

Press Y= . Notice you can graph 10 different graphs. Type the function
        y = (9/5)X+ 32 into Y1.
See Figure 22 or 23.

Trouble Shooting:
The independent variable, in this case C, must be entered as X on the calculator. The dependent variable F will become Y.

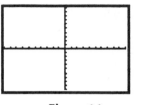

Figure 16

Figure 17

```
WINDOW FORMAT
 Xmin=-10
 Xmax=10
 Xscl=1
 Ymin=-10
 Ymax=10
 Yscl=1
```

Figure 18

X=2.9787234 Y=4.516129

Figure 19

X=2.6   Y=1.2

Figure 20

```
WINDOW FORMAT
 Xmin=-4.7
 Xmax=4.7
 Xscl=1
 Ymin=-3.1
 Ymax=3.1
 Yscl=1
```

Figure 21

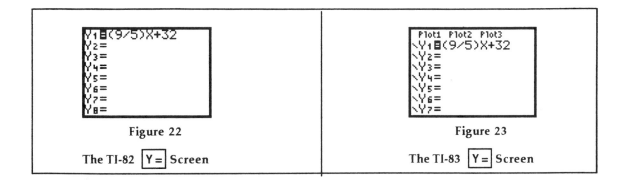

Figure 22

The TI-82 $\boxed{Y=}$ Screen

Figure 23

The TI-83 $\boxed{Y=}$ Screen

A Special Note to TI-83 Users

Plots can be turned ON and OFF from the $\boxed{Y=}$ screen. To turn a plot ON, use the arrow keys to position the cursor on the desired Plot, then press $\boxed{ENTER}$. To turn the plot OFF press $\boxed{ENTER}$ Darkened plots are ON.

2.2.4 The Integer Window

Press $\boxed{WINDOW}$ $\boxed{\nabla}$, then set the window to the settings in Figure 24.

Press $\boxed{GRAPH}$, then $\boxed{TRACE}$. Right arrow to 8° Centigrade to see the equivalent 46.4° Fahrenheit temperature. Left arrow to -10°. See Figures 25 & 26. Notice the pixel jumps are now integers of 2 units. Play with your arrow keys as you TRACE on the function.

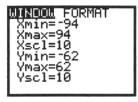

Figure 24

Friendly Windows

As you move in an X direction from pixel to pixel there are 94 jumps across the screen. Like wise there are 62 pixel jumps in a Y direction. To Figure out the horizontal jump Δx, and the vertical jump Δy use the following formulas:

$$\Delta x = \frac{x_{max} - x_{min}}{94}$$

$$\Delta y = \frac{y_{max} - y_{min}}{62}$$

Note: A "Friendly Window" is any window whose distance between Xmax and X min is evenly divisible by 94.

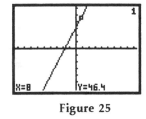

Figure 25

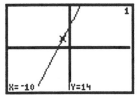

Figure 26

2.2.5 Using Graphs to Determine the Domain of a Function

For most functions the *domain* (possible x values) is all real numbers. However there are functions that are the exception.

Example 3
Use a graph and a table of values to help determine the domain of the function f(x)=1/x.

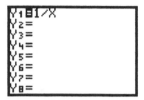

Figure 27

Press [Y=] [CLEAR] . In Y1 type 1/X. See Figure 27.

Press [ZOOM] , select [4:ZDecimal] [TRACE] .

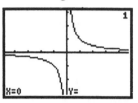

Figure 28

We see a graph in two pieces. TRACE is telling us that there is no Y value associated with x = 0. See Figure 28.

Set up a table of values. Press [2nd] [TblSet] . Begin the table at x = -5 and increment by 1 unit (ΔTbl=1). See Figure 29.

Figure 29

Press [2nd] [TABLE] . Notice that for x =0 we get an ERROR message, which means that y is undefined for x = 0. See Figure 30.

Both the graph and the table confirm that x = 0 is not in the domain of x.

Figure 30

Example 4

Use a graph and a table of values to help determine the domain of f(x) = √x

Press [Y=] [CLEAR] . In Y1 type [2nd] [√] X.

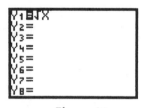

Figure 31

Press [GRAPH] [TRACE] .

The function f(x) appears to be defined for x=0, but not to the left of zero. See Figures 32 and 33.

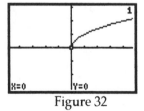

Figure 32

Press 2nd TABLE . Notice that values of x<0 give an ERROR message, which means the function is undefined for x<0. See Figure 34. Use ∇ to see other values of y that are defined by x.

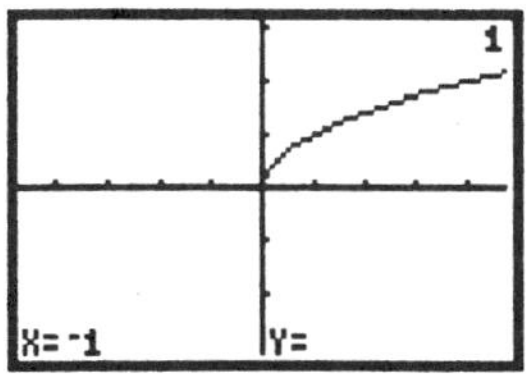

The domain of the square root function is the set of all x such that x ≥ 0.

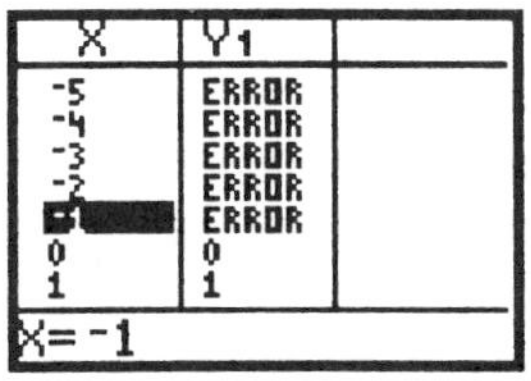

2.2.6 Graphs that are NOT functions.
Example 5
Graph $y^2 = x$.
First solve for y. y has two values: $y = \sqrt{x}$ or $y = -\sqrt{x}$, therefore y is not a function. However, a relationship does exist between x and y. We need to trick the calculator into graphing the relationship.

Let $Y1 = \sqrt{x}$ and $Y2 = -\sqrt{x}$. See Figure 35. Press GRAPH . See Figure 36.

In addition we can tell y is NOT a function of x because it fails the *vertical line test* .

Press 2nd DRAW , select [4:Vertical]. See Figure 37.

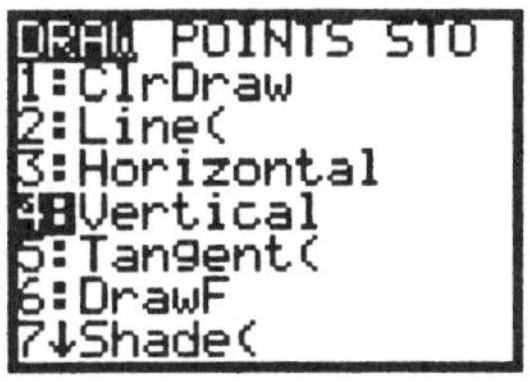

Press GRAPH .

Use the right arrow ▷ to position the vertical line on x = 3. Since a vertical line intersects the graph in two places (i.e. there are two y values for one x value), the graph is NOT a function. See Figure 38.

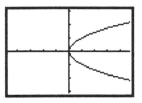

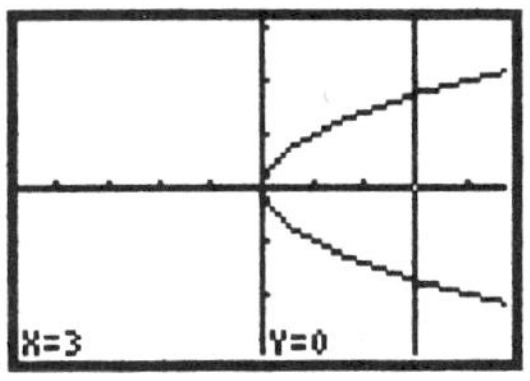

Figure 33

Figure 34

Figure 35

Figure 36

Figure 37

Figure 38

Appendix Section 3
Finding Average Rate of Change on the TI-82 or TI-83[1]. Graphing calculator techniques for chapter 3.

3.1 Finding Average Rate of Change

Example 1:
In 1980 the US Federal debt was 909 billion dollars. In 1990 the Federal debt was 3206 billion dollars. Find the average rate of change. The average rate of change is:

$$\frac{\text{change in debt}}{\text{change in years}} = \frac{3206 - 909}{1990 - 1980} =$$

229.7 billion dollars per year

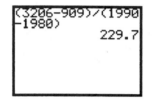

Figure 1

Interpretation:
On average the US Federal debt increased by $229.7 billion/yr. See Figure 1.

3.1.1 Using Lists to Find the Rate of Change

Enter the data from example 1 in List1 and List2, then calculate the average rate of change in List3.

Press [STAT], select [4:ClrList].

CLEAR L1, L2, and L3. See Figure 2.
Enter the data in L1 and L2.

Press [STAT], select [1:Edit]. Remember to

press [ENTER] after every entry.

See Figure 3.
Now you can use the list position to calculate the average rate of change.
Type the following list equation:

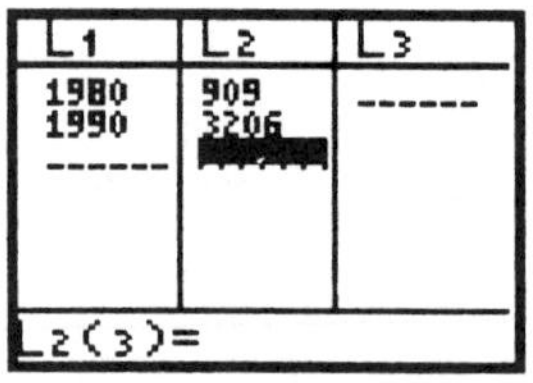

$$\frac{(L2(2) - L2(1))}{(L1(2) - L1(1))}$$

Press: ([2nd] [L2] (2) - [2nd] [L2] (1)) /
([2nd] [L1] (2) - [2nd] [L1] (1))

See Figure 4.
While this may seem tiresome to type, the advantage is that you can change the numbers in the list and then recall the rate of change list equation.

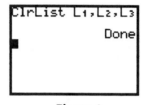

Figure 2

Figure 3

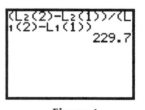

Figure 4

---

[1] The key stroking and menus for the TI-82 and TI-83 are nearly the same. Where they differ, screen images are presented side by side.

Example 2
In 1985 the Federal debt was 1817 billion dollars. Find the average rate of change from 1985 to 1990.

Change L1 and L2 to look like Figure 5.

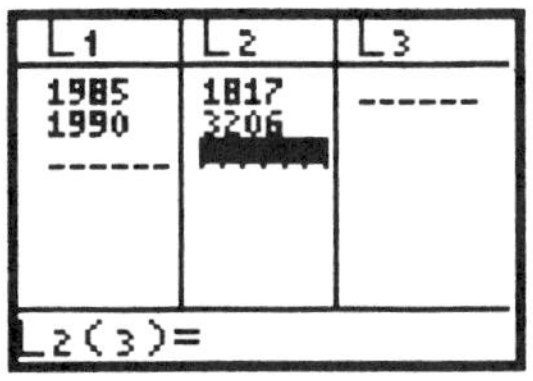

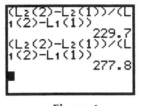

Figure 5

Recall the rate of change equation on the Home Screen.

Press [2nd] [QUIT] [2nd] [ENTRY] [ENTER]

See Figure 6.

Figure 6

Interpretation:
From 1985 to 1990 the Federal debt increase on average $277.8 billion/yr.

3.1.2 Working With Longer Lists

Example 3:
Below is the Federal debt from 1985 to 1990

Year	Billions of $
1985	1817
1986	2120
1987	2346
1988	2601
1989	2868
1990	3206

Calculate the average rate of change for each year.
Enter the above data into L1 and L2.

Press [STAT] [1:Edit]. See Figure 7

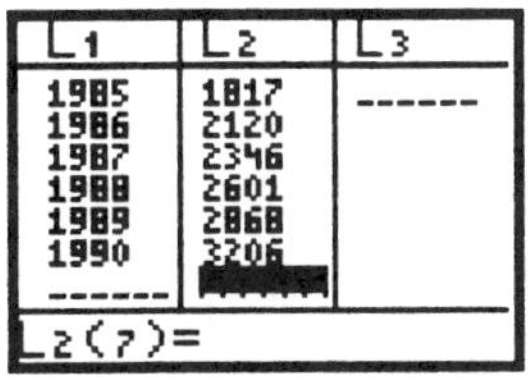

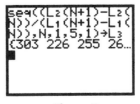

Figure 7

Use a shortcut technique to clear lists: [Δ] to L1 [CLEAR] [∇] [▷] [Δ] to L2 [CLEAR] [∇].

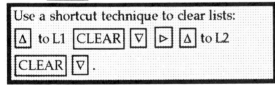

Create the rate of change equation by using the sequence command and store the values to List3.

Figure 8

The rate of change equation for any year would be:

$$\frac{(L2(N+1) - L2(N))}{(L1(N+1) - L1(N)}$$

Press [2nd] [QUIT]. Type the following:

[2nd] [LIST], select [5:seq( ].

Type the commands as in Figure 9 and store to List3.

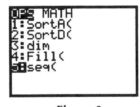

Figure 9

Figure 10 shows the rate of change values stored in L3.

This is a very handy technique for calculating the rate of change for very long lists.

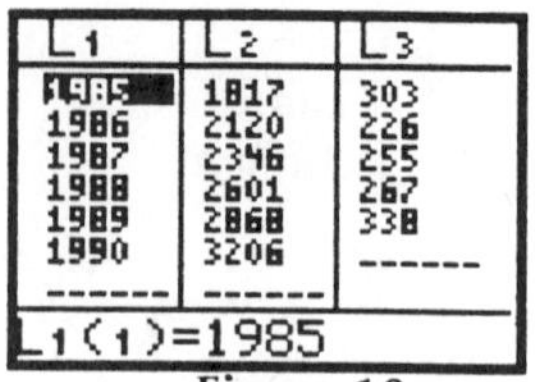

L1 L2 L3
1985 1817 303
1986 2120 226
1987 2346 255
1988 2601 267
1989 2868 338
1990 3206
L1(1)=1985

Figure 10

> **Trouble Shooting:**
>
> The sequence command generates a list of values by evaluating the *expression*, in terms of a *variable*, from *begin* value to *end* value by an *increment* value. Thus the command must include the following:
>
> seq(*expression*,*variable*,*begin*, *end* , *increment*)

3.2 Special TI-83 Techniques for Finding the Average Rate of Change of Lists

The TI-83 has a short cut method for finding the rate of change of a list.

Press [2nd] [LIST] [▷] to <OPS> , select [7:ΔList( ] . See Figure 11.

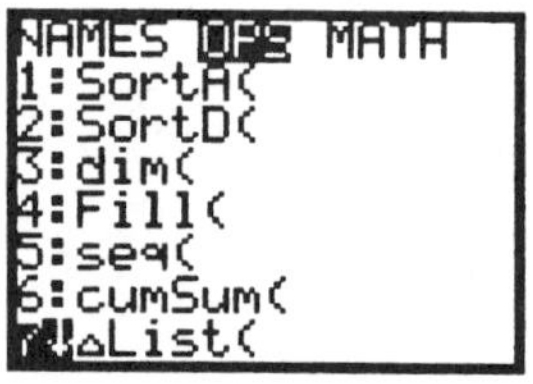

NAMES OPS MATH
1:SortA(
2:SortD(
3:dim(
4:Fill(
5:seq(
6:cumSum(
7:ΔList(

Figure 11 TI-83 Δlist

> ΔList(*list*) generates a list containing the differences between consecutive elements in a named *list* .

ΔList(L2)/ΔList(
L1)→L3
{303 226 255 26...

Figure 12 TI-83 Rate of Change

Calculate the average rate of change, which is the difference of elements in L2 divided by the difference of elements in L1, then store it to L3..
See Figure 12.

Press [STAT] [1:Edit] , to see all three lists.

L1 L2 L3
1985 1817 303
1986 2120 226
1987 2346 255
1988 2601 267
1989 2868 338
1990 3206
L3 ={303,226,255...

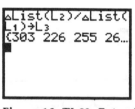

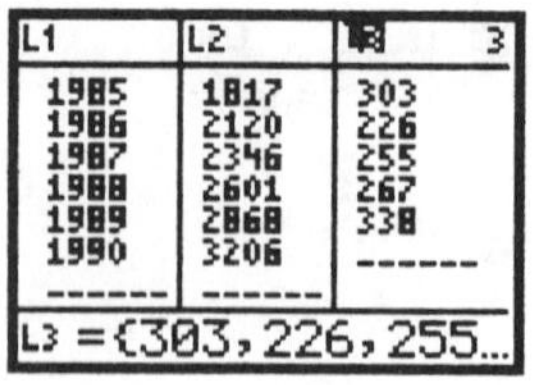

Figure 13 TI -83 Lists

Appendix Section 4
Graphing Linear Functions on the TI-82 or TI-83[1].
Graphing Calculator Techniques for Chapter 4

4.1   Linear Functions
4.1.1   Definition of a Linear Function
The relationship between the variables x and y is said to be linear if for any set of points (x,y) the rate of change of y with respect to x is constant. The rate of change is called **the slope of the line.** Any collection of points (x,y) that have a linear relationship will satisfy an equation of the form $y = mx + b$ where **m** is the rate of change of y with respect to x ($m = slope$).
b is the **y-intercept** (value of y when x = 0).
$y = mx + b$ is called the **slope intercept form.**

An example of a linear equation (function) is :
$$y = 5x + 4.$$
The slope is: $m = 5$ and the y-intercept is: $b = 4.$

If needed, refer back to Appendix Section 2.2 to review the introduction to graphing.

Example 1:
Plot the following points on an xy coordinate plane and determine if they line on a straight line.
(0,2), (1,5), (2,8), (3,11), (5,17), and (7,23)
Organize the data as in Table 1

x	0	1	2	3	5	7
y	2	5	8	11	17	23

Table 1

4.1.2   Plotting Points
You can use the graphing calculator to display the data in list form:
First CLEAR any data in List1 (L1) and List2 (L2)

Press STAT , select [4:ClrList].

Press 2nd L1 , 2nd L2 See Figure 1.

To enter the data points press STAT , select [1:Edit] enter the x- coordinate in L1 pressing ENTER after each entry. Right

arrow ▷ to L2 and enter the y-coordinate data. See Figure 2.

Press 2nd STATPLOT , select [1:Plot 1] and setup the plot as in Figure 3 or 4.

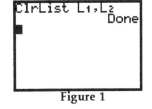
Figure 1

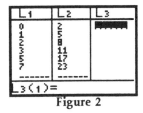
Figure 2

---

[1] The key stroking and menus for the TI-82 and TI-83 are nearly the same. Where they differ screens are presented side by side.

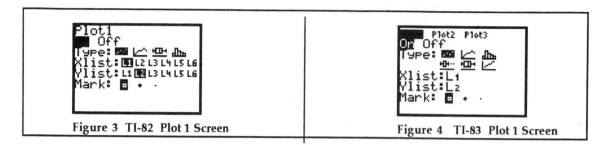

Figure 3 TI-82 Plot 1 Screen

Figure 4   TI-83 Plot 1 Screen

To see the plot:

Press select ZOOM , select [9:ZoomStat].

See Figure 5

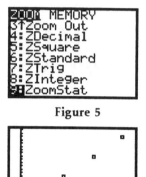

Figure 5

The points appear to lie on a straight line.
See Figure 6.

4.1.3   The Algebraic Representation.
We see from the data that when x = 0 y = 2,
so **b = 2**. As x changes one unit, y changes 3
units. So the rate of change is 3 units, or
m = 3. The linear function in slope intercept
form y = mx + b is :

$$y = 3x + 2$$

Figure 6

Enter this function into Y1.

Press Y =   CLEAR   to clear any old functions.

Press 3 X,T,Θ + 2. See Figure 6 or 7.

Figure 7 TI-82 [Y=]Screen

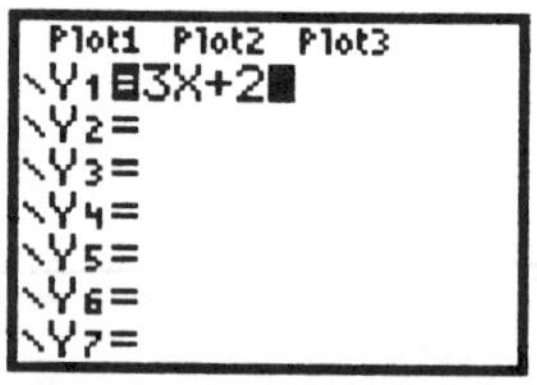

Figure 8 TI-83 [Y=]Screen

Press GRAPH   TRACE

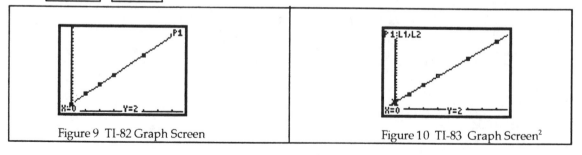

Figure 9 TI-82 Graph Screen

Figure 10 TI-83 Graph Screen[2]

---

[2] The TI-83 graph screen shows the name of the lists being plotted and the equation. From now on only the TI-82 graph screen will be shown.

Verify that the data points are the same as the points on the graph.

Press 2nd TBLSET . Set as in Figure 11.

Figure 11

Compare the function values of x and y to the data points.

Press 2nd TABLE

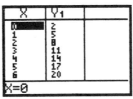

Figure 12

Figure 12 and Table 1 above share the same values. The difference is that the algebraic model can be used to find other values of y when x is given

Use ▽ to find x = 15. We see y = 47

Algebraically:
$$y = 3(15) + 2 = 47$$

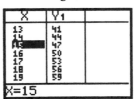

Figure 13

Use △ to find x = - 10 .We see y = -28

Algebraically:
$$y = 3(-10) + 2 = -28$$

See Figures 13 and 14.

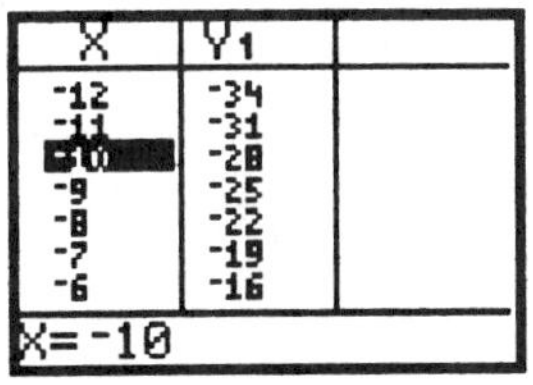

4.2 Graphing Linear Functions
4.2.1 The ZOOM Menu

The graphing calculator has a default viewing window. This window extends in an X direction from -10 to 10 and in a Y direction from -10 to 10.

This is the "standard" window. To set this window quickly:

Press ZOOM 6 . See Figure 15.

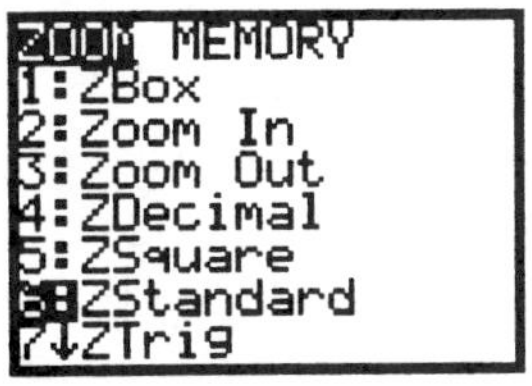

Your old function and plot will show up

Press WINDOW to verify the setting.

See Figure 16.

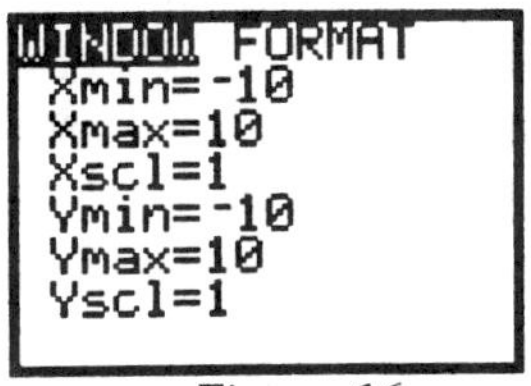

Figure 14

Figure 15

Figure 16

Trouble Shooting:
Before we continue graphing turn OFF all plots. Press 2nd STATPLOT 4 ENTER . After you press ENTER the calculator will tell you *Done*..

Example 2:
Graph the functions:
$Y_1 = x + 5$
$Y_2 = -2x - 3$

Press $\boxed{Y=}$. Use arrows and $\boxed{CLEAR}$ all the old entries. Type the functions.
See Figure 17.

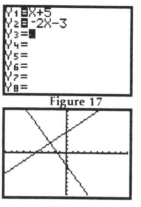

Figure 17

Press $\boxed{GRAPH}$. Your graphs should be the same as Figure 18.

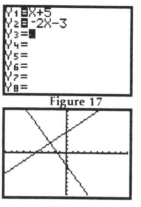

Figure 18

Press [TRACE] and $\boxed{\triangleright}$ $\boxed{\triangleleft}$ to move along the function. Notice X is being incremented in rather ugly decimal values. This is NOT a "friendly" window.

Use $\boxed{\nabla}$ and $\boxed{\Delta}$ to switch between graphs.

Change the window.
Press $\boxed{ZOOM}$ $\boxed{8}$ the "Integer" window.

ZOOM MEMORY
3↑Zoom Out
4:ZDecimal
5:ZSquare
6:ZStandard
7:ZTrig
8▉ZInteger
9:ZoomStat

Figure 19

Trouble Shooting:
You are now given an opportunity to reposition your cursor, if you would like, before the window changes.
Press $\boxed{ENTER}$ to activate the change.

If you didn't move your cursor, your graph will look like Figure 20. These are the graphs of the same functions but the window is now much larger. Press $\boxed{TRACE}$ and notice that X changes by one integer now.

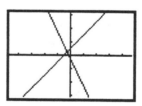

Figure 20

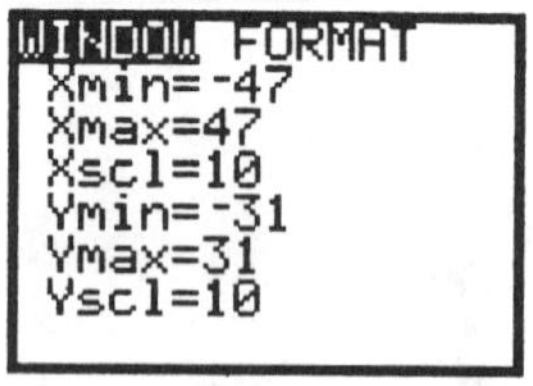

Press $\boxed{WINDOW}$. See Figure 21.
Calculate $X_{max} - X_{min}$. The distance is exactly 94, the same as the number of pixel jumps across the screen. This is a "friendly" window. Refer back to section 2.2.4 for more on *friendly* windows.

WINDOW FORMAT
Xmin=-47
Xmax=47
Xscl=10
Ymin=-31
Ymax=31
Yscl=10

Figure 21 Integer Window

Press $\boxed{ZOOM}$ $\boxed{4}$. This is the "decimal" window. When you press [TRACE] and move the arrows, X now increments by .1, or one decimal point. This is another "friendly" window. See Figure 22.

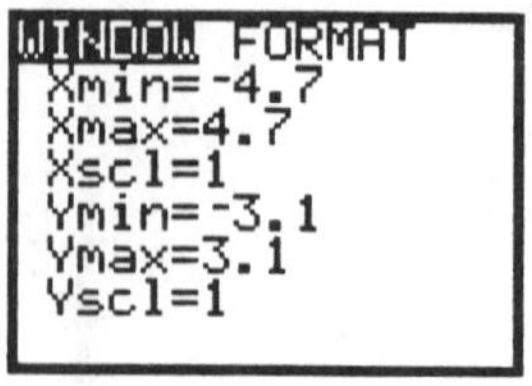

WINDOW FORMAT
Xmin=-4.7
Xmax=4.7
Xscl=1
Ymin=-3.1
Ymax=3.1
Yscl=1

Figure 22 Decimal Window

4.2.2 Two Points Determine a Line

If two points (x1,y1) and (x2,y2) are known, we have enough information to determine the slope or average rate of change using the slope formula

$$m = slope = \frac{(y2) - (y1)}{(x2) - (x1)}.$$

Once the slope is known, we can use $y=mx+b$ to find the equation of the line.

Example 2:

If two points on the graph of a linear equation are (5,-1) and (-15,-13), determine the linear equation associated with these points.

1. Find the slope.

$$m = \frac{y_2 - y_1}{x_2 - x_1} = \frac{(-13) - (-1)}{(-15) - (+5)} = \frac{-12}{-20} = \frac{3}{5}$$

2. Substitute $m = 3/5$ into $y = mx + b$
$$y = (3/5)x + b$$
3. Pick one of the point for values of x and y. (5,-1) means x = 5, and y = -1. Substitute these values into $y = mx + b$ and solve for b

$$-1 = (3/5)(5) + b$$
$$-1 = 3 + b$$
$$b = -4$$

Write the linear function. The linear function through the points (5,-1) and (15,13) is :

$$y = (3/5) x - 4$$

4.2.3 Use the Calculator to Verify the Equation of a Line Between Two Points with a Linear Regression Equation

CLEAR L1 and L2. Use a short cut method:

Press $\boxed{STAT}$ [1:Edit] Use $\boxed{\Delta}$ to position the cursor on the word [L1] press $\boxed{\nabla}$ $\boxed{\triangleright}$.

Repeat for L2 by referring to instructions for clearing lists above.

Example 3:

Find the linear equation that goes through the points (5,-1) and (-15,-13).

Enter the points into L1 and L2 (see Figure 23) then find the linear regression equation.

Press $\boxed{STAT}$ $\boxed{\triangleright}$ to <CALC> , select [5:LinReg(ax+b)]. Press $\boxed{ENTER}$

See Figure 24.

Note: The calculator uses "a" instead of "m". Both expressions represent a linear function: $y = mx + b = ax + b$

In Figure 25 we see the values for the linear equation: a=m=.6 and b= -4.

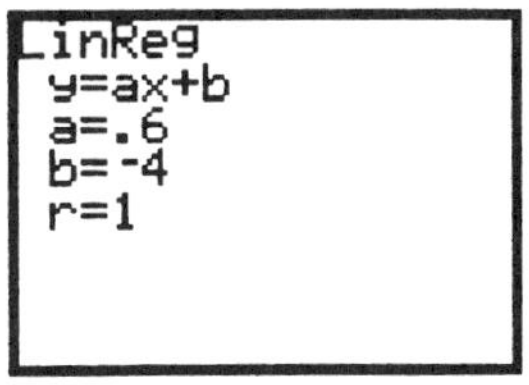

$y = .6x - 4$ is the same as the linear function we determined above:

$$y = (3/5)x - 4 = .6x - 4$$

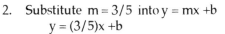

Figure 23

Figure 24

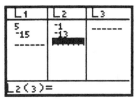

Figure 25

The ratio of y to x is constant (m) for every point on the graph line.

4.3   Special lines

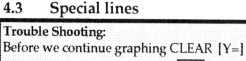

Trouble Shooting:
Before we continue graphing CLEAR [Y=] and Turn OFF all plots. Press 2nd
STATPLOT 4 ENTER . After you press ENTER the calculator will tell you *Done*..

4.3.1   Horizontal Lines
Horizontal lines are parallel to the x-axis and have a slope = 0. For this case, if $m = 0$ in the general equation $y = mx + b$, the equation becomes $y = 0x + b$ or $y = b$.
Examples would be:

Y1 = -3
Y2= 3
Y3 = 2
Y4 = -1

See Figures 26 and 27.

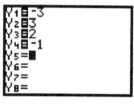

Figure 26

Figure 27

4.3.2   Vertical Lines
Vertical lines are of the form $x = c$ or $x = $ a constant. y can be any number.
Examples would be :

x = 3
x = -4
x = 1

The calculator can draw a vertical line using the DRAW menu. (Since these are NOT functions the calculator must draw them, rather than graphing them)
First CLEAR [Y=] of all functions.

Press ZOOM select [4: ZDecimal]

Press 2nd DRAW select [4:Vertical] use

▷ to position the cursor on x = 3. Press

ENTER .Use ◁   to reposition on x = 1 .

Press ENTER . Use ◁ to reposition on x = -4

press ENTER . You should see three vertical lines as in Figure 29.

4.3.3   Proportional Relationships
The graph of lines which pass through the origin (0,0) have a y-intercept = 0.
For this case, b = 0 or the equation is simply : $y = mx$ or $y/x = m$

Figure 28

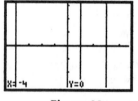

Figure 29

Examples of proportional lines are:
        Y1 = 3x
        Y2= (2/3)x
        Y3= -4 x
See Figure 30

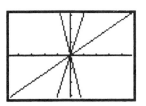

Figure 30

4.3.4 Parallel lines
Parallel lines have the same slope.
Examples:
        Y1= 3x
        Y2= 3x +4
        Y3 = 3x -5
        Y4 = 3x +6,.

Press $\boxed{Y=}$ and enter the above functions .
See Figure 31.

Press $\boxed{ZOOM}$ select [6:ZStandard].

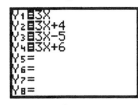

Figure 31

Looking at the graphs of the lines
associated with the above equations you
should note that they are all parallel but
that they have different y-intercepts.

Press $\boxed{TRACE}$ then $\boxed{\nabla}$ to see the intercepts.
See Figure 32.

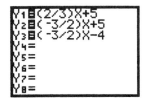

Figure 32

4.3.5 Perpendicular Lines
If the graph of the equation, $y = (2/3)x +5$ is
rotated 90 degrees, the resulting graph line
will have a slope of -3/2. In general, if the
slope of a line is m_1, the slope of a
perpendicular line, m_2, is $-1/m_1$ (the
negative reciprocal of m_1).

> Note: The product of the slopes m_1 and m_2
> will always equal -1.
> If $m_1 m_2 = -1$ then line1 is perpendicular to
> line2.

Examples:
        Y1 = (2/3)x +5
        Y2= (-3/2)x + 5
        Y3= (-3/2)x - 4

Figure 33

Enter the above functions into $\boxed{Y=}$, press

$\boxed{GRAPH}$. We see the graphs in Figure 34.
Due to the scaling the graphs do not look
perpendicular.
To adjust the window so that x and y scaling

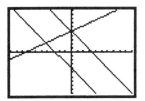

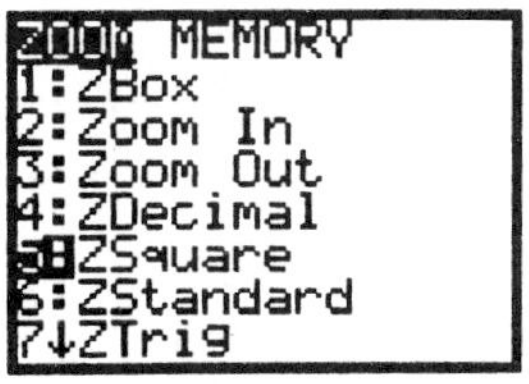

Figure 34

are proportional press $\boxed{ZOOM}$, select

[5:ZSquare] as in Figure 35. Now the
equations look perpendicular. See Figure 36.

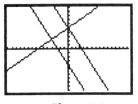

Figure 35

Figure 36

Note: Both Y2 and Y3 are perpendicular to Y1. Y2 is parallel to Y3 because their slopes are equal.

4.3.5 Entering Linear Equations NOT in Function Form

The equation y = mx +b is called the slope-intercept form of a linear equation, or the function form, since the equation is solved for y. A linear relationship may also be described by an equation in the standard form:

Ax + By = C.

Example 4:
Transform the equation **-2x + 3y = 15** into function form so that you can enter it into your calculator and graph it.

$$-2x + 3y = 15$$
$$3y = 2x + 15 \quad \text{add } 2x$$
$$y = (2/3)x + 15/3 \quad \text{divide by } 3$$
$$y = (2/3)x + 5 \quad \text{simplify}$$

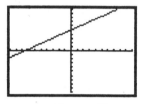

Figure 37

-2x + 3y = 15 is equivalent to
y = (2/3)x + 5 which we graphed above.
See Figure 37, then refer back to Figure 34 .

4.3.5 Fractional Coefficients of Linear Functions

Notice that in all of the examples above when the coefficient of x was a fraction, we enclosed the fraction in parenthesis. Let us see what happens if you do not . The TI-82 and TI-83 behave differently.

Enter the function y=1/2x.

Press ZOOM 4

The TI-82 interprets this as y=1/(2x), while the TI-83 interprets the function as y=(1/2)(x)

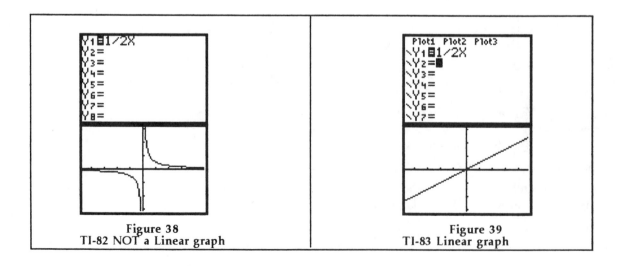

Figure 38
TI-82 NOT a Linear graph

Figure 39
TI-83 Linear graph

Appendix Section 5
Linear Regression Equations on the TI-82 or TI-83[1]
Graphing Calculator Techniques for Chapter 5

5.1   Linking Calculators

This course comes with programs that store data to the graphing calculator. Once the data is stored to lists you can create plots and find regression equations on the data. Your instructor will be able to download the programs from a computer to a calculator. Your calculator came with a cable that allows you to link calculators. Once linked you can receive and send programs as well as data lists.

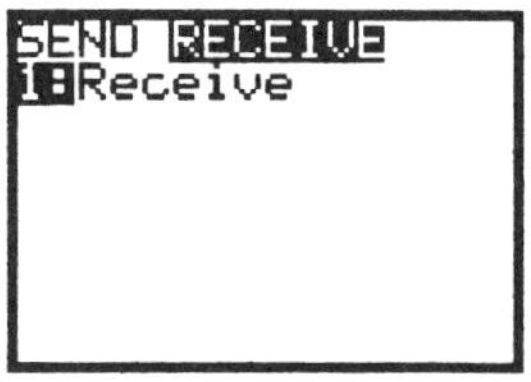

5.1.1   Receiving data:

1. Attach the cable to both calculators. Be sure to push the cable ALL the way in.

2. Press 2nd LINK ▷ to <RECEIVE>.

    Press ENTER . See Figure 1.

3. The receiving calculator must say *Waiting...*

5.1.2   Sending data:

1. Press 2nd LINK , select [3:Select Current...].

2. ∇ to the programs or lists to be sent.

    Press ENTER to select. A small square indicates the selection has been made. See Figure 2.

3. ▷ to <TRANSMIT> press ENTER .
    See Figure 3.

4. Wait for the message *Done..* on the receiving calculator. See Figure 4.

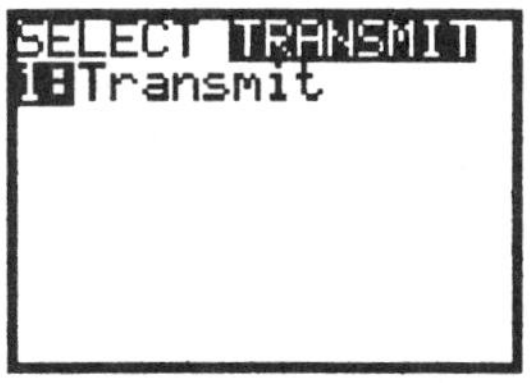

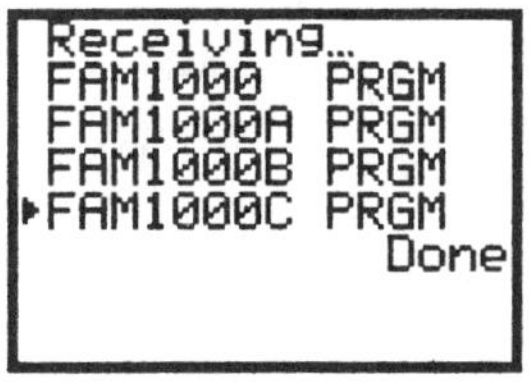

5.1.3   Running a Program

Press PRGM ,then select the program. Press

ENTER . See Figure 5 & 6.

The program has stored the FAM1000 data to List1-List6.

Press STAT select [1:Edit] to see the data in the lists. See Figure 7.

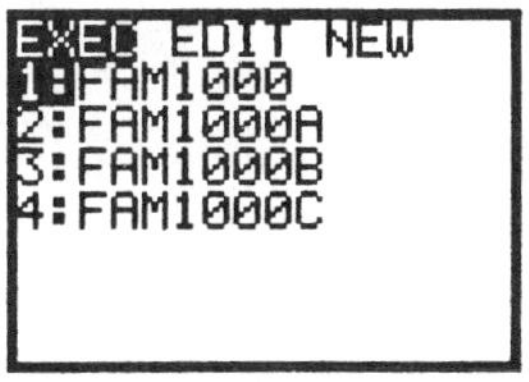

Figure 1

Figure 2

Figure 3

Figure 4

Figure 5

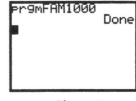

Figure 6

---

[1] The key stroking and menus for the TI-82 and TI-83 are nearly the same. Where they differ, screen images are presented side by side

To see the program press PRGM ▷ to
<EDIT>. Select the program to view. Press
ENTER .

The program will look like Figure 9. The
complete program is below:[2]

```
PROGRAM:FAM1000
ClrList L₁,L₂,L₃,L₄,L₅,L₆
{0,1,2,3,5,6,7,8,9,10,11,12,13,14
,15,16,18,20}→L₁
{12200,15000,11134,8333,8583,1504
1,17112,12094,13153,14321,13230,1
8948,19059,20995,19611,28388,3833
1,50966}→L₂
{0,1,2,5,6,7,8,9,10,11,12,13,14,1
5,16,18,20}→L₃
{10900,15000,11134,18700,17657,20
017,14716,15389,17940,13955,22844
,21092,25184,22415,34279,44429,54
255}→L₄
{0,3,5,6,7,8,9,10,11,12,13,14,15,
16,18,20}→L₅
{13500,8333,3525,8500,11300,6850,
10358,7987,11861,15095,16270,1788
0,15581,21874,32233,32333}→L₆
```

Trouble Shooting:
If you get lost and don't know which menu
you are in, press 2nd QUIT and start over

5.2   Linear Regression Equations.
The FAM1000 program stores in L1 the
Years of Education and in L2 the Mean

Personal Income. Plot the data. Press 2nd

STATPLOT ENTER .

5.2.1   Set up the plot:

1.   Select ON ENTER .

2.   Select Type: choose the scatterplot
      icon, ENTER .

3.   Select Xlist: L1 ENTER .

4.   Select Ylist: L2 ENTER .

5.   Select Mark: □

Trouble Shooting:
CLEAR Y= and turn OFF all other plots.

5.2.2   Draw A Scatter Plot
Press ZOOM , select [9:ZoomStat]. See
Figure 11 & 12.

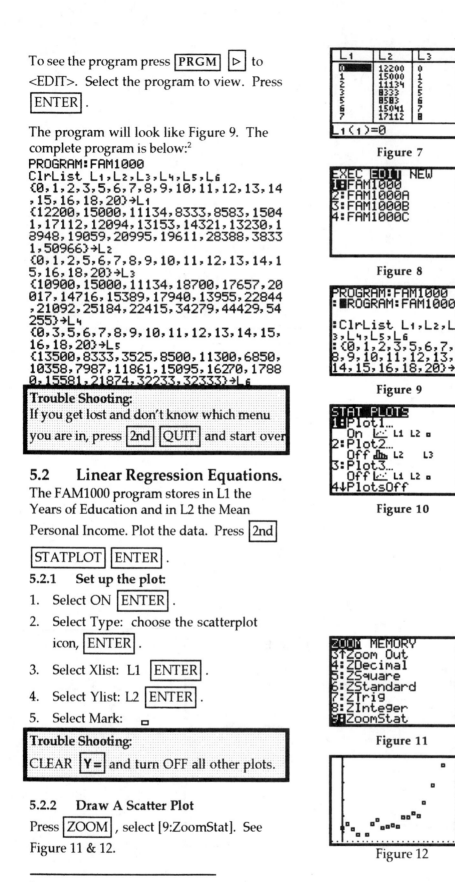

Figure 7

Figure 8

Figure 9

Figure 10

Figure 11

Figure 12

---

[2] At the end of this section you will find the
complete FAM 1000 data and list headings.

5.2.3 Find the Equation

1. Press $\boxed{\text{STAT}}$ $\boxed{\triangleright}$ to <CALC>, select

   [LinReg(ax+b] $\boxed{\text{2nd}}$ $\boxed{\text{L1}}$ $\boxed{,}$ $\boxed{\text{2nd}}$ $\boxed{\text{L2}}$ $\boxed{\text{ENTER}}$

   See Figure 14 or 15, and 13.

2. Enter the equation in Y1. For the TI-82

   press $\boxed{\text{Y=}}$ $\boxed{\text{VARS}}$, select [5:statistics] ,

   $\boxed{\triangleright}$ $\boxed{\triangleright}$ to <EQ>, select [7:RegEQ].

   See Figure 16.

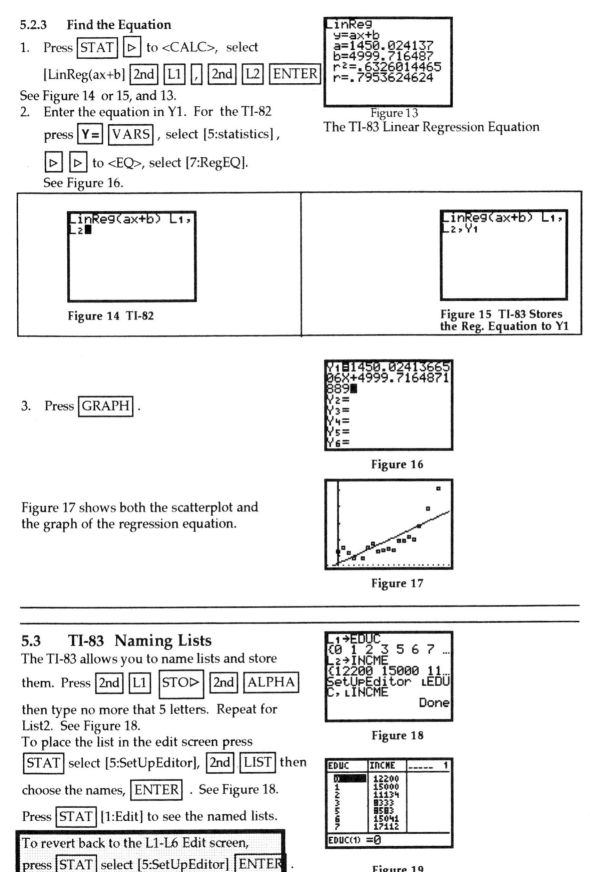

Figure 13
The TI-83 Linear Regression Equation

Figure 14 TI-82

Figure 15 TI-83 Stores
the Reg. Equation to Y1

3. Press $\boxed{\text{GRAPH}}$.

Figure 16

Figure 17 shows both the scatterplot and
the graph of the regression equation.

Figure 17

5.3 TI-83 Naming Lists

The TI-83 allows you to name lists and store

them. Press $\boxed{\text{2nd}}$ $\boxed{\text{L1}}$ $\boxed{\text{STO}\triangleright}$ $\boxed{\text{2nd}}$ $\boxed{\text{ALPHA}}$

then type no more that 5 letters. Repeat for
List2. See Figure 18.

To place the list in the edit screen press

$\boxed{\text{STAT}}$ select [5:SetUpEditor], $\boxed{\text{2nd}}$ $\boxed{\text{LIST}}$ then

choose the names, $\boxed{\text{ENTER}}$. See Figure 18.

Press $\boxed{\text{STAT}}$ [1:Edit] to see the named lists.

To revert back to the L1-L6 Edit screen,
press $\boxed{\text{STAT}}$ select [5:SetUpEditor] $\boxed{\text{ENTER}}$.

Figure 18

Figure 19

The following tables contain information from the FAM 1000 data set described in Chapter 5.

MEAN PERSONAL WAGES
MEN AND WOMEN
Graphing Calculator Program (FAM 1000.82P)

Years of Education L1	Mean P-Wages ($) L2	Years of Education L3	Mean P-Wages ($) Men L4	Years of Education L5	Mean P-Wages ($) Women L6
0	12200	0	10900	0	13500
1	15000	1	15000	3	8333
2	11134	2	11134	5	3525
3	8333	5	18700	6	8500
5	8583	6	17657	7	11300
6	15041	7	20017	8	6850
7	17112	8	14716	9	10358
8	12094	9	15389	10	7987
9	13153	10	17940	11	11861
10	14321	11	13955	12	15095
11	13230	12	22844	13	16270
12	18948	13	21092	14	17880
13	19059	14	25184	15	15581
14	20995	15	22415	16	21874
15	19611	16	34279	18	32233
16	28388	18	44429	20	32333
18	38331	20	54255		
20	50966				

MEAN PERSONAL TOTAL INCOME
MEN AND WOMEN
Graphing Calculator Program (FAM 1000A.82P)

Years of Education L1	Mean P-Total Income ($) L2	Years of Education L3	Mean P-Total Income ($) All Men L2	Years of Education L5	Mean P-Total Income ($) All Individuals L2
0	13500.0	0	10900.0	0	12200.0
3	8333.3	1	15215.0	1	15215.0
5	6078.3	2	11134.0	2	11134.0
6	9046.5	5	18750.0	3	8333.3
7	11606.0	6	19056.2	5	10302.2
8	6850.0	7	20056.3	6	16196.3
9	11118.4	8	14792.3	7	17239.5
10	8740.6	9	16793.4	8	12144.9
11	11909.4	10	18713.8	9	14271.2
12	15850.3	11	14009.4	10	15087.2
13	18039.9	12	24230.3	11	13282.5
14	19748.4	13	22498.3	12	20016.7
15	21877.8	14	26302.7	13	20618.2
16	22940.4	15	23828.4	14	22543.6
18	34540.3	16	36519.4	15	23028.2
20	32833.3	18	46434.9	16	30071.2
		20	59370.2	18	40487.6
				20	55389.7

MEAN PERSONAL TOTAL INCOME
WHITE AND NON-WHITE MEN
Graphing Calculator Program (FAM 1000B.82P)

Years of Education L1	Mean Personal Total Income ($) All Non White Men L2	Years of Education L3	Mean Personal Total Income ($) All White Men L4	Years of Education L5	Mean Personal Total Income ($) All Men L6
6	12750.0	0	10900.0	0	10900.0
7	19000.0	1	15215.0	1	15215.0
8	18505.0	2	11134.0	2	11134.0
10	29850.0	5	18750.0	5	18750.0
11	12784.5	6	20632.7	6	19056.2
12	16312.4	7	20408.3	7	20056.3
13	22371.7	8	12936.0	8	14792.3
14	27928.4	9	16793.4	9	16793.4
15	9366.7	10	16857.8	10	18713.8
16	31478.0	11	14386.2	11	14009.4
18	33586.3	12	25354.1	12	24230.3
		13	22506.7	13	22498.3
		14	26042.6	14	26302.7
		15	25997.6	15	23828.4
		16	36925.0	16	36519.4
		18	48669.4	18	46434.9
		20	59370.2	20	59370.2

MEAN PERSONAL TOTAL INCOME
WHITE AND NON-WHITE WOMEN
Graphing Calculator Program (FAM 1000C.82P)

Years of Education L1	Mean Personal Total Income ($) All Non White Women L2	Years of Education L3	Mean Personal Total Income ($) All White Women L4	Years of Education L5	Mean Personal Total Income ($) All Women L6
6	6908.00	0	13500.00	0	13500.00
7	16000.00	3	8333.33	3	8333.33
9	15750.00	5	6078.25	5	6078.25
10	18304.30	6	9759.33	6	9046.50
12	12758.60	7	7212.00	7	11606.00
13	12512.00	8	6850.00	8	6850.00
14	17140.60	9	9574.50	9	11118.40
15	19500.00	10	6533.54	10	8740.56
16	24201.40	11	11909.40	11	11909.40
18	25042.00	12	16497.80	12	15850.30
		13	18753.10	13	18039.90
		14	20005.50	14	19748.40
		15	22036.30	15	21877.80
		16	22791.00	16	22940.40
		18	34905.60	18	34540.30
		20	32833.30	20	32833.30

Appendix Section 6
Intersecting Lines on the TI-82 or TI-83[1]
Graphing Calculator Techniques for Chapter 6

6.1 Intersecting Lines
Lines that cross each other are said to
intersect. All lines eventually intersect
unless they are parallel lines or are the
same line.
Example 1:
Graph the two functions below and
determine the point of intersection, if it
exists.
$Y1 = 3x + 4$ and $Y2 = 0.5x - 1$

Press $\boxed{Y=}$ then $\boxed{CLEAR}$ all the old

functions. Enter the above equations into Y1
and Y2. See Figure 1.

Press $\boxed{ZOOM}$, select [6: ZStandard] . Press

$\boxed{TRACE}$, then use $\boxed{\triangleleft}$ or $\boxed{\triangleright}$ to estimate the

point of intersection. It looks like the
graphs cross when x is about -2 and y is
about -2. See Figure 2.
6.1.1 The Calculate Menu
Use the calculator to find the point of
intersection.

Press $\boxed{2nd}$ $\boxed{CALC}$, select [5:intersect].

See Figure 3.

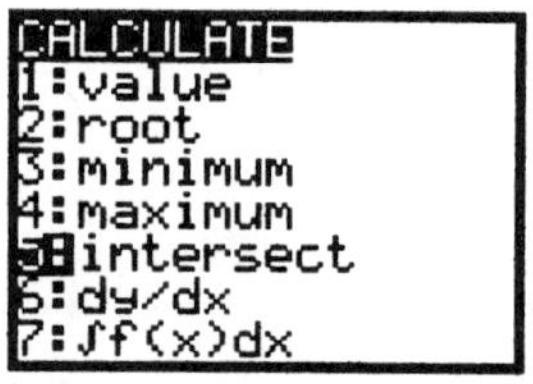

The calculator prompts you:

1. Select the first curve press $\boxed{ENTER}$.

2. Select the second curve press $\boxed{ENTER}$.

3. Using the arrows move the cursor near
   the point of intersection (your guess).

   Press $\boxed{ENTER}$. See Figure 4 & 5.

The point of intersection is (-2,-2).
6.1.2 Algebraic Verification
Check algebraically:
If x = -2 then
$Y1 = 3x + 4$ becomes $Y1 = 3(-2) + 4 = -2$
$Y2 = 0.5x - 1$ becomes $Y2 = 0.5(-2) - 1 = -2$

Both equations have the same y values so
the solution is x = -2 and y=-2 or the point
of intersection is (-2,-2).

Figure 1

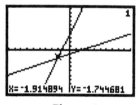

Figure 2

Figure 3

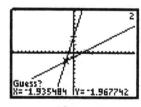

Figure 4

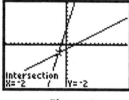

Figure 5

---

[1] The key stroking and menus for the TI-82 and TI-83 are nearly the same. Where they differ, screen images are
presented side by side

6.1.3   Solving Algebraically
Since Y1=Y2 at the point of intersection, you solve algebraically by substitution:

$$3x + 4 = .5x - 1$$
$$2.5x + 4 = -1 \qquad \text{add } -.5x$$
$$2.5x = -5 \qquad \text{add } -4$$
$$x = -5/2.5 = -2 \qquad \text{divide by } 2.5$$

Substitute x=-2 in Y1 and Y2 as above to find y = -2. The solution is (-2,-2).

6.1.4   Adjusting the Window
Example 2:
Find the point of intersection for the system of equations :

$$Y1 = .025x - 25 \quad \text{and}$$
$$Y2 = -2x + 50$$

Enter the expressions into Y1 and Y2. Press GRAPH . See Figures 6 & 7.

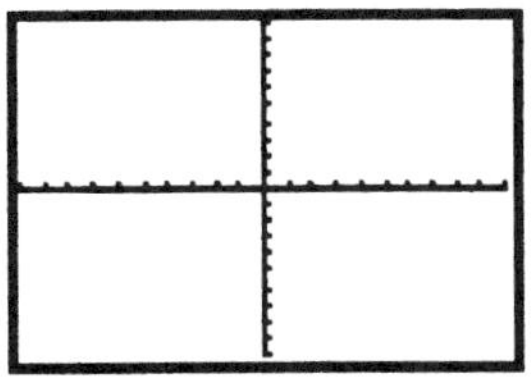

The graphs do not appear! Where are they? Press TRACE it will give you points on the graph. See Figure 8. Using algebra we know that the y-intercepts for the graphs are (0, -25) and (0,50). To see these points we must choose Ymin and Ymax beyond those points. Reset the WINDOW to Figure 9. Now we see that the graphs are getting closer. The graphs are headed to the right.

Press 2nd TABLE for more information. Y1 is climbing very slowly (m=.025) while Y2 is decreasing and some where around x = 37 they are about equal. See Figure 11.

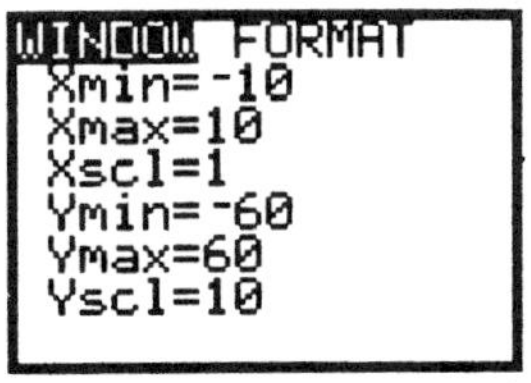

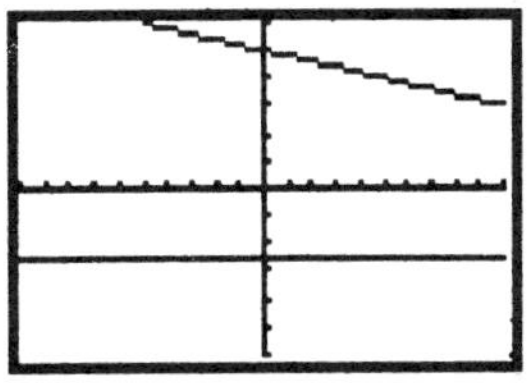

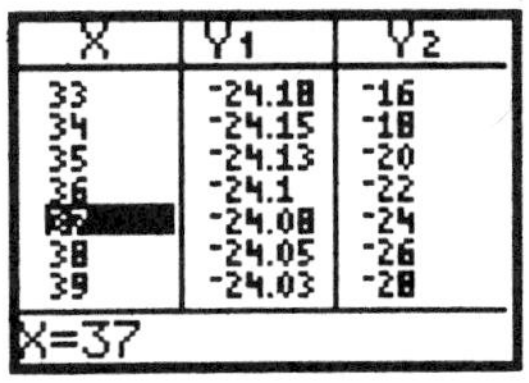

Set your WINDOW so that Xmin=-10 and Xmax=50. Press GRAPH and repeat the CALC Menu steps in 6.1.1. See Figure 12.

Solve Algebraically:
Let Y1=Y2:

$$.025x - 25 = -2x + 50$$
$$2.025x - 25 = 50$$
$$2.025x = 75$$
$$x = 75/2.025 = 37.0370370...$$

$$Y1 = .025(75/2.025) - 25 = -24.07407407...$$
$$Y2 = -2(75/2.025) + 50 = -24.07407407...$$

While the algebra is pretty fast the arithmetic could still use some help from a calculator!

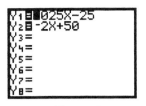

Figure 6

Figure 7

Figure 8

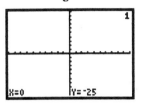

Figure 9

Figure 10

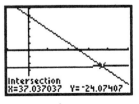

Figure 11

Figure 12

6.2   Piecewise Functions

The graphing calculator can graph piecewise functions very easily. However it is important to understand how the graphing calculator performs a test.

6.2.1   The TEST Menu

The graphing calculator can tell if a statement is true or false using the TEST menu. Notice that you find the equal and inequality symbols here. Press $\boxed{\text{2nd}}$ $\boxed{\text{TEST}}$. See Figure 13.

Go to the Home Screen. Press $\boxed{\text{2nd}}$ $\boxed{\text{QUIT}}$.

Example 3:
Determine if the following are True or False by typing the following:

a. $5 = 5$
b. $5 \le 7$
c. $5 = 4$
d. $5 \ge 7$

See Figures 14 and 15.

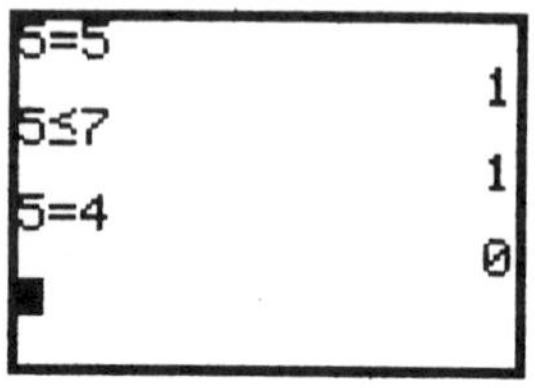

> The calculator is performing a test. It tells you **1** for **True** and **0** for **False**.

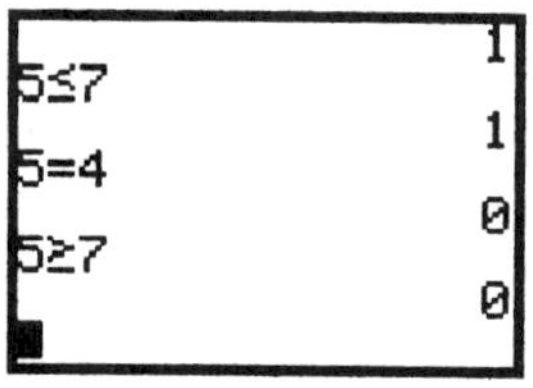

This method can be used to enter a piecewise function (a function defined in pieces) into the calculator.

Example 4:
Graph the piecewise function:

$$y = \begin{cases} x + 2 \text{ for } x > 0 \\ -x - 3 \text{ for } x \le 0 \end{cases}$$

y is defined under two conditions:

1.   when x>0 and
2.   when x≤ 0.

Press $\boxed{\text{Y} =}$ and enter the equations as in Figure 16 .

Since this is a piecewise function, if x>0 choose the function y1 = x + 2 but if x≤0 choose the function y2= -x - 3.
See Figure 17.

> On the calculator we will use the division sign to enter the condition because the calculator will perform a test putting 1 or 0 in the denominator.
> When the denominator is 0 the function is undefined and no points will be drawn.
> When the denominator is 1 the function is plotted.

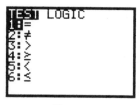

Figure 13

Figure 14

Figure 15

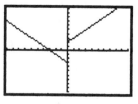

Figure 16

Figure 17

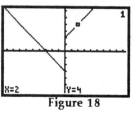

Figure 18

Thus our graph is in two pieces.
When you are on Y1 you are on the graph of
condition one, or y = x+2.

Press TRACE . See Figure 18.

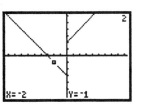

Figure 19

When you are on Y2 you are on the graph of
condition two, or y=-x - 3.

Press ∇ then TRACE to duplicate Figure 19.

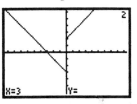

Figure 20

Notice that when the condition no longer
applies, no y value is given. For example if
x=3 condition two no longer applies.
See Figure 20.
To find the value of y for x = 3, you must

switch to condition one. Δ to the graph of

Y1. See Figure 21.

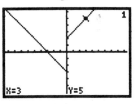

Figure 21

Example 5:
An 8% flat income tax is represented by
f(x). A graduated income tax is represented
by the piecewise function g(x). Graph the
functions.

f(x) = .08x for x ≥ 0

Figure 22

$$g(x) = \begin{cases} 0 & \text{for } 0 \le x \le 20000 \\ .05x(x-20000) & \text{for} \\ & 20000 < x \le 100000 \\ 4000 +.10(x-100000) & \text{for} \\ & x > 100000 \end{cases}$$

Enter the piecewise functions into the
graphing calculator as individual functions
with separate conditions.
Enter the function as shown in Figure 22.
Y1 = .08x /(x≥ 0)
Y2= 0/(x≥0)(x ≤ 20000)
Y3 =. 05(x - 20000) / (x>20000)(x ≤ 100000)
Y4 = 4000+ .1(x-100000) / (x >100000)

Press WINDOW , set as in Figure 23.

Press GRAPH . See Figure 24.

Figure 23

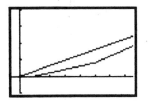

Figure 24

To see the point of intersection change
Xmax to 350000 and Ymax to 30000

Press GRAPH . See Figure 25.

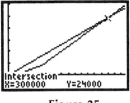

Figure 25

Section 6 Practice Worksheet
Graphing Systems of Linear Equations and Finding Points of Intersection

1) Graph the following two linear equations on your calculator:
    $$y1 = .7x + 15 \quad \text{and} \quad y2 = -.5x - 12,$$
    Graph on the standard viewing screen ([ZOOM] [6]). Then do the following.

    a) Explain why you are having trouble seeing the complete graphs.

    b) Select an appropriate window range so that you can see the x and y-intercepts of both graphs and the point of intersection, if it exists. State your range and scale and hand sketch the graphs in your notebook. Be sure to label the intercepts.

    c) Use trace to estimate the point of intersection.
        Now use the calculator to find the intersection.
            Press: [2nd][CALC][5:intersect]. Select the first curve, press [ENTER].
            Select the second curve, press [ENTER]. Position the cursor near the point of
            intersection, press [ENTER]. What is the intersection?

    d) Use algebra to find the point of intersection. Set y1 = y2, and solve for x. Then substitute back into each equation to find the value of y.

    e) Describe which graph is on top, first to the left and then to the right of the intersection.

2) Graph the following two linear equations from the FAM 1000 data on personal income.
    $$y1 = \text{men's personal wage} \quad = 6325 + 1631 \bullet \text{educ years and}$$
    $$y2 = \text{women's personal wage} = 2\,055 + 1201 \bullet \text{educ years.}$$
    Use your calculator to help with the following:

    a) Graph on the standard viewing screen ([ZOOM] [6]). Explain why you are having trouble seeing the graphs.

    b) Select an appropriate range so that you can see the complete algebraic graph which includes the x and y-intercepts for both graphs, and the point of intersection if it exists. State your range and scale and sketch the graphs. Be sure to label the intercepts.

    c) Use trace to estimate the point of intersection.
        Now use the calculator to find the intersection.

    d) Use algebra to find the point of intersection. Set y1 = y2, and solve for x. Then substitute back into each equation to find the value of y.

    e) Describe which graph is on top, first to the left and then to the right of the intersection.

    f) Now sketch the graphs (above next to the algebraic graph) using values of x that make sense for this real world situation.

Appendix Section 7
Scientific Notation and Exponent Properties on the TI-82 or TI-83[1].   Graphing Calculator Techniques for Chapter 7

7.1   Scientific Notation

Scientific notation expresses large numbers like 3250000000 as 3.25×10^9 and small numbers like 0.00000123586 as 1.23586×10^{-6} , using powers of ten. This can be done on the calculator using either $\boxed{10^x}$, the power of ten key or $\boxed{EE}$, the exponent of ten key.

> Any decimal number can be written in scientific notation using the form $K \times 10^x$, where $1 < K < 10$ and x is an integer. If $x>0$ move x decimal places to the right, if $x<0$ move x decimal places to the left to change back to decimal number form.

Example 1:
Type the following numbers:
- a. 3.25×10^9
- b. 3.25×10^{10}
- c. .00000123589

Press 3.25 $\boxed{2nd}$ $\boxed{10^x}$ 9, then press 3.25 $\boxed{2nd}$ $\boxed{EE}$ 9. Compare the results. See Figure 1&2.

When .00000123589 is typed, it is changed to scientific notation.

> **Trouble Shooting**
> Note: When $x \geq 10$ in 10^x, the number is written in scientific notation.

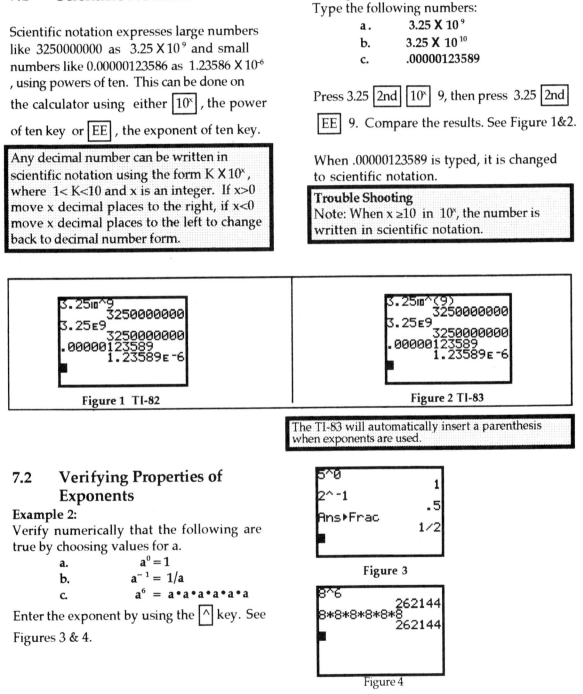

Figure 1 TI-82

Figure 2 TI-83

> The TI-83 will automatically insert a parenthesis when exponents are used.

7.2   Verifying Properties of Exponents

Example 2:
Verify numerically that the following are true by choosing values for a.
- a. $a^0 = 1$
- b. $a^{-1} = 1/a$
- c. $a^6 = a \cdot a \cdot a \cdot a \cdot a \cdot a$

Enter the exponent by using the $\boxed{\wedge}$ key. See Figures 3 & 4.

Figure 3

Figure 4

---

[1] The key stroking and menus for the TI-82 and TI-83 are nearly the same. Where they differ, screen images are presented side by side.

7.2.1   Other Exponent Keys

There are other short cut keys for exponents. The $\boxed{x^2}$ $\boxed{x^{-1}}$ keys are used to paste the exponents without using the $\boxed{\wedge}$ key. See Figure 5.

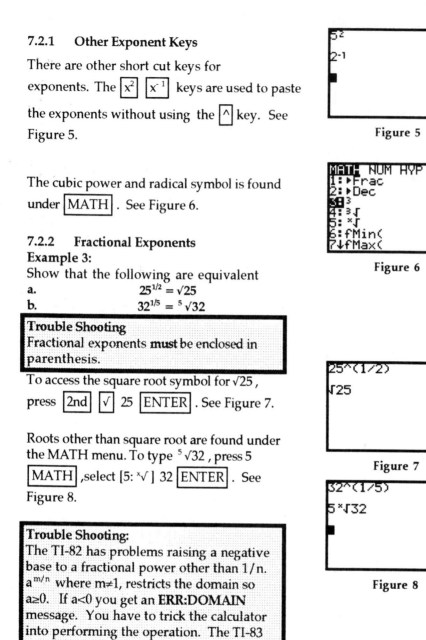

Figure 5

The cubic power and radical symbol is found under $\boxed{\text{MATH}}$. See Figure 6.

Figure 6

7.2.2   Fractional Exponents
Example 3:
Show that the following are equivalent
a. $25^{1/2} = \sqrt{25}$
b. $32^{1/5} = {}^5\sqrt{32}$

Trouble Shooting
Fractional exponents **must** be enclosed in parenthesis.

To access the square root symbol for $\sqrt{25}$, press $\boxed{\text{2nd}}$ $\boxed{\sqrt{}}$ 25 $\boxed{\text{ENTER}}$. See Figure 7.

Roots other than square root are found under the MATH menu. To type ${}^5\sqrt{32}$, press 5 $\boxed{\text{MATH}}$,select [5: $^x\sqrt{}$] 32 $\boxed{\text{ENTER}}$. See Figure 8.

Figure 7

Figure 8

Trouble Shooting:
The TI-82 has problems raising a negative base to a fractional power other than $1/n$. $a^{m/n}$ where $m \neq 1$, restricts the domain so $a \geq 0$. If $a < 0$ you get an **ERR:DOMAIN** message. You have to trick the calculator into performing the operation. The TI-83 does not have this problem. See Figure 9 and 10.

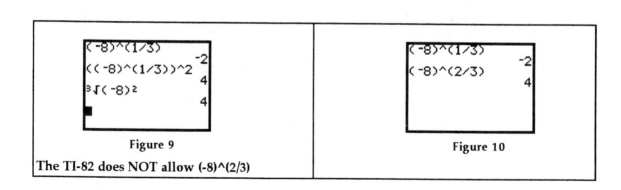

Figure 9

Figure 10

The TI-82 does NOT allow (-8)^(2/3)

Appendix Section 8
Exponential Functions on the TI-82 or TI-83[1]
Graphing Calculator Techniques for Chapter 8

8.1 Exponential Functions

Exponential functions are in the form
$y = b^x$ where $b>0$ and $b \neq 1$ and the power x
can be any real number.

Example 1:

Graph the following:

$$Y1 = 2^x \qquad Y3 = 7^x$$
$$Y2 = 5^x \qquad Y4 = 10^x$$

Set your WINDOW to the decimal scale,
but let Ymax = 15. Type the functions into

$\boxed{Y=}$. The graphs of these functions are

shown in Figure 2.

8.1.1 Exponential Growth

All of these graphs cross the y-axis at (0,1),
because $b^0 = 1$.

These exponential graphs have the same
shape. As x increases y increases slowly at
first and then y increases very rapidly.
This type of change is commonly called
exponential growth.

Figure 4 gives an idea of the growth rate of
$y1 = 2^x$. As x changes from 0 to 10, Y1
changes from 1 to about 1000. But as x
changes from 10 to 20, Y1 does not increase
to 2000, instead Y1 now changes from 1000 to
about 1,000,000! This is a magnitude of 3.

8.1.2 Exponential Decay

Examples 2:

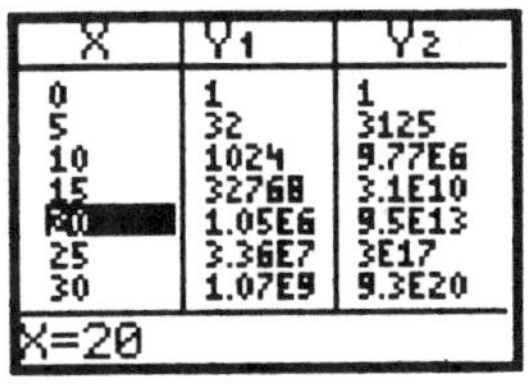

Graph the following:

$$Y1 = 2^{-x} \qquad Y3 = 7^{-x}$$
$$Y2 = 5^{-x} \qquad Y4 = 10^{-x}$$

These exponential functions differ from the
ones above in that a negative now appears
in front of the exponent x. Enter the
functions as in Figure 5. Recall $b^{-x} = 1/b^x$.
The effect on the graphs is shown in Figure
6. The graphs have reversed. Now as x
increases y decreases very rapidly. This is
commonly called exponential decay.
Notice that as x gets larger y gets smaller
and gets very close to zero.
All of these exponential decay graphs
have y-intercept of (1,0).

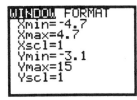

Figure 1

Figure 2

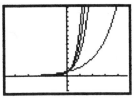

Figure 3

Figure 4

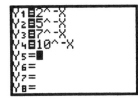

Figure 5

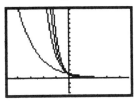

Figure 6

---

[1] The key stroking and menus for the TI-82 and TI-83 are nearly the same. Where they differ, screen images are presented side by side

Appendix Section 9
Power Functions on the TI-82 or TI-83[1]
Graphing Calculator Techniques for Chapter 9

9.1 Power Functions with Positive Integral Powers

A power Function has the form
$$y = kx^p$$
where k and p are constants. The simplest power function is y = x, which has a linear graph. However all other power functions do not look linear.

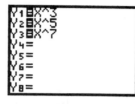

Figure 1

9.1.1 Visualizing Odd Power Functions
Example 1 :
Graph the following and generalize the shape of the graph.

$$Y1 = x^3$$
$$Y2 = x^5$$
$$Y3 = x^7$$

Enter the functions into $\boxed{Y=}$, CLEAR all old functions. Set your WINDOW to Figures 2. Press $\boxed{GRAPH}$. See Figure 3.

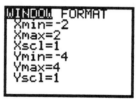

Figure 2

The graphs go through the origin (0,0), have both positive and negative y values and have a "lazy S " shape. As you move left to right the functions are always increasing.

A look at the table of values, in Figure 4, shows that for the odd power functions as x increases to +∞, y increases pretty fast to +∞ . As x decreases to -∞ y decreases to -∞.

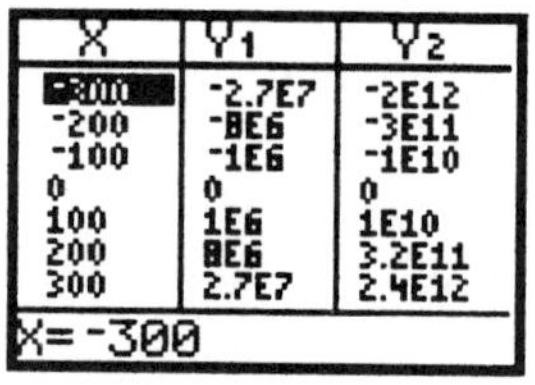

Press $\boxed{ZOOM}$, select [3:Zoom Out]. The calculator gives you a chance to reposition your cursor then press $\boxed{ENTER}$. This shows that all the graphs have a similar shape or "global behavior".

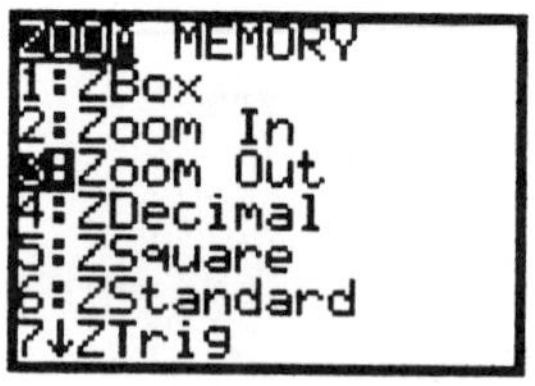

Figure 3

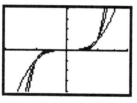

Figure 4

9.1.2 Visualizing Even Power Functions
Example 2:
Graph the following and generalize the shape of the graph.

$$Y1 = x^2$$
$$Y2 = x^4$$
$$Y3 = x^6$$

Figure 5

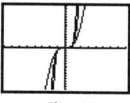

Figure 6

---

[1] The key stroking and menus for the TI-82 and TI-83 are nearly the same. Where they differ, screen images are presented side by side

Enter the functions into $\boxed{Y=}$. Reset the WINDOW to Figure 2.

Press $\boxed{\text{GRAPH}}$. See Figure 7.

The graphs go through the origin (0,0), have only positive y values and have a "U" shape. As you move left to right the functions decrease to zero then increases.

A look at the table of values, in Figure 8, shows that for the even power functions, as x increases to $+\infty$ y increases pretty fast to $+\infty$. As x decreases to $-\infty$ y increases to $+\infty$.

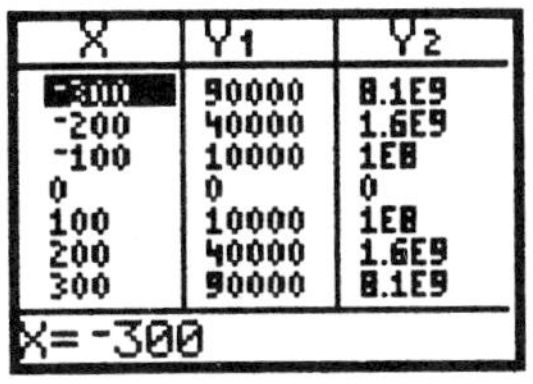

9.1.3 Polynomial Functions

When positive integer power functions are added together you get a polynomial function:

$$y = a_n x^n + a_{n-1} x^{n-1} + ... + a_1 x^a + a_0$$

where a_i is a constant coefficient ($a_n \neq 0$), and n is positive integer power.

Example 3.

The deer population in a national forest was monitored over a 25 year period. The data collected can be modeled by the fifth degree polynomial:

$D(x) = -.125x^5 + 3.125x^4 + 4000$. Graph the function and interpret the graph.

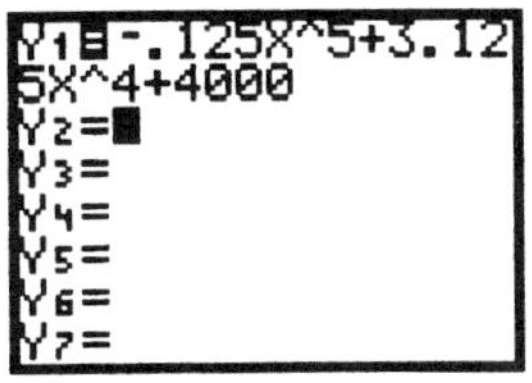

Let Y1= $-.125x^5 + 3.125x^4 + 4000$.
See Figure 9.
Use the table of values to find the appropriate WINDOW. Press $\boxed{\text{2nd}}$ $\boxed{\text{TblSet}}$.

Let TblMin=0 and ΔTbl=1. Press $\boxed{\text{2nd}}$ $\boxed{\text{TABLE}}$
See Figures 10 and 11.
Set the WINDOW as in Figure 12.

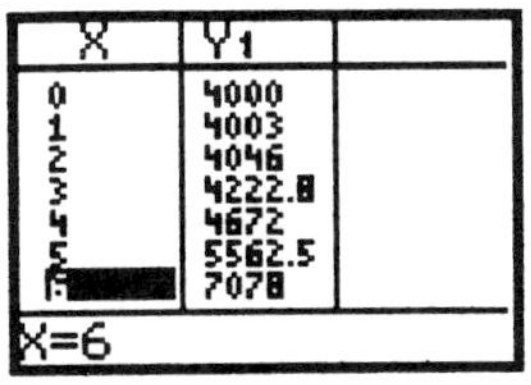

Press $\boxed{\text{GRAPH}}$, then $\boxed{\text{TRACE}}$ to explore the graph.

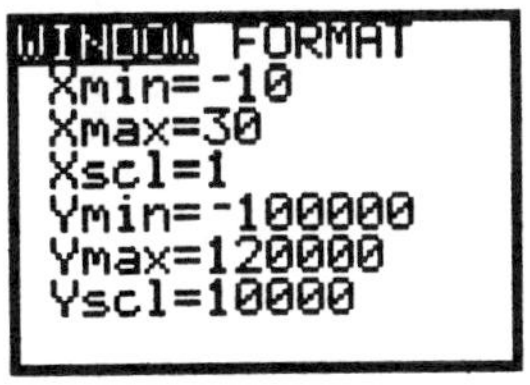

Interpretation:

It took 20 years for the deer population to go from 4000 to a maximum population of about 104,000. Over the next five year period the population decreased back to 4000. The sharp decrease was probably due to lack of food supply due to over population.

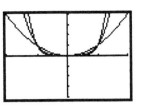

Figure 7

<table>
<thead>
<tr><th>X</th><th>Y1</th><th>Y2</th></tr>
</thead>
<tbody>
<tr><td>-300</td><td>90000</td><td>8.1E9</td></tr>
<tr><td>-200</td><td>40000</td><td>1.6E9</td></tr>
<tr><td>-100</td><td>10000</td><td>1E8</td></tr>
<tr><td>0</td><td>0</td><td>0</td></tr>
<tr><td>100</td><td>10000</td><td>1E8</td></tr>
<tr><td>200</td><td>40000</td><td>1.6E9</td></tr>
<tr><td>300</td><td>90000</td><td>8.1E9</td></tr>
</tbody>
</table>

X=-300

Figure 8

Y1=-.125X^5+3.12
5X^4+4000
Y2=
Y3=
Y4=
Y5=
Y6=
Y7=

Figure 9

<table>
<thead>
<tr><th>X</th><th>Y1</th></tr>
</thead>
<tbody>
<tr><td>0</td><td>4000</td></tr>
<tr><td>1</td><td>4003</td></tr>
<tr><td>2</td><td>4046</td></tr>
<tr><td>3</td><td>4222.8</td></tr>
<tr><td>4</td><td>4672</td></tr>
<tr><td>5</td><td>5562.5</td></tr>
<tr><td>6</td><td>7078</td></tr>
</tbody>
</table>

X=6

Figure 10

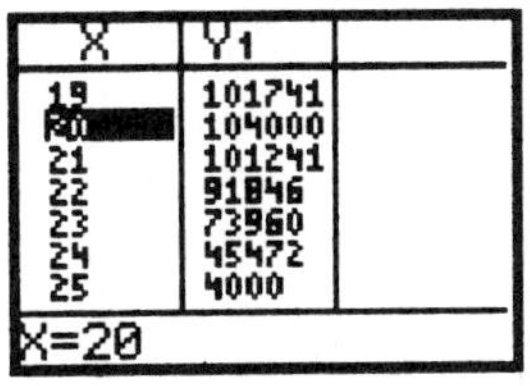

<table>
<thead>
<tr><th>X</th><th>Y1</th></tr>
</thead>
<tbody>
<tr><td>19</td><td>101741</td></tr>
<tr><td>20</td><td>104000</td></tr>
<tr><td>21</td><td>101241</td></tr>
<tr><td>22</td><td>91846</td></tr>
<tr><td>23</td><td>73960</td></tr>
<tr><td>24</td><td>45472</td></tr>
<tr><td>25</td><td>4000</td></tr>
</tbody>
</table>

X=20

Figure 11

WINDOW FORMAT
Xmin=-10
Xmax=30
Xscl=1
Ymin=-100000
Ymax=120000
Yscl=10000

Figure 12

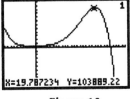

X=19.787234 Y=103889.22

Figure 13

9.1.4 Power Regression Equations.
Example 4:

Below are some data reported on AIDS in women. Find a power regression equation that models the data.

YEAR	AIDS CASES
1	18
2	30
3	36
4	92
5	198
6	360
7	631
8	1016
9	1430

Clear functions from $\boxed{Y=}$. Turn OFF plots

Enter the data in L1 and L2. Press $\boxed{STAT}$ [1:Edit]. See Figure 14.

Press $\boxed{2nd}$ $\boxed{STATPLOT}$ $\boxed{1}$ and set up the plot as in Figure 15.

Press $\boxed{ZOOM}$, select [9:ZoomStat]. See Figure 16 & 17.

Find the regression equation.

Press $\boxed{STAT}$ $\boxed{\triangleright}$ to <CALC>, select [B:PwrReg] $\boxed{2nd}$ $\boxed{L1}$ $\boxed{,}$ $\boxed{2nd}$ $\boxed{L2}$ $\boxed{ENTER}$.

TI-83 NOTE:
Press $\boxed{STAT}$ $\boxed{\triangleright}$ to <CALC>. Select [A:PwrReg], $\boxed{2nd}$ $\boxed{L1}$ $\boxed{,}$ $\boxed{2nd}$ $\boxed{L2}$ $\boxed{:}$ $\boxed{Y1}$ to directly store the equation to Y1.

Put the equation into Y1 either by hand or by pasting the variable. To paste:
press $\boxed{Y=}$ $\boxed{VARS}$, select [5:Statistics] $\boxed{\triangleright}$ $\boxed{\triangleright}$ to <EQ>, select [7:RegEQ].

Now press $\boxed{GRAPH}$. See Figure 20.

The fit is pretty good in the beginning, but not so good at the end.

Example 5.

Figure 18 shows other regression equations available. Find other polynomial regression equations by repeating the steps above.

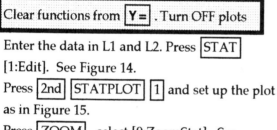

Figure 14

Figure 15

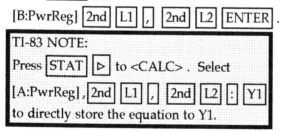

Figure 16

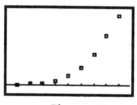

Figure 17

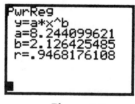

Figure 18

Figure 19

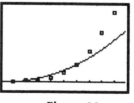

Figure 20

Appendix Section 10
Freefall and Quadratic Regression Equations on the
TI-82 or TI-83[1] Graphing Calculator Techniques for Chapter 10

10.1 The Calculator-Based Laboratory, CBL™

In chapter 10 you will be given the opportunity to explore how the height of an object varies over time when the object is dropped. The CBL™ instructions for this exploration are in the instructors manual and will be shared with you by your instructor.

Be sure to review section 5.1.1 Receiving Data.

The text also comes with a program on FREEFALL data which can be down loaded from a computer to a calculator.
Once the data is in your calculator either by experiment or from the program, do the following.
1. Plot the data
2. Determine a quadratic regression equation for the data.

The Figures to the right might represent the calculator steps you need to perform. Below is the program FREEFALL which stores values to List1-List3. See the next page for a hard copy of the data.

```
PROGRAM:FREEFALL
ClrList L₁,L₂,L₃,L₄,L₅,L₆
{0.0000,0.0167,0.0333,0.0500,0.06
67,0.0833,0.1000,0.1167,0.1333,0.
1500,0.1667,0.1833,0.2000,0.2167,
0.2333,0.2500,0.2667,0.2833,0.300
0,0.3167,0.3333}→L₁
{0.00,1.72,3.75,6.10,8.67,11.58,1
4.71,18.10,21.77,25.71,29.90,34.4
5,39.22,44.22,49.58,55.15,60.99,6
7.11,73.48,80.10,87.05}→L₂
{87.05,85.33,83.30,80.95,78.38,75
.47,72.34,68.95,65.28,61.34,57.15
,52.60,47.83,42.83,37.47,31.90,26
.06,19.94,13.57,6.95,0.00}→L₃
```

Remember to execute the program before you set up your plot. Press [PRGM] ,select
FREEFALL [ENTER] .

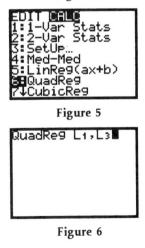

EXEC EDIT NEW
1:FAM1000
2:FAM1000A
3:FAM1000B
4:FAM1000C
5:FREEFALL

Figure 1

L₁	L₂	L₃
0	0	87.05
.0167	1.72	85.33
.0333	3.75	83.3
.05	6.1	80.95
.0667	8.67	78.38
.0833	11.58	75.47
.1	14.71	72.34

L₃(1)=87.05

Figure 2

Plot1
  Off
Type: ▦ ⌐ ⠃⠉ dⅈ
Xlist: L1 L2 L3 L4 L5 L6
Ylist: L1 L2 L3 L4 L5 L6
Mark: □ + ·

Figure 3

Figure 4

EDIT CALC
1:1-Var Stats
2:2-Var Stats
3:SetUp...
4:Med-Med
5:LinReg(ax+b)
6:QuadReg
7↓CubicReg

Figure 5

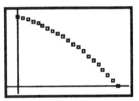
QuadReg L₁,L₃

Figure 6

---

[1] The key stroking and menus for the TI-82 and TI-83 are nearly the same. Where they differ, screen images are presented side by side

The following table summarizes data from a free fall experiment.

The total distance that the object fell, in centimeters, was measured vs. the time of, in seconds. using a spark timer and tape attached to an object experiencing free fall. The position data was obtained as a difference between the total distance fallen and the initial position. 87.05 cm.

Time (t), sec. L1	Total Distance Fallen (D) cm. L2	Position (P), cm. L3
0.0000	0.00	87.05
0.0167	1.72	85.33
0.0333	3.75	83.30
0.0500	6.10	80.95
0.0667	8.67	78.38
0.0833	11.58	75.47
0.1000	14.71	72.34
0.1167	18.10	68.95
0.1333	21.77	65.28
0.1500	25.71	61.34
0.1667	29.90	57.15
0.1833	34.45	52.60
0.2000	39.22	47.83
0.2167	44.22	42.83
0.2333	49.58	37.47
0.2500	55.15	31.90
0.2667	60.99	26.06
0.2833	67.11	19.94
0.3000	73.48	13.57
0.3167	80.10	6.95
0.3333	87.05	0.00

Appendix Section 11
Quadratic Functions on the TI-82 or TI-83[1]
Graphing Calculator Techniques for Chapter 11

11.1   Quadratic Functions

Functions of the form $y = ax^2 + bx + c$ are called a quadratic or second degree equations.
Note that the highest power of x is two. All quadratic equations have the shape of a parabola.

Example 1:
Graph $y = x^2 - x - 6$ and identify a, b, & c.

For this quadratic equation $a = 1$, $b = -1$ and $c = -6$

To graph press $\boxed{Y=}$ CLEAR all the old functions. Into Y1 enter $x^2 - x - 6$. Press $\boxed{ZOOM}$ $\boxed{6}$. See Figure 1

11.1.1   Finding the y-intercept

Press $\boxed{TRACE}$ to find the y-intercept (if $x = 0$ then $y = -6 = c$). See Figure 2. The y-intercept is $(0,-6) = (0,c)$

11.1.2   Finding the x-intercepts / Finding the Roots/ Finding the zeros.

Use $\boxed{TRACE}$ and arrows to estimate where the graph crosses the x-axis. There are two *x-intercepts* (where $y = 0$) also known as *roots* or *zeros*.. The left intercept is around $x = -2$ and the right intercept is around $x = 3$. See Figures 4 and 5.
Use the calculator to find the x-intercept.
Find the Left Intercept:

Press $\boxed{2nd}$ $\boxed{CALC}$ $\boxed{2}$. Follow these steps:

1.   Position the cursor **to the left** of intercept when prompted. Press $\boxed{ENTER}$

2.   Reposition the cursor **to the right** of the x-intercept when prompted. Press $\boxed{ENTER}$.

3.   When prompted GUESS? position the cursor near the x-intercept. Press $\boxed{ENTER}$. See Figure 5.

Figures 8-11 show TI-82 & TI-83 differences.

Figure 1

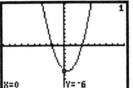

Figure 2

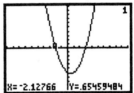

Figure 3

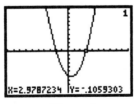

Figure 4

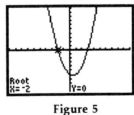

Figure 5

---

[1] The key stroking and menus for the TI-82 and TI-83 are nearly the same. Where they differ, screen images are presented side by side

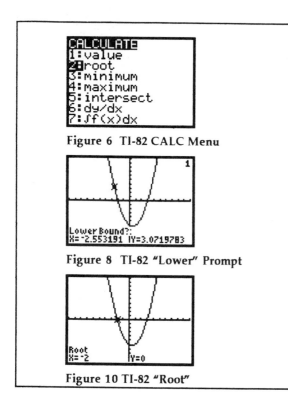

Figure 6 TI-82 CALC Menu

Figure 8 TI-82 "Lower" Prompt

Figure 10 TI-82 "Root"

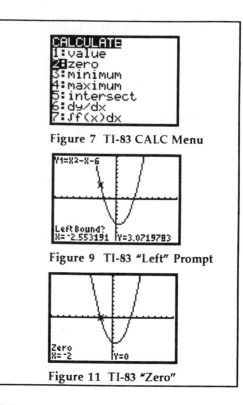

Figure 7 TI-83 CALC Menu

Figure 9 TI-83 "Left" Prompt

Figure 11 TI-83 "Zero"

Find the Right Intercept:
Repeating the process we find that the other x-intercept is (3,0) See Figure 12. There are two roots when x=-2 and x = 3.

11.1.3   Check Roots Algebraically:
$$y = x^2 - x - 6$$
if $x = -2$, $y = (-2)^2 - (-2) - 6 = 0$
if $x = 3$, $y = 3^2 - 3 - 6 = 0$. See Figure 13.

11.1.4   Finding the Vertex of a Quadratic Function

We see from the graph in Figure 12 that the function $y = x^2 - x - 6$ opens upward. This will be true when $a > 0$. As we move left to right the *minimum* point on the graph is called the **vertex**. It is some where between x=0 and x=2. See Figure 12.

Example 2:
Find the vertex for $y = -2x^2 - 12x - 13$.

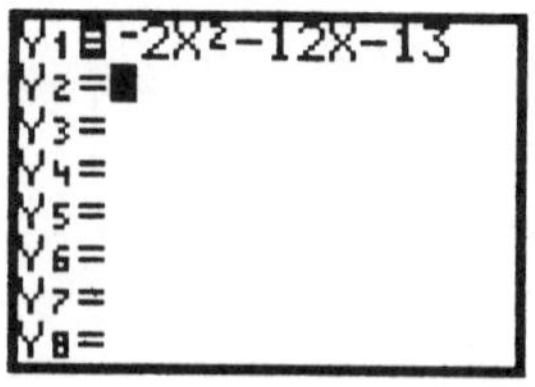

Enter the function into Y1. Notice $a < 0$ so the graph opens downward (Figures 14 and 15). Now the **vertex** is the maximum point on the graph. The vertex appears to be around the point (-3, 5). See Figure 15. Use the calculator to find the vertex.

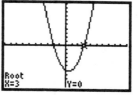

Figure 12

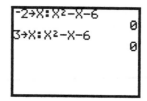

Figure 13

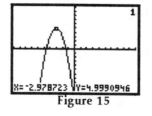

Figure 14

Figure 15

Press 2nd CALC 4 for maximum. Then follow these steps:

1. Position the cursor **to the left** of maximum when prompted. Press ENTER.

2. Reposition the cursor **to the right** of the maximum when prompted. Press ENTER.

3. When prompted GUESS? position the cursor near the maximum. Press ENTER

The vertex is at the point (-3, 5).
See Figures 17-20.

TI-83 Note:
As with the x-intercept the TI-83 prompts for left and right bounds. You can also enter a number guess for the maximum. Only the TI-82 screens are shown to the right.

The vertex is at the point (-3, 5).
See Figure 20
.

11.1.5 The Vertex Form of a Quadratic Function.

We write the vertex as the point (h,k).
Then the quadratic equation is transformed to :

$$y = a(x - h)^2 + k$$

So $y = -2x^2 -12x -13$ becomes
$$y = -2(x - \bar{\ }3)^2 + 5$$
$$= -2(x + 3)^2 + 5$$

Verify algebraically:
$$y = -2(x + 3)^2 + 5$$
$$= -2(x^2 + 6x + 9) + 5$$
$$= -2 x^2 -12x - 18 + 5$$
$$= -2 x^2 -12x - 13$$

Verify Numerically

Enter both equations into Y= . See Figure 21.

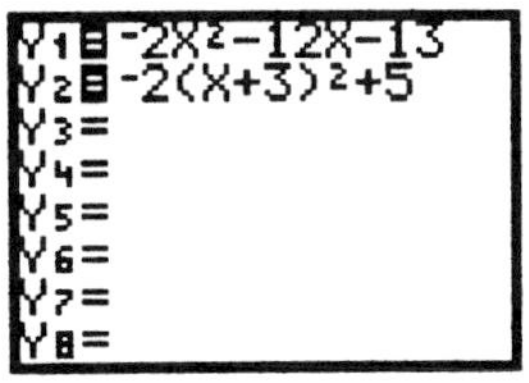

Set up a table of values. Press 2nd TblSet . See Figure 22.

Press 2nd TABLE .

Figure 16

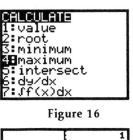

Figure 17

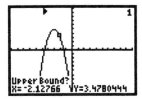

Figure 18

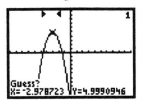

Figure 19

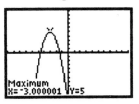

Figure 20

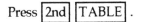

Figure 21

Figure 22

For all values of x , Y1 and Y2 are equivalent. Therefore Y1 = Y2, or
$-2x^2 - 12x - 13 = -2(x + 3)^2 + 5$ and
$ax^2 + bx + c = a(x - h)^2 + k$

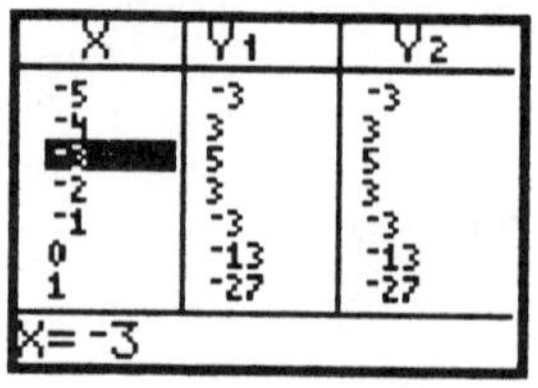

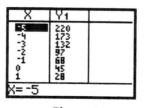

Figure 23

11.1.6 Finding an Appropriate Window.

Example 3 :
Graph $y = 3x^2 - 20x + 45$.

Enter $3x^2 - 20x + 45$ into Y1. CLEAR all other functions.
Press | ZOOM | | 6 | . We see nothing!

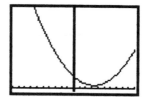

Figure 24

Use the table to get a feel for what happens to y as x increases.
Press | 2nd | | TABLE | . Figure 24 shows that when x = -5 , y = 220 and when x=0, y= 45. No wonder we couldn't see anything on a [-10,10] standard window.
From algebra we know that a = 3 > 0 so the graph opens upward. Use down arrow to find the minimum value of y (the vertex).

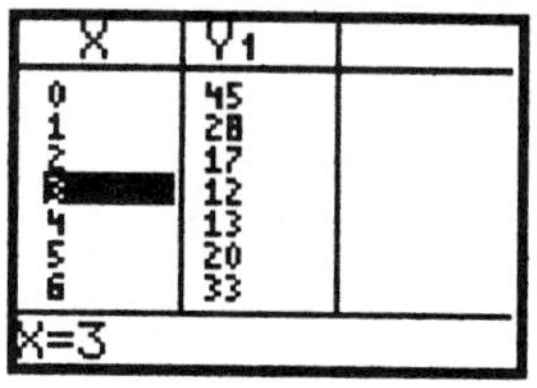

Figure 25

From the table (Figure 25) it looks like the minimum occurs around x = 3. Now we would like to see the Vertex as well as the point (-5,220).

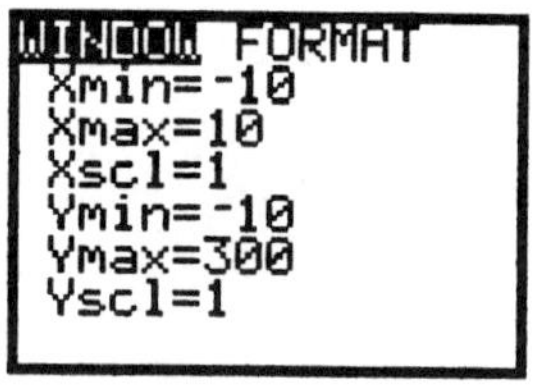

Press | WINDOW | , let Ymax = 300.
See Figure 26.

Figure 26

Press | GRAPH | . See Figure 28. The graph is being cut off on the right.
Press | WINDOW | . Change Xmax to 15.
Press | GRAPH | . Much better!!

11.1.7 A Complete Graph

Try to select a window that displays a complete graph. A complete graph shows the whole shape of the graph , with all its turning points . Also shown are the y-intercept and x-intercept(s) ,if they exist. There are many complete graphs.

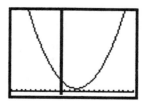

Figure 27

Figure 28

Figure 28 shows a complete graph of
$y = 3x^2 - 20x + 45$.

INDEX